BNi Building News

BNi Building News

GENERAL CONSTRUCTION 2006 COSTBOOK

SIXTEENTH EDITION

BNi Building News
• Los Angeles • Anaheim • Vista
• New England • Washington, D.C.

BNI Publications, Inc.

LOS ANGELES
10801 NATIONAL BLVD. SUITE 100
LOS ANGELES, CA 90064

NEW ENGLAND
PO Box 14527
E. PROVIDENCE, RI 02494

ANAHEIM
1612 S. CLEMENTINE STREET
ANAHEIM, CA 92802

VISTA
990 Park Center Drive, Suite E
Vista, CA 92081

WASHINGTON, D.C.
P.O. Box 1218
Herndon, VA 20172

1-888-BNI-BOOK (1-888-264-2665)
www.bnibooks.com

ISBN 155701-519-8

PREFACE

For the past 50 years, Building News has been dedicated to providing construction professionals with timely and reliable information. Based on this experience, our staff has researched and compiled thousands of up-to-the-minute costs for the **Building News 2006 Costbooks**. This book is an essential reference for contractors, engineers, architects, facilities managers — any construction professional who must provide an estimate or any type of building project.

Whether working up a preliminary estimate or submitting a formal bid, the costs listed here can quickly and easily be tailored to your needs. All costs are based on national averages, while a table of modifiers is provided for regional adjustments. Overhead and profit are included in all costs.

Complete man-hour tables follow the unit costs to provide data on typical durations of specific tasks. This information can be used to schedule projects as well as to determine specific labor costs based on local labor rates.

All data is categorized according to the MASTERFORMAT of the Construction Specifications Institute (CSI). This industry standard provides an all-inclusive checklist to ensure that no element of a project is overlooked. In addition, to make specific items even easier to locate, there is a complete alphabetical index.

This costbook contains an appendix with reference charts and tables taken from an array of sources. Text explains the costs in certain categories and provides helpful pointers that should be taken into account with every estimate.

A section on square foot costs provides an overview of project costs for different building types — commercial, residential, etc. — with summaries of the actual projects. Square foot costs are invaluable for making budget estimates and checking prices when time is a factor.

The "Features in this Book" section presents a clear overview of the many features of this book. Included is an explanation of the data, sample page layout and discussion of how to best use the information in the book.

Of course, all buildings and construction projects are unique. The costs provided in this book are based on averages from well-managed projects with good labor productivity under normal working conditions (eight hours a day). Other circumstances affecting costs such as overtime, unusual working conditions, savings from buying bulk quantities for large projects, and unusual or hidden costs must be factored in as they arise.

The data provided in this book is for estimating purposes only. Check all applicable federal, state and local codes and regulations for specific requirements.

BNi Building News

TABLE OF CONTENTS

CSI MASTERFORMAT

All data in the Costbook pages and Man-Hour tables is organized according to the CSI MASTERFORMAT — the industry standard numbering/classification system. The data is divided into the 16 divisions as shown below. The five digit numbers within each division correspond to the MASTERFORMAT Broadscope designations. Each section is further broken down into MASTERFORMAT Mediumscope designations. These numbers, in most cases, are the same as those used in architectural and engineering specifications.

The construction estimating information in this book is divided into two main sections: Costbook Pages and Man-Hour Tables. Each is organized to the 16 divisions of the CSI MASTERFORMAT as shown below. In addition, there are extensive Construction Reference Tables, Geographic Cost Modifiers, Square Foot Tables and a detailed Index. Sample pages with graphic explanations are included before the Costbook pages, Man-Hour Tables and Square Foot Tables. These explanations, along with the discussions below, will provide a good understanding of what is included in this book and how it can best be used for construction estimating.

GENERAL REQUIREMENTS	*Division 1*
Summary of Work	01010
Allowances	01020
Measurement and Payment	01025
Alternates/Alternatives	01030
Modification Procedures	01035
Coordination	01040
Field Engineering	01050
Regulatory Requirements	01060
Identification Systems	01070
References	01090
Special Project Procedures	01100
Project Meetings	01200
Submittals	01300
Quality Control	01400
Construction Facilities & Temporary Controls	01500
Material and Equipment	01600
Facility Startup/Commissioning	01650
Contract Closeout	01700
Maintenance	01800

SITEWORK	*Division 2*
Subsurface Investigation	02010
Demolition	02050
Site Preparation	02100
Dewatering	02140
Shoring and Underpinning	02150
Excavation Support Systems	02160
Cofferdams	02170
Earthwork	02200
Tunneling	02300
Piles and Caissons	02350
Railroad Work	02450
Marine Work	02480
Paving and Surfacing	02500
Utility Piping Materials	02600
Water Distribution	02660
Fuel and Steam Distribution	02680
Sewerage and Drainage	02700
Restoration and Underground Pipe	02760
Ponds and Reservoirs	02770
Power and Communications	02780
Site Improvements	02800
Landscaping	02900

CONCRETE	*Division 3*
Concrete Formwork	03100
Concrete Reinforcement	03200
Concrete Accessories	03250
Cast-In-Place Concrete	03300
Concrete Curing	03370
Precast Concrete	03400
Cementitious Decks and Toppings	03500
Grout	03600
Concrete Restoration and Cleaning	03700
Mass Concrete	03800

MASONRY	*Division 4*
Mortar and Masonry Grout	04100
Masonry Accessories	04150
Unit Masonry	04200
Stone	04400
Masonry Restoration and Cleaning	04500
Refractories	04550
Corrosion Resistant Masonry	04600
Simulated Masonry	04700

METALS	*Division 5*
Metal Materials	05010
Metal Coatings	05030
Metal Fastening	05050
Structural Metal Framing	05100
Metal Joists	05200
Metal Decking	05300
Cold Formed Metal Framing	05400

FURNISHINGS ***Division 12***

SPECIAL CONSTRUCTION ***Division 13***

CONVEYING SYSTEMS ***Division 14***

MECHANICAL ***Division 15***

ELECTRICAL ***Division 16***

FEATURES IN THIS BOOK

The construction estimating information in this book is divided into two main sections: Costbook Pages and Man-Hour Tables. Each section is organized according to the 16 divisions of the MASTERFORMAT as shown on the previous pages. In addition, there are extensive Supporting Construction Reference tables, Geographic Costs Modifiers, Square Foot tables and a detailed Index.

Sample pages with graphic explanations are included before the Costbook pages and Man-Hour tables. These explanations, along with the discussions below, will provide a good understanding of what is included in this book and how it can best be used in construction estimating.

Material Costs

The material costs used in this book represent national averages for prices that a contractor would expect to pay plus an allowance for freight (if applicable), handling and storage. These costs reflect neither the lowest or highest prices, but rather a typical average cost over time. Periodic fluctuations in availability and in certain commodities (e.g. copper, lumber) can significantly affect local material pricing. In the final estimating and bidding stages of a project when the highest degree of accuracy is required, it is best to check local, current prices.

Labor Costs

Labor costs include the basic wage, plus commonly applicable taxes, insurance and markups for overhead and profit. The labor rates used here to develop the costs are typical average prevailing wage rates. Rates for different trades are used where appropriate for each type of work.

Taxes and insurance which are most often applied to labor rates include employer-paid Social Security/Medicare taxes (FICA), Worker's Compensation insurance, state and federal unemployment taxes, and business insurance. Fixed government rates as well as average allowances are included in the labor costs. However, most of these items vary significantly from state to state and within states. For more specific data, local agencies and sources should be consulted.

Equipment Costs

Costs for various types and pieces of equipment are included in Division 1 - General Requirements and can be included in an estimate when required either as a total "Equipment" category or with specific appropriate trades. Costs for equipment are included when appropriate in the installation costs in the Costbook pages.

Overhead And Profit

Included in the labor costs are allowances for overhead and profit for the contractor/employer whose workers are performing the specific tasks. No cost allowances or fees are included for management of subcontractors by the general contractor or construction manager. These costs, where appropriate, must be added to the costs as listed in the book.

The allowance for overhead is included to account for office overhead, the contractors' typical costs of doing business. These costs normally include in-house office staff salaries and benefits, office rent and operating expenses, professional fees, vehicle costs and other operating costs which are not directly applicable to specific jobs. It should be noted for this book that office overhead as included should be distinguished from project overhead, the General Requirements (CSI Division 1) which are specific to particular projects. Project overhead should be included on an item by item basis for each job.

Depending on the trade, an allowance of 10-15 percent is incorporated into the labor/installation costs to account for typical profit of the installing contractor. See Division 1, General Requirements, for a more detailed review of typical profit allowances.

Adjustments to Costs

The costs as presented in this book attempt to represent national averages. Costs, however, vary among regions, states and even between adjacent localities.

In order to more closely approximate the probable costs for specific locations throughout the U.S., a table of Geographic Cost Modifiers is provided. These adjustment factors are used to modify costs obtained from this book to help account for regional variations of construction costs. Whenever local current costs are known, whether material or equipment prices or labor rates, they should be used if more accuracy is required.

Man-Hour Tables

The man-hour data used to develop the labor costs are listed in the second main section of this book, the "Man-Hour Tables". These productivties represent typical installation labor for thousands of construction items. The data takes into account all activities involved in normal construction under commonly experienced working conditions such as site movement, material handling, start-up, etc. As with the Costbook pages, these items are listed according to the CSI MASTERFORMAT.

Square Foot Tables

Included as an additional reference are Square Foot Tables which list hundreds of actual projects for dozens of building types, each with associated building size and total square foot building cost. This data provides an overview of construction costs by building type. These costs are for actual projects. The variations within similar building types may be due, among other factors, to size, location, quality and specified components, material and processes. Depending upon all such factors, specific building costs can vary significantly and may not necessarily fall within the range of costs as presented.

Editor's Note: The **Building News 2006 Costbooks** are intended to provide accurate, reliable, average costs and typical productivities for thousands of common construction components. The data is developed and compiled from various industry sources, including government, manufacturers, suppliers and working professionals. The intent of the information is to provide assistance and guidelines to construction professionals in estimating. The user should be aware that local conditions, material and labor availability and cost variations, economic considerations, weather, local codes and regulations, etc., all affect the actual cost of construction. These and other such factors must be considered and incorporated into any and all construction estimates.

Sample Costbook Page

In order to best use the information in this book, please review this sample page and read the "Features In This Book" section.

CSI MASTERFORMAT Division

CSI Broadscope Category

CSI Mediumscope Category (First 5 Digits)

Detailed Descriptions

Complete descriptions of items may include information listed above a particular line. Review of the whole category is recommended for a complete description.

03 CONCRETE				
PLACING CONCRETE	**UNIT**	**MAT.**	**INST.**	**TOTAL**
03380.55 SLAB/MAT CONCRETE				
By chute	C.Y.	67.00	5.40	72.40
By crane	"	67.00	18.65	85.65
By pump	"	74.00	16.35	90.35
By hand buggy	"	74.00	13.45	87.45
03380.58 SIDEWALKS				
Walks, cast in place with wire mesh, base not incl.				
4" thick	S.F.	1.00	0.90	1.90
5" thick	"	1.35	1.10	2.45
6" thick	"	1.65	1.35	3.00
03380.60 STAIR CONCRETE				
Stairs				
2500# or 3000# concrete				
By chute	C.Y.	74.00	8.95	82.95
By crane	"	74.00	43.50	117.50
By pump	"	74.00	37.30	111.80
By hand buggy	"	74.00	26.90	100.90
3500# or 4000# concrete				
By chute	C.Y.	78.50	8.95	87.45
By crane	"	78.50	43.50	122.00
By pump	"	78.50	37.30	115.80
By hand buggy	"	78.50	26.90	105.40
5000# concrete				
By chute	C.Y.	83.00	8.95	91.95
By crane	"	83.00	43.50	126.50
By pump	"	83.00	37.30	120.30
By hand buggy	"	83.00	26.90	109.90
03380.65 WALL CONCRETE				
Walls				
2500# or 3000# concrete				
To 4'				
By chute	C.Y.	74.00	7.70	81.70
By crane	"	74.00	43.50	117.50
By pump	"	74.00	40.20	114.20
To 8'				
By crane	C.Y.	74.00	47.50	121.50
By pump	"	74.00	43.50	117.50
To 16'				
By crane	C.Y.	74.00	52.50	126.50
By pump	"	74.00	47.50	121.50
Over 16'				
By crane	C.Y.	74.00	58.00	132.00
By pump	"	74.00	52.50	126.50
3500# or 4000# concrete				
To 4'				
By chute	C.Y.	78.50	7.70	86.20
By crane	"	78.50	43.50	[illegible]
By pump	"	78.50	40.20	118.70
To 8'				
By crane	C.Y.	78.50	47.50	126.00

Material Cost

Material cost represent average contractor prices plus an allowance for freight, handling and storage.

Installation Cost

Installation cost includes basic wage rates, markups for taxes, insurance overhead and profit and also includes equipment costs where appropriate.

Total Cost

The total cost is the sum of material and installation costs. This total represents typical contractors' costs including overhead and profit, but does not include markups for the general contractor or construction management fees.

Unit of Measurement

Each item (and cost) is defined in terms of the common estimating unit. All costs are listed in dollars per unit.

BNi Building News

01 GENERAL

Requirements

01020.10 Allowances

	UNIT	MAT.	INST.	TOTAL
Overhead				
$20,000 project				
Minimum	PCT.			15.00
Average	"			20.00
Maximum	"			40.00
$100,000 project				
Minimum	PCT.			12.00
Average	"			15.00
Maximum	"			25.00
$500,000 project				
Minimum	PCT.			10.00
Average	"			12.00
Maximum	"			20.00
$1,000,000 project				
Minimum	PCT.			6.00
Average	"			10.00
Maximum	"			12.00
$10,000,000 project				
Minimum	PCT.			1.50
Average	"			5.00
Maximum	"			8.00
Profit				
$20,000 project				
Minimum	PCT.			10.00
Average	"			15.00
Maximum	"			25.00
$100,000 project				
Minimum	PCT.			10.00
Average	"			12.00
Maximum	"			20.00
$500,000 project				
Minimum	PCT.			5.00
Average	"			10.00
Maximum	"			15.00
$1,000,000 project				
Minimum	PCT.			3.00
Average	"			8.00
Maximum	"			15.00
Professional fees				
Architectural				
$100,000 project				
Minimum	PCT.			5.00
Average	"			10.00
Maximum	"			20.00
$500,000 project				
Minimum	PCT.			5.00
Average	"			8.00
Maximum	"			12.00
$1,000,000 project				
Minimum	PCT.			3.50
Average	"			7.00
Maximum	"			10.00

01 GENERAL

Requirements

Requirements	UNIT	MAT.	INST.	TOTAL
01020.10 Allowances				
Structural engineering				
Minimum	PCT.			2.00
Average	"			3.00
Maximum	"			5.00
Mechanical engineering				
Minimum	PCT.			4.00
Average	"			5.00
Maximum	"			15.00
Electrical engineering				
Minimum	PCT.			3.00
Average	"			5.00
Maximum	"			12.00
Taxes				
Sales tax				
Minimum	PCT.			4.00
Average	"			5.00
Maximum	"			10.00
Unemployment				
Minimum	PCT.			3.00
Average	"			6.50
Maximum	"			8.00
Social security (FICA)	"			7.85
01050.10 Field Staff				
Superintendent				
Minimum	YEAR			66,462
Average	"			96,778
Maximum	"			139,920
Field engineer				
Minimum	YEAR			52,470
Average	"			78,122
Maximum	"			111,936
Foreman				
Minimum	YEAR			37,312
Average	"			59,466
Maximum	"			88,616
Bookkeeper/timekeeper				
Minimum	YEAR			22,154
Average	"			29,150
Maximum	"			45,474
Watchman				
Minimum	YEAR			14,925
Average	"			18,656
Maximum	"			30,316
01310.10 Scheduling				
Scheduling for				
$100,000 project				
Minimum	PCT.			1.00
Average	"			2.00
Maximum	"			4.00
$500,000 project				

01 GENERAL

Requirements	UNIT	MAT.	INST.	TOTAL
01310.10 Scheduling				
Minimum	PCT.			0.50
Average	"			1.00
Maximum	"			2.00
$1,000,000 project				
Minimum	PCT.			0.30
Average	"			0.80
Maximum	"			1.50
Scheduling software, not including computer				
Minimum	EA.			385.00
Average	"			2,200
Maximum	"			44,000
01330.10 Surveying				
Surveying				
Small crew	DAY		680.00	680.00
Average crew	"		1,030	1,030
Large crew	"		1,350	1,350
Lot lines and boundaries				
Minimum	ACRE		480.00	480.00
Average	"		1,030	1,030
Maximum	"		1,690	1,690
01380.10 Job Requirements				
Job photographs, small jobs				
Minimum	EA.			110.00
Average	"			165.00
Maximum	"			330.00
Large projects				
Minimum	EA.			550.00
Average	"			825.00
Maximum	"			2,200
01410.10 Testing				
Testing concrete, per test				
Minimum	EA.			16.50
Average	"			27.50
Maximum	"			55.00
Soil, per test				
Minimum	EA.			33.00
Average	"			82.50
Maximum	"			220.00
Welding, per test				
Minimum	EA.			16.50
Average	"			27.50
Maximum	"			110.00
01500.10 Temporary Facilities				
Barricades, temporary				
Highway				
Concrete	L.F.	9.95	3.40	13.35
Wood	"	3.25	1.35	4.60
Steel	"	3.45	1.15	4.60
Pedestrian barricades				

01 GENERAL

Requirements	UNIT	MAT.	INST.	TOTAL
01500.10 Temporary Facilities				
Plywood	S.F.	2.30	1.15	3.45
Chain link fence	"	2.60	1.15	3.75
Trailers, general office type, per month				
Minimum	EA.			173.25
Average	"			288.75
Maximum	"			577.50
Crew change trailers, per month				
Minimum	EA.			103.95
Average	"			115.50
Maximum	"			173.25
01505.10 Mobilization				
Equipment mobilization				
Bulldozer				
Minimum	EA.			148.72
Average	"			311.21
Maximum	"			520.52
Backhoe/front-end loader				
Minimum	EA.			89.36
Average	"			155.61
Maximum	"			340.13
Crane, crawler type				
Minimum	EA.			1,637
Average	"			4,021
Maximum	"			8,637
Truck crane				
Minimum	EA.			385.57
Average	"			579.74
Maximum	"			996.98
Pile driving rig				
Minimum	EA.			7,446
Average	"			14,893
Maximum	"			26,807
01525.10 Construction Aids				
Scaffolding/staging, rent per month				
Measured by lineal feet of base				
10' high	L.F.			10.13
20' high	"			18.28
30' high	"			25.59
40' high	"			29.53
50' high	"			35.15
Measured by square foot of surface				
Minimum	S.F.			0.45
Average	"			0.77
Maximum	"			1.39
Safety nets, heavy duty, per job				
Minimum	S.F.			0.30
Average	"			0.36
Maximum	"			0.79
Tarpaulins, fabric, per job				
Minimum	S.F.			0.21

01 GENERAL

Requirements

Requirements	UNIT	MAT.	INST.	TOTAL
01525.10 Construction Aids				
Average	S.F.			0.36
Maximum	"			0.91
01570.10 Signs				
Construction signs, temporary				
Signs, 2' x 4'				
Minimum	EA.			29.49
Average	"			70.76
Maximum	"			248.79
Signs, 4' x 8'				
Minimum	EA.			61.92
Average	"			160.84
Maximum	"			691.08
Signs, 8' x 8'				
Minimum	EA.			79.61
Average	"			249.44
Maximum	"			2,579
01600.10 Equipment				
Air compressor				
60 cfm				
By day	EA.			75.19
By week	"			223.96
By month	"			679.16
300 cfm				
By day	EA.			157.37
By week	"			474.61
By month	"			1,429
600 cfm				
By day	EA.			433.17
By week	"			1,301
By month	"			3,904
Air tools, per compressor, per day				
Minimum	EA.			28.92
Average	"			36.16
Maximum	"			50.61
Generators, 5 kw				
By day	EA.			72.31
By week	"			215.25
By month	"			663.12
Heaters, salamander type, per week				
Minimum	EA.			86.76
Average	"			122.91
Maximum	"			259.37
Pumps, submersible				
50 gpm				
By day	EA.			57.84
By week	"			172.47
By month	"			518.73
100 gpm				
By day	EA.			72.31
By week	"			215.25

Requirements	UNIT	MAT.	INST.	TOTAL
01600.10 Equipment				
By month	EA.			649.75
500 gpm				
By day	EA.			115.67
By week	"			346.27
By month	"			1,040
Diaphragm pump, by week				
Minimum	EA.			101.22
Average	"			172.47
Maximum	"			359.64
Pickup truck				
By day	EA.			108.45
By week	"			316.85
By month	"			982.65
Dump truck				
6 cy truck				
By day	EA.			288.78
By week	"			866.33
By month	"			2,602
10 cy truck				
By day	EA.			359.64
By week	"			1,083
By month	"			3,251
16 cy truck				
By day	EA.			577.56
By week	"			1,734
By month	"			5,205
Backhoe, track mounted				
1/2 cy capacity				
By day	EA.			592.26
By week	"			1,806
By month	"			5,349
1 cy capacity				
By day	EA.			938.53
By week	"			2,818
By month	"			8,457
2 cy capacity				
By day	EA.			1,590
By week	"			4,770
By month	"			14,313
3 cy capacity				
By day	EA.			3,036
By week	"			9,109
By month	"			27,327
Backhoe/loader, rubber tired				
1/2 cy capacity				
By day	EA.			359.64
By week	"			1,083
By month	"			3,251
3/4 cy capacity				
By day	EA.			433.17
By week	"			1,301
By month	"			3,904

Requirements

01600.10 Equipment

	UNIT	MAT.	INST.	TOTAL
Bulldozer				
75 hp				
By day	EA.			505.36
By week	"			1,517
By month	"			4,554
200 hp				
By day	EA.			1,445
By week	"			4,337
By month	"			13,012
400 hp				
By day	EA.			2,169
By week	"			6,506
By month	"			19,519
Cranes, crawler type				
15 ton capacity				
By day	EA.			649.75
By week	"			1,951
By month	"			5,854
25 ton capacity				
By day	EA.			794.14
By week	"			2,384
By month	"			7,155
50 ton capacity				
By day	EA.			1,445
By week	"			4,337
By month	"			13,012
100 ton capacity				
By day	EA.			2,167
By week	"			6,506
By month	"			19,519
Truck mounted, hydraulic				
15 ton capacity				
By day	EA.			612.32
By week	"			1,842
By month	"			5,316
Loader, rubber tired				
1 cy capacity				
By day	EA.			433.17
By week	"			1,301
By month	"			3,904
2 cy capacity				
By day	EA.			649.75
By week	"			2,529
By month	"			7,590
3 cy capacity				
By day	EA.			1,156
By week	"			3,469
By month	"			10,409

01 GENERAL

Requirements	UNIT	MAT.	INST.	TOTAL
01740.10 Bonds				
Performance bonds				
Minimum	PCT.			0.64
Average	"			2.01
Maximum	"			3.18

02 SITEWORK

Soil Tests	UNIT	MAT.	INST.	TOTAL
02010.10 Soil Boring				
Borings, uncased, stable earth				
2-1/2" dia.				
Minimum	L.F.		15.15	15.15
Average	"		22.75	22.75
Maximum	"		36.40	36.40
4" dia.				
Minimum	L.F.		16.55	16.55
Average	"		26.00	26.00
Maximum	"		45.50	45.50
Cased, including samples				
2-1/2" dia.				
Minimum	L.F.		18.20	18.20
Average	"		30.30	30.30
Maximum	"		60.50	60.50
4" dia.				
Minumum	L.F.		36.40	36.40
Average	"		52.00	52.00
Maximum	"		72.50	72.50
Drilling in rock				
No sampling				
Minimum	L.F.		33.10	33.10
Average	"		47.80	47.80
Maximum	"		65.00	65.00
With casing and sampling				
Minimum	L.F.		45.50	45.50
Average	"		60.50	60.50
Maximum	"		91.00	91.00
Test pits				
Light soil				
Minimum	EA.		230.00	230.00
Average	"		300.00	300.00
Maximum	"		610.00	610.00
Heavy soil				
Minimum	EA.		360.00	360.00
Average	"		450.00	450.00
Maximum	"		910.00	910.00

Demolition	UNIT	MAT.	INST.	TOTAL
02060.10 Building Demolition				
Building, complete with disposal				
Wood frame	C.F.		0.26	0.26
Concrete	"		0.40	0.40
Steel frame	"		0.53	0.53
Partition removal				
Concrete block partitions				

Demolition

02060.10 Building Demolition

	UNIT	MAT.	INST.	TOTAL
4" thick	S.F.		1.70	1.70
8" thick	"		2.25	2.25
12" thick	"		3.10	3.10
Brick masonry partitions				
4" thick	S.F.		1.70	1.70
8" thick	"		2.10	2.10
12" thick	"		2.80	2.80
16" thick	"		4.25	4.25
Cast in place concrete partitions				
Unreinforced				
6" thick	S.F.		12.10	12.10
8" thick	"		13.00	13.00
10" thick	"		15.15	15.15
12" thick	"		18.20	18.20
Reinforced				
6" thick	S.F.		14.00	14.00
8" thick	"		18.20	18.20
10" thick	"		20.20	20.20
12" thick	"		24.25	24.25
Terra cotta				
To 6" thick	S.F.		1.70	1.70
Stud partitions				
Metal or wood, with drywall both sides	S.F.		1.70	1.70
Metal studs, both sides, lath and plaster	"		2.25	2.25
Door and frame removal				
Hollow metal in masonry wall				
Single				
2'6"x6'8"	EA.		42.30	42.30
3'x7'	"		56.50	56.50
Double				
3'x7'	EA.		67.50	67.50
4'x8'	"		67.50	67.50
Wood in framed wall				
Single				
2'6"x6'8"	EA.		24.15	24.15
3'x6'8"	"		28.20	28.20
Double				
2'6"x6'8"	EA.		33.80	33.80
3'x6'8"	"		37.60	37.60
Remove for re-use				
Hollow metal	EA.		84.50	84.50
Wood	"		56.50	56.50
Floor removal				
Brick flooring	S.F.		1.35	1.35
Ceramic or quarry tile	"		0.75	0.75
Terrazzo	"		1.50	1.50
Heavy wood	"		0.90	0.90
Residential wood	"		0.97	0.97
Resilient tile or linoleum	"		0.34	0.34
Ceiling removal				
Acoustical tile ceiling				
Adhesive fastened	S.F.		0.34	0.34

Demolition

02060.10 Building Demolition	UNIT	MAT.	INST.	TOTAL
Furred and glued	S.F.		0.28	0.28
Suspended grid	"		0.21	0.21
Drywall ceiling				
Furred and nailed	S.F.		0.38	0.38
Nailed to framing	"		0.34	0.34
Plastered ceiling				
Furred on framing	S.F.		0.85	0.85
Suspended system	"		1.15	1.15
Roofing removal				
Steel frame				
Corrugated metal roofing	S.F.		0.68	0.68
Built-up roof on metal deck	"		1.15	1.15
Wood frame				
Built up roof on wood deck	S.F.		1.05	1.05
Roof shingles	"		0.56	0.56
Roof tiles	"		1.15	1.15
Concrete frame	C.F.		2.25	2.25
Concrete plank	S.F.		1.70	1.70
Built-up roof on concrete	"		0.97	0.97
Cut-outs				
Concrete, elevated slabs, mesh reinforcing				
Under 5 cf	C.F.		33.80	33.80
Over 5 cf	"		28.20	28.20
Bar reinforcing				
Under 5 cf	C.F.		56.50	56.50
Over 5 cf	"		42.30	42.30
Window removal				
Metal windows, trim included				
2'x3'	EA.		33.80	33.80
2'x4'	"		37.60	37.60
2'x6'	"		42.30	42.30
3'x4'	"		42.30	42.30
3'x6'	"		48.30	48.30
3'x8'	"		56.50	56.50
4'x4'	"		56.50	56.50
4'x6'	"		67.50	67.50
4'x8'	"		84.50	84.50
Wood windows, trim included				
2'x3'	EA.		18.80	18.80
2'x4'	"		19.90	19.90
2'x6'	"		21.15	21.15
3'x4'	"		22.55	22.55
3'x6'	"		24.15	24.15
3'x8'	"		26.00	26.00
6'x4'	"		28.20	28.20
6'x6'	"		30.80	30.80
6'x8'	"		33.80	33.80
Walls, concrete, bar reinforcing				
Small jobs	C.F.		22.55	22.55
Large jobs	"		18.80	18.80
Brick walls, not including toothing				
4" thick	S.F.		1.70	1.70

Demolition

02060.10 Building Demolition

	UNIT	MAT.	INST.	TOTAL
8" thick	S.F.		2.10	2.10
12" thick	"		2.80	2.80
16" thick	"		4.25	4.25
Concrete block walls, not including toothing				
4" thick	S.F.		1.90	1.90
6" thick	"		2.00	2.00
8" thick	"		2.10	2.10
10" thick	"		2.40	2.40
12" thick	"		2.80	2.80
Rubbish handling				
Load in dumpster or truck				
Minimum	C.F.		0.75	0.75
Maximum	"		1.15	1.15
For use of elevators, add				
Minimum	C.F.		0.17	0.17
Maximum	"		0.34	0.34
Rubbish hauling				
Hand loaded on trucks, 2 mile trip	C.Y.		27.80	27.80
Machine loaded on trucks, 2 mile trip	"		18.20	18.20

Highway Demolition

02065.10 Pavement Demolition

	UNIT	MAT.	INST.	TOTAL
Bituminous pavement, up to 3" thick				
On streets				
Minimum	S.Y.		5.20	5.20
Average	"		7.25	7.25
Maximum	"		12.10	12.10
On pipe trench				
Minimum	S.Y.		7.25	7.25
Average	"		9.10	9.10
Maximum	"		18.20	18.20
Concrete pavement, 6" thick				
No reinforcement				
Minimum	S.Y.		9.10	9.10
Average	"		12.10	12.10
Maximum	"		18.20	18.20
With wire mesh				
Minimum	S.Y.		14.00	14.00
Average	"		18.20	18.20
Maximum	"		22.75	22.75
With rebars				
Minimum	S.Y.		18.20	18.20
Average	"		22.75	22.75
Maximum	"		30.30	30.30
9" thick				

Highway Demolition

Highway Demolition	UNIT	MAT.	INST.	TOTAL
02065.10 Pavement Demolition				
No reinforcement				
Minimum	S.Y.		12.10	12.10
Average	"		15.15	15.15
Maximum	"		18.20	18.20
With wire mesh				
Minimum	S.Y.		19.15	19.15
Average	"		22.75	22.75
Maximum	"		28.00	28.00
With rebars				
Minimum	S.Y.		24.25	24.25
Average	"		30.30	30.30
Maximum	"		40.40	40.40
12" thick				
No reinforcement				
Minimum	S.Y.		15.15	15.15
Average	"		18.20	18.20
Maximum	"		22.75	22.75
With wire mesh				
Minimum	S.Y.		21.40	21.40
Average	"		26.00	26.00
Maximum	"		33.10	33.10
With rebars				
Minimum	S.Y.		30.30	30.30
Average	"		36.40	36.40
Maximum	"		45.50	45.50
Sidewalk, 4" thick, with disposal				
Minimum	S.Y.		4.35	4.35
Average	"		6.05	6.05
Maximum	"		8.65	8.65
Removal of pavement markings by waterblasting				
Minimum	S.F.		0.14	0.14
Average	"		0.17	0.17
Maximum	"		0.34	0.34
02065.15 Saw Cutting Pavement				
Pavement, bituminous				
2" thick	L.F.		1.40	1.40
3" thick	"		1.75	1.75
4" thick	"		2.15	2.15
5" thick	"		2.30	2.30
6" thick	"		2.50	2.50
Concrete pavement, with wire mesh				
4" thick	L.F.		2.65	2.65
5" thick	"		2.90	2.90
6" thick	"		3.15	3.15
8" thick	"		3.45	3.45
10" thick	"		3.85	3.85
Plain concrete, unreinforced				
4" thick	L.F.		2.30	2.30
5" thick	"		2.65	2.65
6" thick	"		2.90	2.90
8" thick	"		3.15	3.15

02 SITEWORK

Highway Demolition	UNIT	MAT.	INST.	TOTAL
02065.15 Saw Cutting Pavement				
10" thick	L.F.		3.45	3.45
02065.80 Curb & Gutter				
Curb removal				
Concrete, unreinforced				
Minimum	L.F.		3.65	3.65
Average	"		4.55	4.55
Maximum	"		5.70	5.70
Reinforced				
Minimum	L.F.		5.85	5.85
Average	"		6.50	6.50
Maximum	"		7.25	7.25
Combination curb and 2'gutter				
Unreinforced				
Minimum	L.F.		4.80	4.80
Average	"		6.25	6.25
Maximum	"		9.10	9.10
Reinforced				
Minimum	L.F.		7.60	7.60
Average	"		10.10	10.10
Maximum	"		18.20	18.20
Granite curb				
Minimum	L.F.		5.20	5.20
Average	"		6.05	6.05
Maximum	"		7.00	7.00
Asphalt curb				
Minimum	L.F.		3.05	3.05
Average	"		3.65	3.65
Maximum	"		4.35	4.35
02065.85 Guardrails				
Remove standard guardrail				
Steel				
Minimum	L.F.		4.55	4.55
Average	"		6.05	6.05
Maximum	"		9.10	9.10
Wood				
Minimum	L.F.		3.95	3.95
Average	"		4.65	4.65
Maximum	"		7.60	7.60
02075.80 Core Drilling				
Concrete				
6" thick				
3" dia.	EA.		31.30	31.30
4" dia.	"		36.50	36.50
6" dia.	"		43.80	43.80
8" dia.	"		73.00	73.00
8" thick				
3" dia.	EA.		43.80	43.80
4" dia.	"		55.00	55.00
6" dia.	"		62.50	62.50

Highway Demolition

02075.80 Core Drilling	UNIT	MAT.	INST.	TOTAL
8" dia.	EA.		87.50	87.50
10" thick				
3" dia.	EA.		55.00	55.00
4" dia.	"		62.50	62.50
6" dia.	"		73.00	73.00
8" dia.	"		110.00	110.00
12" thick				
3" dia.	EA.		73.00	73.00
4" dia.	"		87.50	87.50
6" dia.	"		110.00	110.00
8" dia.	"		150.00	150.00

Hazardous Waste

02080.10 Asbestos Removal	UNIT	MAT.	INST.	TOTAL
Enclosure using wood studs & poly, install & remove	S.F.	1.05	0.85	1.90
Trailer (change room)	DAY			98.89
Disposal suits (4 suits per man day)	"			40.53
Type C respirator mask, includes hose & filters, per man	"			20.26
Respirator mask & filter, light contamination	"			7.93
Air monitoring test, 12 tests per day				
Off job testing	DAY			1,045
On the job testing	"			1,384
Asbestos vacuum with attachments	EA.			606.32
Hydraspray piston pump	"			775.06
Negative air pressure system	"			707.85
Grade D breathing air equipment	"			1,773
Glove bag, 44" x 60" x 6 mil plastic	"			5.40
40 CY asbestos dumpster				
Weekly rental	EA.			673.53
Pick up/delivery	"			318.89
Asbestos dump fee	"			167.31

02080.12 Duct Insulation Removal	UNIT	MAT.	INST.	TOTAL
Remove duct insulation, duct size				
6" x 12"	L.F.		1.90	1.90
x 18"	"		2.60	2.60
x 24"	"		3.75	3.75
8" x 12"	"		2.80	2.80
x 18"	"		3.10	3.10
x 24"	"		4.25	4.25
12" x 12"	"		2.80	2.80
x 18"	"		3.75	3.75
x 24"	"		4.85	4.85

Hazardous Waste

	UNIT	MAT.	INST.	TOTAL
02080.15 Pipe Insulation Removal				
Removal, asbestos insulation				
2" thick, pipe				
1" to 3" dia.	L.F.		2.80	2.80
4" to 6" dia.	"		3.20	3.20
3" thick				
7" to 8" dia.	L.F.		3.40	3.40
9" to 10" dia.	"		3.55	3.55
11" to 12" dia.	"		3.75	3.75
13" to 14" dia.	"		4.00	4.00
15" to 18" dia.	"		4.25	4.25

Site Demolition

	UNIT	MAT.	INST.	TOTAL
02105.10 Catch Basins/manholes				
Abandon catch basin or manhole (fill with sand)				
Minimum	EA.		230.00	230.00
Average	"		360.00	360.00
Maximum	"		610.00	610.00
Remove and reset frame and cover				
Minimum	EA.		120.00	120.00
Average	"		180.00	180.00
Maximum	"		300.00	300.00
Remove catch basin, to 10' deep				
Masonry				
Minumum	EA.		360.00	360.00
Average	"		450.00	450.00
Maximum	"		610.00	610.00
Concrete				
Minimum	EA.		450.00	450.00
Average	"		610.00	610.00
Maximum	"		730.00	730.00
02105.20 Fences				
Remove fencing				
Chain link, 8' high				
For disposal	L.F.		1.70	1.70
For reuse	"		4.25	4.25
Wood				
4' high	S.F.		1.15	1.15
6' high	"		1.35	1.35
8' high	"		1.70	1.70
Masonry				
8" thick				
4' high	S.F.		3.40	3.40
6' high	"		4.25	4.25
8' high	"		4.85	4.85

02 SITEWORK

Site Demolition	UNIT	MAT.	INST.	TOTAL
02105.20 Fences				
12" thick				
4' high	S.F.		5.65	5.65
6' high	"		6.75	6.75
8' high	"		8.45	8.45
12' high	"		11.30	11.30
02105.30 Hydrants				
Remove fire hydrant				
Minimum	EA.		230.00	230.00
Average	"		300.00	300.00
Maximum	"		450.00	450.00
Remove and reset fire hydrant				
Minimum	EA.		610.00	610.00
Average	"		910.00	910.00
Maximum	"		1,820	1,820
02105.42 Drainage Piping				
Remove drainage pipe, not including excavation				
12" dia.				
Minimum	L.F.		6.05	6.05
Average	"		7.60	7.60
Maximum	"		9.55	9.55
18" dia.				
Minimum	L.F.		8.25	8.25
Average	"		9.55	9.55
Maximum	"		12.10	12.10
24" dia.				
Minimum	L.F.		10.10	10.10
Average	"		12.10	12.10
Maximum	"		15.15	15.15
36" dia.				
Minimum	L.F.		12.10	12.10
Average	"		15.15	15.15
Maximum	"		19.15	19.15
02105.43 Gas Piping				
Remove welded steel pipe, not including excavation				
4" dia.				
Minimum	L.F.		9.10	9.10
Average	"		11.35	11.35
Maximum	"		15.15	15.15
5" dia.				
Minimum	L.F.		15.15	15.15
Average	"		18.20	18.20
Maximum	"		22.75	22.75
6" dia.				
Minimum	L.F.		19.15	19.15
Average	"		22.75	22.75
Maximum	"		30.30	30.30
8" dia.				
Minimum	L.F.		28.00	28.00
Average	"		36.40	36.40

Site Demolition

	UNIT	MAT.	INST.	TOTAL
02105.43 Gas Piping				
Maximum	L.F.		47.80	47.80
10" dia.				
Minimum	L.F.		36.40	36.40
Average	"		45.50	45.50
Maximum	"		60.50	60.50
02105.45 Sanitary Piping				
Remove sewer pipe, not including excavation				
4" dia.				
Minimum	L.F.		5.05	5.05
Average	"		7.25	7.25
Maximum	"		12.10	12.10
6" dia.				
Minimum	L.F.		5.70	5.70
Average	"		8.25	8.25
Maximum	"		15.15	15.15
8" dia.				
Minimum	L.F.		6.05	6.05
Average	"		9.10	9.10
Maximum	"		18.20	18.20
10" dia.				
Minimum	L.F.		6.50	6.50
Average	"		9.55	9.55
Maximum	"		20.20	20.20
12" dia.				
Minimum	L.F.		7.00	7.00
Average	"		10.10	10.10
Maximum	"		22.75	22.75
15" dia.				
Minimum	L.F.		7.60	7.60
Average	"		10.70	10.70
Maximum	"		26.00	26.00
18" dia.				
Minimum	L.F.		8.25	8.25
Average	"		12.10	12.10
Maximum	"		30.30	30.30
24" dia.				
Minimum	L.F.		9.10	9.10
Average	"		15.15	15.15
Maximum	"		36.40	36.40
30" dia.				
Minimum	L.F.		10.10	10.10
Average	"		18.20	18.20
Maximum	"		45.50	45.50
36" dia.				
Minimum	L.F.		12.10	12.10
Average	"		22.75	22.75
Maximum	"		60.50	60.50

02 SITEWORK

Site Demolition

	UNIT	MAT.	INST.	TOTAL
02105.48 Water Piping				
Remove water pipe, not including excavation				
4" dia.				
Minimum	L.F.		7.25	7.25
Average	"		8.25	8.25
Maximum	"		9.55	9.55
6" dia.				
Minimum	L.F.		7.60	7.60
Average	"		8.65	8.65
Maximum	"		10.10	10.10
8" dia.				
Minimum	L.F.		8.25	8.25
Average	"		9.55	9.55
Maximum	"		11.35	11.35
10" dia.				
Minimum	L.F.		8.65	8.65
Average	"		10.10	10.10
Maximum	"		12.10	12.10
12" dia.				
Minimum	L.F.		9.10	9.10
Average	"		10.70	10.70
Maximum	"		13.00	13.00
14" dia.				
Minimum	L.F.		9.55	9.55
Average	"		11.35	11.35
Maximum	"		14.00	14.00
16" dia.				
Minimum	L.F.		10.10	10.10
Average	"		12.10	12.10
Maximum	"		15.15	15.15
18" dia.				
Minimum	L.F.		10.70	10.70
Average	"		13.00	13.00
Maximum	"		16.55	16.55
20" dia.				
Minimum	L.F.		11.35	11.35
Average	"		14.00	14.00
Maximum	"		18.20	18.20
Remove valves				
6"	EA.		91.00	91.00
10"	"		100.00	100.00
14"	"		110.00	110.00
18"	"		150.00	150.00
02105.60 Underground Tanks				
Remove underground storage tank, and backfill				
50 to 250 gals	EA.		610.00	610.00
600 gals	"		610.00	610.00
1000 gals	"		910.00	910.00
4000 gals	"		1,450	1,450
5000 gals	"		1,450	1,450
10,000 gals	"		2,420	2,420
12,000 gals	"		3,030	3,030

Site Demolition

Site Demolition	UNIT	MAT.	INST.	TOTAL
02105.60 Underground Tanks				
15,000 gals	EA.		3,640	3,640
20,000 gals	"		4,550	4,550
02105.66 Septic Tanks				
Remove septic tank				
1000 gals	EA.		150.00	150.00
2000 gals	"		180.00	180.00
5000 gals	"		230.00	230.00
15,000 gals	"		1,820	1,820
25,000 gals	"		2,420	2,420
40,000 gals	"		3,640	3,640
02105.80 Walls, Exterior				
Concrete wall				
Light reinforcing				
6" thick	S.F.		9.10	9.10
8" thick	"		9.55	9.55
10" thick	"		10.10	10.10
12" thick	"		11.35	11.35
Medium reinforcing				
6" thick	S.F.		9.55	9.55
8" thick	"		10.10	10.10
10" thick	"		11.35	11.35
12" thick	"		13.00	13.00
Heavy reinforcing				
6" thick	S.F.		10.70	10.70
8" thick	"		11.35	11.35
10" thick	"		13.00	13.00
12" thick	"		15.15	15.15
Masonry				
No reinforcing				
8" thick	S.F.		4.05	4.05
12" thick	"		4.55	4.55
16" thick	"		5.20	5.20
Horizontal reinforcing				
8" thick	S.F.		4.55	4.55
12" thick	"		4.90	4.90
16" thick	"		5.85	5.85
Vertical reinforcing				
8" thick	S.F.		5.85	5.85
12" thick	"		6.75	6.75
16" thick	"		8.25	8.25
Remove concrete headwall				
15" pipe	EA.		130.00	130.00
18" pipe	"		150.00	150.00
24" pipe	"		170.00	170.00
30" pipe	"		180.00	180.00
36" pipe	"		200.00	200.00
48" pipe	"		260.00	260.00
60" pipe	"		360.00	360.00

02 SITEWORK

Site Demolition

	UNIT	MAT.	INST.	TOTAL
02110.10 Clearing And Grubbing				
Clear wooded area				
Light density	ACRE		4,550	4,550
Medium density	"		6,060	6,060
Heavy density	"		7,270	7,270
02110.50 Tree Cutting & Clearing				
Cut trees and clear out stumps				
9" to 12" dia.	EA.		360.00	360.00
To 24" dia.	"		450.00	450.00
24" dia. and up	"		610.00	610.00
Loading and trucking				
For machine load, per load, round trip				
1 mile	EA.		72.50	72.50
3 mile	"		82.50	82.50
5 mile	"		91.00	91.00
10 mile	"		120.00	120.00
20 mile	"		180.00	180.00
Hand loaded, round trip				
1 mile	EA.		170.00	170.00
3 mile	"		200.00	200.00
5 mile	"		230.00	230.00
10 mile	"		280.00	280.00
20 mile	"		350.00	350.00
Tree trimming for pole line construction				
Light cutting	L.F.		0.91	0.91
Medium cutting	"		1.20	1.20
Heavy cutting	"		1.80	1.80

Dewatering

	UNIT	MAT.	INST.	TOTAL
02144.10 Wellpoint Systems				
Pumping, gas driven, 50' hose				
3" header pipe	DAY		690.00	690.00
6" header pipe	"		870.00	870.00
Wellpoint system per job				
6" header pipe	L.F.	1.60	2.80	4.40
8" header pipe	"	1.95	3.45	5.40
10" header pipe	"	2.70	4.65	7.35
Jetting wellpoint system				
14' long	EA.	52.00	46.30	98.30
18' long	"	61.50	58.00	119.50
Sand filter for wellpoints	L.F.	2.60	1.15	3.75
Replacement of wellpoint components	EA.		13.90	13.90

02 SITEWORK

Shoring And Underpinning

Shoring And Underpinning	UNIT	MAT.	INST.	TOTAL
02162.10 Trench Sheeting				
Closed timber, including pull and salvage, excavation				
8' deep	S.F.	3.10	6.35	9.45
10' deep	"	3.10	6.70	9.80
12' deep	"	3.10	7.05	10.15
14' deep	"	3.10	7.45	10.55
16' deep	"	3.10	7.95	11.05
18' deep	"	3.10	9.05	12.15
20' deep	"	3.10	9.75	12.85
02170.10 Cofferdams				
Cofferdam, steel, driven from shore				
15' deep	S.F.	12.60	15.40	28.00
20' deep	"	12.60	14.40	27.00
25' deep	"	12.60	13.50	26.10
30' deep	"	12.60	12.70	25.30
40' deep	"	12.60	12.00	24.60
Driven from barge				
20' deep	S.F.	12.60	16.60	29.20
30' deep	"	12.60	15.40	28.00
40' deep	"	12.60	14.40	27.00
50' deep	"	12.60	13.50	26.10

Earthwork

Earthwork	UNIT	MAT.	INST.	TOTAL
02210.10 Hauling Material				
Haul material by 10 cy dump truck, round trip distance				
1 mile	C.Y.		3.85	3.85
2 mile	"		4.65	4.65
5 mile	"		6.30	6.30
10 mile	"		6.95	6.95
20 mile	"		7.70	7.70
30 mile	"		9.25	9.25
Site grading, cut & fill, sandy clay, 200' haul, 75 hp dozer	"		2.80	2.80
Spread topsoil by equipment on site	"		3.10	3.10
Site grading (cut and fill to 6") less than 1 acre				
75 hp dozer	C.Y.		4.65	4.65
1.5 cy backhoe/loader	"		6.95	6.95
02210.30 Bulk Excavation				
Excavation, by small dozer				
Large areas	C.Y.		1.40	1.40
Small areas	"		2.30	2.30
Trim banks	"		3.45	3.45
Drag line				
1-1/2 cy bucket				
Sand or gravel	C.Y.		3.05	3.05

Earthwork	UNIT	MAT.	INST.	TOTAL
02210.30 Bulk Excavation				
Light clay	C.Y.		4.05	4.05
Heavy clay	"		4.55	4.55
Unclassified	"		4.85	4.85
2 cy bucket				
Sand or gravel	C.Y.		2.80	2.80
Light clay	"		3.65	3.65
Heavy clay	"		4.05	4.05
Unclassified	"		4.30	4.30
2-1/2 cy bucket				
Sand or gravel	C.Y.		2.60	2.60
Light clay	"		3.30	3.30
Heavy clay	"		3.65	3.65
Unclassified	"		3.85	3.85
3 cy bucket				
Sand or gravel	C.Y.		2.25	2.25
Light clay	"		3.05	3.05
Heavy clay	"		3.30	3.30
Unclassified	"		3.45	3.45
Hydraulic excavator				
1 cy capacity				
Light material	C.Y.		3.05	3.05
Medium material	"		3.65	3.65
Wet material	"		4.55	4.55
Blasted rock	"		5.20	5.20
1-1/2 cy capacity				
Light material	C.Y.		1.20	1.20
Medium material	"		1.60	1.60
Wet material	"		1.95	1.95
Blasted rock	"		2.40	2.40
2 cy capacity				
Light material	C.Y.		1.05	1.05
Medium material	"		1.40	1.40
Wet material	"		1.60	1.60
Blasted rock	"		1.95	1.95
Wheel mounted front-end loader				
7/8 cy capacity				
Light material	C.Y.		2.40	2.40
Medium material	"		2.75	2.75
Wet material	"		3.20	3.20
Blasted rock	"		3.85	3.85
1-1/2 cy capacity				
Light material	C.Y.		1.40	1.40
Medium material	"		1.50	1.50
Wet material	"		1.60	1.60
Blasted rock	"		1.75	1.75
2-1/2 cy capacity				
Light material	C.Y.		1.15	1.15
Medium material	"		1.20	1.20
Wet material	"		1.30	1.30
Blasted rock	"		1.40	1.40
3-1/2 cy capacity				
Light material	C.Y.		1.05	1.05

Earthwork

Earthwork	UNIT	MAT.	INST.	TOTAL
02210.30 Bulk Excavation				
Medium material	C.Y.		1.15	1.15
Wet material	"		1.20	1.20
Blasted rock	"		1.30	1.30
6 cy capacity				
Light material	C.Y.		0.64	0.64
Medium material	"		0.69	0.69
Wet material	"		0.74	0.74
Blasted rock	"		0.80	0.80
Track mounted front-end loader				
1-1/2 cy capacity				
Light material	C.Y.		1.60	1.60
Medium material	"		1.75	1.75
Wet material	"		1.95	1.95
Blasted rock	"		2.15	2.15
2-3/4 cy capacity				
Light material	C.Y.		0.96	0.96
Medium material	"		1.05	1.05
Wet material	"		1.20	1.20
Blasted rock	"		1.40	1.40
02220.10 Borrow				
Borrow fill, F.O.B. at pit				
Sand, haul to site, round trip				
10 mile	C.Y.	11.65	9.65	21.30
20 mile	"	11.65	16.05	27.70
30 mile	"	11.65	24.10	35.75
Place borrow fill and compact				
Less than 1 in 4 slope	C.Y.	11.65	4.80	16.45
Greater than 1 in 4 slope	"	11.65	6.45	18.10
02220.20 Gravel And Stone				
F.O.B. PLANT				
No. 21 crusher run stone	C.Y.			27.97
No. 26 crusher run stone	"			27.20
No. 57 stone	"			28.76
No. 67 gravel	"			13.98
No. 68 stone	"			27.97
No. 78 stone	"			27.97
No. 78 gravel, (pea gravel)	"			16.32
No. 357 or B-3 stone	"			28.76
Structural & foundation backfill				
No. 21 crusher run stone	TON			24.85
No. 26 crusher run stone	"			25.64
No. 57 stone	"			27.97
No. 67 gravel	"			12.44
No. 68 stone	"			23.89
No. 78 stone	"			23.89
No. 78 gravel, (pea gravel)	"			19.43
No. 357 or B-3 stone	"			27.65

02 SITEWORK

Earthwork

02220.40 Building Excavation	UNIT	MAT.	INST.	TOTAL
Structural excavation, unclassified earth				
3/8 cy backhoe	C.Y.		12.85	12.85
3/4 cy backhoe	"		9.65	9.65
1 cy backhoe	"		8.05	8.05
Foundation backfill and compaction by machine	"		19.30	19.30

02220.50 Utility Excavation	UNIT	MAT.	INST.	TOTAL
Trencher, sandy clay, 8" wide trench				
18" deep	L.F.		1.55	1.55
24" deep	"		1.75	1.75
36" deep	"		2.00	2.00
Trench backfill, 95% compaction				
Tamp by hand	C.Y.		21.15	21.15
Vibratory compaction	"		16.90	16.90
Trench backfilling, with borrow sand, place & compact	"	11.75	16.90	28.65

02220.60 Trenching	UNIT	MAT.	INST.	TOTAL
Trenching and continuous footing excavation				
By gradall				
1 cy capacity				
Light soil	C.Y.		2.75	2.75
Medium soil	"		2.95	2.95
Heavy/wet soil	"		3.20	3.20
Loose rock	"		3.50	3.50
Blasted rock	"		3.70	3.70
By hydraulic excavator				
1/2 cy capacity				
Light soil	C.Y.		3.20	3.20
Medium soil	"		3.50	3.50
Heavy/wet soil	"		3.85	3.85
Loose rock	"		4.30	4.30
Blasted rock	"		4.80	4.80
1 cy capacity				
Light soil	C.Y.		2.25	2.25
Medium soil	"		2.40	2.40
Heavy/wet soil	"		2.55	2.55
Loose rock	"		2.75	2.75
Blasted rock	"		2.95	2.95
1-1/2 cy capacity				
Light soil	C.Y.		2.05	2.05
Medium soil	"		2.15	2.15
Heavy/wet soil	"		2.25	2.25
Loose rock	"		2.40	2.40
Blasted rock	"		2.55	2.55
2 cy capacity				
Light soil	C.Y.		1.95	1.95
Medium soil	"		2.05	2.05
Heavy/wet soil	"		2.15	2.15
Loose rock	"		2.25	2.25
Blasted rock	"		2.40	2.40
2-1/2 cy capacity				
Light soil	C.Y.		1.75	1.75

02 SITEWORK

Earthwork

Earthwork	UNIT	MAT.	INST.	TOTAL
02220.60 Trenching				
Medium soil	C.Y.		1.85	1.85
Heavy/wet soil	"		1.95	1.95
Loose rock	"		2.05	2.05
Blasted rock	"		2.15	2.15
Trencher, chain, 1' wide to 4' deep				
Light soil	C.Y.		1.75	1.75
Medium soil	"		2.00	2.00
Heavy soil	"		2.30	2.30
Hand excavation				
Bulk, wheeled 100'				
Normal soil	C.Y.		37.60	37.60
Sand or gravel	"		33.80	33.80
Medium clay	"		48.30	48.30
Heavy clay	"		67.50	67.50
Loose rock	"		84.50	84.50
Trenches, up to 2' deep				
Normal soil	C.Y.		42.30	42.30
Sand or gravel	"		37.60	37.60
Medium clay	"		56.50	56.50
Heavy clay	"		84.50	84.50
Loose rock	"		110.00	110.00
Trenches, to 6' deep				
Normal soil	C.Y.		48.30	48.30
Sand or gravel	"		42.30	42.30
Medium clay	"		67.50	67.50
Heavy clay	"		110.00	110.00
Loose rock	"		170.00	170.00
Backfill trenches				
With compaction				
By hand	C.Y.		28.20	28.20
By 60 hp tracked dozer	"		1.75	1.75
By 200 hp tracked dozer	"		1.05	1.05
By small front-end loader	"		2.00	2.00
Spread dumped fill or gravel, no compaction				
6" layers	S.Y.		1.15	1.15
12" layers	"		1.40	1.40
Compaction in 6" layers				
By hand with air tamper	S.Y.		0.88	0.88
Backfill trenches, sand bedding, no compaction				
By hand	C.Y.	11.55	28.20	39.75
By small front-end loader	"	11.55	2.75	14.30
02220.70 Roadway Excavation				
Roadway excavation				
1/4 mile haul	C.Y.		1.95	1.95
2 mile haul	"		3.20	3.20
5 mile haul	"		4.80	4.80
Excavation of open ditches	"		1.40	1.40
Trim banks, swales or ditches	S.Y.		1.60	1.60
Bulk swale excavation by dragline				
Small jobs	C.Y.		4.55	4.55
Large jobs	"		2.60	2.60

02 SITEWORK

Earthwork	UNIT	MAT.	INST.	TOTAL
02220.70 Roadway Excavation				
Spread base course				
Roll and compact	C.Y.		2.40	2.40
	"		3.20	3.20
02220.71 Base Course				
Base course, crushed stone				
3" thick	S.Y.	2.55	0.48	3.03
4" thick	"	3.40	0.52	3.92
6" thick	"	5.15	0.57	5.72
8" thick	"	6.50	0.64	7.14
10" thick	"	8.05	0.69	8.74
12" thick	"	8.75	0.80	9.55
Base course, bank run gravel				
4" deep	S.Y.	1.85	0.51	2.36
6" deep	"	2.70	0.55	3.25
8" deep	"	3.50	0.60	4.10
10" deep	"	4.55	0.64	5.19
12" deep	"	5.30	0.74	6.04
Prepare and roll sub base				
Minimum	S.Y.		0.48	0.48
Average	"		0.60	0.60
Maximum	"		0.80	0.80
02220.90 Hand Excavation				
Excavation				
To 2' deep				
Normal soil	C.Y.		37.60	37.60
Sand and gravel	"		33.80	33.80
Medium clay	"		42.30	42.30
Heavy clay	"		48.30	48.30
Loose rock	"		56.50	56.50
To 6' deep				
Normal soil	C.Y.		40.00	40.00
Sand and gravel	"		42.30	42.30
Medium clay	"		56.50	56.50
Heavy clay	"		67.50	67.50
Loose rock	"		84.50	84.50
Backfilling foundation without compaction, 6" lifts	"		21.15	21.15
Compaction of backfill around structures or in trench				
By hand with air tamper	C.Y.		24.15	24.15
By hand with vibrating plate tamper	"		22.55	22.55
1 ton roller	"		34.70	34.70
Miscellaneous hand labor				
Trim slopes, sides of excavation	S.F.		0.06	0.06
Trim bottom of excavation	"		0.07	0.07
Excavation around obstructions and services	C.Y.		110.00	110.00
02240.05 Soil Stabilization				
Straw bale secured with rebar	L.F.	1.60	1.15	2.75
Filter barrier, 18" high filter fabric	"	1.60	3.40	5.00
Sediment fence, 36" fabric with 6" mesh	"	3.75	4.25	8.00
Soil stabilization with tar paper, burlap, straw and stakes	S.F.	0.32	0.05	0.37

Earthwork

Earthwork	UNIT	MAT.	INST.	TOTAL
02240.30 Geotextile				
Filter cloth, light reinforcement				
Woven				
12'-6" wide x 50' long	S.F.	0.30	0.05	0.34
Various lengths	"	0.45	0.05	0.49
Non-woven				
14'-8" wide x 430' long	S.F.	0.16	0.05	0.21
Various lengths	"	0.23	0.05	0.28
02270.10 Slope Protection				
Gabions, stone filled				
6" deep	S.Y.	23.05	17.35	40.40
9" deep	"	28.30	19.85	48.15
12" deep	"	37.40	23.15	60.55
18" deep	"	47.90	27.80	75.70
36" deep	"	85.00	46.30	131.30
02270.40 Riprap				
Riprap				
Crushed stone blanket, max size 2-1/2"	TON	25.70	53.00	78.70
Stone, quarry run, 300 lb. stones	"	32.10	48.80	80.90
400 lb. stones	"	32.10	45.30	77.40
500 lb. stones	"	32.10	42.30	74.40
750 lb. stones	"	32.10	39.70	71.80
Dry concrete riprap in bags 3" thick, 80 lb. per bag	BAG	5.15	2.65	7.80
02280.20 Soil Treatment				
Soil treatment, termite control pretreatment				
Under slabs	S.F.	0.14	0.19	0.33
By walls	"	0.14	0.23	0.37
02290.30 Weed Control				
Weed control, bromicil, 15 lb./acre, wettable powder	ACRE	250.00	170.00	420.00
Vegetation control, by application of plant killer	S.Y.	0.02	0.14	0.16
Weed killer, lawns and fields	"	0.21	0.07	0.28

Tunneling

Tunneling	UNIT	MAT.	INST.	TOTAL
02300.10 Pipe Jacking				
Pipe casing, horizontal jacking				
18" dia.	L.F.	75.00	70.50	145.50
21" dia.	"	88.50	75.50	164.00
24" dia.	"	93.00	79.50	172.50
27" dia.	"	100.00	79.50	179.50
30" dia.	"	110.00	83.50	193.50
36" dia.	"	130.00	90.50	220.50
42" dia.	"	190.00	100.00	290.00

Tunneling

Tunneling	UNIT	MAT.	INST.	TOTAL
02300.10 Pipe Jacking				
48" dia.	L.F.	230.00	110.00	340.00

Piles And Caissons

Piles And Caissons	UNIT	MAT.	INST.	TOTAL
02360.50 Prestressed Piling				
Prestressed concrete piling, less than 60' long				
10" sq.	L.F.	9.75	4.50	14.25
12" sq.	"	13.65	4.70	18.35
14" sq.	"	14.20	4.80	19.00
16" sq.	"	17.45	4.90	22.35
18" sq.	"	24.15	5.25	29.40
20" sq.	"	32.70	5.40	38.10
24" sq.	"	41.70	5.55	47.25
More than 60' long				
12" sq.	L.F.	13.90	3.85	17.75
14" sq.	"	15.30	3.90	19.20
16" sq.	"	18.45	4.00	22.45
18" sq.	"	24.20	4.05	28.25
20" sq.	"	32.70	4.15	36.85
24" sq.	"	39.00	4.25	43.25
Straight cylinder, less than 60' long				
12" dia.	L.F.	12.65	4.90	17.55
14" dia.	"	17.15	5.00	22.15
16" dia.	"	20.85	5.15	26.00
18" dia.	"	26.40	5.25	31.65
20" dia.	"	31.30	5.40	36.70
24" dia.	"	38.80	5.55	44.35
More than 60' long				
12" dia.	L.F.	12.65	3.90	16.55
14" dia.	"	17.15	4.00	21.15
16" dia.	"	20.85	4.05	24.90
18" dia.	"	26.40	4.15	30.55
20" dia.	"	31.30	4.25	35.55
24" dia.	"	39.50	4.30	43.80
Concrete sheet piling				
12" thick x 20' long	S.F.	14.50	10.80	25.30
25' long	"	14.50	9.80	24.30
30' long	"	14.50	9.00	23.50
35' long	"	14.50	8.30	22.80
40' long	"	14.50	7.70	22.20
16" thick x 40' long	"	19.85	6.00	25.85
45' long	"	19.85	5.70	25.55
50' long	"	19.85	5.40	25.25
55' long	"	19.85	5.15	25.00
60' long	"	19.85	4.90	24.75

Piles And Caissons

02360.60 Steel Piles

	UNIT	MAT.	INST.	TOTAL
H-section piles				
8x8				
36 lb/ft				
30' long	L.F.	12.95	9.00	21.95
40' long	"	12.95	7.20	20.15
50' long	"	12.95	6.00	18.95
10x10				
42 lb/ft				
30' long	L.F.	14.75	9.00	23.75
40' long	"	14.75	7.20	21.95
50' long	"	14.75	6.00	20.75
57 lb/ft				
30' long	L.F.	18.85	9.00	27.85
40' long	"	18.85	7.20	26.05
50' long	"	18.85	6.00	24.85
12x12				
53 lb/ft				
30' long	L.F.	17.65	9.80	27.45
40' long	"	17.65	7.70	25.35
50' long	"	17.65	6.00	23.65
74 lb/ft				
30' long	L.F.	23.55	9.80	33.35
40' long	"	23.55	7.70	31.25
50' long	"	23.55	6.00	29.55
14x14				
73 lb/ft				
40' long	L.F.	23.55	9.80	33.35
50' long	"	23.55	7.70	31.25
60' long	"	23.55	6.00	29.55
89 lb/ft				
40' long	L.F.	29.50	9.80	39.30
50' long	"	29.50	7.70	37.20
60' long	"	29.50	6.00	35.50
102 lb/ft				
40' long	L.F.	34.20	9.80	44.00
50' long	"	34.20	7.70	41.90
60' long	"	34.20	6.00	40.20
117 lb/ft				
40' long	L.F.	38.90	10.30	49.20
50' long	"	38.90	8.00	46.90
60' long	"	38.90	6.15	45.05
Splice				
8"	EA.	65.00	56.50	121.50
10"	"	70.50	67.50	138.00
12"	"	96.50	67.50	164.00
14"	"	120.00	84.50	204.50
Driving cap				
8"	EA.	36.50	33.80	70.30
10"	"	36.50	42.30	78.80
12"	"	36.50	42.30	78.80
14"	"	36.50	48.30	84.80
Standard point				

Piles And Caissons

	UNIT	MAT.	INST.	TOTAL
02360.60 Steel Piles				
8"	EA.	47.20	33.80	81.00
10"	"	58.50	42.30	100.80
12"	"	83.00	48.30	131.30
14"	"	110.00	56.50	166.50
Heavy duty point				
8"	EA.	54.00	37.60	91.60
10"	"	72.00	48.30	120.30
12"	"	96.00	56.50	152.50
14"	"	120.00	67.50	187.50
Tapered friction piles, with fluted steel casing, up to 50'				
With 4000 psi concrete no reinforcing				
12" dia.	L.F.	16.30	5.40	21.70
14" dia.	"	18.80	5.55	24.35
16" dia.	"	22.55	5.70	28.25
18" dia.	"	25.30	6.35	31.65
02360.65 Steel Pipe Piles				
Concrete filled, 3000# concrete, up to 40'				
8" dia.	L.F.	16.05	7.70	23.75
10" dia.	"	19.70	8.00	27.70
12" dia.	"	24.10	8.30	32.40
14" dia.	"	26.60	8.65	35.25
16" dia.	"	30.10	9.00	39.10
18" dia.	"	36.50	9.40	45.90
Pipe piles, non-filled				
8" dia.	L.F.	12.65	6.00	18.65
10" dia.	"	14.95	6.15	21.10
12" dia.	"	18.40	6.35	24.75
14" dia.	"	19.55	6.75	26.30
16" dia.	"	22.95	6.95	29.90
18" dia.	"	25.30	7.20	32.50
Splice				
8" dia.	EA.	43.40	67.50	110.90
10" dia.	"	46.00	67.50	113.50
12" dia.	"	51.50	84.50	136.00
14" dia.	"	57.50	84.50	142.00
16" dia.	"	76.50	110.00	186.50
18" dia.	"	100.00	110.00	210.00
Standard point				
8" dia.	EA.	55.50	67.50	123.00
10" dia.	"	62.50	67.50	130.00
12" dia.	"	99.00	84.50	183.50
14" dia.	"	140.00	84.50	224.50
16" dia.	"	150.00	110.00	260.00
18" dia.	"	190.00	110.00	300.00
Heavy duty point				
8" dia.	EA.	69.50	84.50	154.00
10" dia.	"	83.50	84.50	168.00
12" dia.	"	110.00	110.00	220.00
14" dia.	"	150.00	110.00	260.00
16" dia.	"	190.00	140.00	330.00
18" dia.	"	250.00	140.00	390.00

Piles And Caissons

	UNIT	MAT.	INST.	TOTAL
02360.70 Steel Sheet Piling				
Steel sheet piling,12" wide				
20' long	S.F.	10.40	10.80	21.20
35' long	"	10.40	7.70	18.10
50' long	"	10.40	5.40	15.80
Over 50' long	"	10.40	4.90	15.30
02360.80 Wood And Timber Piles				
Treated wood piles, 12" butt, 8" tip				
25' long	L.F.	8.35	10.80	19.15
30' long	"	8.90	9.00	17.90
35' long	"	8.90	7.70	16.60
40' long	"	8.90	6.75	15.65
12" butt, 7" tip				
40' long	L.F.	10.10	6.75	16.85
45' long	"	10.10	6.00	16.10
50' long	"	12.25	5.40	17.65
55' long	"	12.25	4.90	17.15
60' long	"	12.25	4.50	16.75
02360.90 Pile Testing				
Pile test				
50 ton to 100 ton	EA.			16,798
To 200 ton	"			24,934
To 300 ton	"			27,559
To 400 ton	"			41,995
To 600 ton	"			48,557
02380.10 Caissons				
Caisson, including 3000# concrete, in stable ground				
18" dia.	L.F.	13.50	21.60	35.10
24" dia.	"	22.55	22.50	45.05
30" dia.	"	35.50	27.00	62.50
36" dia.	"	49.60	30.80	80.40
48" dia.	"	85.00	36.00	121.00
60" dia.	"	140.00	49.10	189.10
72" dia.	"	210.00	60.00	270.00
84" dia.	"	250.00	77.00	327.00
Wet ground, casing required but pulled				
18" dia.	L.F.	13.50	27.00	40.50
24" dia.	"	22.55	30.00	52.55
30" dia.	"	35.50	33.70	69.20
36" dia.	"	49.60	36.00	85.60
48" dia.	"	85.00	45.00	130.00
60" dia.	"	140.00	60.00	200.00
72" dia.	"	210.00	90.00	300.00
84" dia.	"	250.00	130.00	380.00
Soft rock				
18" dia.	L.F.	13.50	77.00	90.50
24" dia.	"	22.55	130.00	152.55
30" dia.	"	35.50	180.00	215.50
36" dia.	"	49.60	270.00	319.60
48" dia.	"	85.00	360.00	445.00

Piles And Caissons

02380.10 Caissons	UNIT	MAT.	INST.	TOTAL
60" dia.	L.F.	140.00	540.00	680.00
72" dia.	"	210.00	600.00	810.00
84" dia.	"	250.00	670.00	920.00

Railroad Work

02450.10 Railroad Work	UNIT	MAT.	INST.	TOTAL
Rail				
90 lb	L.F.	22.15	0.73	22.88
100 lb	"	25.30	0.73	26.03
115 lb	"	28.40	0.73	29.13
132 lb	"	31.60	0.73	32.33
Rail relay				
90 lb	L.F.	8.85	0.73	9.58
100 lb	"	9.80	0.73	10.53
115 lb	"	11.85	0.73	12.58
132 lb	"	14.35	0.73	15.08
New angle bars, per pair				
90 lb	EA.	82.00	0.91	82.91
100 lb	"	88.50	0.91	89.41
115 lb	"	110.00	0.91	110.91
132 lb	"	130.00	0.91	130.91
Angle bar relay				
90 lb	EA.	34.00	0.91	34.91
100 lb	"	34.60	0.91	35.51
115 lb	"	36.30	0.91	37.21
132 lb	"	38.70	0.91	39.61
New tie plates				
90 lb	EA.	11.05	0.65	11.70
100 lb	"	11.55	0.65	12.20
115 lb	"	12.65	0.65	13.30
132 lb	"	13.45	0.65	14.10
Tie plate relay				
90 lb	EA.	3.45	0.65	4.10
100 lb	"	4.90	0.65	5.55
115 lb	"	4.90	0.65	5.55
132 lb	"	6.05	0.65	6.70
Track accessories				
Wooden cross ties, 8'	EA.	44.20	4.55	48.75
Concrete cross ties, 8'	"	270.00	9.10	279.10
Tie plugs, 5"	"	17.40	0.46	17.86
Track bolts and nuts, 1"	"	4.30	0.46	4.76
Lockwashers, 1"	"	1.10	0.30	1.40
Track spikes, 6"	"	1.05	1.80	2.85
Wooden switch ties	B.F.	1.60	0.46	2.06
Rail anchors	EA.	4.35	1.65	6.00

Railroad Work

02450.10 Railroad Work	UNIT	MAT.	INST.	TOTAL
Ballast	TON	12.50	9.10	21.60
Gauge rods	EA.	34.00	7.25	41.25
Compromise splice bars	"	430.00	12.10	442.10
Turnout				
90 lb	EA.	11,850	1,820	13,670
100 lb	"	12,480	1,820	14,300
110 lb	"	13,420	1,820	15,240
115 lb	"	13,740	1,820	15,560
132 lb	"	15,000	1,820	16,820
Turnout relay				
90 lb	EA.	7,580	1,820	9,400
100 lb	"	8,370	1,820	10,190
110 lb	"	8,690	1,820	10,510
115 lb	"	9,160	1,820	10,980
132 lb	"	9,950	1,820	11,770
Railroad track in place, complete				
New rail				
90 lb	L.F.	150.00	18.20	168.20
100 lb	"	150.00	18.20	168.20
110 lb	"	140.00	18.20	158.20
115 lb	"	160.00	18.20	178.20
132 lb	"	160.00	18.20	178.20
Rail relay				
90 lb	L.F.	83.00	18.20	101.20
100 lb	"	91.50	18.20	109.70
110 lb	"	91.50	18.20	109.70
115 lb	"	100.00	18.20	118.20
132 lb	"	100.00	18.20	118.20
No. 8 turnout				
90 lb	EA.	29,690	2,420	32,110
100 lb	"	33,010	2,420	35,430
110 lb	"	36,170	2,420	38,590
115 lb	"	36,960	2,420	39,380
132 lb	"	37,750	2,420	40,170
No. 8 turnout relay				
90 lb	EA.	24,640	2,420	27,060
100 lb	"	24,640	2,420	27,060
110 lb	"	24,640	2,420	27,060
115 lb	"	27,960	2,420	30,380
132 lb	"	27,960	2,420	30,380
Railroad crossings, asphalt, based on 8" thick x 20'				
Including track and approach				
12' roadway	EA.	710.00	450.00	1,160
15' roadway	"	840.00	520.00	1,360
18' roadway	"	980.00	610.00	1,590
21' roadway	"	1,090	730.00	1,820
24' roadway	"	1,230	910.00	2,140
Precast concrete inserts				
12' roadway	EA.	1,030	180.00	1,210
15' roadway	"	1,230	230.00	1,460
18' roadway	"	1,470	300.00	1,770
21' roadway	"	1,890	360.00	2,250

Railroad Work

02450.10 Railroad Work	UNIT	MAT.	INST.	TOTAL
24' roadway	EA.	2,290	400.00	2,690
Molded rubber, with headers				
12' roadway	EA.	6,190	180.00	6,370
15' roadway	"	7,740	230.00	7,970
18' roadway	"	9,160	300.00	9,460
21' roadway	"	10,110	360.00	10,470
24' roadway	"	12,320	400.00	12,720

Paving And Surfacing

02510.20 Asphalt Surfaces	UNIT	MAT.	INST.	TOTAL
Asphalt wearing surface, for flexible pavement				
1" thick	S.Y.	3.10	1.80	4.90
1-1/2" thick	"	4.75	2.15	6.90
2" thick	"	6.30	2.70	9.00
3" thick	"	9.30	3.60	12.90
Binder course				
1-1/2" thick	S.Y.	4.05	2.00	6.05
2" thick	"	5.45	2.45	7.90
3" thick	"	8.15	3.25	11.40
4" thick	"	10.85	3.60	14.45
5" thick	"	13.95	4.00	17.95
6" thick	"	16.65	4.50	21.15
Bituminous sidewalk, no base				
2" thick	S.Y.	6.80	2.15	8.95
3" thick	"	10.10	2.25	12.35

02520.10 Concrete Paving	UNIT	MAT.	INST.	TOTAL
Concrete paving, reinforced, 5000 psi concrete				
6" thick	S.Y.	21.95	16.85	38.80
7" thick	"	24.20	18.00	42.20
8" thick	"	27.30	19.25	46.55
9" thick	"	28.10	20.75	48.85
10" thick	"	32.20	22.50	54.70
11" thick	"	35.80	24.55	60.35
12" thick	"	38.00	27.00	65.00
15" thick	"	47.80	33.70	81.50
Concrete paving, for pipe trench, reinforced				
7" thick	S.Y.	45.10	18.20	63.30
8" thick	"	51.50	20.20	71.70
9" thick	"	54.50	22.75	77.25
10" thick	"	61.50	26.00	87.50
Fibrous concrete				
5" thick	S.Y.	21.50	20.75	42.25
8" thick	"	27.40	22.50	49.90
Roller compacted concrete, (RCC), place and compact				

02 SITEWORK

Paving And Surfacing

02520.10 Concrete Paving

Paving And Surfacing	UNIT	MAT.	INST.	TOTAL
8" thick	S.Y.	26.50	27.00	53.50
12" thick	"	40.30	33.70	74.00
Steel edge forms up to				
12" deep	L.F.	0.66	1.15	1.81
15" deep	"	0.85	1.35	2.20
Paving finishes				
Belt dragged	S.Y.		1.70	1.70
Curing	"	0.32	0.34	0.66

02545.10 Asphalt Repair

Item	UNIT	MAT.	INST.	TOTAL
Coal tar emulsion seal coat, rubber additive, fuel resistant	S.Y.	1.50	0.48	1.98
Bituminous surface treatment, single	"	1.35	0.34	1.69
Double	"	1.80	0.04	1.84
Bituminous prime coat	"	0.91	0.04	0.95
Tack coat	"	0.45	0.03	0.48
Crack sealing, concrete paving	L.F.	0.76	0.23	0.99
Bituminous paving for pipe trench, 4" thick	S.Y.	8.65	12.10	20.75
Polypropylene, nonwoven paving fabric	"	1.15	0.17	1.32
Rubberized asphalt	"	1.80	3.10	4.90
Asphalt slurry seal	"	4.45	2.00	6.45

02580.10 Pavement Markings

Item	UNIT	MAT.	INST.	TOTAL
Pavement line marking, paint				
4" wide	L.F.	0.08	0.09	0.16
6" wide	"	0.10	0.19	0.29
8" wide	"	0.12	0.28	0.40
Reflective paint, 4" wide	"	0.30	0.28	0.58
Airfield markings, retro-reflective				
White	L.F.	0.48	0.28	0.76
Yellow	"	0.52	0.28	0.80
Preformed tape, 4" wide				
Inlaid reflective	L.F.	1.35	0.05	1.40
Reflective paint	"	0.92	0.09	1.01
Thermoplastic				
White	L.F.	0.52	0.17	0.69
Yellow	"	0.52	0.17	0.69
12" wide, thermoplastic, white	"	1.10	0.48	1.58
Directional arrows, reflective preformed tape	EA.	97.50	33.80	131.30
Messages, reflective preformed tape (per letter)	"	48.70	16.90	65.60
Handicap symbol, preformed tape	"	81.00	33.80	114.80
Parking stall painting	"	1.30	6.75	8.05

02 SITEWORK

Utilities

02605.30 Manholes

Utilities	UNIT	MAT.	INST.	TOTAL
Precast sections, 48" dia.				
Base section	EA.	170.00	150.00	320.00
1'0" riser	"	47.20	120.00	167.20
1'4" riser	"	57.50	130.00	187.50
2'8" riser	"	85.00	140.00	225.00
4'0" riser	"	160.00	150.00	310.00
2'8" cone top	"	100.00	180.00	280.00
Precast manholes, 48" dia.				
4' deep	EA.	390.00	360.00	750.00
6' deep	"	590.00	450.00	1,040
7' deep	"	670.00	520.00	1,190
8' deep	"	760.00	610.00	1,370
10' deep	"	850.00	730.00	1,580
Cast-in-place, 48" dia., with frame and cover				
5' deep	EA.	510.00	910.00	1,420
6' deep	"	670.00	1,040	1,710
8' deep	"	970.00	1,210	2,180
10' deep	"	1,130	1,450	2,580
Brick manholes, 48" dia. with cover, 8" thick				
4' deep	EA.	620.00	420.00	1,040
6' deep	"	780.00	470.00	1,250
8' deep	"	1,000	520.00	1,520
10' deep	"	1,250	600.00	1,850
12' deep	"	1,560	700.00	2,260
14' deep	"	1,900	840.00	2,740
Inverts for manholes				
Single channel	EA.	100.00	170.00	270.00
Triple channel	"	120.00	210.00	330.00
Frames and covers, 24" diameter				
300 lb	EA.	330.00	33.80	363.80
400 lb	"	350.00	37.60	387.60
500 lb	"	400.00	48.30	448.30
Watertight, 350 lb	"	420.00	110.00	530.00
For heavy equipment, 1200 lb	"	910.00	170.00	1,080
Steps for manholes				
7" x 9"	EA.	11.40	6.75	18.15
8" x 9"	"	14.45	7.50	21.95
Curb inlet, 4' throat, cast in place				
12"-30" pipe	EA.	320.00	910.00	1,230
36"-48" pipe	"	360.00	1,040	1,400
Raise exist frame and cover, when repaving	"		360.00	360.00

02610.10 Cast Iron Flanged Pipe

Utilities	UNIT	MAT.	INST.	TOTAL
Cast iron flanged sections				
4" pipe, with one bolt set				
3' section	EA.	160.00	16.55	176.55
4' section	"	170.00	18.20	188.20
5' section	"	180.00	20.20	200.20
6' section	"	200.00	22.75	222.75
8' section	"	240.00	26.00	266.00
10' section	"	260.00	36.40	296.40
12' section	"	290.00	60.50	350.50

Utilities

	UNIT	MAT.	INST.	TOTAL
02610.10 Cast Iron Flanged Pipe				
15' section	EA.	340.00	91.00	431.00
18' section	"	370.00	120.00	490.00
6" pipe, with one bolt set				
3' section	EA.	200.00	18.20	218.20
4' section	"	220.00	21.40	241.40
5' section	"	230.00	24.25	254.25
6' section	"	260.00	28.00	288.00
8' section	"	290.00	40.40	330.40
10' section	"	330.00	45.50	375.50
12' section	"	360.00	60.50	420.50
15' section	"	420.00	91.00	511.00
18' section	"	540.00	130.00	670.00
8" pipe, with one bolt set				
3' section	EA.	270.00	22.75	292.75
4' section	"	300.00	26.00	326.00
5' section	"	330.00	30.30	360.30
6' section	"	340.00	36.40	376.40
8' section	"	390.00	52.00	442.00
10' section	"	440.00	60.50	500.50
12' section	"	470.00	91.00	561.00
15' section	"	550.00	120.00	670.00
18' section	"	680.00	150.00	830.00
10" pipe, with one bolt set				
3' section	EA.	350.00	23.30	373.30
4' section	"	400.00	26.70	426.70
5' section	"	430.00	31.30	461.30
6' section	"	470.00	37.90	507.90
8' section	"	520.00	55.00	575.00
10' section	"	560.00	65.00	625.00
12' section	"	640.00	100.00	740.00
15' section	"	700.00	130.00	830.00
18' section	"	860.00	180.00	1,040
12" pipe, with one bolt set				
3' section	EA.	470.00	25.30	495.30
4' section	"	500.00	29.30	529.30
5' section	"	550.00	35.00	585.00
6' section	"	580.00	41.30	621.30
8' section	"	660.00	60.50	720.50
10' section	"	740.00	70.00	810.00
12' section	"	810.00	110.00	920.00
15' section	"	920.00	150.00	1,070
18' section	"	980.00	200.00	1,180
02610.11 Cast Iron Fittings				
Mechanical joint, with 2 bolt kits				
90 deg bend				
4"	EA.	110.00	22.55	132.55
6"	"	140.00	26.00	166.00
8"	"	180.00	33.80	213.80
10"	"	270.00	48.30	318.30
12"	"	410.00	67.50	477.50
14"	"	570.00	84.50	654.50

Utilities	UNIT	MAT.	INST.	TOTAL
02610.11 Cast Iron Fittings				
16"	EA.	650.00	110.00	760.00
45 deg bend				
4"	EA.	97.00	22.55	119.55
6"	"	130.00	26.00	156.00
8"	"	180.00	33.80	213.80
10"	"	230.00	48.30	278.30
12"	"	310.00	67.50	377.50
14"	"	370.00	84.50	454.50
16"	"	490.00	110.00	600.00
Tee, with 3 bolt kits				
4" x 4"	EA.	170.00	33.80	203.80
6" x 6"	"	240.00	42.30	282.30
8" x 8"	"	320.00	56.50	376.50
10" x 10"	"	490.00	84.50	574.50
12" x 12"	"	650.00	110.00	760.00
Wye, with 3 bolt kits				
6" x 6"	EA.	320.00	42.30	362.30
8" x 8"	"	490.00	56.50	546.50
10" x 10"	"	650.00	84.50	734.50
12" x 12"	"	890.00	110.00	1,000
Reducer, with 2 bolt kits				
6" x 4"	EA.	120.00	42.30	162.30
8" x 6"	"	160.00	56.50	216.50
10" x 8"	"	290.00	84.50	374.50
12" x 10"	"	350.00	110.00	460.00
Flanged, 90 deg bend, 125 lb.				
4"	EA.	73.00	28.20	101.20
6"	"	100.00	33.80	133.80
8"	"	140.00	42.30	182.30
10"	"	250.00	56.50	306.50
12"	"	350.00	84.50	434.50
14"	"	700.00	110.00	810.00
16"	"	1,040	110.00	1,150
Tee				
4"	EA.	130.00	42.30	172.30
6"	"	180.00	48.30	228.30
8"	"	270.00	56.50	326.50
10"	"	500.00	67.50	567.50
12"	"	680.00	84.50	764.50
14"	"	1,510	110.00	1,620
16"	"	2,270	170.00	2,440
02610.13 Gate Valves				
Gate valve, (AWWA) mechanical joint, with adjustable box				
4" valve	EA.	650.00	60.50	710.50
6" valve	"	740.00	72.50	812.50
8" valve	"	980.00	91.00	1,071
10" valve	"	1,470	110.00	1,580
12" valve	"	1,960	130.00	2,090
14" valve	"	4,910	150.00	5,060
16" valve	"	6,550	170.00	6,720
18" valve	"	8,190	180.00	8,370

Utilities

	UNIT	MAT.	INST.	TOTAL
02610.13 Gate Valves				
Flanged, with box, post indicator (AWWA)				
4" valve	EA.	590.00	72.50	662.50
6" valve	"	690.00	82.50	772.50
8" valve	"	980.00	100.00	1,080
10" valve	"	1,470	120.00	1,590
12" valve	"	2,130	150.00	2,280
14" valve	"	4,910	180.00	5,090
16" valve	"	6,550	230.00	6,780
02610.15 Water Meters				
Water meter, displacement type				
1"	EA.	240.00	47.70	287.70
1-1/2"	"	830.00	53.00	883.00
2"	"	1,240	59.50	1,300
02610.17 Corporation Stops				
Stop for flared copper service pipe				
3/4"	EA.	26.80	23.85	50.65
1"	"	36.20	26.50	62.70
1-1/4"	"	97.50	31.80	129.30
1-1/2"	"	120.00	39.80	159.80
2"	"	170.00	47.70	217.70
02610.40 Ductile Iron Pipe				
Ductile iron pipe, cement lined, slip-on joints				
4"	L.F.	10.90	5.05	15.95
6"	"	13.25	5.35	18.60
8"	"	17.30	5.70	23.00
10"	"	23.85	6.05	29.90
12"	"	29.40	7.25	36.65
14"	"	37.10	9.10	46.20
16"	"	45.50	10.10	55.60
18"	"	51.00	11.35	62.35
20"	"	58.50	13.00	71.50
Mechanical joint pipe				
4"	L.F.	13.00	7.00	20.00
6"	"	15.55	7.60	23.15
8"	"	20.45	8.25	28.70
10"	"	26.80	9.10	35.90
12"	"	34.00	12.10	46.10
14"	"	42.70	14.00	56.70
16"	"	46.80	16.55	63.35
18"	"	53.00	18.20	71.20
20"	"	61.00	20.20	81.20
Fittings, mechanical joint				
90 degree elbow				
4"	EA.	160.00	22.55	182.55
6"	"	210.00	26.00	236.00
8"	"	300.00	33.80	333.80
10"	"	440.00	48.30	488.30
12"	"	580.00	67.50	647.50
14"	"	910.00	84.50	994.50

Utilities	UNIT	MAT.	INST.	TOTAL
02610.40 Ductile Iron Pipe				
16"	EA.	1,140	110.00	1,250
18"	"	1,700	140.00	1,840
20"	"	1,890	170.00	2,060
45 degree elbow				
4"	EA.	170.00	22.55	192.55
6"	"	190.00	26.00	216.00
8"	"	260.00	33.80	293.80
10"	"	380.00	48.30	428.30
12"	"	550.00	67.50	617.50
14"	"	760.00	84.50	844.50
16"	"	910.00	110.00	1,020
18"	"	1,320	170.00	1,490
20"	"	1,570	170.00	1,740
Tee				
4"x3"	EA.	230.00	42.30	272.30
4"x4"	"	250.00	42.30	292.30
6"x3"	"	280.00	48.30	328.30
6"x4"	"	290.00	48.30	338.30
6"x6"	"	320.00	48.30	368.30
8"x4"	"	400.00	56.50	456.50
8"x6"	"	450.00	56.50	506.50
8"x8"	"	420.00	56.50	476.50
10"x4"	"	530.00	67.50	597.50
10"x6"	"	580.00	67.50	647.50
10"x8"	"	600.00	67.50	667.50
10"x10"	"	680.00	67.50	747.50
12"x4"	"	600.00	84.50	684.50
12"x6"	"	640.00	84.50	724.50
12"x8"	"	700.00	84.50	784.50
12"x10"	"	790.00	84.50	874.50
12"x12"	"	850.00	90.00	940.00
14"x4"	"	1,040	96.50	1,137
14"x6"	"	1,110	96.50	1,207
14"x8"	"	1,130	96.50	1,227
14"x10"	"	1,160	96.50	1,257
14"x12"	"	1,190	100.00	1,290
14"x14"	"	1,180	100.00	1,280
16"x4"	"	1,340	110.00	1,450
16"x6"	"	1,370	110.00	1,480
16"x8"	"	1,220	110.00	1,330
16"x10"	"	1,240	110.00	1,350
16"x12"	"	1,210	110.00	1,320
16"x14"	"	1,280	110.00	1,390
16"x16"	"	1,310	110.00	1,420
18"x6"	"	1,560	120.00	1,680
18"x8"	"	1,590	120.00	1,710
18"x10"	"	1,620	120.00	1,740
18"x12"	"	1,650	120.00	1,770
18"x14"	"	1,840	120.00	1,960
18"x16"	"	1,820	120.00	1,940
18"x18"	"	2,060	120.00	2,180
20"x6"	"	1,990	140.00	2,130

Utilities

02610.40 Ductile Iron Pipe

	UNIT	MAT.	INST.	TOTAL
20"x8"	EA.	2,010	140.00	2,150
20"x10"	"	2,050	140.00	2,190
20"x12"	"	2,080	140.00	2,220
20"x14"	"	2,140	140.00	2,280
20"x16"	"	2,500	140.00	2,640
20"x18"	"	2,610	140.00	2,750
20"x20"	"	2,670	140.00	2,810
Cross				
4"x3"	EA.	280.00	56.50	336.50
4"x4"	"	310.00	56.50	366.50
6"x3"	"	320.00	67.50	387.50
6"x4"	"	340.00	67.50	407.50
6"x6"	"	380.00	67.50	447.50
8"x4"	"	440.00	75.00	515.00
8"x6"	"	470.00	75.00	545.00
8"x8"	"	520.00	75.00	595.00
10"x4"	"	600.00	84.50	684.50
10"x6"	"	640.00	84.50	724.50
10"x8"	"	700.00	84.50	784.50
10"x10"	"	830.00	84.50	914.50
12"x4"	"	780.00	96.50	876.50
12"x6"	"	870.00	96.50	966.50
12"x8"	"	850.00	96.50	946.50
12"x10"	"	980.00	100.00	1,080
12"x12"	"	1,060	100.00	1,160
14"x4"	"	1,000	110.00	1,110
14"x6"	"	1,140	110.00	1,250
14"x8"	"	1,180	110.00	1,290
14"x10"	"	1,270	110.00	1,380
14"x12"	"	1,360	120.00	1,480
14"x14"	"	1,500	120.00	1,620
16"x4"	"	1,310	140.00	1,450
16"x6"	"	1,340	140.00	1,480
16"x8"	"	1,420	140.00	1,560
16"x10"	"	1,510	140.00	1,650
16"x12"	"	1,580	140.00	1,720
16"x14"	"	1,710	140.00	1,850
16"x16"	"	1,820	140.00	1,960
18"x6"	"	1,690	150.00	1,840
18"x8"	"	1,740	150.00	1,890
18"x10"	"	1,820	150.00	1,970
18"x12"	"	1,910	150.00	2,060
18"x14"	"	2,270	150.00	2,420
18"x16"	"	2,430	150.00	2,580
18"x18"	"	2,560	150.00	2,710
20"x6"	"	2,030	160.00	2,190
20"x8"	"	2,080	160.00	2,240
20"x10"	"	2,180	160.00	2,340
20"x12"	"	2,270	160.00	2,430
20"x14"	"	2,390	160.00	2,550
20"x16"	"	2,770	160.00	2,930
20"x18"	"	2,950	170.00	3,120

02 SITEWORK

Utilities	UNIT	MAT.	INST.	TOTAL
02610.40 Ductile Iron Pipe				
20"x20"	EA.	3,130	170.00	3,300
02610.60 Plastic Pipe				
PVC, class 150 pipe				
4" dia.	L.F.	5.30	4.55	9.85
6" dia.	"	10.00	4.90	14.90
8" dia.	"	15.85	5.20	21.05
10" dia.	"	22.50	5.70	28.20
12" dia.	"	33.30	6.05	39.35
Schedule 40 pipe				
1-1/2" dia.	L.F.	2.95	2.00	4.95
2" dia.	"	3.45	2.10	5.55
2-1/2" dia.	"	3.80	2.25	6.05
3" dia.	"	5.50	2.40	7.90
4" dia.	"	7.85	2.80	10.65
6" dia.	"	13.20	3.40	16.60
90 degree elbows				
1"	EA.	1.95	5.65	7.60
1-1/2"	"	3.30	5.65	8.95
2"	"	3.75	6.15	9.90
2-1/2"	"	9.40	6.75	16.15
3"	"	11.35	7.50	18.85
4"	"	31.30	8.45	39.75
6"	"	70.50	11.30	81.80
45 degree elbows				
1"	EA.	2.15	5.65	7.80
1-1/2"	"	3.55	5.65	9.20
2"	"	4.70	6.15	10.85
2-1/2"	"	9.60	6.75	16.35
3"	"	11.65	7.50	19.15
4"	"	34.30	8.45	42.75
6"	"	74.50	11.30	85.80
Tees				
1"	EA.	2.55	6.75	9.30
1-1/2"	"	3.70	6.75	10.45
2"	"	5.10	7.50	12.60
2-1/2"	"	9.20	8.45	17.65
3"	"	12.55	9.65	22.20
4"	"	35.30	11.30	46.60
6"	"	70.50	13.55	84.05
Couplings				
1"	EA.	1.15	5.65	6.80
1-1/2"	"	1.35	5.65	7.00
2"	"	2.35	6.15	8.50
2-1/2"	"	4.70	6.75	11.45
3"	"	8.80	7.50	16.30
4"	"	19.60	8.45	28.05
6"	"	25.50	11.30	36.80
Drainage pipe				
PVC schedule 80				
1" dia.	L.F.	1.80	2.00	3.80
1-1/2" dia.	"	2.20	2.00	4.20

Utilities

02610.60 Plastic Pipe

Utilities	UNIT	MAT.	INST.	TOTAL
ABS, 2" dia.	L.F.	2.80	2.10	4.90
2-1/2" dia.	"	4.00	2.25	6.25
3" dia.	"	4.70	2.40	7.10
4" dia.	"	6.40	2.80	9.20
6" dia.	"	10.60	3.40	14.00
8" dia.	"	14.20	4.80	19.00
10" dia.	"	18.95	5.70	24.65
12" dia.	"	31.00	6.05	37.05
90 degree elbows				
1"	EA.	3.20	5.65	8.85
1-1/2"	"	4.00	5.65	9.65
2"	"	4.80	6.15	10.95
2-1/2"	"	11.50	6.75	18.25
3"	"	11.90	7.50	19.40
4"	"	20.95	8.45	29.40
6"	"	45.90	11.30	57.20
45 degree elbows				
1"	EA.	5.20	5.65	10.85
1-1/2"	"	6.60	5.65	12.25
2"	"	8.20	6.15	14.35
2-1/2"	"	15.35	6.75	22.10
3"	"	16.35	7.50	23.85
4"	"	31.00	8.45	39.45
6"	"	72.00	11.30	83.30
Tees				
1"	EA.	3.40	6.75	10.15
1-1/2"	"	10.80	6.75	17.55
2"	"	13.20	7.50	20.70
2-1/2"	"	15.35	8.45	23.80
3"	"	16.80	9.65	26.45
4"	"	32.00	11.30	43.30
6"	"	63.00	13.55	76.55
Couplings				
1"	EA.	2.80	5.65	8.45
1-1/2"	"	4.80	5.65	10.45
2"	"	7.10	6.15	13.25
2-1/2"	"	14.95	6.75	21.70
3"	"	15.40	7.50	22.90
4"	"	15.95	8.45	24.40
6"	"	27.00	11.30	38.30
Pressure pipe				
PVC, class 200 pipe				
3/4"	L.F.	1.75	1.70	3.45
1"	"	2.00	1.80	3.80
1-1/4"	"	2.20	1.90	4.10
1-1/2"	"	2.65	2.00	4.65
2"	"	2.85	2.10	4.95
2-1/2"	"	4.20	2.25	6.45
3"	"	5.30	2.40	7.70
4"	"	9.20	2.80	12.00
6"	"	17.35	3.40	20.75
8"	"	26.40	5.20	31.60

Utilities

02610.60 Plastic Pipe

	UNIT	MAT.	INST.	TOTAL
90 degree elbows				
3/4"	EA.	0.89	5.65	6.54
1"	"	1.10	5.65	6.75
1-1/4"	"	1.75	5.65	7.40
1-1/2"	"	2.10	5.65	7.75
2"	"	4.20	6.15	10.35
2-1/2"	"	6.05	6.75	12.80
3"	"	13.70	7.50	21.20
4"	"	35.20	8.45	43.65
6"	"	67.00	11.30	78.30
8"	"	130.00	16.90	146.90
45 degree elbows				
3/4"	EA.	1.30	5.65	6.95
1"	"	1.65	5.65	7.30
1-1/4"	"	2.40	5.65	8.05
1-1/2"	"	2.90	5.65	8.55
2"	"	4.10	6.15	10.25
2-1/2"	"	6.70	6.75	13.45
3"	"	15.35	7.50	22.85
4"	"	28.80	8.45	37.25
6"	"	71.00	11.30	82.30
8"	"	140.00	16.90	156.90
Tees				
3/4"	EA.	1.10	6.75	7.85
1"	"	1.45	6.75	8.20
1-1/4"	"	2.05	6.75	8.80
1-1/2"	"	2.85	6.75	9.60
2"	"	4.20	7.50	11.70
2-1/2"	"	6.60	8.45	15.05
3"	"	20.25	9.65	29.90
4"	"	29.00	11.30	40.30
6"	"	100.00	13.55	113.55
8"	"	210.00	18.80	228.80
Couplings				
3/4"	EA.	0.68	5.65	6.33
1"	"	1.00	5.65	6.65
1-1/4"	"	1.30	5.65	6.95
1-1/2"	"	1.40	5.65	7.05
2"	"	2.00	6.15	8.15
2-1/2"	"	4.20	6.75	10.95
3"	"	6.60	7.50	14.10
4"	"	9.30	7.50	16.80
6"	"	29.00	8.45	37.45
8"	"	52.00	11.30	63.30

02610.90 Vitrified Clay Pipe

	UNIT	MAT.	INST.	TOTAL
Vitrified clay pipe, extra strength				
6" dia.	L.F.	3.50	8.25	11.75
8" dia.	"	4.20	8.65	12.85
10" dia.	"	6.45	9.10	15.55
12" dia.	"	9.30	12.10	21.40
15" dia.	"	16.85	18.20	35.05

Utilities

Utilities	UNIT	MAT.	INST.	TOTAL
02610.90 Vitrified Clay Pipe				
18" dia.	L.F.	25.30	20.20	45.50
24" dia.	"	46.40	26.00	72.40
30" dia.	"	79.00	36.40	115.40
36" dia.	"	110.00	52.00	162.00
02630.10 Tapping Saddles & Sleeves				
Tapping saddle, tap size to 2"				
4" saddle	EA.	54.50	16.90	71.40
6" saddle	"	63.50	21.15	84.65
8" saddle	"	74.50	28.20	102.70
10" saddle	"	86.50	33.80	120.30
12" saddle	"	100.00	48.30	148.30
14" saddle	"	120.00	67.50	187.50
Tapping sleeve				
4x4	EA.	570.00	22.55	592.55
6x4	"	740.00	26.00	766.00
6x6	"	750.00	26.00	776.00
8x4	"	770.00	33.80	803.80
8x6	"	790.00	33.80	823.80
10x4	"	1,260	72.50	1,333
10x6	"	1,820	72.50	1,893
10x8	"	1,890	72.50	1,963
10x10	"	1,940	76.00	2,016
12x4	"	1,950	76.00	2,026
12x6	"	1,970	82.50	2,053
12x8	"	2,020	91.00	2,111
12x10	"	2,120	100.00	2,220
12x12	"	2,200	110.00	2,310
Tapping valve, mechanical joint				
4" valve	EA.	540.00	230.00	770.00
6" valve	"	650.00	300.00	950.00
8" valve	"	970.00	450.00	1,420
10" valve	"	1,550	610.00	2,160
12" valve	"	2,730	910.00	3,640
Tap hole in pipe				
4" hole	EA.		42.30	42.30
6" hole	"		67.50	67.50
8" hole	"		110.00	110.00
10" hole	"		140.00	140.00
12" hole	"		170.00	170.00
02640.15 Valve Boxes				
Valve box, adjustable, for valves up to 20"				
3' deep	EA.	140.00	11.30	151.30
4' deep	"	140.00	13.55	153.55
5' deep	"	140.00	16.90	156.90
02640.19 Thrust Blocks				
Thrust block, 3000# concrete				
1/4 c.y.	EA.	95.50	72.00	167.50
1/2 c.y.	"	120.00	86.50	206.50
3/4 c.y.	"	140.00	140.00	280.00

Utilities

Utilities	UNIT	MAT.	INST.	TOTAL
02640.19 Thrust Blocks				
1 c.y.	EA.	180.00	290.00	470.00
02645.10 Fire Hydrants				
Standard, 3 way post, 6" mechanical joint				
2' deep	EA.	1,360	610.00	1,970
4' deep	"	1,460	730.00	2,190
6' deep	"	1,620	910.00	2,530
8' deep	"	1,820	1,040	2,860
02665.10 Chilled Water Systems				
Chilled water pipe, 2" thick insulation, w/casing				
Align and tack weld on sleepers				
1-1/2" dia.	L.F.	15.20	1.65	16.85
3" dia.	"	24.45	2.60	27.05
4" dia.	"	28.10	3.65	31.75
6" dia.	"	32.10	4.55	36.65
8" dia.	"	45.20	5.20	50.40
10" dia.	"	57.50	6.05	63.55
12" dia.	"	68.50	7.25	75.75
14" dia.	"	93.00	7.90	100.90
16" dia.	"	120.00	9.10	129.10
Align and tack weld on trench bottom				
18" dia.	L.F.	120.00	10.10	130.10
20" dia.	"	160.00	11.35	171.35
Preinsulated fittings				
Align and tack weld on sleepers				
Elbows				
1-1/2"	EA.	320.00	29.80	349.80
3"	"	400.00	47.70	447.70
4"	"	520.00	59.50	579.50
6"	"	720.00	79.50	799.50
8"	"	1,020	95.50	1,116
Tees				
1-1/2"	EA.	480.00	31.80	511.80
3"	"	680.00	53.00	733.00
4"	"	520.00	68.00	588.00
6"	"	1,120	95.50	1,216
8"	"	1,490	120.00	1,610
Reducers				
3"	EA.	520.00	39.80	559.80
4"	"	820.00	47.70	867.70
6"	"	1,100	59.50	1,160
8"	"	1,190	79.50	1,270
Anchors, not including concrete				
4"	EA.	220.00	59.50	279.50
6"	"	320.00	59.50	379.50
Align and tack weld on trench bottom				
Elbows				
10"	EA.	1,340	110.00	1,450
12"	"	1,600	130.00	1,730
14"	"	2,020	140.00	2,160
16"	"	2,170	150.00	2,320

Utilities

	UNIT	MAT.	INST.	TOTAL
02665.10 Chilled Water Systems				
18"	EA.	2,420	170.00	2,590
20"	"	2,920	180.00	3,100
Tees				
10"	EA.	2,170	110.00	2,280
12"	"	2,760	130.00	2,890
14"	"	2,920	140.00	3,060
16"	"	3,090	150.00	3,240
18"	"	3,510	170.00	3,680
20"	"	4,010	180.00	4,190
Reducers				
10"	EA.	1,650	76.00	1,726
12"	"	2,010	82.50	2,093
14"	"	2,170	91.00	2,261
16"	"	2,670	100.00	2,770
18"	"	2,920	110.00	3,030
20"	"	3,090	130.00	3,220
Anchors, not including concrete				
10"	EA.	450.00	76.00	526.00
12"	"	500.00	82.50	582.50
14"	"	580.00	91.00	671.00
16"	"	810.00	100.00	910.00
18"	"	1,130	110.00	1,240
20"	"	1,530	130.00	1,660
02670.10 Wells				
Domestic water, drilled and cased				
4" dia.	L.F.	13.75	54.00	67.75
6" dia.	"	15.15	60.00	75.15
8" dia.	"	17.90	67.50	85.40
02685.10 Gas Distribution				
Gas distribution lines				
Polyethylene, 60 psi coils				
1-1/4" dia.	L.F.	1.15	3.20	4.35
1-1/2" dia.	"	1.55	3.40	4.95
2" dia.	"	1.95	4.00	5.95
3" dia.	"	4.15	4.75	8.90
30' pipe lengths				
3" dia.	L.F.	4.05	5.30	9.35
4" dia.	"	6.35	5.95	12.30
6" dia.	"	10.05	7.95	18.00
8" dia.	"	18.65	9.55	28.20
Steel, schedule 40, plain end				
1" dia.	L.F.	4.35	4.00	8.35
2" dia.	"	6.00	4.35	10.35
3" dia.	"	10.35	4.75	15.10
4" dia.	"	12.70	12.10	24.80
5" dia.	"	23.85	13.00	36.85
6" dia.	"	31.80	15.15	46.95
8" dia.	"	39.80	16.55	56.35
Natural gas meters, direct digital reading, threaded				
250 cfh @ 5 lbs	EA.	150.00	95.50	245.50

Utilities

	UNIT	MAT.	INST.	TOTAL
02685.10 Gas Distribution				
425 cfh @ 10 lbs	EA.	380.00	95.50	475.50
800 cfh @ 20 lbs	"	530.00	120.00	650.00
1,000 cfh @ 25 lbs	"	1,570	120.00	1,690
1,400 cfh @ 100 lbs	"	3,620	160.00	3,780
2,300 cfh @ 100 lbs	"	5,040	240.00	5,280
5,000 cfh @ 100 lbs	"	7,400	480.00	7,880
Gas pressure regulators				
Threaded				
3/4"	EA.	65.50	59.50	125.00
1"	"	68.50	79.50	148.00
1-1/4"	"	72.00	79.50	151.50
1-1/2"	"	470.00	79.50	549.50
2"	"	480.00	95.50	575.50
Flanged				
3"	EA.	1,360	120.00	1,480
4"	"	2,030	160.00	2,190
02690.10 Storage Tanks				
Oil storage tank, underground				
Steel				
500 gals	EA.	700.00	230.00	930.00
1,000 gals	"	1,220	300.00	1,520
4,000 gals	"	3,350	610.00	3,960
5,000 gals	"	4,570	910.00	5,480
10,000 gals	"	6,550	1,820	8,370
Fiberglass, double wall				
550 gals	EA.	2,420	300.00	2,720
1,000 gals	"	3,140	300.00	3,440
2,000 gals	"	3,710	450.00	4,160
4,000 gals	"	4,990	910.00	5,900
6,000 gals	"	5,700	1,210	6,910
8,000 gals	"	7,700	1,820	9,520
10,000 gals	"	9,690	2,270	11,960
12,000 gals	"	11,540	3,030	14,570
15,000 gals	"	13,680	4,040	17,720
20,000 gals	"	17,390	4,550	21,940
Above ground				
Steel				
275 gals	EA.	290.00	180.00	470.00
500 gals	"	910.00	300.00	1,210
1,000 gals	"	1,370	360.00	1,730
1,500 gals	"	2,280	450.00	2,730
2,000 gals	"	2,870	610.00	3,480
5,000 gals	"	7,460	910.00	8,370
Fill cap	"	37.80	47.70	85.50
Vent cap	"	37.80	47.70	85.50
Level indicator	"	87.50	47.70	135.20
02695.80 Steam Meters				
In-line turbine, direct reading, 300 lb, flanged				
2"	EA.	4,370	59.50	4,430
3"	"	4,680	79.50	4,760

Utilities

02695.80 Steam Meters	UNIT	MAT.	INST.	TOTAL
4"	EA.	5,150	95.50	5,246
Threaded, 2"				
5" line	EA.	8,270	480.00	8,750
6" line	"	8,430	480.00	8,910
8" line	"	8,580	480.00	9,060
10" line	"	9,050	480.00	9,530
12" line	"	9,210	480.00	9,690
14" line	"	9,520	480.00	10,000
16" line	"	10,140	480.00	10,620

Sewerage And Drainage

02720.10 Catch Basins	UNIT	MAT.	INST.	TOTAL
Standard concrete catch basin				
Cast in place, 3'8" x 3'8", 6" thick wall				
2' deep	EA.	260.00	450.00	710.00
3' deep	"	350.00	450.00	800.00
4' deep	"	460.00	610.00	1,070
5' deep	"	540.00	610.00	1,150
6' deep	"	600.00	730.00	1,330
4'x4', 8" thick wall, cast in place				
2' deep	EA.	280.00	450.00	730.00
3' deep	"	390.00	450.00	840.00
4' deep	"	520.00	610.00	1,130
5' deep	"	600.00	610.00	1,210
6' deep	"	660.00	730.00	1,390
Frames and covers, cast iron				
Round				
24" dia.	EA.	250.00	84.50	334.50
26" dia.	"	280.00	84.50	364.50
28" dia.	"	330.00	84.50	414.50
Rectangular				
23"x23"	EA.	220.00	84.50	304.50
27"x20"	"	270.00	84.50	354.50
24"x24"	"	260.00	84.50	344.50
26"x26"	"	280.00	84.50	364.50
Curb inlet frames and covers				
27"x27"	EA.	530.00	84.50	614.50
24"x36"	"	320.00	84.50	404.50
24"x25"	"	300.00	84.50	384.50
24"x22"	"	260.00	84.50	344.50
20"x22"	"	340.00	84.50	424.50
Airfield catch basin frame and grating, galvanized				
2'x4'	EA.	620.00	84.50	704.50
2'x2'	"	440.00	84.50	524.50

Sewerage And Drainage

02720.40 Storm Drainage

	UNIT	MAT.	INST.	TOTAL
Concrete pipe				
Plain, bell and spigot joint, Class II				
6" pipe	L.F.	5.05	8.25	13.30
8" pipe	"	5.60	9.10	14.70
10" pipe	"	5.75	9.55	15.30
12" pipe	"	6.75	10.10	16.85
15" pipe	"	8.20	10.70	18.90
18" pipe	"	11.10	11.35	22.45
21" pipe	"	13.90	12.10	26.00
24" pipe	"	17.35	13.00	30.35
Reinforced, class III, tongue and groove joint				
12" pipe	L.F.	12.05	10.10	22.15
15" pipe	"	13.55	10.70	24.25
18" pipe	"	15.05	11.35	26.40
21" pipe	"	19.55	12.10	31.65
24" pipe	"	25.50	13.00	38.50
27" pipe	"	30.10	14.00	44.10
30" pipe	"	33.10	15.15	48.25
36" pipe	"	49.60	16.55	66.15
42" pipe	"	61.50	18.20	79.70
48" pipe	"	85.50	20.20	105.70
54" pipe	"	100.00	22.75	122.75
60" pipe	"	110.00	26.00	136.00
66" pipe	"	140.00	30.30	170.30
72" pipe	"	160.00	36.40	196.40
Flared end-section, concrete				
12" pipe	L.F.	54.50	10.10	64.60
15" pipe	"	64.50	10.70	75.20
18" pipe	"	76.50	11.35	87.85
24" pipe	"	89.00	13.00	102.00
30" pipe	"	110.00	15.15	125.15
36" pipe	"	150.00	16.55	166.55
42" pipe	"	160.00	18.20	178.20
48" pipe	"	180.00	20.20	200.20
54" pipe	"	190.00	22.75	212.75
Corrugated metal pipe, coated, paved invert				
16 ga.				
8" pipe	L.F.	8.25	6.05	14.30
10" pipe	"	11.00	6.25	17.25
12" pipe	"	12.35	6.50	18.85
15" pipe	"	15.10	7.00	22.10
18" pipe	"	17.85	7.60	25.45
21" pipe	"	21.95	8.25	30.20
24" pipe	"	26.10	9.10	35.20
30" pipe	"	34.30	10.10	44.40
36" pipe	"	46.70	11.35	58.05
12 ga., 48" pipe	"	83.50	13.00	96.50
10 ga.				
60" pipe	L.F.	100.00	15.15	115.15
72" pipe	"	140.00	18.20	158.20
Galvanized or aluminum, plain				
16 ga.				

Sewerage And Drainage

02720.40 Storm Drainage

Sewerage And Drainage	UNIT	MAT.	INST.	TOTAL
8" pipe	L.F.	6.85	6.05	12.90
10" pipe	"	9.60	6.25	15.85
12" pipe	"	11.00	6.50	17.50
15" pipe	"	13.70	7.00	20.70
18" pipe	"	16.45	7.60	24.05
24" pipe	"	24.70	9.10	33.80
30" pipe	"	32.90	10.10	43.00
36" pipe	"	41.20	11.35	52.55
12 ga., 48" pipe	"	75.50	13.00	88.50
10 ga., 60" pipe	"	100.00	15.15	115.15
Galvanized or aluminum, coated oval arch				
16 ga.				
17" x 13"	L.F.	19.40	8.25	27.65
21" x 15"	"	26.10	9.10	35.20
14 ga.				
28" x 20"	L.F.	36.80	10.10	46.90
35" x 24"	"	55.00	13.00	68.00
12 ga.				
42" x 29"	L.F.	64.00	15.15	79.15
57" x 38"	"	96.00	18.20	114.20
64" x 43"	"	120.00	19.15	139.15
Oval arch culverts, plain				
16 ga.				
17" x 13"	L.F.	11.40	8.25	19.65
21" x 15"	"	16.45	9.10	25.55
14 ga.				
28" x 20"	L.F.	30.60	10.10	40.70
35" x 24"	"	38.40	13.00	51.40
12 ga.				
57" x 38"	L.F.	61.50	15.15	76.65
64" x 43"	"	75.50	18.20	93.70
71" x 47"	"	110.00	19.15	129.15
Nestable corrugated metal pipe				
16 ga.				
10" pipe	L.F.	11.00	6.25	17.25
12" pipe	"	13.70	6.50	20.20
15" pipe	"	17.85	7.00	24.85
18" pipe	"	20.60	7.60	28.20
24" pipe	"	28.80	9.10	37.90
30" pipe	"	35.70	10.10	45.80
14 ga., 36" pipe	"	41.20	11.35	52.55
Headwalls, cast in place, 30 deg wingwall				
12" pipe	EA.	360.00	110.00	470.00
15" pipe	"	440.00	110.00	550.00
18" pipe	"	530.00	120.00	650.00
24" pipe	"	800.00	120.00	920.00
30" pipe	"	960.00	140.00	1,100
36" pipe	"	1,040	220.00	1,260
42" pipe	"	1,300	220.00	1,520
48" pipe	"	1,380	290.00	1,670
54" pipe	"	1,560	360.00	1,920
60" pipe	"	1,870	430.00	2,300

Sewerage And Drainage

Sewerage And Drainage	UNIT	MAT.	INST.	TOTAL
02720.40 Storm Drainage				
4" cleanout for storm drain				
4" pipe	EA.	520.00	42.30	562.30
6" pipe	"	630.00	42.30	672.30
8" pipe	"	880.00	42.30	922.30
Connect new drain line				
To existing manhole	EA.	120.00	110.00	230.00
To new manhole	"	100.00	67.50	167.50
02720.70 Underdrain				
Drain tile, clay				
6" pipe	L.F.	3.35	4.05	7.40
8" pipe	"	5.30	4.25	9.55
12" pipe	"	10.75	4.55	15.30
Porous concrete, standard strength				
6" pipe	L.F.	3.35	4.05	7.40
8" pipe	"	4.85	4.25	9.10
12" pipe	"	9.10	4.55	13.65
15" pipe	"	10.70	5.05	15.75
18" pipe	"	13.95	6.05	20.00
Corrugated metal pipe, perforated type				
6" pipe	L.F.	6.05	4.55	10.60
8" pipe	"	7.15	4.80	11.95
10" pipe	"	8.65	5.05	13.70
12" pipe	"	12.20	5.35	17.55
18" pipe	"	15.10	5.70	20.80
Perforated clay pipe				
6" pipe	L.F.	4.00	5.20	9.20
8" pipe	"	5.40	5.35	10.75
12" pipe	"	9.40	5.50	14.90
Drain tile, concrete				
6" pipe	L.F.	2.75	4.05	6.80
8" pipe	"	4.30	4.25	8.55
12" pipe	"	8.60	4.55	13.15
Perforated rigid PVC underdrain pipe				
4" pipe	L.F.	1.35	3.05	4.40
6" pipe	"	2.45	3.65	6.10
8" pipe	"	3.50	4.05	7.55
10" pipe	"	5.40	4.55	9.95
12" pipe	"	9.05	5.20	14.25
Underslab drainage, crushed stone				
3" thick	S.F.	0.25	0.61	0.86
4" thick	"	0.31	0.70	1.01
6" thick	"	0.33	0.76	1.09
8" thick	"	0.39	0.79	1.18
Plastic filter fabric for drain lines	"	0.12	0.34	0.46
Gravel fill in trench, crushed or bank run, 1/2" to 3/4"	C.Y.	26.40	45.50	71.90
02730.10 Sanitary Sewers				
Clay				
6" pipe	L.F.	4.00	6.05	10.05
8" pipe	"	5.35	6.50	11.85
10" pipe	"	6.70	7.00	13.70

Sewerage And Drainage

	UNIT	MAT.	INST.	TOTAL
02730.10 Sanitary Sewers				
12" pipe	L.F.	10.70	7.60	18.30
PVC				
4" pipe	L.F.	1.85	4.55	6.40
6" pipe	"	3.75	4.80	8.55
8" pipe	"	5.60	5.05	10.65
10" pipe	"	7.45	5.35	12.80
12" pipe	"	11.20	5.70	16.90
Cleanout				
4" pipe	EA.	6.65	42.30	48.95
6" pipe	"	14.65	42.30	56.95
8" pipe	"	43.30	42.30	85.60
Connect new sewer line				
To existing manhole	EA.	77.00	110.00	187.00
To new manhole	"	55.00	67.50	122.50
02740.10 Drainage Fields				
Perforated PVC pipe, for drain field				
4" pipe	L.F.	1.45	4.05	5.50
6" pipe	"	2.65	4.35	7.00
02740.50 Septic Tanks				
Septic tank, precast concrete				
1000 gals	EA.	660.00	300.00	960.00
2000 gals	"	1,240	450.00	1,690
5000 gals	"	5,410	910.00	6,320
25,000 gals	"	21,930	3,640	25,570
40,000 gals	"	35,120	6,060	41,180
Leaching pit, precast concrete, 72" diameter				
3' deep	EA.	480.00	230.00	710.00
6' deep	"	610.00	260.00	870.00
8' deep	"	760.00	300.00	1,060
02760.10 Pipeline Restoration				
Relining existing water main				
6" dia.	L.F.	6.55	27.00	33.55
8" dia.	"	7.40	28.40	35.80
10" dia.	"	8.30	30.00	38.30
12" dia.	"	9.05	31.70	40.75
14" dia.	"	9.70	33.70	43.40
16" dia.	"	10.55	36.00	46.55
18" dia.	"	11.35	38.50	49.85
20" dia.	"	12.50	41.50	54.00
24" dia.	"	13.30	45.00	58.30
36" dia.	"	14.45	54.00	68.45
48" dia.	"	16.10	60.00	76.10
72" dia.	"	20.55	67.50	88.05
Replacing in line gate valves				
6" valve	EA.	710.00	360.00	1,070
8" valve	"	1,100	450.00	1,550
10" valve	"	1,660	540.00	2,200
12" valve	"	2,890	670.00	3,560
16" valve	"	6,560	770.00	7,330

Sewerage And Drainage

02760.10 Pipeline Restoration

	UNIT	MAT.	INST.	TOTAL
18" valve	EA.	9,920	900.00	10,820
20" valve	"	13,640	1,080	14,720
24" valve	"	19,490	1,350	20,840
36" valve	"	53,150	1,800	54,950

Site Improvements

02810.40 Lawn Irrigation

	UNIT	MAT.	INST.	TOTAL
Residential system, complete				
Minimum	ACRE			16,748
Maximum	"			31,878
Commercial system, complete				
Minimum	ACRE			24,255
Maximum	"			40,194
Components				
Pipe				
Schedule 40, PVC				
1/2"	L.F.	0.44	1.80	2.24
3/4"	"	0.63	1.90	2.53
1"	"	0.94	1.95	2.89
1-1/4"	"	1.25	1.95	3.20
1-1/2"	"	1.50	2.00	3.50
2"	"	2.20	2.10	4.30
2-1/2"	"	3.30	2.25	5.55
3"	"	4.50	2.40	6.90
4"	"	6.40	2.80	9.20
6"	"	11.20	3.40	14.60
Fittings				
Tee				
1/2"	EA.	0.75	5.65	6.40
3/4"	"	0.86	5.65	6.51
1"	"	1.55	5.65	7.20
1-1/4"	"	2.80	6.15	8.95
1-1/2"	"	3.20	6.75	9.95
2"	"	4.50	7.50	12.00
2-1/2"	"	16.95	8.45	25.40
3"	"	21.75	9.65	31.40
4"	"	38.80	11.30	50.10
6"	"	160.00	13.55	173.55
El				
1/2"	EA.	0.63	5.20	5.83
3/4"	"	0.71	5.65	6.36
1"	"	1.25	5.65	6.90
1-1/4"	"	2.20	5.65	7.85
1-1/2"	"	2.50	5.65	8.15
2"	"	3.75	6.15	9.90

Site Improvements

02810.40 Lawn Irrigation	UNIT	MAT.	INST.	TOTAL
2-1/2"	EA.	12.30	6.75	19.05
3"	"	15.55	7.50	23.05
4"	"	25.60	8.45	34.05
6"	"	120.00	11.30	131.30
Coupling				
1/2"	EA.	0.47	5.20	5.67
3/4"	"	0.63	5.65	6.28
1"	"	1.10	5.65	6.75
1-1/4"	"	1.40	5.65	7.05
1-1/2"	"	1.50	5.65	7.15
2"	"	2.35	6.15	8.50
2-1/2"	"	5.30	6.75	12.05
3"	"	9.30	7.50	16.80
4"	"	11.65	8.45	20.10
6"	"	49.70	11.30	61.00
45 El				
1/2"	EA.	0.94	5.20	6.14
3/4"	"	1.55	5.65	7.20
1"	"	2.20	5.65	7.85
1-1/4"	"	2.95	5.65	8.60
1-1/2"	"	3.55	5.65	9.20
2"	"	4.65	6.15	10.80
2-1/2"	"	14.00	6.75	20.75
3"	"	20.20	7.50	27.70
4"	"	37.30	8.45	45.75
6"	"	120.00	11.30	131.30
Riser, 1/2" diameter				
2" (close)	EA.	0.54	8.45	8.99
3"	"	0.63	8.45	9.08
4"	"	0.78	9.65	10.43
5"	"	0.86	9.65	10.51
6"	"	1.10	9.65	10.75
3/4" diameter				
2" (close)	EA.	0.54	8.45	8.99
3"	"	0.63	8.45	9.08
4"	"	0.71	9.65	10.36
5"	"	0.94	9.65	10.59
6"	"	1.00	9.65	10.65
1" diameter				
2" (close)	EA.	0.78	8.45	9.23
3"	"	1.00	8.45	9.45
4"	"	1.20	9.65	10.85
5"	"	1.50	9.65	11.15
6"	"	1.50	9.65	11.15
Valve Box				
Concrete, Square				
12" x 22"	EA.	67.00	42.30	109.30
18" x 20"	"	87.00	48.30	135.30
24" x 13"	"	74.50	56.50	131.00
Round				
12"	EA.	29.50	33.80	63.30
Plastic				

Site Improvements

	UNIT	MAT.	INST.	TOTAL
02810.40 Lawn Irrigation				
Square				
12"	EA.	31.10	42.30	73.40
18"	"			
Round				
6"	EA.	7.75	42.30	50.05
10"	"	17.10	42.30	59.40
12"	"	24.85	48.30	73.15
Sprinkler, Pop-Up				
Spray				
2" high	EA.	5.00	56.50	61.50
3" high	"	5.45	56.50	61.95
4" high	"	5.90	67.50	73.40
6" high	"	12.45	67.50	79.95
12" high	"	15.85	67.50	83.35
Rotor				
4" high	EA.	23.30	56.50	79.80
6" high	"	31.10	67.50	98.60
Impact				
Brass	EA.	31.10	56.50	87.60
Plastic	"	15.55	67.50	83.05
Shrub Head				
Spray	EA.	9.30	56.50	65.80
Rotor	"	23.30	67.50	90.80
Time Clocks				
Minimum	EA.	160.00	84.50	244.50
Average	"	310.00	110.00	420.00
Maximum	"	2,800	170.00	2,970
Valves				
Anti-siphon				
Brass				
3/4"	EA.	67.00	56.50	123.50
1"	"	82.50	56.50	139.00
Plastic				
3/4"	EA.	38.80	56.50	95.30
1"	"	46.60	56.50	103.10
Ball Valve				
Plastic				
1/2"	EA.	8.55	56.50	65.05
3/4"	"	9.30	56.50	65.80
1"	"	12.45	56.50	68.95
1-1/2"	"	24.85	67.50	92.35
2"	"	32.60	67.50	100.10
Brass				
1/2"	EA.	8.55	56.50	65.05
3/4"	"	13.20	56.50	69.70
1"	"	21.00	56.50	77.50
1-1/2"	"	38.80	67.50	106.30
2"	"	54.50	67.50	122.00
Gate valves, Brass				
1/2"	EA.	18.65	56.50	75.15
3/4"	"	31.10	56.50	87.60
1"	"	43.50	56.50	100.00

Site Improvements

	UNIT	MAT.	INST.	TOTAL
02810.40 Lawn Irrigation				
1-1/2"	EA.	74.50	67.50	142.00
2"	"	93.00	67.50	160.50
Vacuum Breakers				
Brass				
3/4"	EA.	120.00	110.00	230.00
1"	"	170.00	110.00	280.00
1-1/2"	"	340.00	110.00	450.00
2"	"	480.00	110.00	590.00
Plastic				
3/4"	EA.	93.00	110.00	203.00
1"	"	110.00	110.00	220.00
1-1/2"	"	170.00	110.00	280.00
2"	"	200.00	110.00	310.00
Backflow Preventors, Brass				
3/4"	EA.	310.00	1,130	1,440
1"	"	340.00	1,130	1,470
1-1/2"	"	750.00	1,130	1,880
2"	"	930.00	1,350	2,280
Pressure Regulators, Brass				
3/4"	EA.	77.50	33.80	111.30
1"	"	110.00	33.80	143.80
1-1/2"	"	330.00	37.60	367.60
2"	"	390.00	42.30	432.30
Quick Coupler Valve				
3/4"	EA.	70.00	56.50	126.50
1"	"	100.00	56.50	156.50
02830.10 Chain Link Fence				
Chain link fence, 9 ga., galvanized, with posts 10' o.c.				
4' high	L.F.	5.70	2.40	8.10
5' high	"	7.60	3.10	10.70
6' high	"	8.60	4.25	12.85
7' high	"	9.80	5.20	15.00
8' high	"	11.30	6.75	18.05
For barbed wire with hangers, add				
3 strand	L.F.	2.10	1.70	3.80
6 strand	"	3.50	2.80	6.30
Corner or gate post, 3" post				
4' high	EA.	66.50	11.30	77.80
5' high	"	73.00	12.55	85.55
6' high	"	81.50	14.70	96.20
7' high	"	98.00	16.90	114.90
8' high	"	100.00	18.80	118.80
4" post				
4' high	EA.	110.00	12.55	122.55
5' high	"	130.00	14.70	144.70
6' high	"	150.00	16.90	166.90
7' high	"	160.00	18.80	178.80
8' high	"	180.00	21.15	201.15
Gate with gate posts, galvanized, 3' wide				
4' high	EA.	73.50	84.50	158.00
5' high	"	94.50	110.00	204.50

Site Improvements

02830.10 Chain Link Fence	UNIT	MAT.	INST.	TOTAL
6' high	EA.	110.00	110.00	220.00
7' high	"	140.00	170.00	310.00
8' high	"	150.00	170.00	320.00
Fabric, galvanized chain link, 2" mesh, 9 ga.				
4' high	L.F.	2.60	1.15	3.75
5' high	"	3.15	1.35	4.50
6' high	"	3.30	1.70	5.00
8' high	"	5.05	2.25	7.30
Line post, no rail fitting, galvanized, 2-1/2" dia.				
4' high	EA.	19.60	9.65	29.25
5' high	"	21.30	10.55	31.85
6' high	"	23.35	11.30	34.65
7' high	"	26.50	13.55	40.05
8' high	"	29.60	16.90	46.50
1-7/8" H beam				
4' high	EA.	26.80	9.65	36.45
5' high	"	30.00	10.55	40.55
6' high	"	35.90	11.30	47.20
7' high	"	40.70	13.55	54.25
8' high	"	44.00	16.90	60.90
2-1/4" H beam				
4' high	EA.	19.60	9.65	29.25
5' high	"	24.40	10.55	34.95
6' high	"	28.00	11.30	39.30
7' high	"	32.50	13.55	46.05
8' high	"	37.70	16.90	54.60
Vinyl coated, 9 ga., with posts 10' o.c.				
4' high	L.F.	6.15	2.40	8.55
5' high	"	7.30	3.10	10.40
6' high	"	8.75	4.25	13.00
7' high	"	9.55	5.20	14.75
8' high	"	10.90	6.75	17.65
For barbed wire w/hangers, add				
3 strand	L.F.	2.25	1.70	3.95
6 Strand	"	3.60	2.80	6.40
Corner, or gate post, 4' high				
3" dia.	EA.	81.50	11.30	92.80
4" dia.	"	130.00	11.30	141.30
6" dia.	"	150.00	13.55	163.55
Gate, with posts, 3' wide				
4' high	EA.	89.00	84.50	173.50
5' high	"	110.00	110.00	220.00
6' high	"	120.00	110.00	230.00
7' high	"	140.00	170.00	310.00
8' high	"	160.00	170.00	330.00
Line post, no rail fitting, 2-1/2" dia.				
4' high	EA.	34.20	9.65	43.85
5' high	"	46.50	10.55	57.05
6' high	"	56.00	11.30	67.30
7' high	"	66.00	13.55	79.55
8' high	"	72.00	16.90	88.90
Corner post, no top rail fitting, 4" dia.				

Site Improvements

02830.10 Chain Link Fence

	UNIT	MAT.	INST.	TOTAL
4' high	EA.	130.00	11.30	141.30
5' high	"	150.00	12.55	162.55
6' high	"	180.00	14.70	194.70
7' high	"	190.00	16.90	206.90
8' high	"	210.00	18.80	228.80
Fabric, vinyl, chain link, 2" mesh, 9 ga.				
4' high	L.F.	4.70	1.15	5.85
5' high	"	5.70	1.35	7.05
6' high	"	6.65	1.70	8.35
8' high	"	8.90	2.25	11.15
Swing gates, galvanized, 4' high				
Single gate				
3' wide	EA.	170.00	84.50	254.50
4' wide	"	180.00	84.50	264.50
Double gate				
10' wide	EA.	440.00	140.00	580.00
12' wide	"	470.00	140.00	610.00
14' wide	"	480.00	140.00	620.00
16' wide	"	540.00	140.00	680.00
18' wide	"	580.00	190.00	770.00
20' wide	"	610.00	190.00	800.00
22' wide	"	690.00	190.00	880.00
24' wide	"	700.00	230.00	930.00
26' wide	"	730.00	230.00	960.00
28' wide	"	780.00	270.00	1,050
30' wide	"	830.00	270.00	1,100
5' high				
Single gate				
3' wide	EA.	180.00	110.00	290.00
4' wide	"	200.00	110.00	310.00
Double gate				
10' wide	EA.	470.00	170.00	640.00
12' wide	"	500.00	170.00	670.00
14' wide	"	870.00	170.00	1,040
16' wide	"	570.00	170.00	740.00
18' wide	"	580.00	190.00	770.00
20' wide	"	660.00	190.00	850.00
22' wide	"	690.00	190.00	880.00
24' wide	"	720.00	230.00	950.00
26' wide	"	760.00	230.00	990.00
28' wide	"	840.00	270.00	1,110
30' wide	"	870.00	270.00	1,140
6' high				
Single gate				
3' wide	EA.	200.00	110.00	310.00
4' wide	"	220.00	110.00	330.00
Double gate				
10' wide	EA.	490.00	170.00	660.00
12' wide	"	550.00	170.00	720.00
14' wide	"	590.00	170.00	760.00
16' wide	"	630.00	170.00	800.00
18' wide	"	670.00	190.00	860.00

Site Improvements

02830.10 Chain Link Fence

Site Improvements	UNIT	MAT.	INST.	TOTAL
20' wide	EA.	700.00	190.00	890.00
22' wide	"	750.00	190.00	940.00
24' wide	"	800.00	230.00	1,030
26' wide	"	830.00	230.00	1,060
28' wide	"	900.00	270.00	1,170
30' wide	"	950.00	270.00	1,220
7' high				
Single gate				
3' wide	EA.	230.00	170.00	400.00
4' wide	"	250.00	170.00	420.00
Double gate				
10' wide	EA.	580.00	230.00	810.00
12' wide	"	640.00	230.00	870.00
14' wide	"	690.00	230.00	920.00
16' wide	"	730.00	230.00	960.00
18' wide	"	780.00	270.00	1,050
20' wide	"	830.00	270.00	1,100
22' wide	"	870.00	270.00	1,140
24' wide	"	940.00	340.00	1,280
26' wide	"	1,000	340.00	1,340
28' wide	"	1,050	420.00	1,470
30' wide	"	1,210	420.00	1,630
8' high				
Single gate				
3' wide	EA.	240.00	170.00	410.00
4' wide	"	260.00	170.00	430.00
Double gate				
10' wide	EA.	640.00	230.00	870.00
12' wide	"	680.00	230.00	910.00
14' wide	"	730.00	230.00	960.00
16' wide	"	800.00	230.00	1,030
18' wide	"	830.00	270.00	1,100
20' wide	"	860.00	270.00	1,130
22' wide	"	910.00	270.00	1,180
24' wide	"	1,010	340.00	1,350
26' wide	"	1,040	340.00	1,380
28' wide	"	1,110	420.00	1,530
30' wide	"	1,220	420.00	1,640
Vinyl coated swing gates, 4' high				
Single gate				
3' wide	EA.	260.00	84.50	344.50
4' wide	"	280.00	84.50	364.50
Double gate				
10' wide	EA.	610.00	140.00	750.00
12' wide	"	780.00	140.00	920.00
14' wide	"	830.00	140.00	970.00
16' wide	"	920.00	140.00	1,060
18' wide	"	1,040	190.00	1,230
20' wide	"	1,230	190.00	1,420
22' wide	"	1,360	190.00	1,550
24' wide	"	1,470	230.00	1,700
26' wide	"	1,540	230.00	1,770

Site Improvements

02830.10 Chain Link Fence	UNIT	MAT.	INST.	TOTAL
28' wide	EA.	1,790	270.00	2,060
30' wide	"	1,880	270.00	2,150
5' high				
Single gate				
3' wide	EA.	280.00	110.00	390.00
4' wide	"	320.00	110.00	430.00
Double gate				
10' wide	EA.	830.00	170.00	1,000
12' wide	"	890.00	170.00	1,060
14' wide	"	1,000	170.00	1,170
16' wide	"	1,040	170.00	1,210
18' wide	"	1,250	190.00	1,440
20' wide	"	1,360	190.00	1,550
22' wide	"	1,500	190.00	1,690
24' wide	"	1,700	230.00	1,930
26' wide	"	1,780	230.00	2,010
28' wide	"	1,980	270.00	2,250
30' wide	"	2,140	270.00	2,410
6' high				
Single gate				
3' wide	EA.	300.00	110.00	410.00
4' wide	"	310.00	110.00	420.00
Double gate				
10' wide	EA.	740.00	170.00	910.00
12' wide	"	850.00	170.00	1,020
14' wide	"	980.00	170.00	1,150
16' wide	"	1,140	170.00	1,310
18' wide	"	1,260	190.00	1,450
20' wide	"	1,360	190.00	1,550
22' wide	"	1,480	190.00	1,670
24' wide	"	1,650	230.00	1,880
26' wide	"	1,750	230.00	1,980
28' wide	"	2,080	270.00	2,350
30' wide	"	2,150	270.00	2,420
7' high				
Single gate				
3' wide	EA.	330.00	170.00	500.00
4' wide	"	420.00	170.00	590.00
Double gate				
10' wide	EA.	830.00	230.00	1,060
12' wide	"	990.00	230.00	1,220
14' wide	"	1,120	230.00	1,350
16' wide	"	1,250	230.00	1,480
18' wide	"	1,420	270.00	1,690
20' wide	"	1,620	270.00	1,890
22' wide	"	1,800	270.00	2,070
24' wide	"	1,970	340.00	2,310
26' wide	"	2,140	340.00	2,480
28' wide	"	2,300	420.00	2,720
30' wide	"	2,480	420.00	2,900
8' high				
Single gate				

Site Improvements

Site Improvements	UNIT	MAT.	INST.	TOTAL
02830.10 Chain Link Fence				
3' wide	EA.	350.00	170.00	520.00
4' wide	"	420.00	170.00	590.00
Double gate				
10' wide	EA.	850.00	230.00	1,080
12' wide	"	990.00	230.00	1,220
14' wide	"	1,130	230.00	1,360
16' wide	"	800.00	230.00	1,030
18' wide	"	1,430	270.00	1,700
20' wide	"	1,660	270.00	1,930
22' wide	"	1,910	270.00	2,180
24' wide	"	2,030	340.00	2,370
28' wide	"	2,370	340.00	2,710
30' wide	"	2,540	420.00	2,960
Motor operator for gates, no wiring	"			4,889
Drilling fence post holes				
In soil				
By hand	EA.		16.90	16.90
By machine auger	"		10.95	10.95
In rock				
By jackhammer	EA.		150.00	150.00
By rock drill	"		43.80	43.80
Aluminum privacy slats, installed vertically	S.F.	0.86	0.85	1.70
Post hole, dig by hand	EA.		22.55	22.55
Set fence post in concrete	"	8.25	16.90	25.15
02830.70 Recreational Courts				
Walls, galvanized steel				
8' high	L.F.	11.35	6.75	18.10
10' high	"	13.50	7.50	21.00
12' high	"	15.50	8.90	24.40
Vinyl coated				
8' high	L.F.	10.90	6.75	17.65
10' high	"	13.35	7.50	20.85
12' high	"	14.80	8.90	23.70
Gates, galvanized steel				
Single, 3' transom				
3'x7'	EA.	270.00	170.00	440.00
4'x7'	"	280.00	190.00	470.00
5'x7'	"	380.00	230.00	610.00
6'x7'	"	420.00	270.00	690.00
Double, 3' transom				
10'x7'	EA.	650.00	680.00	1,330
12'x7'	"	830.00	750.00	1,580
14'x7'	"	990.00	850.00	1,840
Double, no transom				
10'x10'	EA.	700.00	560.00	1,260
12'x10'	"	830.00	680.00	1,510
14'x10'	"	960.00	750.00	1,710
Vinyl coated				
Single, 3' transom				
3'x7'	EA.	520.00	170.00	690.00
4'x7'	"	570.00	190.00	760.00

Site Improvements

	UNIT	MAT.	INST.	TOTAL
02830.70 Recreational Courts				
5'x7'	EA.	570.00	230.00	800.00
6'x7'	"	580.00	270.00	850.00
Double, 3'				
10'x7'	EA.	1,540	680.00	2,220
12'x7'	"	1,580	750.00	2,330
14'x7'	"	1,700	850.00	2,550
Double, no transom				
10'x10'	EA.	1,530	560.00	2,090
12'x10'	"	1,550	680.00	2,230
14'x10'	"	1,700	750.00	2,450
Baseball backstop, regulation				
Galvanized	EA.			6,626
Vinyl coated	"			9,228
Softball backstop, regulation				
14' high				
Galvanized	EA.			6,153
Vinyl coated	"			9,150
18' high				
Galvanized	EA.			7,257
Vinyl coated	"			10,490
20' high				
Galvanized	EA.			8,597
Vinyl coated	"			12,384
22' high				
Galvanized	EA.			9,940
Vinyl coated	"			14,514
24' high				
Galvanized	EA.			12,084
Vinyl coated	"			19,879
Wire and miscellaneous metal fences				
Chicken wire, post 4' o.c.				
2" mesh				
4' high	L.F.	1.30	1.70	3.00
6' high	"	1.45	2.25	3.70
Galvanized steel				
12 gauge, 2" by 4" mesh, posts 5' o.c.				
3' high	L.F.	2.10	1.70	3.80
5' high	"	3.00	2.10	5.10
14 gauge, 1" by 2" mesh, posts 5' o.c.				
3' high	L.F.	1.75	1.70	3.45
5' high	"	2.85	2.10	4.95
02840.30 Guardrails				
Pipe bollard, steel pipe, concrete filled, painted				
6" dia.	EA.	150.00	28.20	178.20
8" dia.	"	230.00	42.30	272.30
12" dia.	"	370.00	110.00	480.00
Corrugated steel, guardrail, galvanized	L.F.	21.15	3.05	24.20
End section, wrap around or flared	EA.	59.00	33.80	92.80
Timber guardrail, 4" x 8"	L.F.	26.50	2.25	28.75
Guard rail, 3 cables, 3/4" dia.				
Steel posts	L.F.	11.25	9.10	20.35

02 SITEWORK

Site Improvements

	UNIT	MAT.	INST.	TOTAL
02840.30 Guardrails				
Wood posts	L.F.	12.85	7.25	20.10
Steel box beam				
6" x 6"	L.F.	46.60	10.10	56.70
6" x 8"	"	49.60	11.35	60.95
Concrete posts	EA.	30.80	16.90	47.70
Barrel type impact barrier	"	390.00	33.80	423.80
Light shield, 6' high	L.F.	25.50	6.75	32.25
02840.40 Parking Barriers				
Timber, treated, 4' long				
4" x 4"	EA.	16.30	28.20	44.50
6" x 6"	"	30.30	33.80	64.10
Precast concrete, 6' long, with dowels				
12" x 6"	EA.	29.80	16.90	46.70
12" x 8"	"	32.60	18.80	51.40
02840.60 Signage				
Traffic signs				
Reflectorized signs per OSHA standards, including post				
Stop, 24"x24"	EA.	58.00	22.55	80.55
Yield, 30" triangle	"	33.90	22.55	56.45
Speed limit, 12"x18"	"	38.80	22.55	61.35
Directional, 12"x18"	"	53.50	22.55	76.05
Exit, 12"x18"	"	53.50	22.55	76.05
Entry, 12"x18"	"	53.50	22.55	76.05
Warning, 24"x24"	"	68.00	22.55	90.55
Informational, 12"x18"	"	24.25	22.55	46.80
Handicap parking, 12"x18"	"	25.90	22.55	48.45
02860.40 Recreational Facilities				
Bleachers, outdoor, portable, per seat				
10 tiers				
Minimum	EA.	30.40	11.35	41.75
Maximum	"	59.00	15.15	74.15
20 tiers				
Minimum	EA.	37.10	10.70	47.80
Maximum	"	71.00	14.00	85.00
Grandstands, fixed, wood seat, steel frame, per seat				
15 tiers				
Minimum	EA.	52.50	18.20	70.70
Maximum	"	89.50	30.30	119.80
30 tiers				
Minimum	EA.	54.00	16.55	70.55
Maximum	"	110.00	26.00	136.00
Seats				
Seat backs only				
Fiberglass	EA.	29.80	3.40	33.20
Steel and wood seat	"	43.10	3.40	46.50
Seat restoration, fiberglass on wood				
Seats	EA.	21.55	6.75	28.30
Plain bench, no backs	"	13.25	2.80	16.05
Benches				

02 SITEWORK

Site Improvements

02860.40 Recreational Facilities

Site Improvements	UNIT	MAT.	INST.	TOTAL
Park, precast concrete with backs				
4' long	EA.	830.00	110.00	940.00
8' long	"	1,820	170.00	1,990
Fiberglass, with backs				
4' long	EA.	660.00	84.50	744.50
8' long	"	1,260	110.00	1,370
Wood, with backs and fiberglass supports				
4' long	EA.	370.00	84.50	454.50
8' long	"	380.00	110.00	490.00
Steel frame, 6' long				
All steel	EA.	310.00	84.50	394.50
Hardwood boards	"	220.00	84.50	304.50
Players bench (no back), steel frame, fir seat, 10' long	"	240.00	110.00	350.00
Backstops				
Handball or squash court, outdoor				
Wood	EA.			30,316
Masonry	"			27,401
Soccer goal posts	PAIR			2,449
Running track				
Gravel and cinders over stone base	S.Y.	7.55	4.55	12.10
Rubber-cork base resilient pavement	"	10.60	36.40	47.00
For colored surfaces, add	"	6.70	3.65	10.35
Colored rubberized asphalt	"	14.20	45.50	59.70
Artificial resilient mat over asphalt	"	33.50	91.00	124.50
Tennis courts				
Bituminous pavement, 2-1/2" thick	S.Y.	10.75	11.35	22.10
Colored sealer, acrylic emulsion				
3 coats	S.Y.	5.60	2.25	7.85
For 2 color seal coating, add	"	1.00	0.34	1.34
For preparing old courts, add	"	2.40	0.23	2.63
Net, nylon, 42' long	EA.	330.00	42.30	372.30
Paint markings on asphalt, 2 coats	"	84.50	340.00	424.50
Complete court with fence, etc., bituminous				
Minimum	EA.			13,206
Average	"			22,727
Maximum	"			32,249
Clay court				
Minimum	EA.			13,743
Average	"			18,733
Maximum	"			28,409
Playground equipment				
Basketball backboard				
Minimum	EA.	580.00	84.50	664.50
Maximum	"	1,100	96.50	1,197
Bike rack, 10' long	"	440.00	67.50	507.50
Golf shelter, fiberglass	"	2,230	84.50	2,315
Ground socket for movable posts				
Minimum	EA.	100.00	21.15	121.15
Maximum	"	200.00	21.15	221.15
Horizontal monkey ladder, 14' long	"	680.00	56.50	736.50
Posts, tether ball	"	330.00	16.90	346.90
Multiple purpose, 10' long	"	340.00	33.80	373.80

Site Improvements

02860.40 Recreational Facilities

	UNIT	MAT.	INST.	TOTAL
See-saw, steel				
Minimum	EA.	560.00	140.00	700.00
Average	"	1,030	170.00	1,200
Maximum	"	1,580	230.00	1,810
Slide				
Minimum	EA.	1,000	270.00	1,270
Maximum	"	1,740	310.00	2,050
Swings, plain seats				
8' high				
Minimum	EA.	740.00	230.00	970.00
Maximum	"	1,400	260.00	1,660
12' high				
Minimum	EA.	1,140	260.00	1,400
Maximum	"	2,050	380.00	2,430

02870.10 Prefabricated Planters

	UNIT	MAT.	INST.	TOTAL
Concrete precast, circular				
24" dia., 18" high	EA.	260.00	33.80	293.80
42" dia., 30" high	"	350.00	42.30	392.30
Fiberglass, circular				
36" dia., 27" high	EA.	490.00	16.90	506.90
60" dia., 39" high	"	1,120	18.80	1,139
Tapered, circular				
24" dia., 36" high	EA.	380.00	15.40	395.40
40" dia., 36" high	"	630.00	16.90	646.90
Square				
2' by 2', 17" high	EA.	320.00	15.40	335.40
4' by 4', 39" high	"	1,120	18.80	1,139
Rectangular				
4' by 1', 18" high	EA.	360.00	16.90	376.90

Landscaping

02910.10 Shrub & Tree Maintenance

	UNIT	MAT.	INST.	TOTAL
Moving shrubs on site				
12" ball	EA.		42.30	42.30
24" ball	"		56.50	56.50
3' high	"		33.80	33.80
4' high	"		37.60	37.60
5' high	"		42.30	42.30
18" spread	"		48.30	48.30
30" spread	"		56.50	56.50
Moving trees on site				
24" ball	EA.		91.00	91.00
48" ball	"		120.00	120.00
Trees				

02 SITEWORK

Landscaping

Landscaping	UNIT	MAT.	INST.	TOTAL
02910.10 Shrub & Tree Maintenance				
3' high	EA.		36.40	36.40
6' high	"		40.40	40.40
8' high	"		45.50	45.50
10' high	"		60.50	60.50
Palm trees				
7' high	EA.		45.50	45.50
10' high	"		60.50	60.50
20' high	"		180.00	180.00
40' high	"		360.00	360.00
Guying trees				
4" dia.	EA.	7.25	16.90	24.15
8" dia.	"	7.25	21.15	28.40
02920.10 Topsoil				
Spread topsoil, with equipment				
Minimum	C.Y.		9.65	9.65
Maximum	"		12.05	12.05
By hand				
Minimum	C.Y.		33.80	33.80
Maximum	"		42.30	42.30
Area preparation for seeding (grade, rake and clean)				
Square yard	S.Y.		0.27	0.27
By acre	ACRE		1,350	1,350
Remove topsoil and stockpile on site				
4" deep	C.Y.		8.05	8.05
6" deep	"		7.40	7.40
Spreading topsoil from stock pile				
By loader	C.Y.		8.75	8.75
By hand	"		96.50	96.50
Top dress by hand	S.Y.		0.96	0.96
Place imported top soil				
By loader				
4" deep	S.Y.		0.96	0.96
6" deep	"		1.05	1.05
By hand				
4" deep	S.Y.		3.75	3.75
6" deep	"		4.25	4.25
Plant bed preparation, 18" deep				
With backhoe/loader	S.Y.		2.40	2.40
By hand	"		5.65	5.65
02930.30 Seeding				
Mechanical seeding, 175 lb/acre				
By square yard	S.Y.	0.60	0.09	0.69
By acre	ACRE	2,390	440.00	2,830
450 lb/acre				
By square yard	S.Y.	0.96	0.11	1.07
By acre	ACRE	3,650	550.00	4,200
Seeding by hand, 10 lb per 100 s.y.				
By square yard	S.Y.	0.64	0.11	0.75
By acre	ACRE	2,530	560.00	3,090
Reseed disturbed areas	S.F.	0.57	0.17	0.74

Landscaping

	UNIT	MAT.	INST.	TOTAL
02950.10 Plants				
Euonymus coloratus, 18" (Purple Wintercreeper)	EA.	1.80	5.65	7.45
Hedera Helix, 2-1/4" pot (English ivy)	"	0.74	5.65	6.39
Liriope muscari, 2" clumps	"	3.20	3.40	6.60
Santolina, 12"	"	3.65	3.40	7.05
Vinca major or minor, 3" pot	"	0.59	3.40	3.98
Cortaderia argentia, 2 gallon (Pampas Grass)	"	11.55	3.40	14.95
Ophiopogan japonicus, 1 quart (4" pot)	"	3.20	3.40	6.60
Ajuga reptans, 2-3/4" pot (carpet bugle)	"	0.59	3.40	3.98
Pachysandra terminalis, 2-3/4" pot (Japanese Spurge)	"	0.81	3.40	4.21
02950.30 Shrubs				
Juniperus conferia litoralis, 18"-24" (Shore Juniper)	EA.	26.80	13.55	40.35
Horizontalis plumosa, 18"-24" (Andorra Juniper)	"	28.40	13.55	41.95
Sabina tamar-iscfolia-tamarix juniper, 18"-24"	"	28.40	13.55	41.95
Chin San Jose, 18"-24" (San Jose Juniper)	"	28.40	13.55	41.95
Sargenti, 18"-24" (Sargent's Juniper)	"	26.80	13.55	40.35
Nandina domestica, 18"-24" (Heavenly Bamboo)	"	17.95	13.55	31.50
Raphiolepis Indica Springtime, 18"-24" (Indian Hawthorn	"	19.35	13.55	32.90
Osmanthus Heterophyllus Gulftide, 18"-24" (Osmanthus)	"	20.65	13.55	34.20
Ilex Cornuta Burfordi Nana, 18"-24" (Dwarf Burford Holly)	"	23.55	13.55	37.10
Glabra, 18"-24" (Inkberry Holly)	"	22.15	13.55	35.70
Azalea, Indica types, 18"-24"	"	25.00	13.55	38.55
Kurume types, 18"-24"	"	28.00	13.55	41.55
Berberis Julianae, 18"-24" (Wintergreen Barberry)	"	16.40	13.55	29.95
Pieris Japonica Japanese, 18"-24" (Japanese Pieris)	"	16.40	13.55	29.95
Ilex Cornuta Rotunda, 18"-24" (Dwarf Chinese Holly)	"	19.40	13.55	32.95
Juniperus Horizontalis Plumosa, 24"-30" (Andorra Juniper)	"	17.90	16.90	34.80
Rhodopendrow Hybrids, 24"-30"	"	47.70	16.90	64.60
Aucuba Japonica Varigata, 24"-30" (Gold Dust Aucuba)	"	16.30	16.90	33.20
Ilex Crenata Willow Leaf, 24"-30" (Japanese Holly)	"	17.90	16.90	34.80
Cleyera Japonica, 30"-36" (Japanese Cleyera)	"	20.95	21.15	42.10
Pittosporum Tobira, 30"-36"	"	23.90	21.15	45.05
Prumus Laurocerasus, 30"-36"	"	44.90	21.15	66.05
Ilex Cornuta Burfordi, 30"-36" (Burford Holly)	"	23.90	21.15	45.05
Abelia Grandiflora, 24"-36" (Yew Podocarpus)	"	16.40	16.90	33.30
Podocarpos Macrophylla, 24"-36" (Yew Podocarpus)	"	26.80	16.90	43.70
Pyracantha Coccinea Lalandi, 3'-4' (Firethorn)	"	15.30	21.15	36.45
Photinia Frazieri, 3'-4' (Red Photinia)	"	24.25	21.15	45.40
Forsythia Suspensa, 3'-4' (Weeping Forsythia)	"	15.30	21.15	36.45
Camellia Japonica, 3'-4' (Common Camellia)	"	26.90	21.15	48.05
Juniperus Chin Torulosa, 3'-4' (Hollywood Juniper)	"	28.60	21.15	49.75
Cupressocyparis Leylandi, 3'-4'	"	24.10	21.15	45.25
Ilex Opaca Fosteri, 5'-6' (Foster's Holly)	"	97.00	28.20	125.20
Opaca, 5'-6' (American Holly)	"	140.00	28.20	168.20
Nyrica Cerifera, 4'-5' (Southern Wax Myrtles)	"	30.40	24.15	54.55
Ligustrum Japonicum, 4'-5' (Japanese Privet)	"	23.90	24.15	48.05
02950.60 Trees				
Cornus Florida, 5'-6' (White flowering Dogwood)	EA.	67.50	28.20	95.70
Prunus Serrulata Kwanzan, 6'-8' (Kwanzan Cherry)	"	74.50	33.80	108.30
Caroliniana, 6'-8' (Carolina Cherry Laurel)	"	88.50	33.80	122.30
Cercis Canadensis, 6'-8' (Eastern Redbud)	"	61.00	33.80	94.80

Landscaping

02950.60 Trees

	UNIT	MAT.	INST.	TOTAL
Koelreuteria Paniculata, 8'-10' (Goldenrain Tree)	EA.	100.00	42.30	142.30
Acer Platanoides, 1-3/4"-2" (11'-13') (Norway Maple)	"	140.00	56.50	196.50
Rubrum, 1-3/4"-2" (11'-13') (Red Maple)	"	100.00	56.50	156.50
Saccharum, 1-3/4"-2" (Sugar Maple)	"	190.00	56.50	246.50
Fraxinus Pennsylvanica, 1-3/4"-2" (Laneolata-Green Ash)	"	90.00	56.50	146.50
Celtis Occidentalis, 1-3/4"-2" (American Hackberry)	"	130.00	56.50	186.50
Glenditsia Triacantos Inermis, 2"	"	120.00	56.50	176.50
Prunus Cerasifera 'Thundercloud', 6'-8'	"	71.00	33.80	104.80
Yeodensis, 6'-8' (Yoshino Cherry)	"	75.50	33.80	109.30
Lagerstroemia Indica, 8'-10' (Crapemyrtle)	"	120.00	42.30	162.30
Crataegus Phaenopyrum, 8'-10' (Washington Hawthorn)	"	190.00	42.30	232.30
Quercus Borealis, 1-3/4"-2" (Northern Red Oak)	"	110.00	56.50	166.50
Quercus Acutissima, 1-3/4"-2" (8'-10') (Sawtooth Oak)	"	100.00	56.50	156.50
Saliz Babylonica, 1-3/4"-2" (Weeping Willow)	"	52.00	56.50	108.50
Tilia Cordata Greenspire, 1-3/4"-2" (10'-12')	"	230.00	56.50	286.50
Malus, 2"-2-1/2" (8'-10') (Flowering Crabapple)	"	110.00	56.50	166.50
Platanus Occidentalis, (12'-14')	"	170.00	67.50	237.50
Pyrus Calleryana Bradford, 2"-2-1/2" (Bradford Pear)	"	130.00	56.50	186.50
Quercus Palustris, 2"-2-1/2" (12'-14') (Pin Oak)	"	150.00	56.50	206.50
Phellos, 2-1/2"-3" (Willow Oak)	"	160.00	67.50	227.50
Nigra, 2"-2-1/2" (Water Oak)	"	140.00	56.50	196.50
Magnolia Soulangeana, 4'-5' (Saucer Magnolia)	"	81.00	28.20	109.20
Grandiflora, 6'-8' (Southern Magnolia)	"	110.00	33.80	143.80
Cedrus Deodara, 10'-12' (Deodare Cedar)	"	180.00	56.50	236.50
Gingko Biloba, 10'-12' (2"-2-1/2") (Maidenhair Tree)	"	170.00	56.50	226.50
Pinus Thunbergi, 5'-6' (Japanese Black Pine)	"	68.00	28.20	96.20
Strobus, 6'-8' (White Pine)	"	75.50	33.80	109.30
Taeda, 6'-8' (Loblolly Pine)	"	63.50	33.80	97.30
Quercus Virginiana, 2"-2-1/2" (Live Oak)	"	160.00	67.50	227.50

02970.10 Fertilizing

	UNIT	MAT.	INST.	TOTAL
Fertilizing (23#/1000 sf)				
By square yard	S.Y.	0.03	0.11	0.14
By acre	ACRE	120.00	550.00	670.00
Liming (70#/1000 sf)				
By square yard	S.Y.	0.03	0.15	0.17
By acre	ACRE	120.00	730.00	850.00

02980.10 Landscape Accessories

	UNIT	MAT.	INST.	TOTAL
Steel edging, 3/16" x 4"	L.F.	0.54	0.42	0.96
Landscaping stepping stones, 15"x15", white	EA.	4.80	1.70	6.50
Wood chip mulch	C.Y.	40.30	22.55	62.85
2" thick	S.Y.	2.50	0.68	3.18
4" thick	"	4.65	0.97	5.62
6" thick	"	7.00	1.25	8.25
Gravel mulch, 3/4" stone	C.Y.	31.00	33.80	64.80
White marble chips, 1" deep	S.F.	0.62	0.34	0.95
Peat moss				
2" thick	S.Y.	2.90	0.75	3.65
4" thick	"	5.60	1.15	6.75
6" thick	"	8.50	1.40	9.90
Landscaping timbers, treated lumber				

Landscaping

02980.10 Landscape Accessories	UNIT	MAT.	INST.	TOTAL
4" x 4"	L.F.	1.20	1.15	2.35
6" x 6"	"	2.40	1.20	3.60
8" x 8"	"	3.90	1.40	5.30

Formwork

	UNIT	MAT.	INST.	TOTAL
03110.05 Beam Formwork				
Beam forms, job built				
Beam bottoms				
1 use	S.F.	4.30	7.20	11.50
2 uses	"	2.55	6.85	9.40
3 uses	"	1.95	6.65	8.60
4 uses	"	1.60	6.35	7.95
5 uses	"	1.45	6.20	7.65
Beam sides				
1 use	S.F.	3.10	4.80	7.90
2 uses	"	1.85	4.55	6.40
3 uses	"	1.60	4.35	5.95
4 uses	"	1.45	4.10	5.55
5 uses	"	1.30	3.95	5.25
03110.10 Box Culvert Formwork				
Box culverts, job built				
6' x 6'				
1 use	S.F.	2.95	4.35	7.30
2 uses	"	1.60	4.10	5.70
3 uses	"	1.35	3.95	5.30
4 uses	"	1.10	3.75	4.85
5 uses	"	1.00	3.60	4.60
8' x 12'				
1 use	S.F.	2.95	3.60	6.55
2 uses	"	1.60	3.45	5.05
3 uses	"	1.35	3.35	4.70
4 uses	"	1.10	3.20	4.30
5 uses	"	1.00	3.10	4.10
03110.15 Column Formwork				
Column, square forms, job built				
8" x 8" columns				
1 use	S.F.	3.40	8.65	12.05
2 uses	"	1.85	8.30	10.15
3 uses	"	1.55	8.00	9.55
4 uses	"	1.40	7.75	9.15
5 uses	"	1.20	7.45	8.65
12" x 12" columns				
1 use	S.F.	3.10	7.85	10.95
2 uses	"	1.70	7.60	9.30
3 uses	"	1.35	7.35	8.70
4 uses	"	1.20	7.10	8.30
5 uses	"	1.00	6.85	7.85
16" x 16" columns				
1 use	S.F.	2.95	7.20	10.15
2 uses	"	1.55	7.00	8.55
3 uses	"	1.25	6.75	8.00
4 uses	"	1.15	6.55	7.70
5 uses	"	0.93	6.35	7.28
24" x 24" columns				
1 use	S.F.	2.95	6.65	9.60
2 uses	"	1.35	6.45	7.80

03 CONCRETE

Formwork

	UNIT	MAT.	INST.	TOTAL
03110.15 Column Formwork				
3 uses	S.F.	1.10	6.25	7.35
4 uses	"	0.93	6.10	7.03
5 uses	"	0.86	5.95	6.81
36" x 36" columns				
1 use	S.F.	2.95	6.20	9.15
2 uses	"	1.40	6.00	7.40
3 uses	"	1.15	5.85	7.00
4 uses	"	1.00	5.70	6.70
5 uses	"	0.93	5.55	6.48
Round fiber forms, 1 use				
10" dia.	L.F.	3.95	8.65	12.60
12" dia.	"	4.90	8.85	13.75
14" dia.	"	6.40	9.20	15.60
16" dia.	"	8.40	9.60	18.00
18" dia.	"	13.75	10.30	24.05
24" dia.	"	16.80	11.10	27.90
30" dia.	"	25.20	12.00	37.20
36" dia.	"	31.30	13.10	44.40
42" dia.	"	57.50	14.45	71.95
03110.18 Curb Formwork				
Curb forms				
Straight, 6" high				
1 use	L.F.	2.00	4.35	6.35
2 uses	"	1.20	4.10	5.30
3 uses	"	0.89	3.95	4.84
4 uses	"	0.80	3.75	4.55
5 uses	"	0.72	3.60	4.32
Curved, 6" high				
1 use	L.F.	2.15	5.40	7.55
2 uses	"	1.35	5.10	6.45
3 uses	"	1.05	4.80	5.85
4 uses	"	0.94	4.60	5.54
5 uses	"	0.89	4.40	5.29
03110.20 Elevated Slab Formwork				
Elevated slab formwork				
Slab, with drop panels				
1 use	S.F.	3.55	3.45	7.00
2 uses	"	2.05	3.35	5.40
3 uses	"	1.60	3.20	4.80
4 uses	"	1.40	3.10	4.50
5 uses	"	1.25	3.00	4.25
Floor slab, hung from steel beams				
1 use	S.F.	2.85	3.35	6.20
2 uses	"	1.55	3.20	4.75
3 uses	"	1.45	3.10	4.55
4 uses	"	1.25	3.00	4.25
5 uses	"	1.05	2.90	3.95
Floor slab, with pans or domes				
1 use	S.F.	5.10	3.95	9.05
2 uses	"	3.40	3.75	7.15

Formwork

Formwork	UNIT	MAT.	INST.	TOTAL
03110.20 Elevated Slab Formwork				
3 uses	S.F.	3.10	3.60	6.70
4 uses	"	2.90	3.45	6.35
5 uses	"	2.55	3.35	5.90
Equipment curbs, 12" high				
1 use	L.F.	2.65	4.35	7.00
2 uses	"	1.70	4.10	5.80
3 uses	"	1.45	3.95	5.40
4 uses	"	1.35	3.75	5.10
5 uses	"	1.10	3.60	4.70
03110.25 Equipment Pad Formwork				
Equipment pad, job built				
1 use	S.F.	3.45	5.40	8.85
2 uses	"	2.10	5.10	7.20
3 uses	"	1.65	4.80	6.45
4 uses	"	1.30	4.55	5.85
5 uses	"	1.05	4.35	5.40
03110.35 Footing Formwork				
Wall footings, job built, continuous				
1 use	S.F.	1.65	4.35	6.00
2 uses	"	1.15	4.10	5.25
3 uses	"	0.95	3.95	4.90
4 uses	"	0.83	3.75	4.58
5 uses	"	0.73	3.60	4.33
Column footings, spread				
1 use	S.F.	1.70	5.40	7.10
2 uses	"	1.30	5.10	6.40
3 uses	"	0.92	4.80	5.72
4 uses	"	0.77	4.55	5.32
5 uses	"	0.70	4.35	5.05
03110.50 Grade Beam Formwork				
Grade beams, job built				
1 use	S.F.	2.55	4.35	6.90
2 uses	"	1.45	4.10	5.55
3 uses	"	1.10	3.95	5.05
4 uses	"	0.93	3.75	4.68
5 uses	"	0.78	3.60	4.38
03110.53 Pile Cap Formwork				
Pile cap forms, job built				
Square				
1 use	S.F.	2.90	5.40	8.30
2 uses	"	1.70	5.10	6.80
3 uses	"	1.35	4.80	6.15
4 uses	"	1.20	4.55	5.75
5 uses	"	0.98	4.35	5.33
Triangular				
1 use	S.F.	3.10	6.20	9.30
2 uses	"	2.05	5.75	7.80
3 uses	"	1.65	5.40	7.05
4 uses	"	1.35	5.10	6.45

03 CONCRETE

Formwork	UNIT	MAT.	INST.	TOTAL
03110.53 Pile Cap Formwork				
5 uses	S.F.	1.05	4.80	5.85
03110.55 Slab/mat Formwork				
Mat foundations, job built				
1 use	S.F.	2.55	5.40	7.95
2 uses	"	1.45	5.10	6.55
3 uses	"	1.10	4.80	5.90
4 uses	"	0.92	4.55	5.47
5 uses	"	0.73	4.35	5.08
Edge forms				
6" high				
1 use	L.F.	2.55	3.95	6.50
2 uses	"	1.45	3.75	5.20
3 uses	"	1.10	3.60	4.70
4 uses	"	0.92	3.45	4.37
5 uses	"	0.73	3.35	4.08
12" high				
1 use	L.F.	2.40	4.35	6.75
2 uses	"	1.35	4.10	5.45
3 uses	"	1.00	3.95	4.95
4 uses	"	0.83	3.75	4.58
5 uses	"	0.68	3.60	4.28
Formwork for openings				
1 use	S.F.	3.50	8.65	12.15
2 uses	"	2.00	7.85	9.85
3 uses	"	1.65	7.20	8.85
4 uses	"	1.30	6.65	7.95
5 uses	"	1.05	6.20	7.25
03110.60 Stair Formwork				
Stairway forms, job built				
1 use	S.F.	4.35	8.65	13.00
2 uses	"	2.45	7.85	10.30
3 uses	"	1.90	7.20	9.10
4 uses	"	1.75	6.65	8.40
5 uses	"	1.45	6.20	7.65
Stairs, elevated				
1 use	S.F.	5.25	8.65	13.90
2 uses	"	2.80	7.20	10.00
3 uses	"	2.45	6.20	8.65
4 uses	"	2.10	5.75	7.85
5 uses	"	1.75	5.40	7.15
03110.65 Wall Formwork				
Wall forms, exterior, job built				
Up to 8' high wall				
1 use	S.F.	2.80	4.35	7.15
2 uses	"	1.55	4.10	5.65
3 uses	"	1.35	3.95	5.30
4 uses	"	1.20	3.75	4.95
5 uses	"	1.05	3.60	4.65
Over 8' high wall				

Formwork

03110.65 Wall Formwork

	UNIT	MAT.	INST.	TOTAL
1 use	S.F.	3.10	5.40	8.50
2 uses	"	1.75	5.10	6.85
3 uses	"	1.60	4.80	6.40
4 uses	"	1.45	4.55	6.00
5 uses	"	1.30	4.35	5.65
Over 16' high wall				
1 use	S.F.	3.25	6.20	9.45
2 uses	"	1.95	5.75	7.70
3 uses	"	1.75	5.40	7.15
4 uses	"	1.60	5.10	6.70
5 uses	"	1.45	4.80	6.25
Radial wall forms				
1 use	S.F.	3.05	6.65	9.70
2 uses	"	1.80	6.20	8.00
3 uses	"	1.70	5.75	7.45
4 uses	"	1.50	5.40	6.90
5 uses	"	1.35	5.10	6.45
Retaining wall forms				
1 use	S.F.	2.60	4.80	7.40
2 uses	"	1.40	4.55	5.95
3 uses	"	1.20	4.35	5.55
4 uses	"	1.05	4.10	5.15
5 uses	"	0.89	3.95	4.84
Radial retaining wall forms				
1 use	S.F.	2.75	7.20	9.95
2 uses	"	1.70	6.65	8.35
3 uses	"	1.45	6.20	7.65
4 uses	"	1.40	5.75	7.15
5 uses	"	1.20	5.40	6.60
Column pier and pilaster				
1 use	S.F.	3.10	8.65	11.75
2 uses	"	1.85	7.85	9.70
3 uses	"	1.70	7.20	8.90
4 uses	"	1.55	6.65	8.20
5 uses	"	1.40	6.20	7.60
Interior wall forms				
Up to 8' high				
1 use	S.F.	2.80	3.95	6.75
2 uses	"	1.55	3.75	5.30
3 uses	"	1.40	3.60	5.00
4 uses	"	1.20	3.45	4.65
5 uses	"	1.00	3.35	4.35
Over 8' high				
1 use	S.F.	3.10	4.80	7.90
2 uses	"	1.75	4.55	6.30
3 uses	"	1.60	4.35	5.95
4 uses	"	1.45	4.10	5.55
5 uses	"	1.30	3.95	5.25
Over 16' high				
1 use	S.F.	3.25	5.40	8.65
2 uses	"	1.95	5.10	7.05
3 uses	"	1.75	4.80	6.55

Formwork	UNIT	MAT.	INST.	TOTAL
03110.65 Wall Formwork				
4 uses	S.F.	1.60	4.55	6.15
5 uses	"	1.45	4.35	5.80
Radial wall forms				
1 use	S.F.	3.05	5.75	8.80
2 uses	"	1.80	5.40	7.20
3 uses	"	1.70	5.10	6.80
4 uses	"	1.50	4.80	6.30
5 uses	"	1.35	4.55	5.90
Curved wall forms, 24" sections				
1 use	S.F.	2.90	8.65	11.55
2 uses	"	1.75	7.85	9.60
3 uses	"	1.60	7.20	8.80
4 uses	"	1.45	6.65	8.10
5 uses	"	1.30	6.20	7.50
PVC form liner, per side, smooth finish				
1 use	S.F.	5.30	3.60	8.90
2 uses	"	2.90	3.45	6.35
3 uses	"	2.40	3.35	5.75
4 uses	"	1.90	3.10	5.00
5 uses	"	1.50	2.90	4.40
03110.90 Miscellaneous Formwork				
Keyway forms (5 uses)				
2 x 4	L.F.	0.19	2.15	2.33
2 x 6	"	0.28	2.40	2.68
Bulkheads				
Walls, with keyways				
2 piece	L.F.	3.15	3.95	7.10
3 piece	"	3.90	4.35	8.25
Elevated slab, with keyway				
2 piece	L.F.	3.60	3.60	7.20
3 piece	"	5.20	3.95	9.15
Ground slab, with keyway				
2 piece	L.F.	3.90	3.10	7.00
3 piece	"	5.20	3.35	8.55
Chamfer strips				
Wood				
1/2" wide	L.F.	0.24	0.96	1.20
3/4" wide	"	0.31	0.96	1.27
1" wide	"	0.40	0.96	1.36
PVC				
1/2" wide	L.F.	0.71	0.96	1.67
3/4" wide	"	0.78	0.96	1.74
1" wide	"	1.00	0.96	1.96
Radius				
1"	L.F.	1.05	1.05	2.10
1-1/2"	"	1.80	1.05	2.85
Reglets				
Galvanized steel, 24 ga.	L.F.	1.10	1.75	2.85
Metal formwork				
Straight edge forms				
4" high	L.F.	0.15	2.70	2.85

03 CONCRETE

Formwork

	UNIT	MAT.	INST.	TOTAL
03110.90 Miscellaneous Formwork				
6" high	L.F.	0.17	2.90	3.07
8" high	"	0.19	3.10	3.29
12" high	"	0.24	3.35	3.59
16" high	"	0.29	3.60	3.89
Curb form, S-shape				
12" x				
1'-6"	L.F.	0.29	6.20	6.49
2'	"	0.36	5.75	6.11
2'-6"	"	0.41	5.40	5.81
3'	"	0.46	4.80	5.26

Reinforcement

	UNIT	MAT.	INST.	TOTAL
03210.05 Beam Reinforcing				
Beam-girders				
#3 - #4	TON	1,100	1,200	2,300
#5 - #6	"	970.00	960.00	1,930
#7 - #8	"	920.00	800.00	1,720
#9 - #10	"	920.00	690.00	1,610
#11 - #12	"	920.00	640.00	1,560
#13 - #14	"	920.00	600.00	1,520
Galvanized				
#3 - #4	TON	1,870	1,200	3,070
#5 - #6	"	1,770	960.00	2,730
#7 - #8	"	1,700	800.00	2,500
#9 - #10	"	1,700	690.00	2,390
#11 - #12	"	1,700	640.00	2,340
#13 - #14	"	1,700	600.00	2,300
Epoxy coated				
#3 - #4	TON	1,620	1,370	2,990
#5 - #6	"	1,520	1,070	2,590
#7 - #8	"	1,470	870.00	2,340
#9 - #10	"	1,470	740.00	2,210
#11 - #12	"	1,470	690.00	2,160
#13 - #14	"	1,470	640.00	2,110
Bond Beams				
#3 - #4	TON	1,100	1,600	2,700
#5 - #6	"	970.00	1,200	2,170
#7 - #8	"	920.00	1,070	1,990
Galvanized				
#3 - #4	TON	1,790	1,600	3,390
#5 - #6	"	1,770	1,200	2,970
#7 - #8	"	1,700	1,070	2,770
Epoxy coated				
#3 - #4	TON	1,620	1,920	3,540
#5 - #6	"	1,520	1,370	2,890

Reinforcement	UNIT	MAT.	INST.	TOTAL
03210.05 Beam Reinforcing				
#7 - #8	TON	1,470	1,200	2,670
03210.10 Box Culvert Reinforcing				
Box culverts				
#3 - #4	TON	1,100	600.00	1,700
#5 - #6	"	970.00	530.00	1,500
#7 - #8	"	920.00	480.00	1,400
#9 - #10	"	920.00	440.00	1,360
#11 - #12	"	920.00	400.00	1,320
Galvanized				
#3 - #4	TON	1,790	600.00	2,390
#5 - #6	"	1,770	530.00	2,300
#7 - #8	"	1,700	480.00	2,180
#9 - #10	"	1,700	440.00	2,140
#11 - #12	"	1,700	400.00	2,100
Epoxy coated				
#3 - #4	TON	1,620	640.00	2,260
#5 - #6	"	1,520	570.00	2,090
#7 - #8	"	1,470	510.00	1,980
#9 - #10	"	1,470	460.00	1,930
#11 - #12	"	1,470	420.00	1,890
03210.15 Column Reinforcing				
Columns				
#3 - #4	TON	1,100	1,370	2,470
#5 - #6	"	970.00	1,070	2,040
#7 - #8	"	920.00	960.00	1,880
#9 - #10	"	920.00	870.00	1,790
#11 - #12	"	920.00	800.00	1,720
#13 - #14	"	920.00	740.00	1,660
#15 - #16	"	920.00	690.00	1,610
Galvanized				
#3 - #4	TON	1,870	1,370	3,240
#5 - #6	"	1,770	1,070	2,840
#7 - #8	"	1,700	960.00	2,660
#9 - #10	"	1,700	870.00	2,570
#11 - #12	"	1,700	800.00	2,500
#13 - #14	"	1,700	740.00	2,440
#15 - #16	"	1,700	690.00	2,390
Epoxy coated				
#3 - #4	TON	1,620	1,600	3,220
#5 - #6	"	1,520	1,200	2,720
#7 - #8	"	1,470	1,070	2,540
#9 - #10	"	1,470	960.00	2,430
#11 - #12	"	1,470	870.00	2,340
#13 - #14	"	1,470	800.00	2,270
#15 - #16	"	1,470	740.00	2,210
Spirals				
8" to 24" dia.	TON	1,700	1,200	2,900
24" to 48" dia.	"	1,700	1,070	2,770
48" to 84" dia.	"	1,870	960.00	2,830

Reinforcement

Reinforcement	UNIT	MAT.	INST.	TOTAL
03210.20 Elevated Slab Reinforcing				
Elevated slab				
#3 - #4	TON	1,100	600.00	1,700
#5 - #6	"	970.00	530.00	1,500
#7 - #8	"	920.00	480.00	1,400
#9 - #10	"	920.00	440.00	1,360
#11 - #12	"	920.00	400.00	1,320
Galvanized				
#3 - #4	TON	1,790	600.00	2,390
#5 - #6	"	1,770	530.00	2,300
#7 - #8	"	1,700	480.00	2,180
#9 - #10	"	1,700	440.00	2,140
#11 - #12	"	1,700	400.00	2,100
Epoxy coated				
#3 - #4	TON	1,620	640.00	2,260
#5 - #6	"	1,520	570.00	2,090
#7 - #8	"	1,470	510.00	1,980
#9 - #10	"	1,470	460.00	1,930
#11 - #12	"	1,470	420.00	1,890
03210.25 Equip. Pad Reinforcing				
Equipment pad				
#3 - #4	TON	1,100	960.00	2,060
#5 - #6	"	970.00	870.00	1,840
#7 - #8	"	920.00	800.00	1,720
#9 - #10	"	920.00	740.00	1,660
#11 - #12	"	920.00	690.00	1,610
03210.35 Footing Reinforcing				
Footings				
Grade 50				
#3 - #4	TON	1,100	800.00	1,900
#5 - #6	"	970.00	690.00	1,660
#7 - #8	"	920.00	600.00	1,520
#9 - #10	"	920.00	530.00	1,450
Grade 60				
#3 - #4	TON	1,100	800.00	1,900
#5 - #6	"	970.00	690.00	1,660
#7 - #8	"	920.00	600.00	1,520
#9 - #10	"	920.00	530.00	1,450
Grade 70				
#3 - #4	TON	1,100	800.00	1,900
#5 - #6	"	970.00	690.00	1,660
#7 - #8	"	920.00	600.00	1,520
#9 - #10	"	920.00	530.00	1,450
#11- #12	"	920.00	480.00	1,400
Straight dowels, 24" long				
1" dia. (#8)	EA.	3.40	4.80	8.20
3/4" dia. (#6)	"	3.10	4.80	7.90
5/8" dia. (#5)	"	2.65	4.00	6.65
1/2" dia. (#4)	"	2.00	3.45	5.45

Reinforcement

	UNIT	MAT.	INST.	TOTAL
03210.45 Foundation Reinforcing				
Foundations				
#3 - #4	TON	1,100	800.00	1,900
#5 - #6	"	970.00	690.00	1,660
#7 - #8	"	920.00	600.00	1,520
#9 - #10	"	920.00	530.00	1,450
#11 - #12	"	920.00	480.00	1,400
Galvanized				
#3 - #4	TON	1,870	800.00	2,670
#5 - #6	"	1,770	690.00	2,460
#7 - #8	"	1,700	600.00	2,300
#9 - #10	"	1,700	530.00	2,230
#11 - #12	"	1,700	480.00	2,180
Epoxy Coated				
#3 - #4	TON	1,620	870.00	2,490
#5 - #6	"	1,520	740.00	2,260
#7 - #8	"	1,470	640.00	2,110
#9 - #10	"	1,470	570.00	2,040
#11 - #12	"	1,470	510.00	1,980
03210.50 Grade Beam Reinforcing				
Grade beams				
#3 - #4	TON	1,100	740.00	1,840
#5 - #6	"	970.00	640.00	1,610
#7 - #8	"	920.00	570.00	1,490
#9 - #10	"	920.00	510.00	1,430
#11 - #12	"	920.00	460.00	1,380
Galvanized				
#3 - #4	TON	1,870	740.00	2,610
#5 - #6	"	1,770	640.00	2,410
#7 - #8	"	1,700	570.00	2,270
#9 - #10	"	1,700	510.00	2,210
#11 - #12	"	1,700	460.00	2,160
Epoxy coated				
#3 - #4	TON	1,620	800.00	2,420
#5 - #6	"	1,520	690.00	2,210
#7 - #8	"	1,470	600.00	2,070
#9 - #10	"	1,470	530.00	2,000
#11 - #12	"	1,470	480.00	1,950
03210.53 Pile Cap Reinforcing				
Pile caps				
#3 - #4	TON	1,100	1,200	2,300
#5 - #6	"	970.00	1,070	2,040
#7 - #8	"	920.00	960.00	1,880
#9 - #10	"	920.00	870.00	1,790
#11 - #12	"	920.00	800.00	1,720
Galvanized				
#3 - #4	TON	1,870	1,200	3,070
#5 - #6	"	1,770	1,070	2,840
#7 - #8	"	1,700	960.00	2,660
#9 - #10	"	1,700	870.00	2,570
#11 - #12	"	1,700	800.00	2,500

Reinforcement

Reinforcement	UNIT	MAT.	INST.	TOTAL
03210.53 Pile Cap Reinforcing				
Epoxy coated				
#3 - #4	TON	1,620	1,370	2,990
#5 - #6	"	1,520	1,200	2,720
#7 - #8	"	1,470	1,070	2,540
#9 - #10	"	1,470	960.00	2,430
#11 - #12	"	1,470	870.00	2,340
03210.55 Slab/mat Reinforcing				
Bars, slabs				
#3 - #4	TON	1,100	800.00	1,900
#5 - #6	"	970.00	690.00	1,660
#7 - #8	"	920.00	600.00	1,520
#9 - #10	"	920.00	530.00	1,450
#11 - #12	"	920.00	480.00	1,400
Galvanized				
#3 - #4	TON	1,870	800.00	2,670
#5 - #6	"	1,770	690.00	2,460
#7 - #8	"	1,700	600.00	2,300
#9 - #10	"	1,700	530.00	2,230
#11 - #12	"	1,700	480.00	2,180
Epoxy coated				
#3 - #4	TON	1,620	870.00	2,490
#5 - #6	"	1,520	740.00	2,260
#7 - #8	"	1,470	640.00	2,110
#9 - #10	"	1,470	570.00	2,040
#11 - #12	"	1,470	510.00	1,980
Wire mesh, slabs				
Galvanized				
4x4				
W1.4xW1.4	S.F.	0.27	0.32	0.59
W2.0xW2.0	"	0.35	0.34	0.70
W2.9xW2.9	"	0.50	0.37	0.87
W4.0xW4.0	"	0.74	0.40	1.14
6x6				
W1.4xW1.4	S.F.	0.26	0.24	0.50
W2.0xW2.0	"	0.35	0.27	0.62
W2.9xW2.9	"	0.48	0.28	0.76
W4.0xW4.0	"	0.51	0.32	0.83
Standard				
2x2				
W.9xW.9	S.F.	0.27	0.32	0.59
4x4				
W1.4xW1.4	S.F.	0.18	0.32	0.50
W2.0xW2.0	"	0.24	0.34	0.58
W2.9xW2.9	"	0.32	0.37	0.69
W4.0xW4.0	"	0.50	0.40	0.90
6x6				
W1.4xW1.4	S.F.	0.11	0.24	0.35
W2.0xW2.0	"	0.16	0.27	0.43
W2.9xW2.9	"	0.24	0.28	0.52
W4.0xW4.0	"	0.34	0.32	0.66

Reinforcement

Reinforcement	UNIT	MAT.	INST.	TOTAL
03210.60 Stair Reinforcing				
Stairs				
#3 - #4	TON	1,100	960.00	2,060
#5 - #6	"	970.00	800.00	1,770
#7 - #8	"	920.00	690.00	1,610
#9 - #10	"	920.00	600.00	1,520
Galvanized				
#3 - #4	TON	1,870	960.00	2,830
#5 - #6	"	1,770	800.00	2,570
#7 - #8	"	1,700	690.00	2,390
#9 - #10	"	1,700	600.00	2,300
Epoxy coated				
#3 - #4	TON	1,620	1,070	2,690
#5 - #6	"	1,520	870.00	2,390
#7 - #8	"	1,470	740.00	2,210
#9 - #10	"	1,470	640.00	2,110
03210.65 Wall Reinforcing				
Walls				
#3 - #4	TON	1,100	690.00	1,790
#5 - #6	"	970.00	600.00	1,570
#7 - #8	"	920.00	530.00	1,450
#9 - #10	"	920.00	480.00	1,400
Galvanized				
#3 - #4	TON	1,870	690.00	2,560
#5 - #6	"	1,770	600.00	2,370
#7 - #8	"	1,700	530.00	2,230
#9 - #10	"	1,700	480.00	2,180
Epoxy coated				
#3 - #4	TON	1,620	740.00	2,360
#5 - #6	"	1,520	640.00	2,160
#7 - #8	"	1,470	570.00	2,040
#9 - #10	"	1,470	510.00	1,980
Masonry wall (horizontal)				
#3 - #4	TON	1,100	1,920	3,020
#5 - #6	"	970.00	1,600	2,570
Galvanized				
#3 - #4	TON	1,870	1,920	3,790
#5 - #6	"	1,770	1,600	3,370
Masonry wall (vertical)				
#3 - #4	TON	1,100	2,400	3,500
#5 - #6	"	970.00	1,920	2,890
Galvanized				
#3 - #4	TON	1,870	2,400	4,270
#5 - #6	"	1,770	1,920	3,690

Accessories

03250.40 Concrete Accessories

Accessories	UNIT	MAT.	INST.	TOTAL
Expansion joint, poured				
Asphalt				
1/2" x 1"	L.F.	0.55	0.68	1.23
1" x 2"	"	1.70	0.74	2.44
Liquid neoprene, cold applied				
1/2" x 1"	L.F.	2.05	0.69	2.74
1" x 2"	"	8.00	0.75	8.75
Polyurethane, 2 parts				
1/2" x 1"	L.F.	1.90	1.15	3.05
1" x 2"	"	7.75	1.25	9.00
Rubberized asphalt, cold				
1/2" x 1"	L.F.	0.47	0.68	1.15
1" x 2"	"	1.45	0.74	2.19
Hot, fuel resistant				
1/2" x 1"	L.F.	0.88	0.68	1.55
1" x 2"	"	4.30	0.74	5.04
Expansion joint, premolded, in slabs				
Asphalt				
1/2" x 6"	L.F.	0.62	0.85	1.47
1" x 12"	"	1.05	1.15	2.20
Cork				
1/2" x 6"	L.F.	1.35	0.85	2.20
1" x 12"	"	5.50	1.15	6.65
Neoprene sponge				
1/2" x 6"	L.F.	1.90	0.85	2.75
1" x 12"	"	6.85	1.15	8.00
Polyethylene foam				
1/2" x 6"	L.F.	0.71	0.85	1.56
1" x 12"	"	3.40	1.15	4.55
Polyurethane foam				
1/2" x 6"	L.F.	0.94	0.85	1.78
1" x 12"	"	2.10	1.15	3.25
Polyvinyl chloride foam				
1/2" x 6"	L.F.	2.00	0.85	2.85
1" x 12"	"	7.30	1.15	8.45
Rubber, gray sponge				
1/2" x 6"	L.F.	3.25	0.85	4.10
1" x 12"	"	13.75	1.15	14.90
Asphalt felt control joints or bond breaker, screed joints				
4" slab	L.F.	0.88	0.68	1.55
6" slab	"	1.10	0.75	1.85
8" slab	"	1.50	0.85	2.35
10" slab	"	2.10	0.97	3.07
Keyed cold expansion and control joints, 24 ga.				
4" slab	L.F.	0.77	2.10	2.87
5" slab	"	0.97	2.10	3.07
6" slab	"	1.10	2.25	3.35
8" slab	"	1.40	2.40	3.80
10" slab	"	1.55	2.60	4.15
Waterstops				
Polyvinyl chloride				
Ribbed				

Accessories

03250.40 Concrete Accessories	UNIT	MAT.	INST.	TOTAL
3/16" thick x				
4" wide	L.F.	1.25	1.70	2.95
6" wide	"	1.55	1.90	3.45
1/2" thick x				
9" wide	L.F.	4.25	2.10	6.35
Ribbed with center bulb				
3/16" thick x 9" wide	L.F.	3.60	2.10	5.70
3/8" thick x 9" wide	"	4.40	2.10	6.50
Dumbbell type, 3/8" thick x 6" wide	"	4.40	1.90	6.30
Plain, 3/8" thick x 9" wide	"	5.45	2.10	7.55
Center bulb, 3/8" thick x 9" wide	"	7.30	2.10	9.40
Rubber				
Flat dumbbell				
3/8" thick x				
6" wide	L.F.	7.10	1.90	9.00
9" wide	"	11.15	2.10	13.25
Center bulb				
3/8" thick x				
6" wide	L.F.	6.60	1.90	8.50
9" wide	"	13.40	2.10	15.50
Vapor barrier				
4 mil polyethylene	S.F.	0.03	0.11	0.14
6 mil polyethylene	"	0.04	0.11	0.16
Gravel porous fill, under floor slabs, 3/4" stone	C.Y.	20.45	56.50	76.95
Reinforcing accessories				
Beam bolsters				
1-1/2" high, plain	L.F.	0.41	0.48	0.89
Galvanized	"	0.51	0.48	0.99
3" high				
Plain	L.F.	0.96	0.60	1.56
Galvanized	"	1.05	0.60	1.65
Slab bolsters				
1" high				
Plain	L.F.	0.28	0.24	0.51
Galvanized	"	0.39	0.24	0.63
2" high				
Plain	L.F.	0.39	0.27	0.65
Galvanized	"	0.44	0.27	0.70
Chairs, high chairs				
3" high				
Plain	EA.	0.45	1.20	1.65
Galvanized	"	0.59	1.20	1.79
5" high				
Plain	EA.	0.59	1.25	1.84
Galvanized	"	0.72	1.25	1.97
8" high				
Plain	EA.	1.20	1.35	2.55
Galvanized	"	1.55	1.35	2.90
12" high				
Plain	EA.	2.55	1.60	4.15
Galvanized	"	3.00	1.60	4.60
Continuous, high chair				

Accessories

03250.40 Concrete Accessories	UNIT	MAT.	INST.	TOTAL
3" high				
Plain	L.F.	0.52	0.32	0.85
Galvanized	"	0.64	0.32	0.96
5" high				
Plain	L.F.	0.79	0.34	1.13
Galvanized	"	1.05	0.34	1.39
8" high				
Plain	L.F.	1.05	0.37	1.42
Galvanized	"	1.40	0.37	1.77
12" high				
Plain	L.F.	2.30	0.40	2.70
Galvanized	"	2.70	0.40	3.10

Cast-in-place Concrete

03300.10 Concrete Admixtures	UNIT	MAT.	INST.	TOTAL
Concrete admixtures				
Water reducing admixture	GAL			13.71
Set retarder	"			24.57
Air entraining agent	"			8.36

03350.10 Concrete Finishes	UNIT	MAT.	INST.	TOTAL
Floor finishes				
Broom	S.F.		0.48	0.48
Screed	"		0.42	0.42
Darby	"		0.42	0.42
Steel float	"		0.56	0.56
Granolithic topping				
1/2" thick	S.F.	0.24	1.55	1.79
1" thick	"	0.45	1.70	2.15
2" thick	"	0.77	1.90	2.67
Wall finishes				
Burlap rub, with cement paste	S.F.	0.06	0.56	0.62
Float finish	"	0.08	0.85	0.93
Etch with acid	"	0.29	0.56	0.85
Sandblast				
Minimum	S.F.	0.06	0.88	0.93
Maximum	"	0.36	0.88	1.24
Bush hammer				
Green concrete	S.F.		1.70	1.70
Cured concrete	"		2.60	2.60
Break ties and patch holes	"		0.68	0.68
Carborundum				
Dry rub	S.F.		1.15	1.15
Wet rub	"		1.70	1.70
Floor hardeners				

03 CONCRETE

Cast-in-place Concrete

	UNIT	MAT.	INST.	TOTAL
03350.10 Concrete Finishes				
Metallic				
Light service	S.F.	0.25	0.42	0.68
Heavy service	"	0.84	0.56	1.40
Non-metallic				
Light service	S.F.	0.13	0.42	0.56
Heavy service	"	0.59	0.56	1.16
Rusticated concrete finish				
Beveled edge	L.F.	0.30	1.90	2.20
Square edge	"	0.40	2.40	2.80
Solid board concrete finish				
Standard	S.F.	0.75	2.80	3.55
Rustic	"	0.66	3.40	4.06
03360.10 Pneumatic Concrete				
Pneumatic applied concrete (gunite)				
2" thick	S.F.	4.20	2.25	6.45
3" thick	"	5.20	3.05	8.25
4" thick	"	6.25	3.65	9.90
Finish surface				
Minimum	S.F.		2.15	2.15
Maximum	"		4.35	4.35
03370.10 Curing Concrete				
Sprayed membrane				
Slabs	S.F.	0.04	0.07	0.11
Walls	"	0.07	0.09	0.15
Curing paper				
Slabs	S.F.	0.07	0.09	0.16
Walls	"	0.07	0.10	0.17
Burlap				
7.5 oz.	S.F.	0.06	0.11	0.17
12 oz.	"	0.08	0.12	0.20

Placing Concrete

	UNIT	MAT.	INST.	TOTAL
03380.05 Beam Concrete				
Beams and girders				
2500# or 3000# concrete				
By crane	C.Y.	87.00	65.50	152.50
By pump	"	87.00	59.50	146.50
By hand buggy	"	87.00	33.80	120.80
3500# or 4000# concrete				
By crane	C.Y.	91.50	65.50	157.00
By pump	"	91.50	59.50	151.00
By hand buggy	"	91.50	33.80	125.30
5000# concrete				

Placing Concrete

	UNIT	MAT.	INST.	TOTAL
03380.05 Beam Concrete				
By crane	C.Y.	95.50	65.50	161.00
By pump	"	95.50	59.50	155.00
By hand buggy	"	95.50	33.80	129.30
Bond beam, 3000# concrete				
By pump				
8" high				
4" wide	L.F.	0.24	1.30	1.54
6" wide	"	0.58	1.50	2.07
8" wide	"	0.74	1.65	2.39
10" wide	"	0.97	1.80	2.77
12" wide	"	1.30	2.05	3.35
16" high				
8" wide	L.F.	1.80	2.05	3.85
10" wide	"	2.40	2.35	4.75
12" wide	"	3.20	2.75	5.95
By crane				
8" high				
4" wide	L.F.	0.30	1.45	1.75
6" wide	"	0.58	1.55	2.12
8" wide	"	0.74	1.65	2.39
10" wide	"	0.97	1.80	2.77
12" wide	"	1.30	2.05	3.35
16" high				
8" wide	L.F.	1.80	2.05	3.85
10" wide	"	2.40	2.20	4.60
12" wide	"	3.20	2.50	5.70
03380.15 Column Concrete				
Columns				
2500# or 3000# concrete				
By crane	C.Y.	87.00	59.50	146.50
By pump	"	87.00	54.50	141.50
3500# or 4000# concrete				
By crane	C.Y.	91.50	59.50	151.00
By pump	"	91.50	54.50	146.00
5000# concrete				
By crane	C.Y.	95.50	59.50	155.00
By pump	"	95.50	54.50	150.00
03380.20 Elevated Slab Concrete				
Elevated slab				
2500# or 3000# concrete				
By crane	C.Y.	87.00	32.80	119.80
By pump	"	87.00	25.20	112.20
By hand buggy	"	87.00	33.80	120.80
3500# or 4000# concrete				
By crane	C.Y.	91.50	32.80	124.30
By pump	"	91.50	25.20	116.70
By hand buggy	"	91.50	33.80	125.30
5000# concrete				
By crane	C.Y.	95.50	32.80	128.30
By pump	"	95.50	25.20	120.70

Placing Concrete

	UNIT	MAT.	INST.	TOTAL
03380.20 Elevated Slab Concrete				
By hand buggy	C.Y.	95.50	33.80	129.30
Topping				
2500# or 3000# concrete				
By crane	C.Y.	87.00	32.80	119.80
By pump	"	87.00	25.20	112.20
By hand buggy	"	87.00	33.80	120.80
3500# or 4000# concrete				
By crane	C.Y.	91.50	32.80	124.30
By pump	"	91.50	25.20	116.70
By hand buggy	"	91.50	33.80	125.30
5000# concrete				
By crane	C.Y.	95.50	32.80	128.30
By pump	"	95.50	25.20	120.70
By hand buggy	"	95.50	33.80	129.30
03380.25 Equipment Pad Concrete				
Equipment pad				
2500# or 3000# concrete				
By chute	C.Y.	87.00	11.30	98.30
By pump	"	87.00	46.90	133.90
By crane	"	87.00	54.50	141.50
3500# or 4000# concrete				
By chute	C.Y.	91.50	11.30	102.80
By pump	"	91.50	46.90	138.40
By crane	"	91.50	54.50	146.00
5000# concrete				
By chute	C.Y.	95.50	11.30	106.80
By pump	"	95.50	46.90	142.40
By crane	"	95.50	54.50	150.00
03380.35 Footing Concrete				
Continuous footing				
2500# or 3000# concrete				
By chute	C.Y.	87.00	11.30	98.30
By pump	"	87.00	41.00	128.00
By crane	"	87.00	46.90	133.90
3500# or 4000# concrete				
By chute	C.Y.	91.50	11.30	102.80
By pump	"	91.50	41.00	132.50
By crane	"	91.50	46.90	138.40
5000# concrete				
By chute	C.Y.	95.50	11.30	106.80
By pump	"	95.50	41.00	136.50
By crane	"	95.50	46.90	142.40
Spread footing				
2500# or 3000# concrete				
Under 5 cy				
By chute	C.Y.	87.00	11.30	98.30
By pump	"	87.00	43.70	130.70
By crane	"	87.00	50.50	137.50
Over 5 cy				
By chute	C.Y.	87.00	8.45	95.45

Placing Concrete

	UNIT	MAT.	INST.	TOTAL
03380.35 Footing Concrete				
By pump	C.Y.	87.00	38.60	125.60
By crane	"	87.00	43.70	130.70
3500# or 4000# concrete				
Under 5 c.y.				
By chute	C.Y.	91.50	11.30	102.80
By pump	"	91.50	43.70	135.20
By crane	"	91.50	50.50	142.00
Over 5 c.y.				
By chute	C.Y.	91.50	8.45	99.95
By pump	"	91.50	38.60	130.10
By crane	"	91.50	43.70	135.20
5000# concrete				
Under 5 c.y.				
By chute	C.Y.	95.50	11.30	106.80
By pump	"	95.50	43.70	139.20
By crane	"	95.50	50.50	146.00
Over 5 c.y.				
By chute	C.Y.	95.50	8.45	103.95
By pump	"	95.50	38.60	134.10
By crane	"	95.50	43.70	139.20
03380.50 Grade Beam Concrete				
Grade beam				
2500# or 3000# concrete				
By chute	C.Y.	87.00	11.30	98.30
By crane	"	87.00	46.90	133.90
By pump	"	87.00	41.00	128.00
By hand buggy	"	87.00	33.80	120.80
3500# or 4000# concrete				
By chute	C.Y.	91.50	11.30	102.80
By crane	"	91.50	46.90	138.40
By pump	"	91.50	41.00	132.50
By hand buggy	"	91.50	33.80	125.30
5000# concrete				
By chute	C.Y.	95.50	11.30	106.80
By crane	"	95.50	46.90	142.40
By pump	"	95.50	41.00	136.50
By hand buggy	"	95.50	33.80	129.30
03380.53 Pile Cap Concrete				
Pile cap				
2500# or 3000 concrete				
By chute	C.Y.	87.00	11.30	98.30
By crane	"	87.00	54.50	141.50
By pump	"	87.00	46.90	133.90
By hand buggy	"	87.00	33.80	120.80
3500# or 4000# concrete				
By chute	C.Y.	91.50	11.30	102.80
By crane	"	91.50	54.50	146.00
By pump	"	91.50	46.90	138.40
By hand buggy	"	91.50	33.80	125.30
5000# concrete				

Placing Concrete	UNIT	MAT.	INST.	TOTAL
03380.53 Pile Cap Concrete				
By chute	C.Y.	95.50	11.30	106.80
By crane	"	95.50	54.50	150.00
By pump	"	95.50	46.90	142.40
By hand buggy	"	95.50	33.80	129.30
03380.55 Slab/mat Concrete				
Slab on grade				
2500# or 3000# concrete				
By chute	C.Y.	87.00	8.45	95.45
By crane	"	87.00	27.30	114.30
By pump	"	87.00	23.45	110.45
By hand buggy	"	87.00	22.55	109.55
3500# or 4000# concrete				
By chute	C.Y.	91.50	8.45	99.95
By crane	"	91.50	27.30	118.80
By pump	"	91.50	23.45	114.95
By hand buggy	"	91.50	22.55	114.05
5000# concrete				
By chute	C.Y.	95.50	8.45	103.95
By crane	"	95.50	27.30	122.80
By pump	"	95.50	23.45	118.95
By hand buggy	"	95.50	22.55	118.05
Foundation mat				
2500# or 3000# concrete, over 20 cy				
By chute	C.Y.	87.00	6.75	93.75
By crane	"	87.00	23.45	110.45
By pump	"	87.00	20.50	107.50
By hand buggy	"	87.00	16.90	103.90
03380.58 Sidewalks				
Walks, cast in place with wire mesh, base not incl.				
4" thick	S.F.	1.20	1.15	2.35
5" thick	"	1.60	1.35	2.95
6" thick	"	2.00	1.70	3.70
03380.60 Stair Concrete				
Stairs				
2500# or 3000# concrete				
By chute	C.Y.	87.00	11.30	98.30
By crane	"	87.00	54.50	141.50
By pump	"	87.00	46.90	133.90
By hand buggy	"	87.00	33.80	120.80
3500# or 4000# concrete				
By chute	C.Y.	91.50	11.30	102.80
By crane	"	91.50	54.50	146.00
By pump	"	91.50	46.90	138.40
By hand buggy	"	91.50	33.80	125.30
5000# concrete				
By chute	C.Y.	95.50	11.30	106.80
By crane	"	95.50	54.50	150.00
By pump	"	95.50	46.90	142.40
By hand buggy	"	95.50	33.80	129.30

Placing Concrete

03380.65 Wall Concrete	UNIT	MAT.	INST.	TOTAL
Walls				
2500# or 3000# concrete				
To 4'				
By chute	C.Y.	87.00	9.65	96.65
By crane	"	87.00	54.50	141.50
By pump	"	87.00	50.50	137.50
To 8'				
By crane	C.Y.	87.00	59.50	146.50
By pump	"	87.00	54.50	141.50
To 16'				
By crane	C.Y.	87.00	65.50	152.50
By pump	"	87.00	59.50	146.50
Over 16'				
By crane	C.Y.	87.00	73.00	160.00
By pump	"	87.00	65.50	152.50
3500# or 4000# concrete				
To 4'				
By chute	C.Y.	91.50	9.65	101.15
By crane	"	91.50	54.50	146.00
By pump	"	91.50	50.50	142.00
To 8'				
By crane	C.Y.	91.50	59.50	151.00
By pump	"	91.50	54.50	146.00
To 16'				
By crane	C.Y.	91.50	65.50	157.00
By pump	"	91.50	59.50	151.00
Over 16'				
By crane	C.Y.	91.50	73.00	164.50
By pump	"	91.50	65.50	157.00
5000# concrete				
To 4'				
By chute	C.Y.	95.50	9.65	105.15
By crane	"	95.50	54.50	150.00
By pump	"	95.50	50.50	146.00
To 8'				
By crane	C.Y.	95.50	59.50	155.00
By pump	"	95.50	54.50	150.00
To 16'				
By crane	C.Y.	95.50	65.50	161.00
By pump	"	95.50	59.50	155.00
Filled block (CMU)				
3000# concrete, by pump				
4" wide	S.F.	0.33	2.35	2.67
6" wide	"	0.75	2.75	3.50
8" wide	"	1.15	3.30	4.45
10" wide	"	1.55	3.85	5.40
12" wide	"	1.95	4.70	6.65
Pilasters, 3000# concrete	C.F.	4.55	65.50	70.05
Wall cavity, 2" thick, 3000# concrete	S.F.	0.84	2.20	3.04

Precast Concrete

Precast Concrete	UNIT	MAT.	INST.	TOTAL
03400.10 Precast Beams				
Prestressed, double tee, 24" deep, 8' wide				
35' span				
115 psf	S.F.	8.45	0.90	9.35
140 psf	"	8.95	0.90	9.85
40' span				
80 psf	S.F.	8.10	0.96	9.06
143 psf	"	8.65	0.96	9.61
45' span				
50 psf	S.F.	7.75	0.83	8.58
70 psf	"	8.30	0.83	9.13
100 psf	"	8.45	0.83	9.28
130 psf	"	9.30	0.83	10.13
50' span				
75 psf	S.F.	7.75	0.75	8.50
100 psf	"	8.45	0.75	9.20
Precast beams, girders and joists				
1000 lb/lf live load				
10' span	L.F.	71.50	18.00	89.50
20' span	"	75.50	10.80	86.30
30' span	"	94.50	9.00	103.50
3000 lb/lf live load				
10' span	L.F.	75.00	18.00	93.00
20' span	"	84.50	10.80	95.30
30' span	"	110.00	9.00	119.00
5000 lb/lf live load				
10' span	L.F.	76.50	18.00	94.50
20' span	"	100.00	10.80	110.80
30' span	"	130.00	9.00	139.00
03400.20 Precast Columns				
Prestressed concrete columns				
10" x 10"				
10' long	EA.	200.00	110.00	310.00
15' long	"	310.00	110.00	420.00
20' long	"	410.00	120.00	530.00
25' long	"	530.00	130.00	660.00
30' long	"	630.00	130.00	760.00
12" x 12"				
20' long	EA.	570.00	130.00	700.00
25' long	"	690.00	150.00	840.00
30' long	"	880.00	150.00	1,030
16" x 16"				
20' long	EA.	880.00	130.00	1,010
25' long	"	1,290	150.00	1,440
30' long	"	1,530	150.00	1,680
20" x 20"				
20' long	EA.	1,570	140.00	1,710
25' long	"	2,150	150.00	2,300
30' long	"	2,530	160.00	2,690
24" x 24"				
20' long	EA.	2,410	150.00	2,560
25' long	"	2,920	160.00	3,080

Precast Concrete

Precast Concrete	UNIT	MAT.	INST.	TOTAL
03400.20 Precast Columns				
30' long	EA.	3,610	170.00	3,780
28" x 28"				
20' long	EA.	3,270	170.00	3,440
25' long	"	3,950	180.00	4,130
30' long	"	4,900	190.00	5,090
32" x 32"				
20' long	EA.	4,130	180.00	4,310
25' long	"	5,330	190.00	5,520
30' long	"	6,110	210.00	6,320
36" x 36"				
20' long	EA.	5,160	190.00	5,350
25' long	"	6,450	210.00	6,660
30' long	"	7,740	220.00	7,960
03400.30 Precast Slabs				
Prestressed flat slab				
6" thick, 4' wide				
20' span				
80 psf	S.F.	12.30	2.25	14.55
110 psf	"	12.40	2.25	14.65
25' span				
80 psf	S.F.	12.90	2.15	15.05
Cored slab				
6" thick, 4' wide				
20' span				
80 psf	S.F.	6.45	2.25	8.70
100 psf	"	6.60	2.25	8.85
130 psf	"	6.75	2.25	9.00
8" thick, 4' wide				
25' span				
70 psf	S.F.	6.45	2.15	8.60
125 psf	"	6.90	2.15	9.05
170 psf	"	6.80	2.15	8.95
30' span				
70 psf	S.F.	6.45	1.80	8.25
90 psf	"	6.90	1.80	8.70
35' span				
70 psf	S.F.	6.80	1.70	8.50
10" thick, 4' wide				
30' span				
75 psf	S.F.	6.80	1.80	8.60
100 psf	"	6.95	1.80	8.75
130 psf	"	7.15	1.80	8.95
35' span				
60 psf	S.F.	6.95	1.70	8.65
80 psf	"	7.15	1.70	8.85
120 psf	"	7.45	1.70	9.15
40' span				
65 psf	S.F.	7.45	1.35	8.80
Slabs, roof and floor members, 4' wide				
6" thick, 25' span	S.F.	6.00	2.15	8.15
8" thick, 30' span	"	6.90	1.65	8.55

Precast Concrete

Precast Concrete	UNIT	MAT.	INST.	TOTAL
03400.30 Precast Slabs				
10" thick, 40' span	S.F.	8.45	1.45	9.90
Tee members				
Multiple tee, roof and floor				
Minimum	S.F.	7.80	1.35	9.15
Maximum	"	9.80	2.70	12.50
Double tee wall member				
Minimum	S.F.	7.15	1.55	8.70
Maximum	"	9.05	3.00	12.05
Single tee				
Short span, roof members				
Minimum	S.F.	8.05	1.65	9.70
Maximum	"	9.95	3.35	13.30
Long span, roof members				
Minimum	S.F.	10.20	1.35	11.55
Maximum	"	12.10	2.70	14.80
03400.40 Precast Walls				
Wall panel, 8' x 20'				
Gray cement				
Liner finish				
4" wall	S.F.	9.45	1.55	11.00
5" wall	"	10.15	1.60	11.75
6" wall	"	11.60	1.65	13.25
8" wall	"	12.00	1.70	13.70
Sandblast finish				
4" wall	S.F.	10.85	1.55	12.40
5" wall	"	11.90	1.60	13.50
6" wall	"	13.00	1.65	14.65
8" wall	"	13.60	1.70	15.30
White cement				
Liner finish				
4" wall	S.F.	11.45	1.55	13.00
5" wall	"	12.15	1.60	13.75
6" wall	"	13.30	1.65	14.95
8" wall	"	14.20	1.70	15.90
Sandblast finish				
4" wall	S.F.	12.30	1.55	13.85
5" wall	"	13.00	1.60	14.60
6" wall	"	13.30	1.65	14.95
8" wall	"	14.75	1.70	16.45
Double tee wall panel, 24" deep				
Gray cement				
Liner finish	S.F.	6.80	1.80	8.60
Sandblast finish	"	8.55	1.80	10.35
White cement				
Form liner finish	S.F.	9.55	1.80	11.35
Sandblast finish	"	12.30	1.80	14.10
Partition panels				
4" wall	S.F.	10.20	1.80	12.00
5" wall	"	11.05	1.80	12.85
6" wall	"	12.10	1.80	13.90
8" wall	"	13.10	1.80	14.90

03 CONCRETE

Precast Concrete

	UNIT	MAT.	INST.	TOTAL
03400.40 Precast Walls				
Cladding panels				
4" wall	S.F.	10.10	1.95	12.05
5" wall	"	11.05	1.95	13.00
6" wall	"	12.20	1.95	14.15
8" wall	"	12.95	1.95	14.90
Sandwich panel, 2.5" cladding panel, 2" insulation				
5" wall	S.F.	14.90	1.95	16.85
6" wall	"	15.65	1.95	17.60
8" wall	"	16.50	1.95	18.45
03400.90 Precast Specialties				
Precast concrete, coping, 4' to 8' long				
12" wide	L.F.	6.80	4.55	11.35
10" wide	"	6.05	5.20	11.25
Splash block, 30"x12"x4"	EA.	10.10	30.30	40.40
Stair unit, per riser	"	66.50	30.30	96.80
Sun screen and trellis, 8' long, 12" high				
4" thick blades	EA.	72.50	22.75	95.25
5" thick blades	"	89.50	22.75	112.25
6" thick blades	"	110.00	24.25	134.25
8" thick blades	"	140.00	24.25	164.25
Bearing pads for precast members, 2" wide strips				
1/8" thick	L.F.	0.24	0.14	0.38
1/4" thick	"	0.33	0.14	0.47
1/2" thick	"	0.36	0.14	0.49
3/4" thick	"	0.71	0.15	0.87
1" thick	"	0.75	0.17	0.92
1-1/2" thick	"	0.92	0.17	1.09

Cementitous Toppings

	UNIT	MAT.	INST.	TOTAL
03550.10 Concrete Toppings				
Gypsum fill				
2" thick	S.F.	1.30	0.34	1.64
2-1/2" thick	"	1.50	0.35	1.85
3" thick	"	1.85	0.36	2.21
3-1/2" thick	"	2.10	0.36	2.46
4" thick	"	2.45	0.41	2.86
Formboard				
Mineral fiber board				
1" thick	S.F.	1.25	0.85	2.10
1-1/2" thick	"	3.35	0.97	4.32
Cement fiber board				
1" thick	S.F.	0.99	1.15	2.14
1-1/2" thick	"	1.25	1.30	2.55
Glass fiber board				

03 CONCRETE

Cementitous Toppings

03550.10 Concrete Toppings	UNIT	MAT.	INST.	TOTAL
1" thick	S.F.	1.50	0.85	2.35
1-1/2" thick	"	2.05	0.97	3.02
Poured deck				
Vermiculite or perlite				
1 to 4 mix	C.Y.	140.00	54.50	194.50
1 to 6 mix	"	120.00	50.50	170.50
Vermiculite or perlite				
2" thick				
1 to 4 mix	S.F.	1.30	0.35	1.65
1 to 6 mix	"	0.94	0.31	1.25
3" thick				
1 to 4 mix	S.F.	1.75	0.51	2.26
1 to 6 mix	"	1.40	0.47	1.87
Concrete plank, lightweight				
2" thick	S.F.	6.65	2.70	9.35
2-1/2" thick	"	6.80	2.70	9.50
3-1/2" thick	"	7.10	3.00	10.10
4" thick	"	7.35	3.00	10.35
Channel slab, lightweight, straight				
2-3/4" thick	S.F.	5.35	2.70	8.05
3-1/2" thick	"	5.60	2.70	8.30
3-3/4" thick	"	5.95	2.70	8.65
4-3/4" thick	"	7.55	3.00	10.55
Gypsum plank				
2" thick	S.F.	2.50	2.70	5.20
3" thick	"	2.60	2.70	5.30
Cement fiber, T and G planks				
1" thick	S.F.	1.40	2.45	3.85
1-1/2" thick	"	1.45	2.45	3.90
2" thick	"	1.70	2.70	4.40
2-1/2" thick	"	1.80	2.70	4.50
3" thick	"	2.35	2.70	5.05
3-1/2" thick	"	2.65	3.00	5.65
4" thick	"	3.05	3.00	6.05

Grout

03600.10 Grouting	UNIT	MAT.	INST.	TOTAL
Grouting for bases				
Nonshrink				
Metallic grout				
1" deep	S.F.	5.75	8.75	14.50
2" deep	"	10.90	9.75	20.65
Non-metallic grout				
1" deep	S.F.	4.30	8.75	13.05
2" deep	"	8.25	9.75	18.00

03 CONCRETE

Grout

Grout	UNIT	MAT.	INST.	TOTAL
03600.10 Grouting				
Fluid type				
Non-metallic				
1" deep	S.F.	4.30	8.75	13.05
2" deep	"	7.95	9.75	17.70
Grouting for joints				
Portland cement grout (1 cement to 3 sand, by volume)				
1/2" joint thickness				
6" wide joints	L.F.	0.10	1.45	1.55
8" wide joints	"	0.15	1.75	1.90
1" joint thickness				
4" wide joints	L.F.	0.15	1.35	1.50
6" wide joints	"	0.26	1.50	1.75
8" wide joints	"	0.31	1.85	2.16
Nonshrink, nonmetallic grout				
1/2" joint thickness				
4" wide joint	L.F.	0.71	1.25	1.96
6" wide joint	"	1.00	1.45	2.45
8" wide joint	"	1.35	1.75	3.10
1" joint thickness				
4" wide joint	L.F.	1.35	1.35	2.70
6" wide joint	"	2.05	1.50	3.55
8" wide joint	"	2.75	1.85	4.60

Concrete Restoration

Concrete Restoration	UNIT	MAT.	INST.	TOTAL
03730.10 Concrete Repair				
Epoxy grout floor patch, 1/4" thick	S.F.	5.05	3.40	8.45
Grout, epoxy, 2 component system	C.F.			243.84
Epoxy sand	BAG			16.51
Epoxy modifier	GAL			105.93
Epoxy gel grout	S.F.	2.45	33.80	36.25
Injection valve, 1 way, threaded plastic	EA.	6.75	6.75	13.50
Grout crack seal, 2 component	C.F.	570.00	33.80	603.80
Grout, non shrink	"	58.00	33.80	91.80
Concrete, epoxy modified				
Sand mix	C.F.	92.00	13.55	105.55
Gravel mix	"	70.50	12.55	83.05
Concrete repair				
Soffit repair				
16" wide	L.F.	2.95	6.75	9.70
18" wide	"	3.05	7.05	10.10
24" wide	"	3.60	7.50	11.10
30" wide	"	4.15	8.05	12.20
32" wide	"	4.45	8.45	12.90
Edge repair				
2" spall	L.F.	1.40	8.45	9.85

Concrete Restoration

03730.10 Concrete Repair	UNIT	MAT.	INST.	TOTAL
3" spall	L.F.	1.40	8.90	10.30
4" spall	"	1.45	9.15	10.60
6" spall	"	1.55	9.40	10.95
8" spall	"	1.60	9.95	11.55
9" spall	"	1.65	11.30	12.95
Crack repair, 1/8" crack	"	2.70	3.40	6.10
Reinforcing steel repair				
1 bar, 4 ft				
#4 bar	L.F.	0.40	6.00	6.40
#5 bar	"	0.55	6.00	6.55
#6 bar	"	0.68	6.40	7.08
#8 bar	"	1.20	6.40	7.60
#9 bar	"	1.55	6.85	8.40
#11 bar	"	2.40	6.85	9.25
Form fabric, nylon				
18" diameter	L.F.			10.28
20" diameter	"			10.46
24" diameter	"			17.19
30" diameter	"			17.66
36" diameter	"			20.26
Pile repairs				
Polyethylene wrap				
30 mil thick				
60" wide	S.F.	11.05	11.30	22.35
72" wide	"	12.00	13.55	25.55
60 mil thick				
60" wide	S.F.	13.30	11.30	24.60
80" wide	"	15.35	15.40	30.75
Pile spall, average repair 3'				
18" x 18"	EA.	34.50	28.20	62.70
20" x 20"	"	46.00	33.80	79.80

Mortar And Grout

04100.10 Masonry Grout	UNIT	MAT.	INST.	TOTAL
Grout, non shrink, non-metallic, trowelable	C.F.	11.40	1.20	12.60
Grout door frame, hollow metal				
Single	EA.	16.60	45.50	62.10
Double	"	24.85	47.80	72.65
Grout-filled concrete block (CMU)				
4" wide	S.F.	0.48	1.50	1.98
6" wide	"	1.00	1.65	2.65
8" wide	"	1.75	1.80	3.55
12" wide	"	2.50	1.90	4.40
Grout-filled individual CMU cells				
4" wide	L.F.	0.26	0.91	1.16
6" wide	"	0.52	0.91	1.43
8" wide	"	0.69	0.91	1.60
10" wide	"	0.91	1.05	1.95
12" wide	"	1.05	1.05	2.10
Bond beams or lintels, 8" deep				
6" thick	L.F.	1.05	1.50	2.55
8" thick	"	1.35	1.65	3.00
10" thick	"	1.65	1.80	3.45
12" thick	"	1.95	2.05	4.00
Cavity walls				
2" thick	S.F.	1.15	2.20	3.35
3" thick	"	1.70	2.20	3.90
4" thick	"	2.25	2.35	4.60
6" thick	"	3.40	2.75	6.15

04150.10 Masonry Accessories	UNIT	MAT.	INST.	TOTAL
Foundation vents	EA.	23.85	16.75	40.60
Bar reinforcing				
Horizontal				
#3 - #4	Lb.	0.47	1.70	2.17
#5 - #6	"	0.44	1.40	1.84
Vertical				
#3 - #4	Lb.	0.47	2.10	2.57
#5 - #6	"	0.44	1.70	2.14
Horizontal joint reinforcing				
Truss type				
4" wide, 6" wall	L.F.	0.18	0.17	0.35
6" wide, 8" wall	"	0.18	0.18	0.36
8" wide, 10" wall	"	0.22	0.18	0.40
10" wide, 12" wall	"	0.22	0.19	0.41
12" wide, 14" wall	"	0.27	0.20	0.47
Ladder type				
4" wide, 6" wall	L.F.	0.13	0.17	0.30
6" wide, 8" wall	"	0.14	0.18	0.32
8" wide, 10" wall	"	0.16	0.18	0.34
10" wide, 12" wall	"	0.18	0.18	0.36
Rectangular wall ties				
3/16" dia., galvanized				
2" x 6"	EA.	0.20	0.70	0.89
2" x 8"	"	0.21	0.70	0.91
2" x 10"	"	0.23	0.70	0.93

04 MASONRY

Mortar And Grout

04150.10 Masonry Accessories	UNIT	MAT.	INST.	TOTAL
2" x 12"	EA.	0.27	0.70	0.97
4" x 6"	"	0.22	0.84	1.06
4" x 8"	"	0.25	0.84	1.08
4" x 10"	"	0.32	0.84	1.16
4" x 12"	"	0.38	0.84	1.21
1/4" dia., galvanized				
2" x 6"	EA.	0.36	0.70	1.06
2" x 8"	"	0.40	0.70	1.10
2" x 10"	"	0.45	0.70	1.15
2" x 12"	"	0.52	0.70	1.22
4" x 6"	"	0.42	0.84	1.25
4" x 8"	"	0.45	0.84	1.29
4" x 10"	"	0.52	0.84	1.36
4" x 12"	"	0.55	0.84	1.38
"Z" type wall ties, galvanized				
6" long				
1/8" dia.	EA.	0.20	0.70	0.89
3/16" dia.	"	0.21	0.70	0.91
1/4" dia.	"	0.22	0.70	0.92
8" long				
1/8" dia.	EA.	0.21	0.70	0.91
3/16" dia.	"	0.22	0.70	0.92
1/4" dia.	"	0.23	0.70	0.93
10" long				
1/8" dia.	EA.	0.22	0.70	0.92
3/16" dia.	"	0.25	0.70	0.94
1/4" dia.	"	0.29	0.70	0.98
Dovetail anchor slots				
Galvanized steel, filled				
24 ga.	L.F.	0.52	1.05	1.57
20 ga.	"	0.65	1.05	1.70
16 oz. copper, foam filled	"	1.30	1.05	2.35
Dovetail anchors				
16 ga.				
3-1/2" long	EA.	0.16	0.70	0.85
5-1/2" long	"	0.18	0.70	0.88
12 ga.				
3-1/2" long	EA.	0.20	0.70	0.89
5-1/2" long	"	0.40	0.70	1.10
Dovetail, triangular galvanized ties, 12 ga.				
3" x 3"	EA.	0.36	0.70	1.06
5" x 5"	"	0.39	0.70	1.09
7" x 7"	"	0.44	0.70	1.14
7" x 9"	"	0.47	0.70	1.17
Brick anchors				
Corrugated, 3-1/2" long				
16 ga.	EA.	0.14	0.70	0.84
12 ga.	"	0.25	0.70	0.94
Non-corrugated, 3-1/2" long				
16 ga.	EA.	0.20	0.70	0.89
12 ga.	"	0.35	0.70	1.05
Cavity wall anchors, corrugated, galvanized				

04 MASONRY

Mortar And Grout

	UNIT	MAT.	INST.	TOTAL

04150.10 Masonry Accessories

	UNIT	MAT.	INST.	TOTAL
5" long				
16 ga.	EA.	0.44	0.70	1.14
12 ga.	"	0.65	0.70	1.35
7" long				
28 ga.	EA.	0.48	0.70	1.18
24 ga.	"	0.61	0.70	1.31
22 ga.	"	0.62	0.70	1.32
16 ga.	"	0.70	0.70	1.40
Mesh ties, 16 ga., 3" wide				
8" long	EA.	0.58	0.70	1.28
12" long	"	0.65	0.70	1.35
20" long	"	0.90	0.70	1.59
24" long	"	0.99	0.70	1.68

04150.20 Masonry Control Joints

	UNIT	MAT.	INST.	TOTAL
Control joint, cross shaped PVC	L.F.	4.00	1.05	5.05
Closed cell joint filler				
1/2"	L.F.	0.60	1.05	1.65
3/4"	"	1.10	1.05	2.15
Rubber, for				
4" wall	L.F.	4.30	1.05	5.35
6" wall	"	6.75	1.10	7.85
8" wall	"	7.95	1.15	9.10
PVC, for				
4" wall	L.F.	3.30	1.05	4.35
6" wall	"	4.15	1.10	5.25
8" wall	"	4.70	1.15	5.85

04150.50 Masonry Flashing

	UNIT	MAT.	INST.	TOTAL
Through-wall flashing				
5 oz. coated copper	S.F.	4.15	3.50	7.65
0.030" elastomeric	"	0.81	2.80	3.61

Unit Masonry

04210.10 Brick Masonry

	UNIT	MAT.	INST.	TOTAL
Standard size brick, running bond				
Face brick, red (6.4/sf)				
Veneer	S.F.	3.15	7.00	10.15
Cavity wall	"	3.15	6.00	9.15
9" solid wall	"	6.05	11.95	18.00
Common brick (6.4/sf)				
Select common for veneers	S.F.	3.55	7.00	10.55
Back-up				
4" thick	S.F.	3.25	5.25	8.50
8" thick	"	6.35	8.40	14.75

Unit Masonry

	UNIT	MAT.	INST.	TOTAL
04210.10 Brick Masonry				
Firewall				
12" thick	S.F.	9.75	13.95	23.70
16" thick	"	12.85	19.05	31.90
Glazed brick (7.4/sf)				
Veneer	S.F.	10.35	7.60	17.95
Buff or gray face brick (6.4/sf)				
Veneer	S.F.	3.50	7.00	10.50
Cavity wall	"	3.50	6.00	9.50
Jumbo or oversize brick (3/sf)				
4" veneer	S.F.	4.05	4.20	8.25
4" back-up	"	4.05	3.50	7.55
8" back-up	"	9.45	6.00	15.45
12" firewall	"	13.50	10.45	23.95
16" firewall	"	20.00	13.95	33.95
Norman brick, red face, (4.5/sf)				
4" veneer	S.F.	4.75	5.25	10.00
Cavity wall	"	4.75	4.65	9.40
Chimney, standard brick, including flue				
16" x 16"	L.F.	18.90	41.90	60.80
16" x 20"	"	22.35	41.90	64.25
16" x 24"	"	26.50	41.90	68.40
20" x 20"	"	28.90	52.50	81.40
20" x 24"	"	37.20	52.50	89.70
20" x 32"	"	46.90	60.00	106.90
Window sill, face brick on edge	"	2.80	10.45	13.25
04210.20 Structural Tile				
Structural glazed tile				
6T series, 5-1/2" x 12"				
Glazed on one side				
2" thick	S.F.	7.30	4.20	11.50
4" thick	"	9.15	4.20	13.35
6" thick	"	13.35	4.65	18.00
8" thick	"	17.60	5.25	22.85
Glazed on two sides				
4" thick	S.F.	14.05	5.25	19.30
6" thick	"	21.10	6.00	27.10
Special shapes				
Group 1	S.F.	10.55	8.40	18.95
Group 2	"	13.35	8.40	21.75
Group 3	"	14.75	8.40	23.15
Group 4	"	25.70	8.40	34.10
Group 5	"	30.90	8.40	39.30
Fire rated				
4" thick, 1 hr rating	S.F.	19.70	4.20	23.90
6" thick, 2 hr rating	"	23.90	4.65	28.55
8W series, 8" x 16"				
Glazed on one side				
2" thick	S.F.	7.75	2.80	10.55
4" thick	"	8.45	2.80	11.25
6" thick	"	12.65	3.20	15.85
8" thick	"	14.05	3.20	17.25

Unit Masonry

	UNIT	MAT.	INST.	TOTAL
04210.20 Structural Tile				
Glazed on two sides				
4" thick	S.F.	14.05	3.50	17.55
6" thick	"	16.90	4.20	21.10
8" thick	"	23.90	4.20	28.10
Special shapes				
Group 1	S.F.	12.65	6.00	18.65
Group 2	"	15.45	6.00	21.45
Group 3	"	16.90	6.00	22.90
Group 4	"	28.10	6.00	34.10
Group 5	"	70.50	6.00	76.50
Fire rated				
4" thick, 1 hr rating	S.F.	21.10	6.00	27.10
6" thick, 2 hr rating	"	25.30	6.00	31.30
04210.60 Pavers, Masonry				
Brick walk laid on sand, sand joints				
Laid flat, (4.5 per sf)	S.F.	3.25	4.65	7.90
Laid on edge, (7.2 per sf)	"	5.10	7.00	12.10
Precast concrete patio blocks				
2" thick				
Natural	S.F.	2.50	1.40	3.90
Colors	"	3.50	1.40	4.90
Exposed aggregates, local aggregate				
Natural	S.F.	2.95	1.40	4.35
Colors	"	3.90	1.40	5.30
Granite or limestone aggregate	"	5.35	1.40	6.75
White tumblestone aggregate	"	4.20	1.40	5.60
Stone pavers, set in mortar				
Bluestone				
1" thick				
Irregular	S.F.	2.70	10.45	13.15
Snapped rectangular	"	4.10	8.40	12.50
1-1/2" thick, random rectangular	"	4.75	10.45	15.20
2" thick, random rectangular	"	5.60	11.95	17.55
Slate				
Natural cleft				
Irregular, 3/4" thick	S.F.	2.70	11.95	14.65
Random rectangular				
1-1/4" thick	S.F.	5.85	10.45	16.30
1-1/2" thick	"	6.60	11.65	18.25
Granite blocks				
3" thick, 3" to 6" wide				
4" to 12" long	S.F.	6.30	13.95	20.25
6" to 15" long	"	3.65	11.95	15.60
Crushed stone, white marble, 3" thick	"	1.50	0.68	2.18
04220.10 Concrete Masonry Units				
Hollow, load bearing				
4"	S.F.	1.50	3.10	4.60
6"	"	1.90	3.20	5.10
8"	"	2.45	3.50	5.95
10"	"	3.30	3.80	7.10

Unit Masonry

04220.10 Concrete Masonry Units

	UNIT	MAT.	INST.	TOTAL
12"	S.F.	3.45	4.20	7.65
Solid, load bearing				
4"	S.F.	2.05	3.10	5.15
6"	"	2.45	3.20	5.65
8"	"	2.80	3.50	6.30
10"	"	3.80	3.80	7.60
12"	"	4.25	4.20	8.45
Back-up block, 8" x 16"				
2"	S.F.	1.00	2.40	3.40
4"	"	1.45	2.45	3.90
6"	"	1.80	2.60	4.40
8"	"	2.10	2.80	4.90
10"	"	2.85	3.00	5.85
12"	"	2.95	3.20	6.15
Foundation wall, 8" x 16"				
6"	S.F.	1.95	3.00	4.95
8"	"	2.15	3.20	5.35
10"	"	3.30	3.50	6.80
12"	"	3.65	3.80	7.45
Solid				
6"	S.F.	2.70	3.20	5.90
8"	"	3.35	3.50	6.85
10"	"	4.05	3.80	7.85
12"	"	4.95	4.20	9.15
Exterior, styrofoam inserts, standard weight, 8" x 16"				
6"	S.F.	3.70	3.20	6.90
8"	"	3.20	3.50	6.70
10"	"	4.55	3.80	8.35
12"	"	4.80	4.20	9.00
Lightweight				
6"	S.F.	2.95	3.20	6.15
8"	"	3.60	3.50	7.10
10"	"	4.55	3.80	8.35
12"	"	4.80	4.20	9.00
Acoustical slotted block				
4"	S.F.	3.20	3.80	7.00
6"	"	3.90	3.80	7.70
8"	"	5.85	4.20	10.05
Filled cavities				
4"	S.F.	4.05	4.65	8.70
6"	"	5.00	4.95	9.95
8"	"	6.70	5.25	11.95
Hollow, split face				
4"	S.F.	3.05	3.10	6.15
6"	"	3.55	3.20	6.75
8"	"	3.20	3.50	6.70
10"	"	4.20	3.80	8.00
12"	"	4.45	4.20	8.65
Split rib profile				
4"	S.F.	2.25	3.80	6.05
6"	"	2.55	3.80	6.35
8"	"	2.80	4.20	7.00

Unit Masonry

04220.10 Concrete Masonry Units	UNIT	MAT.	INST.	TOTAL
10"	S.F.	3.05	4.20	7.25
12"	"	3.35	4.20	7.55
High strength block, 3500 psi				
2"	S.F.	1.45	3.10	4.55
4"	"	1.80	3.20	5.00
6"	"	2.15	3.20	5.35
8"	"	2.45	3.50	5.95
10"	"	2.85	3.80	6.65
12"	"	3.40	4.20	7.60
Solar screen concrete block				
4" thick				
6" x 6"	S.F.	4.30	9.30	13.60
8" x 8"	"	5.15	8.40	13.55
12" x 12"	"	3.00	6.45	9.45
8" thick				
8" x 16"	S.F.	3.20	6.00	9.20
Glazed block				
Cove base, glazed 1 side, 2"	L.F.	9.85	4.65	14.50
4"	"	9.85	4.65	14.50
6"	"	10.40	5.25	15.65
8"	"	11.15	5.25	16.40
Single face				
2"	S.F.	7.30	3.50	10.80
4"	"	7.90	3.50	11.40
6"	"	8.00	3.80	11.80
8"	"	8.30	4.20	12.50
10"	"	8.60	4.65	13.25
12"	"	8.85	4.95	13.80
Double face				
4"	S.F.	11.70	4.40	16.10
6"	"	12.65	4.65	17.30
8"	"	13.00	5.25	18.25
Corner or bullnose				
2"	EA.	9.85	5.25	15.10
4"	"	12.65	6.00	18.65
6"	"	15.45	6.00	21.45
8"	"	16.85	7.00	23.85
10"	"	18.25	7.60	25.85
12"	"	19.70	8.40	28.10
Gypsum unit masonry				
Partition blocks (12"x30")				
Solid				
2"	S.F.	0.90	1.70	2.60
Hollow				
3"	S.F.	0.91	1.70	2.61
4"	"	1.05	1.75	2.80
6"	"	1.10	1.90	3.00
Vertical reinforcing				
4' o.c., add 5% to labor				
2'8" o.c., add 15% to labor				
Interior partitions, add 10% to labor				

04 MASONRY

Unit Masonry	UNIT	MAT.	INST.	TOTAL
04220.90 Bond Beams & Lintels				
Bond beam, no grout or reinforcement				
8" x 16" x				
4" thick	L.F.	1.90	3.20	5.10
6" thick	"	2.35	3.35	5.70
8" thick	"	2.85	3.50	6.35
10" thick	"	3.00	3.65	6.65
12" thick	"	3.00	3.80	6.80
Beam lintel, no grout or reinforcement				
8" x 16" x				
10" thick	L.F.	5.25	4.20	9.45
12" thick	"	6.70	4.65	11.35
Precast masonry lintel				
6 lf, 8" high x				
4" thick	L.F.	5.00	7.00	12.00
6" thick	"	6.40	7.00	13.40
8" thick	"	7.25	7.60	14.85
10" thick	"	8.65	7.60	16.25
10 lf, 8" high x				
4" thick	L.F.	6.30	4.20	10.50
6" thick	"	7.75	4.20	11.95
8" thick	"	8.65	4.65	13.30
10" thick	"	11.70	4.65	16.35
Steel angles and plates				
Minimum	Lb.	0.44	0.60	1.03
Maximum	"	0.60	1.05	1.65
Various size angle lintels				
1/4" stock				
3" x 3"	L.F.	3.05	2.60	5.65
3" x 3-1/2"	"	3.20	2.60	5.80
3/8" stock				
3" x 4"	L.F.	4.95	2.60	7.55
3-1/2" x 4"	"	5.15	2.60	7.75
4" x 4"	"	5.60	2.60	8.20
5" x 3-1/2"	"	5.85	2.60	8.45
6" x 3-1/2"	"	6.00	2.60	8.60
1/2" stock				
6" x 4"	L.F.	7.45	2.60	10.05
04240.10 Clay Tile				
Hollow clay tile, for back-up, 12" x 12"				
Scored face				
Load bearing				
4" thick	S.F.	4.20	3.00	7.20
6" thick	"	5.15	3.10	8.25
8" thick	"	6.20	3.20	9.40
10" thick	"	6.90	3.35	10.25
12" thick	"	8.10	3.50	11.60
Non-load bearing				
3" thick	S.F.	3.95	2.90	6.85
4" thick	"	4.05	3.00	7.05
6" thick	"	4.20	3.10	7.30
8" thick	"	5.40	3.20	8.60

04 MASONRY

Unit Masonry

04240.10 Clay Tile	UNIT	MAT.	INST.	TOTAL
12" thick	S.F.	5.95	3.50	9.45
Partition, 12" x 12"				
In walls				
3" thick	S.F.	3.40	3.50	6.90
4" thick	"	3.95	3.50	7.45
6" thick	"	4.35	3.65	8.00
8" thick	"	5.70	3.80	9.50
10" thick	"	6.75	4.00	10.75
12" thick	"	7.20	4.20	11.40
Clay tile floors				
4" thick	S.F.	4.20	2.35	6.55
6" thick	"	5.15	2.45	7.60
8" thick	"	6.20	2.60	8.80
10" thick	"	6.90	2.80	9.70
12" thick	"	8.10	3.00	11.10
Terra cotta				
Coping, 10" or 12" wide, 3" thick	L.F.	13.55	8.40	21.95

04270.10 Glass Block	UNIT	MAT.	INST.	TOTAL
Glass block, 4" thick				
6" x 6"	S.F.	30.90	13.95	44.85
8" x 8"	"	26.70	10.45	37.15
12" x 12"	"	26.70	8.40	35.10

04295.10 Parging/masonry Plaster	UNIT	MAT.	INST.	TOTAL
Parging				
1/2" thick	S.F.	0.57	2.80	3.37
3/4" thick	"	0.72	3.50	4.22
1" thick	"	0.99	4.20	5.19

Stone

04400.10 Stone	UNIT	MAT.	INST.	TOTAL
Rubble stone				
Walls set in mortar				
8" thick	S.F.	13.00	10.45	23.45
12" thick	"	15.70	16.75	32.45
18" thick	"	20.85	20.95	41.80
24" thick	"	26.10	27.90	54.00
Dry set wall				
8" thick	S.F.	14.65	7.00	21.65
12" thick	"	16.55	10.45	27.00
18" thick	"	22.85	13.95	36.80
24" thick	"	27.90	16.75	44.65
Cut stone				
Imported marble				

Stone	UNIT	MAT.	INST.	TOTAL
04400.10 Stone				
Facing panels				
3/4" thick	S.F.	34.70	16.75	51.45
1-1/2" thick	"	52.00	19.05	71.05
2-1/4" thick	"	60.50	23.25	83.75
Base				
1" thick				
4" high	L.F.	15.70	20.95	36.65
6" high	"	19.25	20.95	40.20
Columns, solid				
Plain faced	C.F.	100.00	280.00	380.00
Fluted	"	280.00	280.00	560.00
Flooring, travertine, minimum	S.F.	14.30	6.45	20.75
Average	"	19.10	8.40	27.50
Maximum	"	28.60	9.30	37.90
Domestic marble				
Facing panels				
7/8" thick	S.F.	31.40	16.75	48.15
1-1/2" thick	"	47.10	19.05	66.15
2-1/4" thick	"	57.00	23.25	80.25
Stairs				
12" treads	L.F.	28.10	20.95	49.05
6" risers	"	20.85	13.95	34.80
Thresholds, 7/8" thick, 3' long, 4" to 6" wide				
Plain	EA.	26.80	34.90	61.70
Beveled	"	29.60	34.90	64.50
Window sill				
6" wide, 2" thick	L.F.	13.50	16.75	30.25
Stools				
5" wide, 7/8" thick	L.F.	19.85	16.75	36.60
Limestone panels up to 12' x 5', smooth finish				
2" thick	S.F.	19.10	7.25	26.35
3" thick	"	22.25	7.25	29.50
4" thick	"	31.70	7.25	38.95
Miscellaneous limestone items				
Steps, 14" wide, 6" deep	L.F.	66.50	27.90	94.40
Coping, smooth finish	C.F.	92.50	13.95	106.45
Sills, lintels, jambs, smooth finish	"	92.50	16.75	109.25
Granite veneer facing panels, polished				
7/8" thick				
Black	S.F.	37.20	16.75	53.95
Gray	"	32.30	16.75	49.05
Base				
4" high	L.F.	17.35	8.40	25.75
6" high	"	20.85	9.30	30.15
Curbing, straight, 6" x 16"	"	19.20	30.30	49.50
Radius curbs, radius over 5'	"	23.50	40.40	63.90
Ashlar veneer				
4" thick, random	S.F.	31.80	16.75	48.55
Pavers, 4" x 4" split				
Gray	S.F.	31.30	8.40	39.70
Pink	"	30.60	8.40	39.00
Black	"	30.30	8.40	38.70

Stone

Stone	UNIT	MAT.	INST.	TOTAL
04400.10 Stone				
Slate, panels				
1" thick	S.F.	22.25	16.75	39.00
2" thick	"	30.20	19.05	49.25
Sills or stools				
1" thick				
6" wide	L.F.	10.35	16.75	27.10
10" wide	"	16.85	18.20	35.05
2" thick				
6" wide	L.F.	16.95	19.05	36.00
10" wide	"	28.10	20.95	49.05

Masonry Restoration

Masonry Restoration	UNIT	MAT.	INST.	TOTAL
04520.10 Restoration And Cleaning				
Masonry cleaning				
Washing brick				
Smooth surface	S.F.	0.38	0.70	1.07
Rough surface	"	0.52	0.93	1.45
Steam clean masonry				
Smooth face				
Minimum	S.F.		0.55	0.55
Maximum	"		0.80	0.80
Rough face				
Minimum	S.F.		0.73	0.73
Maximum	"		1.10	1.10
Sandblast masonry				
Minimum	S.F.	0.46	0.88	1.33
Maximum	"	0.68	1.45	2.13
Pointing masonry				
Brick	S.F.	1.20	1.70	2.90
Concrete block	"	0.53	1.20	1.73
Cut and repoint				
Brick				
Minimum	S.F.	0.33	2.10	2.42
Maximum	"	0.56	4.20	4.76
Stone work	L.F.	1.50	3.20	4.70
Cut and recaulk				
Oil base caulks	L.F.	1.15	2.80	3.95
Butyl caulks	"	1.00	2.80	3.80
Polysulfides and acrylics	"	1.95	2.80	4.75
Silicones	"	2.30	2.80	5.10
Cement and sand grout on walls, to 1/8" thick				
Minimum	S.F.	0.57	1.70	2.27
Maximum	"	1.45	2.10	3.55
Brick removal and replacement				
Minimum	EA.	0.60	5.25	5.85

Masonry Restoration	UNIT	MAT.	INST.	TOTAL
04520.10 Restoration And Cleaning				
Average	EA.	0.78	7.00	7.78
Maximum	"	1.60	20.95	22.55
04550.10 Refractories				
Flue liners				
Rectangular				
8" x 12"	L.F.	5.95	7.00	12.95
12" x 12"	"	7.45	7.60	15.05
12" x 18"	"	13.40	8.40	21.80
16" x 16"	"	14.90	9.30	24.20
18" x 18"	"	17.85	9.95	27.80
20" x 20"	"	29.80	10.45	40.25
24" x 24"	"	35.70	11.95	47.65
Round				
18" dia.	L.F.	29.80	9.95	39.75
24" dia.	"	41.70	11.95	53.65

05 METALS

Metal Fastening

Metal Fastening	UNIT	MAT.	INST.	TOTAL
05050.10 Structural Welding				
Welding				
Single pass				
1/8"	L.F.	0.52	2.40	2.92
3/16"	"	0.99	3.20	4.19
1/4"	"	1.35	4.00	5.35
Miscellaneous steel shapes				
Plain	Lb.	0.99	0.10	1.08
Galvanized	"	1.50	0.16	1.66
Plates				
Plain	Lb.	0.91	0.12	1.03
Galvanized	"	1.40	0.19	1.59
05050.90 Metal Anchors				
Anchor bolts				
3/8" x				
8" long	EA.			2.92
10" long	"			3.20
12" long	"			3.50
1/2" x				
8" long	EA.			4.38
10" long	"			4.67
12" long	"			5.12
18" long	"			5.55
5/8" x				
8" long	EA.			4.08
10" long	"			4.52
12" long	"			4.83
18" long	"			5.12
24" long	"			5.55
3/4" x				
8" long	EA.			5.83
12" long	"			6.57
18" long	"			9.04
24" long	"			11.96
7/8" x				
8" long	EA.			5.83
12" long	"			6.57
18" long	"			9.04
24" long	"			11.96
1" x				
12" long	EA.			11.67
18" long	"			14.59
24" long	"			17.50
36" long	"			26.25
Expansion shield				
1/4"	EA.			0.87
3/8"	"			1.47
1/2"	"			2.92
5/8"	"			4.24
3/4"	"			5.12
1"	"			7.02
Non-drilling anchor				

05 METALS

Metal Fastening

	UNIT	MAT.	INST.	TOTAL
05050.90 Metal Anchors				
1/4"	EA.			0.56
3/8"	"			0.69
1/2"	"			1.06
5/8"	"			1.73
3/4"	"			2.96
Self-drilling anchor				
1/4"	EA.			1.47
5/16"	"			1.82
3/8"	"			2.19
1/2"	"			2.92
5/8"	"			5.55
3/4"	"			7.30
7/8"	"			10.22
Add 25% for galvanized anchor bolts				
Channel door frame, with anchors	Lb.	1.45	0.53	1.98
Corner guard angle, with anchors	"	1.30	0.80	2.10
05050.95 Metal Lintels				
Lintels, steel				
Plain	Lb.	1.05	1.20	2.25
Galvanized	"	1.55	1.20	2.75
05120.10 Structural Steel				
Beams and girders, A-36				
Welded	TON	1,740	540.00	2,280
Bolted	"	1,700	490.00	2,190
Columns				
Pipe				
6" dia.	Lb.	1.15	0.54	1.69
12" dia.	"	0.99	0.45	1.44
Purlins and girts				
Welded	TON	2,170	900.00	3,070
Bolted	"	2,140	770.00	2,910
Column base plates				
Up to 150 lb each	Lb.	1.25	0.32	1.57
Over 150 lb each	"	1.00	0.40	1.40
Structural pipe				
3" to 5" o.d.	TON	2,340	1,080	3,420
6" to 12" o.d.	"	2,170	770.00	2,940
Structural tube				
6" square				
Light sections	TON	2,720	1,080	3,800
Heavy sections	"	2,540	770.00	3,310
6" wide rectangular				
Light sections	TON	2,720	900.00	3,620
Heavy sections	"	2,540	670.00	3,210
Greater than 6" wide rectangular				
Light sections	TON	2,980	900.00	3,880
Heavy sections	"	2,810	670.00	3,480
Miscellaneous structural shapes				
Steel angle	TON	1,930	1,350	3,280
Steel plate	"	2,190	900.00	3,090

Metal Fastening

	UNIT	MAT.	INST.	TOTAL
05120.10 Structural Steel				
Trusses, field welded				
60 lb/lf	TON	2,810	670.00	3,480
100 lb/lf	"	2,460	540.00	3,000
150 lb/lf	"	2,310	450.00	2,760
Bolted				
60 lb/lf	TON	2,770	600.00	3,370
100 lb/lf	"	2,420	490.00	2,910
150 lb/lf	"	2,300	420.00	2,720
Add for galvanizing	"			551.14
05200.10 Metal Joists				
Joist				
DLH series	TON	1,210	360.00	1,570
K series	"	1,210	360.00	1,570
LH series	"	1,210	360.00	1,570
05300.10 Metal Decking				
Roof, 1-1/2" deep, non-composite				
16 ga.				
Primed	S.F.	2.75	0.90	3.65
Galvanized	"	2.90	0.90	3.80
18 ga.				
Primed	S.F.	2.30	0.90	3.20
Galvanized	"	2.55	0.90	3.45
20 ga.				
Primed	S.F.	1.70	0.90	2.60
Galvanized	"	2.00	0.90	2.90
22 ga.				
Primed	S.F.	1.40	0.90	2.30
Galvanized	"	1.55	0.90	2.45
Open type decking, galvanized				
1-1/2" deep				
18 ga.	S.F.	2.20	0.90	3.10
20 ga.	"	1.70	0.90	2.60
22 ga.	"	1.45	0.90	2.35
3" deep				
16 ga.	S.F.	3.35	0.98	4.33
18 ga.	"	3.20	0.98	4.18
20 ga.	"	2.45	0.98	3.43
22 ga.	"	2.35	0.98	3.33
4-1/2" deep				
16 ga.	S.F.	6.10	1.10	7.20
18 ga.	"	5.05	1.10	6.15
6" deep				
16 ga.	S.F.	7.05	1.20	8.25
18 ga.	"	6.10	1.20	7.30
7-1/2" deep				
16 ga.	S.F.	7.65	1.25	8.90
18 ga.	"	7.20	1.25	8.45
Cellular type				
1-1/2" deep, galvanized				
18-18 ga.	S.F.	5.50	1.10	6.60

05 METALS

Metal Fastening

05300.10 Metal Decking	UNIT	MAT.	INST.	TOTAL
22-18 ga.	S.F.	5.05	1.10	6.15
3" deep, galvanized				
16-16 ga.	S.F.	7.90	1.20	9.10
18-16 ga.	"	7.40	1.20	8.60
18-18 ga.	"	6.75	1.20	7.95
20-18 ga.	"	6.00	1.20	7.20
4-1/2" deep, galvanized				
16-16 ga.	S.F.	11.80	1.30	13.10
18-16 ga.	"	10.85	1.30	12.15
18-18 ga.	"	10.40	1.30	11.70
20-18 ga.	"	9.45	1.30	10.75
Composite deck, non-cellular, galvanized				
1-1/2" deep				
18 ga.	S.F.	1.85	0.98	2.83
20 ga.	"	1.70	0.98	2.68
22 ga.	"	1.40	0.98	2.38
3" deep				
18 ga.	S.F.	2.60	1.05	3.65
20 ga.	"	1.70	1.05	2.75
22 ga.	"	1.40	1.05	2.45
Slab form floor deck				
9/16" deep				
28 ga.	S.F.	0.73	0.98	1.72
1-5/16" deep				
24 ga.	S.F.	1.15	1.00	2.15
22 ga.	"	1.30	1.00	2.30

Cold Formed Framing

05410.10 Metal Framing	UNIT	MAT.	INST.	TOTAL
Furring channel, galvanized				
Beams and columns, 3/4"				
12" o.c.	S.F.	0.47	4.80	5.27
16" o.c.	"	0.38	4.35	4.73
Walls, 3/4"				
12" o.c.	S.F.	0.45	2.40	2.85
16" o.c.	"	0.36	2.00	2.36
24" o.c.	"	0.26	1.60	1.86
1-1/2"				
12" o.c.	S.F.	0.58	2.40	2.98
16" o.c.	"	0.47	2.00	2.47
24" o.c.	"	0.36	1.60	1.96
Stud, load bearing				
16" o.c.				
16 ga.				
2-1/2"	S.F.	1.20	2.15	3.35

Cold Formed Framing

05410.10 Metal Framing	UNIT	MAT.	INST.	TOTAL
3-5/8"	S.F.	1.55	2.15	3.70
4"	"	1.70	2.15	3.85
6"	"	2.45	2.40	4.85
18 ga.				
2-1/2"	S.F.	1.05	2.15	3.20
3-5/8"	"	1.40	2.15	3.55
4"	"	1.55	2.15	3.70
6"	"	1.95	2.40	4.35
8"	"	2.15	2.40	4.55
20 ga.				
2-1/2"	S.F.	0.89	2.15	3.04
3-5/8"	"	1.05	2.15	3.20
4"	"	1.20	2.15	3.35
6"	"	1.40	2.40	3.80
8"	"	1.55	2.40	3.95
24" o.c.				
16 ga.				
2-1/2"	S.F.	0.98	1.85	2.82
3-5/8"	"	1.30	1.85	3.15
4"	"	1.45	1.85	3.30
6"	"	2.15	2.00	4.15
8"	"	2.30	2.00	4.30
18 ga.				
2-1/2"	S.F.	0.82	1.85	2.67
3-5/8"	"	1.15	1.85	3.00
4"	"	1.30	1.85	3.15
6"	"	1.45	2.00	3.45
8"	"	1.65	2.00	3.65
20 ga.				
2-1/2"	S.F.	0.74	1.85	2.59
3-5/8"	"	0.95	1.85	2.80
4"	"	1.10	1.85	2.95
6"	"	1.20	2.00	3.20
8"	"	1.45	2.00	3.45

Metal Fabrications

05510.10 Stairs	UNIT	MAT.	INST.	TOTAL
Stock unit, steel, complete, per riser				
Tread				
3'-6" wide	EA.	110.00	60.00	170.00
4' wide	"	130.00	68.50	198.50
5' wide	"	150.00	80.00	230.00
Metal pan stair, cement filled, per riser				
3'-6" wide	EA.	120.00	48.10	168.10
4' wide	"	140.00	53.50	193.50

Metal Fabrications

Metal Fabrications	UNIT	MAT.	INST.	TOTAL
05510.10 Stairs				
5' wide	EA.	150.00	60.00	210.00
Landing, steel pan	S.F.	47.20	12.00	59.20
Cast iron tread, steel stringers, stock units, per riser				
Tread				
3'-6" wide	EA.	210.00	60.00	270.00
4' wide	"	240.00	68.50	308.50
5' wide	"	290.00	80.00	370.00
Stair treads, abrasive, 12" x 3'-6"				
Cast iron				
3/8"	EA.	120.00	24.05	144.05
1/2"	"	140.00	24.05	164.05
Cast aluminum				
5/16"	EA.	140.00	24.05	164.05
3/8"	"	150.00	24.05	174.05
1/2"	"	170.00	24.05	194.05
05515.10 Ladders				
Ladder, 18" wide				
With cage	L.F.	83.00	32.10	115.10
Without cage	"	53.00	24.05	77.05
05520.10 Railings				
Railing, pipe				
1-1/4" diameter, welded steel				
2-rail				
Primed	L.F.	16.65	9.60	26.25
Galvanized	"	21.15	9.60	30.75
3-rail				
Primed	L.F.	21.15	12.00	33.15
Galvanized	"	27.20	12.00	39.20
Wall mounted, single rail, welded steel				
Primed	L.F.	7.05	7.40	14.45
Galvanized	"	10.60	7.40	18.00
1-1/2" diameter, welded steel				
2-rail				
Primed	L.F.	18.15	9.60	27.75
Galvanized	"	22.70	9.60	32.30
3-rail				
Primed	L.F.	22.70	12.00	34.70
Galvanized	"	28.60	12.00	40.60
Wall mounted, single rail, welded steel				
Primed	L.F.	8.30	7.40	15.70
Galvanized	"	11.35	7.40	18.75
2" diameter, welded steel				
2-rail				
Primed	L.F.	22.70	10.70	33.40
Galvanized	"	28.70	10.70	39.40
3-rail				
Primed	L.F.	28.70	13.75	42.45
Galvanized	"	34.80	13.75	48.55
Wall mounted, single rail, welded steel				
Primed	L.F.	9.10	8.00	17.10

Metal Fabrications

Metal Fabrications	UNIT	MAT.	INST.	TOTAL
05520.10 Railings				
Galvanized	L.F.	11.35	8.00	19.35
05530.10 Metal Grating				
Floor plate, checkered, steel				
1/4"				
Primed	S.F.	6.90	0.69	7.59
Galvanized	"	9.15	0.69	9.84
3/8"				
Primed	S.F.	9.15	0.74	9.89
Galvanized	"	12.25	0.74	12.99
Aluminum grating, pressure-locked bearing bars				
3/4" x 1/8"	S.F.	14.70	1.20	15.90
1" x 1/8"	"	16.20	1.20	17.40
1-1/4" x 1/8"	"	20.60	1.20	21.80
1-1/4" x 3/16"	"	22.05	1.20	23.25
1-1/2" x 1/8"	"	23.55	1.20	24.75
1-3/4" x 3/16"	"	26.50	1.20	27.70
Miscellaneous expenses				
Cutting				
Minimum	L.F.		3.20	3.20
Maximum	"		4.80	4.80
Banding				
Minimum	L.F.		8.00	8.00
Maximum	"		9.60	9.60
Toe plates				
Minimum	L.F.		9.60	9.60
Maximum	"		12.00	12.00
Steel grating, primed				
3/4" x 1/8"	S.F.	6.50	1.60	8.10
1" x 1/8"	"	6.75	1.60	8.35
1-1/4" x 1/8"	"	7.20	1.60	8.80
1-1/4" x 3/16"	"	8.80	1.60	10.40
1-1/2" x 1/8"	"	8.05	1.60	9.65
1-3/4" x 3/16"	"	11.50	1.60	13.10
Galvanized				
3/4" x 1/8"	S.F.	8.05	1.60	9.65
1" x 1/8"	"	8.60	1.60	10.20
1-1/4" x 1/8"	"	9.35	1.60	10.95
1-1/4" x 3/16"	"	11.10	1.60	12.70
1-1/2" x 1/8"	"	9.95	1.60	11.55
1-3/4" x 3/16"	"	14.55	1.60	16.15
Miscellaneous expenses				
Cutting				
Minimum	L.F.		3.45	3.45
Maximum	"		5.35	5.35
Banding				
Minimum	L.F.		8.75	8.75
Maximum	"		10.70	10.70
Toe plates				
Minimum	L.F.		10.70	10.70
Maximum	"		13.75	13.75

05 METALS

Metal Fabrications

Metal Fabrications	UNIT	MAT.	INST.	TOTAL
05540.10 Castings				
Miscellaneous castings				
Light sections	Lb.	1.45	0.96	2.41
Heavy sections	"	1.25	0.69	1.94
Manhole covers and frames				
Regular, city type				
18" dia.				
100 lb	EA.	300.00	96.00	396.00
24" dia.				
200 lb	EA.	310.00	96.00	406.00
300 lb	"	320.00	110.00	430.00
400 lb	"	330.00	110.00	440.00
26" dia., 475 lb	"	400.00	120.00	520.00
30" dia., 600 lb	"	470.00	140.00	610.00
8" square, 75 lb	"	130.00	19.25	149.25
24" square				
126 lb	EA.	290.00	96.00	386.00
500 lb	"	470.00	120.00	590.00
Watertight type				
20" dia., 200 lb	EA.	270.00	120.00	390.00
24" dia., 350 lb	"	450.00	160.00	610.00
Steps, cast iron				
7" x 9"	EA.	12.80	9.60	22.40
8" x 9"	"	15.35	10.70	26.05
Manhole covers and frames, aluminum				
12" x 12"	EA.	61.50	19.25	80.75
18" x 18"	"	63.50	19.25	82.75
24" x 24"	"	69.50	24.05	93.55
Corner protection				
Steel angle guard with anchors				
2" x 2" x 3/16"	L.F.	13.30	6.85	20.15
2" x 3" x 1/4"	"	15.10	6.85	21.95
3" x 3" x 5/16"	"	17.50	6.85	24.35
3" x 4" x 5/16"	"	21.10	7.40	28.50
4" x 4" x 5/16"	"	21.90	7.40	29.30

Misc. Fabrications

Misc. Fabrications	UNIT	MAT.	INST.	TOTAL
05580.10 Metal Specialties				
Kick plate				
4" high x 1/4" thick				
Primed	L.F.	6.40	9.60	16.00
Galvanized	"	7.30	9.60	16.90
6" high x 1/4" thick				
Primed	L.F.	7.00	10.70	17.70
Galvanized	"	8.30	10.70	19.00

Misc. Fabrications

Misc. Fabrications	UNIT	MAT.	INST.	TOTAL
05700.10 Ornamental Metal				
Railings, vertical square bars, 6" o.c., with shaped top rails				
Steel	L.F.	69.50	24.05	93.55
Aluminum	"	95.00	24.05	119.05
Bronze	"	150.00	32.10	182.10
Stainless steel	"	140.00	32.10	172.10
Laminated metal or wood handrails with metal supports				
2-1/2" round or oval shape	L.F.	130.00	24.05	154.05
Grilles and louvers				
Fixed type louvers				
4 through 10 sf	S.F.	24.75	8.00	32.75
Over 10 sf	"	29.30	6.00	35.30
Movable type louvers				
4 through 10 sf	S.F.	29.30	8.00	37.30
Over 10 sf	"	32.40	6.00	38.40
Aluminum louvers				
Residential use, fixed type, with screen				
8" x 8"	EA.	15.15	24.05	39.20
12" x 12"	"	16.70	24.05	40.75
12" x 18"	"	20.00	24.05	44.05
14" x 24"	"	28.80	24.05	52.85
18" x 24"	"	32.40	24.05	56.45
30" x 24"	"	44.20	26.70	70.90
05800.10 Expansion Control				
Expansion joints with covers, floor assembly type				
With 1" space				
Aluminum	L.F.	22.55	8.00	30.55
Bronze	"	47.10	8.00	55.10
Stainless steel	"	36.50	8.00	44.50
With 2" space				
Aluminum	L.F.	27.90	8.00	35.90
Bronze	"	49.50	8.00	57.50
Stainless steel	"	43.50	8.00	51.50
Ceiling and wall assembly type				
With 1" space				
Aluminum	L.F.	14.45	9.60	24.05
Bronze	"	49.50	9.60	59.10
Stainless steel	"	44.60	9.60	54.20
With 2" space				
Aluminum	L.F.	15.95	9.60	25.55
Bronze	"	53.50	9.60	63.10
Stainless steel	"	47.10	9.60	56.70
Exterior roof and wall, aluminum				
Roof to roof				
With 1" space	L.F.	42.20	8.00	50.20
With 2" space	"	45.60	8.00	53.60
Roof to wall				
With 1" space	L.F.	32.80	8.75	41.55
With 2" space	"	39.00	8.75	47.75
Flat wall to wall				
With 1" space	L.F.	15.95	8.00	23.95
With 2" space	"	17.35	8.00	25.35

Misc. Fabrications

05800.10 Expansion Control	UNIT	MAT.	INST.	TOTAL
Corner to flat wall				
With 1" space	L.F.	19.15	9.60	28.75
With 2" in space	"	19.45	9.60	29.05

06 WOOD AND PLASTICS

Fasteners And Adhesives

06050.10 Accessories	UNIT	MAT.	INST.	TOTAL
Column/post base, cast aluminum				
4" x 4"	EA.	2.75	10.80	13.55
6" x 6"	"	6.00	10.80	16.80
Bridging, metal, per pair				
12" o.c.	EA.	1.00	4.35	5.35
16" o.c.	"	1.00	3.95	4.95
Anchors				
Bolts, threaded two ends, with nuts and washers				
1/2" dia.				
4" long	EA.	1.35	2.70	4.05
7-1/2" long	"	1.65	2.70	4.35
3/4" dia.				
7-1/2" long	EA.	2.60	2.70	5.30
15" long	"	3.35	2.70	6.05
Framing anchors				
10 gauge	EA.	0.71	3.60	4.31
Bolts, carriage				
1/4 x 4	EA.	1.35	4.35	5.70
5/16 x 6	"	1.45	4.55	6.00
3/8 x 6	"	1.65	4.55	6.20
1/2 x 6	"	1.65	4.55	6.20
Joist and beam hangers				
18 ga.				
2 x 4	EA.	0.66	4.35	5.01
2 x 6	"	0.87	4.35	5.22
2 x 8	"	1.00	4.35	5.35
2 x 10	"	1.15	4.80	5.95
2 x 12	"	1.35	5.40	6.75
16 ga.				
3 x 6	EA.	2.40	4.80	7.20
3 x 8	"	2.50	4.80	7.30
3 x 10	"	2.85	5.10	7.95
3 x 12	"	3.10	5.75	8.85
3 x 14	"	3.35	6.20	9.55
4 x 6	"	2.70	4.80	7.50
4 x 8	"	3.10	4.80	7.90
4 x 10	"	3.90	5.10	9.00
4 x 12	"	4.40	5.75	10.15
4 x 14	"	4.75	6.20	10.95
Rafter anchors, 18 ga., 1-1/2" wide				
5-1/4" long	EA.	0.73	3.60	4.33
10-3/4" long	"	1.15	3.60	4.75
Shear plates				
2-5/8" dia.	EA.	2.05	3.35	5.40
4" dia.	"	4.35	3.60	7.95
Sill anchors				
Embedded in concrete	EA.	1.85	4.35	6.20
Split rings				
2-1/2" dia.	EA.	1.45	4.80	6.25
4" dia.	"	2.55	5.40	7.95
Strap ties, 14 ga., 1-3/8" wide				
12" long	EA.	1.55	3.60	5.15

06 WOOD AND PLASTICS

Fasteners And Adhesives

06050.10 Accessories	UNIT	MAT.	INST.	TOTAL
18" long	EA.	1.70	3.95	5.65
24" long	"	2.10	4.35	6.45
36" long	"	3.20	4.80	8.00
Toothed rings				
2-5/8" dia.	EA.	1.25	7.20	8.45
4" dia.	"	1.45	8.65	10.10

Rough Carpentry

06110.10 Blocking	UNIT	MAT.	INST.	TOTAL
Steel construction				
Walls				
2x4	L.F.	0.63	2.90	3.53
2x6	"	0.89	3.35	4.24
2x8	"	1.25	3.60	4.85
2x10	"	1.75	3.95	5.70
2x12	"	2.30	4.35	6.65
Ceilings				
2x4	L.F.	0.63	3.35	3.98
2x6	"	0.89	3.95	4.84
2x8	"	1.25	4.35	5.60
2x10	"	1.75	4.80	6.55
2x12	"	2.30	5.40	7.70
Wood construction				
Walls				
2x4	L.F.	0.63	2.40	3.03
2x6	"	0.89	2.70	3.59
2x8	"	1.25	2.90	4.15
2x10	"	1.75	3.10	4.85
2x12	"	2.30	3.35	5.65
Ceilings				
2x4	L.F.	0.63	2.70	3.33
2x6	"	0.89	3.10	3.99
2x8	"	1.25	3.35	4.60
2x10	"	1.75	3.60	5.35
2x12	"	2.30	3.95	6.25

06110.20 Ceiling Framing	UNIT	MAT.	INST.	TOTAL
Ceiling joists				
12" o.c.				
2x4	S.F.	0.76	1.05	1.81
2x6	"	1.10	1.10	2.20
2x8	"	1.55	1.15	2.70
2x10	"	2.35	1.20	3.55
2x12	"	3.15	1.25	4.40
16" o.c.				

Rough Carpentry

06110.20 Ceiling Framing

	UNIT	MAT.	INST.	TOTAL
2x4	S.F.	0.63	0.83	1.46
2x6	"	0.95	0.87	1.81
2x8	"	1.30	0.90	2.20
2x10	"	1.95	0.94	2.89
2x12	"	2.55	0.98	3.53
24" o.c.				
2x4	S.F.	0.50	0.69	1.18
2x6	"	0.76	0.72	1.48
2x8	"	1.00	0.76	1.76
2x10	"	1.55	0.80	2.35
2x12	"	2.10	0.85	2.95
Headers and nailers				
2x4	L.F.	0.63	1.40	2.03
2x6	"	0.89	1.45	2.34
2x8	"	1.25	1.55	2.80
2x10	"	1.75	1.65	3.40
2x12	"	2.30	1.80	4.10
Sister joists for ceilings				
2x4	L.F.	0.63	3.10	3.73
2x6	"	0.89	3.60	4.49
2x8	"	1.25	4.35	5.60
2x10	"	1.75	5.40	7.15
2x12	"	2.30	7.20	9.50

06110.30 Floor Framing

	UNIT	MAT.	INST.	TOTAL
Floor joists				
12" o.c.				
2x6	S.F.	1.10	0.87	1.97
2x8	"	1.55	0.88	2.43
2x10	"	2.35	0.90	3.25
2x12	"	3.15	0.94	4.09
2x14	"	3.45	0.90	4.35
3x6	"	2.50	0.92	3.42
3x8	"	3.30	0.94	4.24
3x10	"	4.15	0.98	5.13
3x12	"	4.95	1.05	6.00
3x14	"	5.70	1.10	6.80
4x6	"	3.30	0.90	4.20
4x8	"	4.25	0.94	5.19
4x10	"	5.45	0.98	6.43
4x12	"	6.60	1.05	7.65
4x14	"	7.70	1.10	8.80
16" o.c.				
2x6	S.F.	0.95	0.72	1.67
2x8	"	1.30	0.73	2.03
2x10	"	1.95	0.75	2.70
2x12	"	2.55	0.77	3.32
2x14	"	3.15	0.80	3.95
3x6	"	2.10	0.75	2.85
3x8	"	2.75	0.77	3.52
3x10	"	3.45	0.80	4.25
3x12	"	4.10	0.83	4.93

Rough Carpentry

	UNIT	MAT.	INST.	TOTAL
06110.30 Floor Framing				
3x14	S.F.	4.85	0.87	5.72
4x6	"	2.75	0.75	3.50
4x8	"	3.70	0.77	4.47
4x10	"	4.60	0.80	5.40
4x12	"	5.45	0.83	6.28
4x14	"	6.45	0.87	7.32
Sister joists for floors				
2x4	L.F.	0.63	2.70	3.33
2x6	"	0.89	3.10	3.99
2x8	"	1.25	3.60	4.85
2x10	"	1.75	4.35	6.10
2x12	"	2.30	5.40	7.70
3x6	"	2.10	4.35	6.45
3x8	"	2.75	4.80	7.55
3x10	"	3.45	5.40	8.85
3x12	"	4.10	6.20	10.30
4x6	"	2.70	4.35	7.05
4x8	"	3.70	4.80	8.50
4x10	"	4.60	5.40	10.00
4x12	"	5.45	6.20	11.65
06110.40 Furring				
Furring, wood strips				
Walls				
On masonry or concrete walls				
1x2 furring				
12" o.c.	S.F.	0.29	1.35	1.64
16" o.c.	"	0.25	1.25	1.50
24" o.c.	"	0.20	1.15	1.35
1x3 furring				
12" o.c.	S.F.	0.42	1.35	1.77
16" o.c.	"	0.36	1.25	1.61
24" o.c.	"	0.29	1.15	1.44
On wood walls				
1x2 furring				
12" o.c.	S.F.	0.29	0.96	1.25
16" o.c.	"	0.25	0.87	1.11
24" o.c.	"	0.20	0.79	0.99
1x3 furring				
12" o.c.	S.F.	0.42	0.96	1.38
16" o.c.	"	0.36	0.87	1.23
24" o.c.	"	0.29	0.79	1.08
Ceilings				
On masonry or concrete ceilings				
1x2 furring				
12" o.c.	S.F.	0.29	2.40	2.69
16" o.c.	"	0.25	2.15	2.40
24" o.c.	"	0.20	1.95	2.15
1x3 furring				
12" o.c.	S.F.	0.42	2.40	2.82
16" o.c.	"	0.36	2.15	2.51
24" o.c.	"	0.29	1.95	2.24

06 WOOD AND PLASTICS

Rough Carpentry

	UNIT	MAT.	INST.	TOTAL
06110.40 Furring				
On wood ceilings				
1x2 furring				
12" o.c.	S.F.	0.29	1.60	1.89
16" o.c.	"	0.25	1.45	1.70
24" o.c.	"	0.20	1.30	1.50
1x3				
12" o.c.	S.F.	0.42	1.60	2.02
16" o.c.	"	0.36	1.45	1.81
24" o.c.	"	0.29	1.30	1.59
06110.50 Roof Framing				
Roof framing				
Rafters, gable end				
0-2 pitch (flat to 2-in-12)				
12" o.c.				
2x4	S.F.	0.76	0.90	1.66
2x6	"	1.10	0.94	2.04
2x8	"	1.55	0.98	2.53
2x10	"	2.35	1.05	3.40
2x12	"	3.15	1.10	4.25
16" o.c.				
2x6	S.F.	0.95	0.77	1.72
2x8	"	1.30	0.80	2.10
2x10	"	1.95	0.83	2.78
2x12	"	2.55	0.87	3.42
24" o.c.				
2x6	S.F.	0.76	0.66	1.41
2x8	"	1.00	0.68	1.68
2x10	"	1.55	0.70	2.25
2x12	"	2.10	0.72	2.82
4-6 pitch (4-in-12 to 6-in-12)				
12" o.c.				
2x4	S.F.	0.80	0.94	1.74
2x6	"	1.20	0.98	2.18
2x8	"	1.70	1.05	2.75
2x10	"	2.55	1.10	3.65
2x12	"	3.40	1.15	4.55
16" o.c.				
2x6	S.F.	0.99	0.80	1.79
2x8	"	1.45	0.83	2.28
2x10	"	2.10	0.87	2.97
2x12	"	2.80	0.90	3.70
24" o.c.				
2x6	S.F.	0.80	0.68	1.48
2x8	"	1.10	0.70	1.80
2x10	"	1.65	0.75	2.40
2x12	"	2.25	0.83	3.08
8-12 pitch (8-in-12 to 12-in-12)				
12" o.c.				
2x4	S.F.	0.83	0.98	1.81
2x6	"	1.30	1.05	2.35
2x8	"	1.80	1.10	2.90

Rough Carpentry

06110.50 Roof Framing

	UNIT	MAT.	INST.	TOTAL
2x10	S.F.	2.70	1.15	3.85
2x12	"	3.55	1.20	4.75
16" o.c.				
2x6	S.F.	1.10	0.83	1.93
2x8	"	1.50	0.87	2.37
2x10	"	2.25	0.90	3.15
2x12	"	3.00	0.94	3.94
24" o.c.				
2x6	S.F.	0.83	0.70	1.53
2x8	"	1.15	0.72	1.87
2x10	"	1.80	0.75	2.55
2x12	"	2.35	0.77	3.12
Ridge boards				
2x6	L.F.	0.89	2.15	3.04
2x8	"	1.25	2.40	3.65
2x10	"	1.75	2.70	4.45
2x12	"	2.30	3.10	5.40
Hip rafters				
2x6	L.F.	0.89	1.55	2.44
2x8	"	1.25	1.60	2.85
2x10	"	1.75	1.65	3.40
2x12	"	2.30	1.75	4.05
Jack rafters				
4-6 pitch (4-in-12 to 6-in-12)				
16" o.c.				
2x6	S.F.	1.00	1.25	2.25
2x8	"	1.45	1.30	2.75
2x10	"	2.10	1.40	3.50
2x12	"	2.80	1.45	4.25
24" o.c.				
2x6	S.F.	0.80	0.98	1.78
2x8	"	1.10	1.00	2.10
2x10	"	1.65	1.05	2.70
2x12	"	2.25	1.10	3.35
8-12 pitch (8-in-12 to 12-in-12)				
16" o.c.				
2x6	S.F.	1.55	1.35	2.90
2x8	"	2.20	1.40	3.60
2x10	"	3.30	1.45	4.75
2x12	"	4.40	1.50	5.90
24" o.c.				
2x6	S.F.	1.30	1.05	2.35
2x8	"	1.80	1.05	2.85
2x10	"	2.70	1.10	3.80
2x12	"	3.55	1.10	4.65
Sister rafters				
2x4	L.F.	0.63	3.10	3.73
2x6	"	0.89	3.60	4.49
2x8	"	1.25	4.35	5.60
2x10	"	1.75	5.40	7.15
2x12	"	2.30	7.20	9.50
Fascia boards				

Rough Carpentry

Rough Carpentry	UNIT	MAT.	INST.	TOTAL
06110.50 Roof Framing				
2x4	L.F.	0.63	2.15	2.78
2x6	"	0.89	2.15	3.04
2x8	"	1.25	2.40	3.65
2x10	"	1.75	2.40	4.15
2x12	"	2.30	2.70	5.00
Cant strips				
Fiber				
3x3	L.F.	0.31	1.25	1.56
4x4	"	0.41	1.30	1.71
Wood				
3x3	L.F.	1.70	1.30	3.00
06110.60 Sleepers				
Sleepers, over concrete				
12" o.c.				
1x2	S.F.	0.26	0.98	1.25
1x3	"	0.36	1.05	1.41
2x4	"	0.76	1.20	1.96
2x6	"	1.10	1.25	2.35
16" o.c.				
1x2	S.F.	0.22	0.87	1.08
1x3	"	0.31	0.87	1.17
2x4	"	0.63	1.05	1.68
2x6	"	0.95	1.10	2.05
06110.65 Soffits				
Soffit framing				
2x3	L.F.	0.50	3.10	3.59
2x4	"	0.63	3.35	3.98
2x6	"	0.89	3.60	4.49
2x8	"	1.25	3.95	5.20
06110.70 Wall Framing				
Framing wall, studs				
12" o.c.				
2x3	S.F.	0.54	0.80	1.34
2x4	"	0.76	0.80	1.56
2x6	"	1.10	0.87	1.97
2x8	"	1.55	0.90	2.45
16" o.c.				
2x3	S.F.	0.47	0.68	1.14
2x4	"	0.63	0.68	1.30
2x6	"	0.89	0.72	1.61
2x8	"	1.30	0.75	2.05
24" o.c.				
2x3	S.F.	0.35	0.59	0.93
2x4	"	0.51	0.59	1.09
2x6	"	0.76	0.62	1.37
2x8	"	1.00	0.64	1.64
Plates, top or bottom				
2x3	L.F.	0.50	1.25	1.74
2x4	"	0.63	1.35	1.98

06 WOOD AND PLASTICS

Rough Carpentry

	UNIT	MAT.	INST.	TOTAL
06110.70 Wall Framing				
2x6	L.F.	0.89	1.45	2.34
2x8	"	1.25	1.55	2.80
Headers, door or window				
2x6				
Single				
3' long	EA.	2.85	21.65	24.50
6' long	"	5.70	27.10	32.80
Double				
3' long	EA.	5.70	24.05	29.75
6' long	"	11.35	30.90	42.25
2x8				
Single				
4' long	EA.	5.25	27.10	32.35
8' long	"	10.55	33.30	43.85
Double				
4' long	EA.	10.55	30.90	41.45
8' long	"	21.10	39.30	60.40
2x10				
Single				
5' long	EA.	9.80	33.30	43.10
10' long	"	19.60	43.30	62.90
Double				
5' long	EA.	19.60	36.10	55.70
10' long	"	39.20	43.30	82.50
2x12				
Single				
6' long	EA.	15.20	33.30	48.50
12' long	"	30.40	43.30	73.70
Double				
6' long	EA.	30.40	39.30	69.70
12' long	"	61.00	48.10	109.10
06115.10 Floor Sheathing				
Sub-flooring, plywood, CDX				
1/2" thick	S.F.	0.69	0.54	1.23
5/8" thick	"	0.83	0.62	1.45
3/4" thick	"	0.99	0.72	1.71
Structural plywood				
1/2" thick	S.F.	0.77	0.54	1.31
5/8" thick	"	0.91	0.62	1.53
3/4" thick	"	1.10	0.67	1.77
Board type subflooring				
1x6				
Minimum	S.F.	1.05	0.96	2.01
Maximum	"	1.40	1.10	2.50
1x8				
Minimum	S.F.	1.25	0.91	2.16
Maximum	"	1.40	1.00	2.40
1x10				
Minimum	S.F.	1.45	0.87	2.32
Maximum	"	1.50	0.96	2.46
Underlayment				

Rough Carpentry

Rough Carpentry	UNIT	MAT.	INST.	TOTAL
06115.10 Floor Sheathing				
Hardboard, 1/4" tempered	S.F.	0.50	0.54	1.04
Plywood, CDX				
3/8" thick	S.F.	0.62	0.54	1.16
1/2" thick	"	0.71	0.58	1.29
5/8" thick	"	0.85	0.62	1.47
3/4" thick	"	1.00	0.67	1.67
06115.20 Roof Sheathing				
Sheathing				
Plywood, CDX				
3/8" thick	S.F.	0.62	0.56	1.17
1/2" thick	"	0.71	0.58	1.29
5/8" thick	"	0.85	0.62	1.47
3/4" thick	"	1.00	0.67	1.67
Structural plywood				
3/8" thick	S.F.	0.68	0.56	1.24
1/2" thick	"	0.77	0.58	1.35
5/8" thick	"	0.92	0.62	1.54
3/4" thick	"	1.10	0.67	1.77
06115.30 Wall Sheathing				
Sheathing				
Plywood, CDX				
3/8" thick	S.F.	0.62	0.64	1.26
1/2" thick	"	0.71	0.67	1.37
5/8" thick	"	0.85	0.72	1.57
3/4" thick	"	1.00	0.79	1.79
Waferboard				
3/8" thick	S.F.	0.54	0.64	1.18
1/2" thick	"	0.66	0.67	1.33
5/8" thick	"	0.80	0.72	1.52
3/4" thick	"	0.88	0.79	1.66
Structural plywood				
3/8" thick	S.F.	0.68	0.64	1.32
1/2" thick	"	0.77	0.67	1.44
5/8" thick	"	0.92	0.72	1.65
3/4" thick	"	1.10	0.79	1.89
Gypsum, 1/2" thick	"	0.65	0.67	1.32
Asphalt impregnated fiberboard, 1/2" thick	"	0.66	0.67	1.33
06125.10 Wood Decking				
Decking, T&G solid				
Cedar				
3" thick	S.F.	7.90	1.10	9.00
4" thick	"	9.55	1.15	10.70
Fir				
3" thick	S.F.	3.30	1.10	4.40
4" thick	"	4.05	1.15	5.20
Southern yellow pine				
3" thick	S.F.	3.25	1.25	4.50
4" thick	"	3.50	1.35	4.85
White pine				

06 WOOD AND PLASTICS

Rough Carpentry

06125.10 Wood Decking

	UNIT	MAT.	INST.	TOTAL
3" thick	S.F.	4.10	1.10	5.20
4" thick	"	5.40	1.15	6.55

06130.10 Heavy Timber

	UNIT	MAT.	INST.	TOTAL
Mill framed structures				
Beams to 20' long				
Douglas fir				
6x8	L.F.	5.85	5.45	11.30
6x10	"	6.90	5.65	12.55
6x12	"	8.25	6.05	14.30
6x14	"	9.90	6.30	16.20
6x16	"	10.75	6.55	17.30
8x10	"	9.10	5.65	14.75
8x12	"	10.75	6.05	16.80
8x14	"	12.40	6.30	18.70
8x16	"	14.15	6.55	20.70
Southern yellow pine				
6x8	L.F.	4.65	5.45	10.10
6x10	"	5.60	5.65	11.25
6x12	"	7.20	6.05	13.25
6x14	"	8.25	6.30	14.55
6x16	"	9.10	6.55	15.65
8x10	"	7.65	5.65	13.30
8x12	"	9.25	6.05	15.30
8x14	"	10.60	6.30	16.90
8x16	"	12.25	6.55	18.80
Columns to 12' high				
Douglas fir				
6x6	L.F.	4.20	8.20	12.40
8x8	"	7.20	8.20	15.40
10x10	"	12.55	9.10	21.65
12x12	"	15.55	9.10	24.05
Southern yellow pine				
6x6	L.F.	3.60	8.20	11.80
8x8	"	6.05	8.20	14.25
10x10	"	9.45	9.10	18.55
12x12	"	13.15	9.10	22.25
Posts, treated				
4x4	L.F.	1.95	1.75	3.70
6x6	"	4.55	2.15	6.70

06190.20 Wood Trusses

	UNIT	MAT.	INST.	TOTAL
Truss, fink, 2x4 members				
3-in-12 slope				
24' span	EA.	73.00	46.90	119.90
26' span	"	84.50	46.90	131.40
28' span	"	87.50	49.70	137.20
30' span	"	93.50	49.70	143.20
34' span	"	100.00	53.00	153.00
38' span	"	120.00	53.00	173.00
5-in-12 slope				
24' span	EA.	74.50	48.20	122.70

Rough Carpentry

06190.20 Wood Trusses

Rough Carpentry	UNIT	MAT.	INST.	TOTAL
28' span	EA.	87.50	49.70	137.20
30' span	"	100.00	51.50	151.50
32' span	"	110.00	51.50	161.50
40' span	"	150.00	54.50	204.50
Gable, 2x4 members				
5-in-12 slope				
24' span	EA.	100.00	48.20	148.20
26' span	"	110.00	48.20	158.20
28' span	"	120.00	49.70	169.70
30' span	"	130.00	51.50	181.50
32' span	"	130.00	51.50	181.50
36' span	"	140.00	53.00	193.00
40' span	"	150.00	54.50	204.50
King post type, 2x4 members				
4-in-12 slope				
16' span	EA.	61.50	44.30	105.80
18' span	"	65.50	45.60	111.10
24' span	"	70.00	48.20	118.20
26' span	"	76.00	48.20	124.20
30' span	"	96.50	51.50	148.00
34' span	"	100.00	51.50	151.50
38' span	"	120.00	53.00	173.00
42' span	"	150.00	56.50	206.50

Finish Carpentry

06200.10 Finish Carpentry

Finish Carpentry	UNIT	MAT.	INST.	TOTAL
Mouldings and trim				
Apron, flat				
9/16 x 2	L.F.	1.25	2.15	3.40
9/16 x 3-1/2	"	2.30	2.30	4.60
Base				
Colonial				
7/16 x 2-1/4	L.F.	1.45	2.15	3.60
7/16 x 3	"	1.75	2.15	3.90
7/16 x 3-1/4	"	1.90	2.15	4.05
9/16 x 3	"	1.85	2.30	4.15
9/16 x 3-1/4	"	1.95	2.30	4.25
11/16 x 2-1/4	"	1.75	2.40	4.15
Ranch				
7/16 x 2-1/4	L.F.	1.50	2.15	3.65
7/16 x 3-1/4	"	1.95	2.15	4.10
9/16 x 2-1/4	"	1.65	2.30	3.95
9/16 x 3	"	1.95	2.30	4.25
9/16 x 3-1/4	"	1.95	2.30	4.25
Casing				

Finish Carpentry

06200.10 Finish Carpentry

Finish Carpentry	UNIT	MAT.	INST.	TOTAL
11/16 x 2-1/2	L.F.	1.50	1.95	3.45
11/16 x 3-1/2	"	1.95	2.05	4.00
Chair rail				
9/16 x 2-1/2	L.F.	1.55	2.15	3.70
9/16 x 3-1/2	"	2.35	2.15	4.50
Closet pole				
1-1/8" dia.	L.F.	1.80	2.90	4.70
1-5/8" dia.	"	2.10	2.90	5.00
Cove				
9/16 x 1-3/4	L.F.	1.20	2.15	3.35
11/16 x 2-3/4	"	1.65	2.15	3.80
Crown				
9/16 x 1-5/8	L.F.	1.80	2.90	4.70
9/16 x 2-5/8	"	2.00	3.35	5.35
11/16 x 3-5/8	"	2.40	3.60	6.00
11/16 x 4-1/4	"	3.35	3.95	7.30
11/16 x 5-1/4	"	4.30	4.35	8.65
Drip cap				
1-1/16 x 1-5/8	L.F.	1.70	2.15	3.85
Glass bead				
3/8 x 3/8	L.F.	0.49	2.70	3.19
1/2 x 9/16	"	0.62	2.70	3.32
5/8 x 5/8	"	0.65	2.70	3.35
3/4 x 3/4	"	0.81	2.70	3.51
Half round				
1/2	L.F.	0.61	1.75	2.36
5/8	"	0.81	1.75	2.56
3/4	"	0.99	1.75	2.74
Lattice				
1/4 x 7/8	L.F.	0.48	1.75	2.23
1/4 x 1-1/8	"	0.50	1.75	2.25
1/4 x 1-3/8	"	0.53	1.75	2.28
1/4 x 1-3/4	"	0.62	1.75	2.37
1/4 x 2	"	0.71	1.75	2.46
Ogee molding				
5/8 x 3/4	L.F.	0.99	2.15	3.14
11/16 x 1-1/8	"	1.50	2.15	3.65
11/16 x 1-3/8	"	1.80	2.15	3.95
Parting bead				
3/8 x 7/8	L.F.	0.78	2.70	3.48
Quarter round				
1/4 x 1/4	L.F.	0.23	1.75	1.98
3/8 x 3/8	"	0.36	1.75	2.11
1/2 x 1/2	"	0.49	1.75	2.24
11/16 x 11/16	"	0.52	1.90	2.42
3/4 x 3/4	"	0.58	1.90	2.48
1-1/16 x 1-1/16	"	0.72	1.95	2.67
Railings, balusters				
1-1/8 x 1-1/8	L.F.	2.60	4.35	6.95
1-1/2 x 1-1/2	"	3.05	3.95	7.00
Screen moldings				
1/4 x 3/4	L.F.	0.63	3.60	4.23

Finish Carpentry

	UNIT	MAT.	INST.	TOTAL
06200.10 Finish Carpentry				
5/8 x 5/16	L.F.	0.79	3.60	4.39
Shoe				
7/16 x 11/16	L.F.	0.79	1.75	2.54
Sash beads				
1/2 x 3/4	L.F.	0.98	3.60	4.58
1/2 x 7/8	"	1.15	3.60	4.75
1/2 x 1-1/8	"	1.25	3.95	5.20
5/8 x 7/8	"	1.25	3.95	5.20
Stop				
5/8 x 1-5/8				
Colonial	L.F.	0.62	2.70	3.32
Ranch	"	0.61	2.70	3.31
Stools				
11/16 x 2-1/4	L.F.	2.80	4.80	7.60
11/16 x 2-1/2	"	2.95	4.80	7.75
11/16 x 5-1/4	"	3.15	5.40	8.55
Exterior trim, casing, select pine, 1x3	"	2.10	2.15	4.25
Douglas fir				
1x3	L.F.	1.10	2.15	3.25
1x4	"	1.50	2.15	3.65
1x6	"	1.80	2.40	4.20
1x8	"	2.60	2.70	5.30
Cornices, white pine, #2 or better				
1x2	L.F.	0.68	2.15	2.83
1x4	"	1.00	2.15	3.15
1x6	"	1.60	2.40	4.00
1x8	"	1.80	2.55	4.35
1x10	"	2.45	2.70	5.15
1x12	"	3.05	2.90	5.95
Shelving, pine				
1x8	L.F.	2.15	3.35	5.50
1x10	"	2.60	3.45	6.05
1x12	"	3.45	3.60	7.05
Plywood shelf, 3/4", with edge band, 12" wide	"	2.35	4.35	6.70
Adjustable shelf, and rod, 12" wide				
3' to 4' long	EA.	15.90	10.80	26.70
5' to 8' long	"	30.00	14.45	44.45
Prefinished wood shelves with brackets and supports				
8" wide				
3' long	EA.	47.10	10.80	57.90
4' long	"	54.00	10.80	64.80
6' long	"	79.00	10.80	89.80
10" wide				
3' long	EA.	52.00	10.80	62.80
4' long	"	74.00	10.80	84.80
6' long	"	87.50	10.80	98.30
06220.10 Millwork				
Countertop, laminated plastic				
25" x 7/8" thick				
Minimum	L.F.	16.20	10.80	27.00
Average	"	29.20	14.45	43.65

Finish Carpentry

06220.10 Millwork

	UNIT	MAT.	INST.	TOTAL
Maximum	L.F.	48.60	17.30	65.90
25" x 1-1/4" thick				
Minimum	L.F.	19.45	14.45	33.90
Average	"	32.40	17.30	49.70
Maximum	"	52.00	21.65	73.65
Add for cutouts	EA.		27.10	27.10
Backsplash, 4" high, 7/8" thick	L.F.	16.20	8.65	24.85
Plywood, sanded, A-C				
1/4" thick	S.F.	1.25	1.45	2.70
3/8" thick	"	1.35	1.55	2.90
1/2" thick	"	1.50	1.65	3.15
A-D				
1/4" thick	S.F.	1.15	1.45	2.60
3/8" thick	"	1.30	1.55	2.85
1/2" thick	"	1.45	1.65	3.10
Base cabinets, 34-1/2" high, 24" deep, hardwood, no tops				
Minimum	L.F.	62.00	17.30	79.30
Average	"	110.00	21.65	131.65
Maximum	"	170.00	28.90	198.90
Wall cabinets				
Minimum	L.F.	49.50	14.45	63.95
Average	"	79.50	17.30	96.80
Maximum	"	160.00	21.65	181.65

Wood Treatment

06300.10 Wood Treatment

	UNIT	MAT.	INST.	TOTAL
Creosote preservative treatment				
8 lb/cf	B.F.			0.40
10 lb/cf	"			0.51
Salt preservative treatment				
Oil borne				
Minimum	B.F.			0.30
Maximum	"			0.48
Water borne				
Minimum	B.F.			0.23
Maximum	"			0.37
Fire retardant treatment				
Minimum	B.F.			0.50
Maximum	"			0.59
Kiln dried, softwood, add to framing costs				
1" thick	B.F.			0.17
2" thick	"			0.26
3" thick	"			0.35
4" thick	"			0.45

06 WOOD AND PLASTICS

Architectural Woodwork

	UNIT	MAT.	INST.	TOTAL
06420.10 Panel Work				
Hardboard, tempered, 1/4" thick				
Natural faced	S.F.	0.66	1.10	1.75
Plastic faced	"	1.10	1.25	2.35
Pegboard, natural	"	0.76	1.10	1.86
Plastic faced	"	0.99	1.25	2.24
Untempered, 1/4" thick				
Natural faced	S.F.	0.60	1.10	1.70
Plastic faced	"	1.05	1.25	2.30
Pegboard, natural	"	0.63	1.10	1.73
Plastic faced	"	0.96	1.25	2.21
Plywood unfinished, 1/4" thick				
Birch				
Natural	S.F.	1.45	1.45	2.90
Select	"	1.90	1.45	3.35
Knotty pine	"	1.90	1.45	3.35
Cedar (closet lining)				
Standard boards T&G	S.F.	1.85	1.45	3.30
Particle board	"	1.45	1.45	2.90
Plywood, prefinished, 1/4" thick, premium grade				
Birch veneer	S.F.	2.60	1.75	4.35
Cherry veneer	"	2.90	1.75	4.65
Chestnut veneer	"	5.20	1.75	6.95
Lauan veneer	"	0.93	1.75	2.68
Mahogany veneer	"	2.95	1.75	4.70
Oak veneer (red)	"	2.95	1.75	4.70
Pecan veneer	"	3.50	1.75	5.25
Rosewood veneer	"	4.95	1.75	6.70
Teak veneer	"	6.35	1.75	8.10
Walnut veneer	"	4.35	1.75	6.10
06430.10 Stairwork				
Risers, 1x8, 42" wide				
White oak	EA.	21.25	21.65	42.90
Pine	"	15.05	21.65	36.70
Treads, 1-1/16" x 9-1/2" x 42"				
White oak	EA.	30.40	27.10	57.50
06440.10 Columns				
Column, hollow, round wood				
12" diameter				
10' high	EA.	610.00	60.50	670.50
12' high	"	750.00	65.00	815.00
14' high	"	900.00	72.50	972.50
16' high	"	1,120	91.00	1,211
24" diameter				
16' high	EA.	2,550	91.00	2,641
18' high	"	2,890	95.50	2,986
20' high	"	3,570	95.50	3,666
22' high	"	3,740	100.00	3,840
24' high	"	4,080	100.00	4,180

Moisture Protection

07100.10 Waterproofing

	UNIT	MAT.	INST.	TOTAL
Membrane waterproofing, elastomeric				
Butyl				
1/32" thick	S.F.	0.87	1.35	2.22
1/16" thick	"	1.15	1.40	2.55
Butyl with nylon				
1/32" thick	S.F.	1.00	1.35	2.35
1/16" thick	"	1.20	1.40	2.60
Neoprene				
1/32" thick	S.F.	1.40	1.35	2.75
1/16" thick	"	2.35	1.40	3.75
Neoprene with nylon				
1/32" thick	S.F.	1.55	1.35	2.90
1/16" thick	"	2.50	1.40	3.90
Bituminous membrane waterproofing, asphalt felt, 15 lb.				
One ply	S.F.	0.52	0.85	1.36
Two ply	"	0.61	1.05	1.66
Three ply	"	0.76	1.20	1.96
Four ply	"	0.86	1.40	2.26
Five ply	"	0.93	1.80	2.73
Modified asphalt membrane waterproofing, fibrous asphalt				
One ply	S.F.	1.20	1.40	2.60
Two ply	"	1.80	1.70	3.50
Three ply	"	2.00	1.90	3.90
Four ply	"	2.40	2.25	4.65
Five ply	"	2.85	2.70	5.55
Asphalt coated protective board				
1/8" thick	S.F.	0.67	0.85	1.51
1/4" thick	"	0.93	0.85	1.78
3/8" thick	"	1.00	0.85	1.85
1/2" thick	"	1.25	0.89	2.14
Cement protective board				
3/8" thick	S.F.	2.30	1.15	3.45
1/2" thick	"	3.20	1.15	4.35
Fluid applied, neoprene				
50 mil	S.F.	2.75	1.15	3.90
90 mil	"	3.80	1.15	4.95
Tab extended polyurethane				
.050" thick	S.F.	1.50	0.85	2.35
Fluid applied rubber based polyurethane				
6 mil	S.F.	0.79	1.05	1.84
15 mil	"	1.50	0.85	2.35
Bentonite waterproofing, panels				
3/16" thick	S.F.	1.35	0.85	2.20
1/4" thick	"	1.50	0.85	2.35
5/8" thick	"	2.25	0.89	3.14
Granular admixtures, trowel on, 3/8" thick	"	1.85	0.85	2.70
Metallic oxide waterproofing, iron compound, troweled				
5/8" thick	S.F.	1.35	0.85	2.20
3/4" thick	"	1.65	0.97	2.62

07 THERMAL AND MOISTURE

Moisture Protection

Moisture Protection	UNIT	MAT.	INST.	TOTAL
07150.10 Dampproofing				
Silicone dampproofing, sprayed on				
Concrete surface				
1 coat	S.F.	0.55	0.19	0.74
2 coats	"	0.90	0.26	1.16
Concrete block				
1 coat	S.F.	0.55	0.23	0.77
2 coats	"	0.90	0.31	1.21
Brick				
1 coat	S.F.	0.64	0.26	0.90
2 coats	"	0.98	0.34	1.32
07160.10 Bituminous Dampproofing				
Building paper, asphalt felt				
15 lb	S.F.	0.12	1.35	1.47
30 lb	"	0.23	1.40	1.63
Asphalt dampproofing, troweled, cold, primer plus				
1 coat	S.F.	0.78	1.15	1.93
2 coats	"	1.15	1.70	2.85
3 coats	"	1.75	2.10	3.85
Fibrous asphalt dampproofing, hot troweled, primer plus				
1 coat	S.F.	1.20	1.35	2.55
2 coats	"	1.50	1.90	3.40
3 coats	"	1.65	2.40	4.05
Asphaltic paint dampproofing, per coat				
Brush on	S.F.	0.23	0.48	0.71
Spray on	"	0.40	0.38	0.77
07190.10 Vapor Barriers				
Vapor barrier, polyethylene				
2 mil	S.F.	0.01	0.17	0.18
6 mil	"	0.03	0.17	0.20
8 mil	"	0.04	0.19	0.23
10 mil	"	0.07	0.19	0.25

Insulation

Insulation	UNIT	MAT.	INST.	TOTAL
07210.10 Batt Insulation				
Ceiling, fiberglass, unfaced				
3-1/2" thick, R11	S.F.	0.35	0.40	0.75
6" thick, R19	"	0.46	0.45	0.91
9" thick, R30	"	0.91	0.52	1.43
Suspended ceiling, unfaced				
3-1/2" thick, R11	S.F.	0.35	0.38	0.73
6" thick, R19	"	0.46	0.42	0.89
9" thick, R30	"	0.91	0.48	1.39
Crawl space, unfaced				

07 THERMAL AND MOISTURE

Insulation

	UNIT	MAT.	INST.	TOTAL
07210.10 Batt Insulation				
3-1/2" thick, R11	S.F.	0.35	0.52	0.87
6" thick, R19	"	0.46	0.56	1.03
9" thick, R30	"	0.91	0.62	1.53
Wall, fiberglass				
Paper backed				
2" thick, R7	S.F.	0.23	0.36	0.59
3" thick, R8	"	0.25	0.38	0.62
4" thick, R11	"	0.41	0.40	0.81
6" thick, R19	"	0.61	0.42	1.04
Foil backed, 1 side				
2" thick, R7	S.F.	0.53	0.36	0.89
3" thick, R11	"	0.56	0.38	0.93
4" thick, R14	"	0.59	0.40	0.98
6" thick, R21	"	0.61	0.42	1.04
Foil backed, 2 sides				
2" thick, R7	S.F.	0.60	0.40	1.00
3" thick, R11	"	0.76	0.42	1.19
4" thick, R14	"	0.90	0.45	1.35
6" thick, R21	"	0.97	0.48	1.45
Unfaced				
2" thick, R7	S.F.	0.34	0.36	0.70
3" thick, R9	"	0.38	0.38	0.76
4" thick, R11	"	0.41	0.40	0.81
6" thick, R19	"	0.53	0.42	0.95
Mineral wool batts				
Paper backed				
2" thick, R6	S.F.	0.57	0.36	0.93
4" thick, R12	"	0.68	0.38	1.06
6" thick, R19	"	0.79	0.42	1.21
Fasteners, self adhering, attached to ceiling deck				
2-1/2" long	EA.	0.18	0.56	0.74
4-1/2" long	"	0.20	0.62	0.82
Capped, self-locking washers for fastening insulation	"	0.18	0.34	0.51
07210.20 Board Insulation				
Insulation, rigid				
Fiberglass, roof				
0.75" thick, R2.78	S.F.	0.33	0.31	0.63
1.06" thick, R4.17	"	0.50	0.32	0.83
1.31" thick, R5.26	"	0.67	0.34	1.00
1.63" thick, R6.67	"	0.83	0.36	1.19
2.25" thick, R8.33	"	0.91	0.38	1.29
Composite board, roof				
1-1/2" thick, R6.67	S.F.	1.10	0.34	1.44
1-5/8" thick, R7.69	"	1.15	0.36	1.51
2" thick, R10.0	"	2.10	0.38	2.48
2-1/4" thick, R12.50	"	2.30	0.40	2.70
2-1/2" thick, R14.29	"	2.50	0.42	2.92
2-3/4" thick, R16.67	"	2.75	0.45	3.20
3-1/4" thick, R20.00	"	3.55	0.48	4.03
Perlite board, roof				
1.00" thick, R2.78	S.F.	0.69	0.28	0.98

Insulation

07210.20 Board Insulation

Description	UNIT	MAT.	INST.	TOTAL
1.50" thick, R4.17	S.F.	1.10	0.29	1.39
2.00" thick, R5.92	"	1.35	0.31	1.66
2.50" thick, R6.67	"	1.65	0.32	1.97
3.00" thick, R8.33	"	2.05	0.34	2.39
4.00" thick, R10.00	"	2.30	0.36	2.66
5.25" thick, R14.29	"	2.50	0.38	2.88
Rigid urethane				
Roof				
1" thick, R6.67	S.F.	0.94	0.28	1.22
1.20" thick, R8.33	"	1.05	0.29	1.34
1.50" thick, R11.11	"	1.30	0.29	1.59
2" thick, R14.29	"	1.65	0.31	1.96
2.25" thick, R16.67	"	2.15	0.32	2.47
Wall				
1" thick, R6.67	S.F.	0.94	0.36	1.29
1.5" thick, R11.11	"	1.30	0.38	1.68
2" thick, R14.29	"	1.70	0.40	2.10
Polystyrene				
Roof				
1.0" thick, R4.17	S.F.	0.88	0.28	1.17
1.5" thick, R6.26	"	1.50	0.29	1.79
2.0" thick, R8.33	"	1.65	0.31	1.96
Wall				
1.0" thick, R4.17	S.F.	0.88	0.36	1.24
1.5" thick, R6.26	"	1.50	0.38	1.88
2.0" thick, R8.33	"	1.65	0.40	2.05
Rigid board insulation, deck				
Mineral fiberboard				
1" thick, R3.0	S.F.	0.54	0.28	0.83
2" thick, R5.26	"	1.20	0.31	1.51
Fiberglass				
1" thick, R4.3	S.F.	0.93	0.28	1.21
2" thick, R8.5	"	1.35	0.31	1.66
Polystyrene				
1" thick, R5.4	S.F.	0.88	0.28	1.17
2" thick, R10.8	"	1.65	0.31	1.96
Urethane				
.75" thick, R5.4	S.F.	0.86	0.28	1.14
1" thick, R6.4	"	1.00	0.28	1.28
1.5" thick, R10.7	"	1.20	0.29	1.49
2" thick, R14.3	"	1.40	0.31	1.71
Foamglass				
1" thick, R1.8	S.F.	2.80	0.28	3.08
2" thick, R5.26	"	3.55	0.31	3.86
Wood fiber				
1" thick, R3.85	S.F.	3.15	0.28	3.43
2" thick, R7.7	"	3.75	0.31	4.06
Particle board				
3/4" thick, R2.08	S.F.	0.83	0.28	1.11
1" thick, R2.77	"	0.84	0.28	1.12
2" thick, R5.50	"	0.87	0.31	1.17

Insulation	UNIT	MAT.	INST.	TOTAL
07210.60 Loose Fill Insulation				
Blown-in type				
Fiberglass				
5" thick, R11	S.F.	0.33	0.28	0.61
6" thick, R13	"	0.38	0.34	0.72
9" thick, R19	"	0.46	0.48	0.95
Rockwool, attic application				
6" thick, R13	S.F.	0.59	0.34	0.92
8" thick, R19	"	0.71	0.42	1.13
10" thick, R22	"	0.86	0.52	1.38
12" thick, R26	"	1.05	0.56	1.61
15" thick, R30	"	1.30	0.68	1.98
Poured type				
Fiberglass				
1" thick, R4	S.F.	0.30	0.21	0.51
2" thick, R8	"	0.56	0.24	0.80
3" thick, R12	"	0.82	0.28	1.10
4" thick, R16	"	1.05	0.34	1.39
Mineral wool				
1" thick, R3	S.F.	0.33	0.21	0.54
2" thick, R6	"	0.61	0.24	0.85
3" thick, R9	"	0.93	0.28	1.21
4" thick, R12	"	1.05	0.34	1.39
Vermiculite or perlite				
2" thick, R4.8	S.F.	0.69	0.24	0.94
3" thick, R7.2	"	0.99	0.28	1.28
4" thick, R9.6	"	1.30	0.34	1.64
Masonry, poured vermiculite or perlite				
4" block	S.F.	0.53	0.17	0.70
6" block	"	0.67	0.21	0.88
8" block	"	1.00	0.24	1.24
10" block	"	1.20	0.26	1.46
12" block	"	1.35	0.28	1.63
07210.70 Sprayed Insulation				
Foam, sprayed on				
Polystyrene				
1" thick, R4	S.F.	0.67	0.34	1.01
2" thick, R8	"	1.35	0.45	1.80
Urethane				
1" thick, R4	S.F.	0.61	0.34	0.94
2" thick, R8	"	1.25	0.45	1.70
07250.10 Fireproofing				
Sprayed on				
1" thick				
On beams	S.F.	0.65	0.75	1.40
On columns	"	0.66	0.68	1.34
On decks				
Flat surface	S.F.	0.66	0.34	1.00
Fluted surface	"	0.84	0.42	1.26
1-1/2" thick				
On beams	S.F.	1.15	0.97	2.12

07 THERMAL AND MOISTURE

Insulation

	UNIT	MAT.	INST.	TOTAL
07250.10 Fireproofing				
On columns	S.F.	1.30	0.85	2.15
On decks				
Flat surface	S.F.	0.99	0.42	1.41
Fluted surface	"	1.15	0.56	1.71

Shingles And Tiles

	UNIT	MAT.	INST.	TOTAL
07310.10 Asphalt Shingles				
Standard asphalt shingles, strip shingles				
210 lb/square	SQ.	55.00	41.60	96.60
235 lb/square	"	58.50	46.20	104.70
240 lb/square	"	60.00	52.00	112.00
260 lb/square	"	86.00	59.50	145.50
300 lb/square	"	94.50	69.50	164.00
385 lb/square	"	130.00	83.00	213.00
Roll roofing, mineral surface				
90 lb	SQ.	30.60	29.70	60.30
110 lb	"	51.00	34.60	85.60
140 lb	"	52.50	41.60	94.10
07310.30 Metal Shingles				
Aluminum, .020" thick				
Plain	SQ.	230.00	83.00	313.00
Colors	"	250.00	83.00	333.00
Steel, galvanized				
26 ga.				
Plain	SQ.	240.00	83.00	323.00
Colors	"	300.00	83.00	383.00
24 ga.				
Plain	SQ.	250.00	83.00	333.00
Colors	"	310.00	83.00	393.00
Porcelain enamel, 22 ga.				
Minimum	SQ.	710.00	100.00	810.00
Average	"	820.00	100.00	920.00
Maximum	"	920.00	100.00	1,020
07310.60 Slate Shingles				
Slate shingles				
Pennsylvania				
Ribbon	SQ.	590.00	210.00	800.00
Clear	"	760.00	210.00	970.00
Vermont				
Black	SQ.	640.00	210.00	850.00
Gray	"	710.00	210.00	920.00
Green	"	730.00	210.00	940.00
Red	"	1,300	210.00	1,510

07 THERMAL AND MOISTURE

Shingles And Tiles

Shingles And Tiles	UNIT	MAT.	INST.	TOTAL
07310.60 Slate Shingles				
Replacement shingles				
Small jobs	EA.	11.35	13.85	25.20
Large jobs	S.F.	8.95	6.95	15.90
07310.70 Wood Shingles				
Wood shingles, on roofs				
White cedar, #1 shingles				
4" exposure	SQ.	230.00	140.00	370.00
5" exposure	"	210.00	100.00	310.00
#2 shingles				
4" exposure	SQ.	160.00	140.00	300.00
5" exposure	"	140.00	100.00	240.00
Resquared and rebutted				
4" exposure	SQ.	210.00	140.00	350.00
5" exposure	"	170.00	100.00	270.00
On walls				
White cedar, #1 shingles				
4" exposure	SQ.	230.00	210.00	440.00
5" exposure	"	210.00	170.00	380.00
6" exposure	"	170.00	140.00	310.00
#2 shingles				
4" exposure	SQ.	160.00	210.00	370.00
5" exposure	"	140.00	170.00	310.00
6" exposure	"	120.00	140.00	260.00
Add for fire retarding	"			107.74
07310.80 Wood Shakes				
Shakes, hand split, 24" red cedar, on roofs				
5" exposure	SQ.	240.00	210.00	450.00
7" exposure	"	220.00	170.00	390.00
9" exposure	"	210.00	140.00	350.00
On walls				
6" exposure	SQ.	220.00	210.00	430.00
8" exposure	"	210.00	170.00	380.00
10" exposure	"	200.00	140.00	340.00
Add for fire retarding	"			74.00

Roofing And Siding

Roofing And Siding	UNIT	MAT.	INST.	TOTAL
07410.10 Manufactured Roofs				
Aluminum roof panels, for structural steel framing				
Corrugated				
Unpainted finish				
.024"	S.F.	1.60	1.05	2.65
.030"	"	1.85	1.05	2.90
Painted finish				

Roofing And Siding

	UNIT	MAT.	INST.	TOTAL
07410.10 Manufactured Roofs				
.024"	S.F.	2.00	1.05	3.05
.030"	"	2.45	1.05	3.50
V-beam				
Unpainted finish				
.032"	S.F.	2.10	1.05	3.15
.040"	"	2.50	1.05	3.55
.050"	"	3.15	1.05	4.20
Painted finish				
.032"	S.F.	2.70	1.05	3.75
.040"	"	3.25	1.05	4.30
.050"	"	3.85	1.05	4.90
Steel roof panels, for structural steel framing				
Corrugated, painted				
18 ga.	S.F.	4.90	1.05	5.95
20 ga.	"	4.60	1.05	5.65
22 ga.	"	4.05	1.05	5.10
Box rib, painted				
18 ga.	S.F.	5.75	1.10	6.85
20 ga.	"	4.75	1.10	5.85
22 ga.	"	4.20	1.10	5.30
4" rib, painted				
18 ga.	S.F.	6.60	1.15	7.75
20 ga.	"	5.60	1.15	6.75
22 ga.	"	5.05	1.15	6.20
Standing seam roof				
2" high seam, painted				
22 ga.	S.F.	6.40	1.65	8.05
24 ga.	"	6.25	1.65	7.90
26 ga.	"	6.10	1.65	7.75
07410.30 Manufactured Walls				
Sandwich panels with 1-1/2" fiberglass insulation				
Galvanized 18 ga. steel interior panels				
Exterior panels				
16 ga. aluminum	S.F.	6.10	6.40	12.50
18 ga. galvanized steel	"	8.65	6.40	15.05
20 ga. painted steel	"	8.80	6.40	15.20
20 ga. stainless steel	"	8.95	6.40	15.35
Metal liner panels, 1-3/8" thick, 24" wide				
Galvanized				
22 ga.	S.F.	3.20	1.60	4.80
20 ga.	"	3.55	1.60	5.15
18 ga.	"	4.35	1.60	5.95
Primed				
22 ga.	S.F.	2.50	1.60	4.10
20 ga.	"	2.90	1.60	4.50
18 ga.	"	3.55	1.60	5.15
07440.10 Aggregate Coated Panels				
Dryvit type system				
1" thick	S.F.	2.45	1.60	4.05
1-1/2" thick	"	2.55	1.70	4.25

Roofing And Siding

Roofing And Siding	UNIT	MAT.	INST.	TOTAL
07440.10 Aggregate Coated Panels				
2" thick	S.F.	2.70	2.00	4.70
07460.10 Metal Siding Panels				
Aluminum siding panels				
Corrugated				
Plain finish				
.024"	S.F.	1.40	1.90	3.30
.032"	"	1.65	1.90	3.55
Painted finish				
.024"	S.F.	1.75	1.90	3.65
.032"	"	2.05	1.90	3.95
V. beam				
Plain finish				
.032"	S.F.	1.90	1.90	3.80
.040"	"	2.20	1.90	4.10
.050"	"	2.80	1.90	4.70
Painted finish				
.032"	S.F.	2.35	1.90	4.25
.040"	"	2.80	1.90	4.70
.050"	"	3.60	1.90	5.50
4" rib				
Plain finish				
.032"	S.F.	1.85	2.20	4.05
.040"	"	2.00	2.20	4.20
.050"	"	2.45	2.20	4.65
Painted finish				
.032"	S.F.	2.30	2.20	4.50
.040"	"	2.45	2.20	4.65
.050"	"	2.80	2.20	5.00
Steel siding panels				
Corrugated				
22 ga.	S.F.	1.65	3.20	4.85
24 ga.	"	1.50	3.20	4.70
26 ga.	"	1.35	3.20	4.55
Box rib				
20 ga.	S.F.	3.00	3.20	6.20
22 ga.	"	2.50	3.20	5.70
24 ga.	"	2.20	3.20	5.40
26 ga.	"	1.55	3.20	4.75
07460.50 Plastic Siding				
Horizontal vinyl siding, solid				
8" wide				
Standard	S.F.	0.98	1.65	2.63
Insulated	"	1.20	1.65	2.85
10" wide				
Standard	S.F.	1.00	1.55	2.55
Insulated	"	1.20	1.55	2.75
Vinyl moldings for doors and windows	L.F.	0.63	1.75	2.38

Roofing And Siding

Roofing And Siding	UNIT	MAT.	INST.	TOTAL
07460.60 Plywood Siding				
Rough sawn cedar, 3/8" thick	S.F.	1.75	1.45	3.20
Fir, 3/8" thick	"	0.97	1.45	2.42
Texture 1-11, 5/8" thick				
Cedar	S.F.	2.35	1.55	3.90
Fir	"	1.65	1.55	3.20
Redwood	"	2.55	1.50	4.05
Southern Yellow Pine	"	1.30	1.55	2.85
07460.70 Steel Siding				
Ribbed, sheets, galvanized				
22 ga.	S.F.	1.65	1.90	3.55
24 ga.	"	1.50	1.90	3.40
26 ga.	"	1.15	1.90	3.05
28 ga.	"	1.00	1.90	2.90
Primed				
24 ga.	S.F.	2.00	1.90	3.90
26 ga.	"	1.40	1.90	3.30
28 ga.	"	1.15	1.90	3.05
07460.80 Wood Siding				
Beveled siding, cedar				
A grade				
1/2 x 6	S.F.	3.35	2.15	5.50
1/2 x 8	"	3.45	1.75	5.20
3/4 x 10	"	4.45	1.45	5.90
Clear				
1/2 x 6	S.F.	3.75	2.15	5.90
1/2 x 8	"	3.85	1.75	5.60
3/4 x 10	"	5.15	1.45	6.60
B grade				
1/2 x 6	S.F.	3.60	2.15	5.75
1/2 x 8	"	4.10	17.30	21.40
3/4 x 10	"	3.85	1.45	5.30
Board and batten				
Cedar				
1x6	S.F.	4.20	2.15	6.35
1x8	"	3.80	1.75	5.55
1x10	"	3.45	1.55	5.00
1x12	"	3.10	1.40	4.50
Pine				
1x6	S.F.	1.25	2.15	3.40
1x8	"	1.25	1.75	3.00
1x10	"	1.20	1.55	2.75
1x12	"	1.10	1.40	2.50
Redwood				
1x6	S.F.	5.45	2.15	7.60
1x8	"	5.10	1.75	6.85
1x10	"	4.70	1.55	6.25
1x12	"	4.35	1.40	5.75
Tongue and groove				
Cedar				
1x4	S.F.	4.70	2.40	7.10

Roofing And Siding

07460.80 Wood Siding	UNIT	MAT.	INST.	TOTAL
1x6	S.F.	4.55	2.30	6.85
1x8	"	4.25	2.15	6.40
1x10	"	4.20	2.05	6.25
Pine				
1x4	S.F.	1.40	2.40	3.80
1x6	"	1.35	2.30	3.65
1x8	"	1.25	2.15	3.40
1x10	"	1.20	2.05	3.25
Redwood				
1x4	S.F.	5.00	2.40	7.40
1x6	"	4.80	2.30	7.10
1x8	"	4.65	2.15	6.80
1x10	"	4.45	2.05	6.50

Membrane Roofing

07510.10 Built-up Asphalt Roofing	UNIT	MAT.	INST.	TOTAL
Built-up roofing, asphalt felt, including gravel				
2 ply	SQ.	40.00	100.00	140.00
3 ply	"	54.50	140.00	194.50
4 ply	"	78.00	170.00	248.00
Walkway, for built-up roofs				
3' x 3' x				
1/2" thick	S.F.	3.80	1.40	5.20
3/4" thick	"	5.00	1.40	6.40
1" thick	"	5.55	1.40	6.95
Roof bonds				
Asphalt felt				
10 yrs	SQ.			29.22
20 yrs	"			33.03
Cant strip, 4" x 4"				
Treated wood	L.F.	1.90	1.20	3.10
Foamglass	"	0.88	1.05	1.93
Mineral fiber	"	0.38	1.05	1.43
New gravel for built-up roofing, 400 lb/sq	SQ.	27.40	83.00	110.40
Roof gravel (ballast)	C.Y.	18.40	210.00	228.40
Aluminum coating, top surfacing, for built-up roofing	SQ.	29.00	69.50	98.50
Remove 4-ply built-up roof (includes gravel)	"		210.00	210.00
Remove & replace gravel, includes flood coat	"	39.40	140.00	179.40

07530.10 Single-ply Roofing	UNIT	MAT.	INST.	TOTAL
Elastic sheet roofing				
Neoprene, 1/16" thick	S.F.	2.25	0.52	2.77
EPDM rubber				
45 mil	S.F.	1.20	0.52	1.72
60 mil	"	1.65	0.52	2.17

Membrane Roofing

07530.10 Single-ply Roofing

Description	UNIT	MAT.	INST.	TOTAL
PVC				
45 mil	S.F.	1.70	0.52	2.22
60 mil	"	2.00	0.52	2.52
Flashing				
Pipe flashing, 90 mil thick				
1" pipe	EA.	24.75	10.40	35.15
2" pipe	"	26.60	10.40	37.00
3" pipe	"	26.80	10.95	37.75
4" pipe	"	29.00	10.95	39.95
5" pipe	"	31.00	11.55	42.55
6" pipe	"	33.80	11.55	45.35
8" pipe	"	38.50	12.20	50.70
10" pipe	"	44.00	13.85	57.85
12" pipe	"	54.00	13.85	67.85
Neoprene flashing, 60 mil thick strip				
6" wide	L.F.	1.45	3.45	4.90
12" wide	"	2.85	5.20	8.05
18" wide	"	4.15	6.95	11.10
24" wide	"	5.45	10.40	15.85
Adhesives				
Mastic sealer, applied at joints only				
1/4" bead	L.F.	0.09	0.21	0.30
Fluid applied roofing				
Urethane, 2 components, elastomeric top membrane				
1" thick	S.F.	2.45	0.69	3.14
Vinyl liquid roofing, 2 coats, 2 mils per coat	"	3.85	0.59	4.44
Silicone roofing, 2 coats sprayed, 16 mil per coat	"	2.90	0.69	3.59
Inverted roof system				
Insulated membrane with coarse gravel ballast				
3 ply with 2" polystyrene	S.F.	5.25	0.69	5.94
Ballast, 3/4" through 1-1/2" dia. river gravel, 100lb/sf	"	0.35	41.60	41.95
Walkway for membrane roofs, 1/2" thick	"	1.95	1.40	3.35

Flashing And Sheet Metal

07610.10 Metal Roofing

Description	UNIT	MAT.	INST.	TOTAL
Sheet metal roofing, copper, 16 oz, batten seam	SQ.	680.00	280.00	960.00
Standing seam	"	660.00	260.00	920.00
Aluminum roofing, natural finish				
Corrugated, on steel frame				
.0175" thick	SQ.	110.00	120.00	230.00
.0215" thick	"	140.00	120.00	260.00
.024" thick	"	170.00	120.00	290.00
.032" thick	"	210.00	120.00	330.00
V-beam, on steel frame				
.032" thick	SQ.	200.00	120.00	320.00

Flashing And Sheet Metal

07610.10 Metal Roofing	UNIT	MAT.	INST.	TOTAL
.040" thick	SQ.	220.00	120.00	340.00
.050" thick	"	270.00	120.00	390.00
Ridge cap				
.019" thick	L.F.	3.25	1.40	4.65
Corrugated galvanized steel roofing, on steel frame				
28 ga.	SQ.	110.00	120.00	230.00
26 ga.	"	120.00	120.00	240.00
24 ga.	"	140.00	120.00	260.00
22 ga.	"	160.00	120.00	280.00
26 ga., factory insulated with 1" polystyrene	"	310.00	170.00	480.00
Ridge roll				
10" wide	L.F.	1.70	1.40	3.10
20" wide	"	3.45	1.65	5.10

07620.10 Flashing And Trim	UNIT	MAT.	INST.	TOTAL
Counter flashing				
Aluminum, .032"	S.F.	1.20	4.15	5.35
Stainless steel, .015"	"	3.85	4.15	8.00
Copper				
16 oz.	S.F.	4.65	4.15	8.80
20 oz.	"	5.85	4.15	10.00
24 oz.	"	7.00	4.15	11.15
32 oz.	"	8.65	4.15	12.80
Valley flashing				
Aluminum, .032"	S.F.	1.30	2.60	3.90
Stainless steel, .015	"	4.10	2.60	6.70
Copper				
16 oz.	S.F.	4.65	2.60	7.25
20 oz.	"	5.85	3.45	9.30
24 oz.	"	7.00	2.60	9.60
32 oz.	"	8.65	2.60	11.25
Base flashing				
Aluminum, .040"	S.F.	2.20	3.45	5.65
Stainless steel, .018"	"	4.90	3.45	8.35
Copper				
16 oz.	S.F.	4.65	3.45	8.10
20 oz.	"	5.85	2.60	8.45
24 oz.	"	7.00	3.45	10.45
32 oz.	"	8.65	3.45	12.10
Waterstop, "T" section, 22 ga.				
1-1/2" x 3"	L.F.	2.25	2.10	4.35
2" x 2"	"	2.45	2.10	4.55
4" x 3"	"	2.75	2.10	4.85
6" x 4"	"	2.90	2.10	5.00
8" x 4"	"	3.60	2.10	5.70
Scupper outlets				
10" x 10" x 4"	EA.	23.35	10.40	33.75
22" x 4" x 4"	"	28.80	10.40	39.20
8" x 8" x 5"	"	23.35	10.40	33.75
Flashing and trim, aluminum				
.019" thick	S.F.	1.10	2.95	4.05
.032" thick	"	1.30	2.95	4.25

Flashing And Sheet Metal

07620.10 Flashing And Trim

Flashing And Sheet Metal	UNIT	MAT.	INST.	TOTAL
.040" thick	S.F.	2.25	3.20	5.45
Neoprene sheet flashing, .060" thick	"	1.75	2.60	4.35
Copper, paper backed				
2 oz.	S.F.	1.50	4.15	5.65
5 oz.	"	1.60	4.15	5.75
Drainage boots, roof, cast iron				
2 x 3	L.F.	43.30	5.20	48.50
3 x 4	"	54.00	5.20	59.20
4 x 5	"	77.50	5.55	83.05
4 x 6	"	74.00	5.55	79.55
5 x 7	"	88.50	5.95	94.45
Pitch pocket, copper, 16 oz.				
4 x 4	EA.	79.00	10.40	89.40
6 x 6	"	86.00	10.40	96.40
8 x 8	"	93.00	10.40	103.40
8 x 10	"	100.00	10.40	110.40
8 x 12	"	120.00	10.40	130.40
Reglets, copper 10 oz.	L.F.	3.20	2.75	5.95
Stainless steel, .020"	"	2.40	2.75	5.15
Gravel stop				
Aluminum, .032"				
4"	L.F.	4.15	1.40	5.55
6"	"	4.90	1.40	6.30
8"	"	7.45	1.60	9.05
10"	"	8.70	1.60	10.30
Copper, 16 oz.				
4"	L.F.	7.25	1.40	8.65
6"	"	9.30	1.40	10.70
8"	"	11.70	1.60	13.30
10"	"	13.05	1.60	14.65

07620.20 Gutters And Downspouts

Flashing And Sheet Metal	UNIT	MAT.	INST.	TOTAL
Copper gutter and downspout				
Downspouts, 16 oz. copper				
Round				
3" dia.	L.F.	7.00	2.75	9.75
4" dia.	"	8.75	2.75	11.50
Rectangular, corrugated				
2" x 3"	L.F.	6.80	2.60	9.40
3" x 4"	"	8.40	2.60	11.00
Rectangular, flat surface				
2" x 3"	L.F.	7.75	2.75	10.50
3" x 4"	"	11.00	2.75	13.75
Lead-coated copper downspouts				
Round				
3" dia.	L.F.	8.75	2.60	11.35
4" dia.	"	10.45	2.95	13.40
Rectangular, corrugated				
2" x 3"	L.F.	6.10	2.75	8.85
3" x 4"	"	7.35	2.75	10.10
Rectangular, plain				
2" x 3"	L.F.	5.80	2.75	8.55

Flashing And Sheet Metal

07620.20 Gutters And Downspouts

	UNIT	MAT.	INST.	TOTAL
3" x 4"	L.F.	6.90	2.75	9.65
Gutters, 16 oz. copper				
Half round				
4" wide	L.F.	6.30	4.15	10.45
5" wide	"	7.70	4.60	12.30
Type K				
4" wide	L.F.	7.00	4.15	11.15
5" wide	"	7.35	4.60	11.95
Lead-coated copper gutters				
Half round				
4" wide	L.F.	7.35	4.15	11.50
6" wide	"	9.95	4.60	14.55
Type K				
4" wide	L.F.	8.40	4.15	12.55
5" wide	"	9.05	4.60	13.65
Aluminum gutter and downspout				
Downspouts				
2" x 3"	L.F.	0.87	2.75	3.62
3" x 4"	"	1.15	2.95	4.10
4" x 5"	"	1.30	3.20	4.50
Round				
3" dia.	L.F.	1.40	2.75	4.15
4" dia.	"	1.80	2.95	4.75
Gutters, stock units				
4" wide	L.F.	1.45	4.35	5.80
5" wide	"	1.55	4.60	6.15
Galvanized steel gutter and downspout				
Downspouts, round corrugated				
3" dia.	L.F.	1.30	2.75	4.05
4" dia.	"	1.45	2.75	4.20
5" dia.	"	1.75	2.95	4.70
6" dia.	"	2.25	2.95	5.20
Rectangular				
2" x 3"	L.F.	1.30	2.75	4.05
3" x 4"	"	2.05	2.60	4.65
4" x 4"	"	2.55	2.60	5.15
Gutters, stock units				
5" wide				
Plain	L.F.	1.45	4.60	6.05
Painted	"	1.55	4.60	6.15
6" wide				
Plain	L.F.	1.90	4.90	6.80
Painted	"	2.25	4.90	7.15

Roofing Specialties

07700.10 Manufactured Specialties

	UNIT	MAT.	INST.	TOTAL
Moisture relief vent				
Aluminum	EA.	17.00	5.95	22.95
Copper	"	37.30	5.95	43.25
Expansion joint				
Aluminum				
Opening to 2.5"	L.F.	9.00	2.95	11.95
Opening to 3.5"	"	10.25	3.20	13.45
Copper, 16 oz.				
Opening to 2.5"	L.F.	18.20	2.95	21.15
Opening to 3.5"	"	22.85	3.20	26.05
Butyl or neoprene				
4" wide				
16 oz. copper bellows	L.F.	17.30	3.45	20.75
28 ga. stainless steel bellows	"	15.80	3.45	19.25
6" wide				
Copper bellows	L.F.	17.40	3.80	21.20
Stainless steel				
Opening to 2.5"	L.F.	16.40	2.95	19.35
Opening to 3.5"	"	20.00	3.20	23.20
Smoke vent, 48" x 48"				
Aluminum	EA.	800.00	100.00	900.00
Galvanized steel	"	800.00	100.00	900.00
Heat/smoke vent, 48" x 96"				
Aluminum	EA.	2,450	140.00	2,590
Galvanized steel	"	2,450	140.00	2,590
Ridge vent strips				
Mill finish	L.F.	3.45	2.75	6.20
Connectors	EA.	16.20	10.40	26.60
End cap	"	33.30	11.85	45.15
Soffit vents				
Mill finish				
2-1/2" wide	L.F.	0.43	1.65	2.08
3" wide	"	0.50	1.65	2.15
6" wide	"	0.89	1.65	2.54
Roof hatches				
Steel, plain, primed				
2'6" x 3'0"	EA.	560.00	100.00	660.00
2'6" x 4'6"	"	830.00	140.00	970.00
2'6" x 8'0"	"	1,280	210.00	1,490
Galvanized steel				
2'6" x 3'0"	EA.	580.00	100.00	680.00
2'6" x 4'6"	"	870.00	140.00	1,010
2'6" x 8'0"	"	1,360	210.00	1,570
Aluminum				
2'6" x 3'0"	EA.	610.00	100.00	710.00
2'6" x 4'6"	"	920.00	140.00	1,060
2'6" x 8'0"	"	1,700	210.00	1,910
Ceiling access doors				
Swing up model, metal frame				
Steel door				
2'6" x 2'6"	EA.	420.00	41.60	461.60
2'6" x 3'0"	"	450.00	41.60	491.60

Roofing Specialties

07700.10 Manufactured Specialties

Description	UNIT	MAT.	INST.	TOTAL
Aluminum door				
2'6" x 2'6"	EA.	450.00	41.60	491.60
2'6" x 3'0"	"	490.00	41.60	531.60
Swing down model, metal frame				
Steel door				
2'6" x 2'6"	EA.	400.00	41.60	441.60
2'6" x 3'0"	"	440.00	41.60	481.60
Aluminum door				
2'6" x 2'6"	EA.	420.00	41.60	461.60
2'6" x 3'0"	"	450.00	41.60	491.60
Gravity ventilators, with curb, base, damper and screen				
Stationary siphon				
6" dia.	EA.	32.80	27.70	60.50
12" dia.	"	50.00	27.70	77.70
24" dia.	"	210.00	41.60	251.60
36" dia.	"	380.00	41.60	421.60
Wind driven spinner				
6" dia.	EA.	50.00	27.70	77.70
12" dia.	"	67.00	27.70	94.70
24" dia.	"	250.00	41.60	291.60
36" dia.	"	510.00	41.60	551.60
Stationary mushroom				
16" dia.	EA.	310.00	41.60	351.60
30" dia.	"	700.00	52.00	752.00
36" dia.	"	900.00	69.50	969.50
42" dia.	"	1,340	83.00	1,423

Skylights

07810.10 Plastic Skylights

Description	UNIT	MAT.	INST.	TOTAL
Single thickness, not including mounting curb				
2' x 4'	EA.	310.00	52.00	362.00
4' x 4'	"	410.00	69.50	479.50
5' x 5'	"	560.00	100.00	660.00
6' x 8'	"	1,180	140.00	1,320
Double thickness, not including mounting curb				
2' x 4'	EA.	400.00	52.00	452.00
4' x 4'	"	510.00	69.50	579.50
5' x 5'	"	750.00	100.00	850.00
6' x 8'	"	1,310	140.00	1,450
Metal framed skylights				
Translucent panels, 2-1/2" thick	S.F.	36.60	4.15	40.75
Continuous vaults, 8' wide				
Single glazed	S.F.	49.30	5.20	54.50
Double glazed	"	79.50	5.95	85.45

Joint Sealers

07920.10 Caulking

Joint Sealers	UNIT	MAT.	INST.	TOTAL
Caulk exterior, two component				
1/4 x 1/2	L.F.	0.34	2.15	2.49
3/8 x 1/2	"	0.52	2.40	2.92
1/2 x 1/2	"	0.71	2.70	3.41
Caulk interior, single component				
1/4 x 1/2	L.F.	0.23	2.05	2.28
3/8 x 1/2	"	0.33	2.30	2.63
1/2 x 1/2	"	0.42	2.55	2.97
Butyl rubber fillers				
1/4" x 1/4"	L.F.	0.65	0.87	1.52
1/2" x 1/2"	"	0.95	1.45	2.40
1/2" x 3/4"	"	1.45	1.75	3.20
3/4" x 3/4"	"	1.90	1.75	3.65
1" x 1"	"	2.25	1.90	4.15
Seals, "O" ring type cord				
1/4" dia.	L.F.	0.72	1.10	1.82
1/2" dia.	"	2.15	1.15	3.30
1" dia.	"	7.55	1.20	8.75
1-1/4" dia.	"	9.55	1.25	10.80
1-1/2" dia.	"	12.75	1.35	14.10
1-3/4" dia.	"	18.75	1.40	20.15
2" dia.	"	24.30	1.45	25.75
Polyvinyl chloride, closed cell				
1/4" x 2"	L.F.	0.50	1.55	2.04
1/4" x 6"	"	1.55	1.95	3.50
Silicon foam penetration seal				
1/4" x 1/2"	L.F.	0.18	0.54	0.72
1/2" x 1/2"	"	0.31	0.72	1.03
1/2" x 3/4"	"	0.48	0.87	1.35
3/4" x 3/4"	"	0.72	1.10	1.82
1/8" x 1"	"	0.18	0.54	0.72
1/8" x 3"	"	0.48	0.87	1.35
1/4" x 3"	"	0.95	1.10	2.05
1/4" x 6"	"	1.95	1.45	3.40
1/2" x 6"	"	3.85	3.35	7.20
1/2" x 9"	"	5.75	5.40	11.15
1/2" x 12"	"	7.60	7.85	15.45
Oil base sealants and caulking				
1/4" x 1/4"	L.F.	0.04	1.10	1.14
1/4" x 3/8"	"	0.09	1.10	1.18
1/4" x 1/2"	"	0.10	1.15	1.25
3/8" x 3/8"	"	0.11	1.25	1.36
3/8" x 1/2"	"	0.14	1.30	1.43
3/8" x 5/8"	"	0.24	1.40	1.64
3/8" x 3/4"	"	0.27	1.50	1.77
1/2" x 1/2"	"	0.24	1.65	1.89
1/2" x 5/8"	"	0.30	1.90	2.20
1/2" x 3/4"	"	0.34	2.15	2.49
1/2" x 7/8"	"	0.40	2.20	2.60
1/2" x 1"	"	0.44	2.30	2.74
3/4" x 3/4"	"	0.51	2.35	2.86
1" x 1"	"	0.89	2.40	3.29

Joint Sealers

07920.10 Caulking

Joint Sealers	UNIT	MAT.	INST.	TOTAL
Polyurethane compounds				
1/4" x 1/4"	L.F.	0.17	1.10	1.27
1/4" x 3/8"	"	0.31	1.10	1.41
1/4" x 1/2"	"	0.39	1.15	1.54
3/8" x 3/8"	"	0.43	1.25	1.68
3/8" x 1/2"	"	0.55	1.30	1.85
3/8" x 5/8"	"	0.69	1.40	2.09
3/8" x 3/4"	"	0.83	1.50	2.33
1/2" x 1/2"	"	0.80	1.65	2.45
1/2" x 5/8"	"	0.87	1.90	2.77
1/2" x 3/4"	"	0.99	2.15	3.14
1/2" x 7/8"	"	1.25	2.20	3.45
1/2" x 1"	"	1.60	2.40	4.00
3/4" x 3/4"	"	1.65	2.35	4.00
3/4" x 1"	"	1.75	2.40	4.15
Backer rod, polyethylene				
1/4"	L.F.	0.04	1.10	1.14
1/2"	"	0.09	1.15	1.23
3/4"	"	0.10	1.20	1.30
1"	"	0.13	1.25	1.38

08 DOORS AND WINDOWS

Metal

	UNIT	MAT.	INST.	TOTAL
08110.10 Metal Doors				
Flush hollow metal, standard duty, 20 ga., 1-3/8" thick				
2-6 x 6-8	EA.	190.00	48.10	238.10
2-8 x 6-8	"	230.00	48.10	278.10
3-0 x 6-8	"	240.00	48.10	288.10
1-3/4" thick				
2-6 x 6-8	EA.	230.00	48.10	278.10
2-8 x 6-8	"	240.00	48.10	288.10
3-0 x 6-8	"	260.00	48.10	308.10
2-6 x 7-0	"	250.00	48.10	298.10
2-8 x 7-0	"	260.00	48.10	308.10
3-0 x 7-0	"	270.00	48.10	318.10
Heavy duty, 20 ga., unrated, 1-3/4"				
2-8 x 6-8	EA.	240.00	48.10	288.10
3-0 x 6-8	"	260.00	48.10	308.10
2-8 x 7-0	"	260.00	48.10	308.10
3-0 x 7-0	"	280.00	48.10	328.10
3-4 x 7-0	"	340.00	48.10	388.10
18 ga., 1-3/4", unrated door				
2-0 x 7-0	EA.	270.00	48.10	318.10
2-4 x 7-0	"	280.00	48.10	328.10
2-6 x 7-0	"	290.00	48.10	338.10
2-8 x 7-0	"	300.00	48.10	348.10
3-0 x 7-0	"	310.00	48.10	358.10
3-4 x 7-0	"	360.00	48.10	408.10
2", unrated door				
2-0 x 7-0	EA.	280.00	54.00	334.00
2-4 x 7-0	"	300.00	54.00	354.00
2-6 x 7-0	"	300.00	54.00	354.00
2-8 x 7-0	"	320.00	54.00	374.00
3-0 x 7-0	"	330.00	54.00	384.00
3-4 x 7-0	"	400.00	54.00	454.00
Galvanized metal door				
3-0 x 7-0	EA.	410.00	54.00	464.00
For lead lining in doors	"			842.28
For sound attenuation	"			39.12
Vision glass				
8" x 8"	EA.	89.00	54.00	143.00
8" x 48"	"	130.00	54.00	184.00
Fixed metal louver	"	140.00	43.30	183.30
For fire rating, add				
3 hr door	EA.			155.12
1-1/2 hr door	"			65.61
3/4 hr door	"			32.80
1' extra height, add to material, 20%				
1'6" extra height, add to material, 60%				
For dutch doors with shelf, add to material, 100%				
08110.40 Metal Door Frames				
Hollow metal, stock, 18 ga., 4-3/4" x 1-3/4"				
2-0 x 7-0	EA.	120.00	54.00	174.00
2-4 x 7-0	"	120.00	54.00	174.00
2-6 x 7-0	"	130.00	54.00	184.00

Metal	UNIT	MAT.	INST.	TOTAL
08110.40 Metal Door Frames				
2-8 x 7-0	EA.	130.00	54.00	184.00
3-0 x 7-0	"	130.00	54.00	184.00
4-0 x 7-0	"	140.00	72.00	212.00
5-0 x 7-0	"	140.00	72.00	212.00
6-0 x 7-0	"	150.00	72.00	222.00
16 ga., 6-3/4" x 1-3/4"				
2-0 x 7-0	EA.	120.00	60.00	180.00
2-4 x 7-0	"	130.00	60.00	190.00
2-6 x 7-0	"	130.00	60.00	190.00
2-8 x 7-0	"	130.00	60.00	190.00
3-0 x 7-0	"	140.00	60.00	200.00
4-0 x 7-0	"	140.00	80.00	220.00
6-0 x 7-0	"	160.00	80.00	240.00
Transom frame				
3-4 x 1-6	EA.	79.00	60.00	139.00
3-8 x 1-6	"	79.00	60.00	139.00
6-4 x 1-6	"	110.00	60.00	170.00
Transom sash				
3-0 x 1-4	EA.	530.00	60.00	590.00
3-4 x 1-4	"	530.00	60.00	590.00
6-0 x 1-4	"	600.00	60.00	660.00
1' extension of frame, add	"			16.68
14 ga. frame, add	"			16.68
For fire rating, add				
3 hour	EA.			45.51
1-1/2 hour	"			36.40
3/4 hour	"			31.85
Lead lining in frame, add	"			98.59
Sidelights, complete				
1-0 x 7-2	EA.	310.00	60.00	370.00
1-4 x 7-2	"	350.00	60.00	410.00
1-0 x 8-8	"	360.00	60.00	420.00
1-6 x 8-8	"	380.00	60.00	440.00
16 ga., 4-3/4" x 1-3/4"				
2-0 x 7-0	EA.	97.00	60.00	157.00
2-4 x 7-0	"	98.50	60.00	158.50
2-6 x 7-0	"	100.00	60.00	160.00
2-8 x 7-0	"	100.00	60.00	160.00
3-0 x 7-0	"	100.00	60.00	160.00
4-0 x 7-0	"	110.00	80.00	190.00
6-0 x 7-0	"	110.00	80.00	190.00
Transom frame				
3-4 x 1-6	EA.	79.00	60.00	139.00
3-8 x 1-6	"	79.00	60.00	139.00
6-4 x 1-6	"	110.00	60.00	170.00
Transom sash				
3-0 x 1-4	EA.	510.00	60.00	570.00
3-4 x 1-4	"	510.00	60.00	570.00
6-0 x 1-4	"	590.00	60.00	650.00
1' extension of door frame, add	"			16.68
14 ga., metal frame, add	"			16.68
For fire rating, add				

Metal

08110.40 Metal Door Frames

	UNIT	MAT.	INST.	TOTAL
3 hour	EA.			45.49
1-1/2 hour	"			36.40
3/4 hour	"			31.85
Lead lining in frame, add	"			95.54
Sidelights, complete				
1-0 x 7-2	EA.	300.00	60.00	360.00
1-4 x 7-2	"	320.00	60.00	380.00
1-0 x 8-8	"	350.00	60.00	410.00
1-4 x 8-8	"	360.00	60.00	420.00
16 ga., 5-3/4" x 1-3/4"				
2-0 x 7-0	EA.	98.50	54.00	152.50
2-4 x 7-0	"	100.00	54.00	154.00
2-6 x 7-0	"	100.00	54.00	154.00
2-8 x 7-0	"	110.00	54.00	164.00
3-0 x 7-0	"	110.00	54.00	164.00
4-0 x 7-0	"	120.00	72.00	192.00
5-0 x 7-0	"	120.00	72.00	192.00
6-0 x 7-0	"	130.00	72.00	202.00
Mullions, vertical				
5-1/4" x 1-3/4"	L.F.	11.35	5.40	16.75
5-1/4" x 2"	"	14.10	5.40	19.50
Horizontal				
5-1/4" x 1-3/4"	L.F.	11.35	5.40	16.75
5-1/4" x 2"	"	14.10	5.40	19.50

08120.10 Aluminum Doors

	UNIT	MAT.	INST.	TOTAL
Aluminum doors, commercial				
Narrow stile				
2-6 x 7-0	EA.	560.00	240.00	800.00
3-0 x 7-0	"	590.00	240.00	830.00
3-6 x 7-0	"	600.00	240.00	840.00
Pair				
5-0 x 7-0	EA.	930.00	480.00	1,410
6-0 x 7-0	"	940.00	480.00	1,420
7-0 x 7-0	"	990.00	480.00	1,470
Wide stile				
2-6 x 7-0	EA.	860.00	240.00	1,100
3-0 x 7-0	"	890.00	240.00	1,130
3-6 x 7-0	"	920.00	240.00	1,160
Pair				
5-0 x 7-0	EA.	1,580	480.00	2,060
6-0 x 7-0	"	1,650	480.00	2,130
7-0 x 7-0	"	1,680	480.00	2,160

08 DOORS AND WINDOWS

Wood And Plastic	UNIT	MAT.	INST.	TOTAL
08210.10 Wood Doors				
Solid core, 1-3/8" thick				
Birch faced				
2-4 x 7-0	EA.	150.00	54.00	204.00
2-8 x 7-0	"	150.00	54.00	204.00
3-0 x 7-0	"	160.00	54.00	214.00
3-4 x 7-0	"	170.00	54.00	224.00
2-4 x 6-8	"	140.00	54.00	194.00
2-6 x 6-8	"	150.00	54.00	204.00
2-8 x 6-8	"	160.00	54.00	214.00
3-0 x 6-8	"	160.00	54.00	214.00
Lauan faced				
2-4 x 6-8	EA.	130.00	54.00	184.00
2-8 x 6-8	"	130.00	54.00	184.00
3-0 x 6-8	"	140.00	54.00	194.00
3-4 x 6-8	"	140.00	54.00	194.00
Tempered hardboard faced				
2-4 x 7-0	EA.	130.00	54.00	184.00
2-8 x 7-0	"	140.00	54.00	194.00
3-0 x 7-0	"	150.00	54.00	204.00
3-4 x 7-0	"	160.00	54.00	214.00
Hollow core, 1-3/8" thick				
Birch faced				
2-4 x 7-0	EA.	76.50	54.00	130.50
2-8 x 7-0	"	81.00	54.00	135.00
3-0 x 7-0	"	84.00	54.00	138.00
3-4 x 7-0	"	92.00	54.00	146.00
Lauan faced				
2-4 x 6-8	EA.	48.30	54.00	102.30
2-6 x 6-8	"	51.50	54.00	105.50
2-8 x 6-8	"	54.50	54.00	108.50
3-0 x 6-8	"	59.00	54.00	113.00
3-4 x 6-8	"	62.50	54.00	116.50
Tempered hardboard faced				
2-4 x 7-0	EA.	51.50	54.00	105.50
2-6 x 7-0	"	54.50	54.00	108.50
2-8 x 7-0	"	59.00	54.00	113.00
3-0 x 7-0	"	62.50	54.00	116.50
3-4 x 7-0	"	65.50	54.00	119.50
Solid core, 1-3/4" thick				
Birch faced				
2-4 x 7-0	EA.	170.00	54.00	224.00
2-6 x 7-0	"	180.00	54.00	234.00
2-8 x 7-0	"	190.00	54.00	244.00
3-0 x 7-0	"	200.00	54.00	254.00
3-4 x 7-0	"	200.00	54.00	254.00
Lauan faced				
2-4 x 7-0	EA.	150.00	54.00	204.00
2-6 x 7-0	"	170.00	54.00	224.00
2-8 x 7-0	"	180.00	54.00	234.00
3-4 x 7-0	"	190.00	54.00	244.00
3-0 x 7-0	"	200.00	54.00	254.00
Tempered hardboard faced				

Wood And Plastic	UNIT	MAT.	INST.	TOTAL
08210.10 Wood Doors				
2-4 x 7-0	EA.	150.00	54.00	204.00
2-6 x 7-0	"	170.00	54.00	224.00
2-8 x 7-0	"	180.00	54.00	234.00
3-0 x 7-0	"	190.00	54.00	244.00
3-4 x 7-0	"	200.00	54.00	254.00
Hollow core, 1-3/4" thick				
Birch faced				
2-4 x 7-0	EA.	78.00	54.00	132.00
2-6 x 7-0	"	81.00	54.00	135.00
2-8 x 7-0	"	85.50	54.00	139.50
3-0 x 7-0	"	88.50	54.00	142.50
3-4 x 7-0	"	93.50	54.00	147.50
Lauan faced				
2-4 x 6-8	EA.	59.00	54.00	113.00
2-6 x 6-8	"	62.50	54.00	116.50
2-8 x 6-8	"	65.50	54.00	119.50
3-0 x 6-8	"	68.50	54.00	122.50
3-4 x 6-8	"	73.00	54.00	127.00
Tempered hardboard				
2-4 x 7-0	EA.	59.00	54.00	113.00
2-6 x 7-0	"	62.50	54.00	116.50
2-8 x 7-0	"	65.50	54.00	119.50
3-0 x 7-0	"	68.50	54.00	122.50
3-4 x 7-0	"	73.00	54.00	127.00
Add-on, louver	"	20.75	43.30	64.05
Glass	"	72.50	43.30	115.80
Exterior doors, 3-0 x 7-0 x 2-1/2", solid core				
Carved				
One face	EA.	960.00	110.00	1,070
Two faces	"	1,410	110.00	1,520
Closet doors, 1-3/4" thick				
Bi-fold or bi-passing, includes frame and trim				
Paneled				
4-0 x 6-8	EA.	320.00	72.00	392.00
6-0 x 6-8	"	380.00	72.00	452.00
Louvered				
4-0 x 6-8	EA.	220.00	72.00	292.00
6-0 x 6-8	"	280.00	72.00	352.00
Flush				
4-0 x 6-8	EA.	190.00	72.00	262.00
6-0 x 6-8	"	230.00	72.00	302.00
Primed				
4-0 x 6-8	EA.	200.00	72.00	272.00
6-0 x 6-8	"	240.00	72.00	312.00
08210.90 Wood Frames				
Frame, interior, pine				
2-6 x 6-8	EA.	38.50	62.00	100.50
2-8 x 6-8	"	40.00	62.00	102.00
3-0 x 6-8	"	41.50	62.00	103.50
5-0 x 6-8	"	43.00	62.00	105.00
6-0 x 6-8	"	46.00	62.00	108.00

08 DOORS AND WINDOWS

Wood And Plastic	UNIT	MAT.	INST.	TOTAL
08210.90 Wood Frames				
2-6 x 7-0	EA.	41.50	62.00	103.50
2-8 x 7-0	"	46.00	62.00	108.00
3-0 x 7-0	"	47.40	62.00	109.40
5-0 x 7-0	"	52.00	86.50	138.50
6-0 x 7-0	"	53.50	86.50	140.00
Exterior, custom, with threshold, including trim				
Walnut				
3-0 x 7-0	EA.	250.00	110.00	360.00
6-0 x 7-0	"	290.00	110.00	400.00
Oak				
3-0 x 7-0	EA.	220.00	110.00	330.00
6-0 x 7-0	"	280.00	110.00	390.00
Pine				
2-4 x 7-0	EA.	90.50	86.50	177.00
2-6 x 7-0	"	93.50	86.50	180.00
2-8 x 7-0	"	95.00	86.50	181.50
3-0 x 7-0	"	100.00	86.50	186.50
3-4 x 7-0	"	110.00	86.50	196.50
6-0 x 7-0	"	200.00	140.00	340.00
08300.10 Special Doors				
Vault door and frame, class 5, steel	EA.	4,680	430.00	5,110
Overhead door, coiling insulated				
Chain gear, no frame, 12' x 12'	EA.	2,500	540.00	3,040
Aluminum, bronze glass panels, 12-9 x 13-0	"	2,810	430.00	3,240
Garage door, flush insulated metal, primed, 9-0 x 7-0	"	880.00	140.00	1,020
Sliding metal fire doors, motorized, fusible link, 3 hr.				
3-0 x 6-8	EA.	3,200	870.00	4,070
3-8 x 6-8	"	3,200	870.00	4,070
4-0 x 8-0	"	3,440	870.00	4,310
5-0 x 8-0	"	3,520	870.00	4,390
Metal clad doors, including electric motor				
Light duty				
Minimum	S.F.	34.30	7.20	41.50
Maximum	"	56.00	17.30	73.30
Heavy duty				
Minimum	S.F.	67.00	21.65	88.65
Maximum	"	110.00	27.10	137.10
Hangar doors, based on 150' openings				
To 20' high	S.F.	39.00	6.55	45.55
20' to 40' high	"	35.90	4.10	40.00
40' to 60' high	"	39.00	2.75	41.75
60' to 80' high	"	45.20	1.65	46.85
Over 80' high	"	61.00	1.30	62.30
Counter doors, (roll-up shutters), standard, manual				
Opening, 4' high				
4' wide	EA.	1,140	360.00	1,500
6' wide	"	1,550	360.00	1,910
8' wide	"	1,800	390.00	2,190
10' wide	"	1,970	540.00	2,510
14' wide	"	2,460	540.00	3,000
6' high				

Wood And Plastic

08300.10 Special Doors	UNIT	MAT.	INST.	TOTAL
4' wide	EA.	1,400	360.00	1,760
6' wide	"	1,800	390.00	2,190
8' wide	"	1,950	430.00	2,380
10' wide	"	2,180	540.00	2,720
14' wide	"	2,340	620.00	2,960
For stainless steel, add to material, 40%				
For motor operator, add	EA.			1,247
Service doors, (roll up shutters), standard, manual				
Opening				
8' high x 8' wide	EA.	1,460	240.00	1,700
10' high x 10' wide	"	1,720	360.00	2,080
12' high x 12' wide	"	2,030	540.00	2,570
14' high x 14' wide	"	2,650	720.00	3,370
16' high x 14' wide	"	3,590	720.00	4,310
20' high x 14' wide	"	5,080	1,080	6,160
24' high x 16' wide	"	6,880	960.00	7,840
For motor operator				
Up to 12-0 x 12-0, add	EA.			1,407
Over 12-0 x 12-0, add	"			1,802
Roll-up doors				
13-0 high x 14-0 wide	EA.	1,250	620.00	1,870
12-0 high x 14-0 wide	"	1,560	620.00	2,180
Top coiling grilles, manually operated, steel or aluminum				
Opening, 4' high x				
4' wide	EA.	1,470	170.00	1,640
6' wide	"	1,510	170.00	1,680
8' wide	"	1,740	240.00	1,980
12' wide	"	2,030	240.00	2,270
16' wide	"	2,340	360.00	2,700
6' high x				
4' wide	EA.	1,560	360.00	1,920
6' wide	"	1,720	390.00	2,110
8' wide	"	1,790	430.00	2,220
12' wide	"	2,180	480.00	2,660
16' wide	"	2,810	620.00	3,430
Side coiling grilles, manually operated, aluminum				
Opening, 8' high x				
18' wide	EA.	10,920	4,100	15,020
24' wide	"	14,040	4,690	18,730
12' high x				
12' wide	EA.	10,920	4,100	15,020
18' wide	"	14,040	4,690	18,730
24' wide	"	18,720	5,470	24,190
Accordion folding doors, tracks and fittings included				
Vinyl covered, 2 layers	S.F.	11.45	17.30	28.75
Woven mahogany and vinyl	"	14.75	17.30	32.05
Economy vinyl	"	9.85	17.30	27.15
Rigid polyvinyl chloride	"	16.40	17.30	33.70
Sectional wood overhead doors, frames not included				
Commercial grade, heavy duty, 1-3/4" thick, manual				
8' x 8'	EA.	640.00	360.00	1,000
10' x 10'	"	1,120	390.00	1,510

Wood And Plastic	UNIT	MAT.	INST.	TOTAL
08300.10 Special Doors				
12' x 12'	EA.	1,440	430.00	1,870
Chain hoist				
12' x 16' high	EA.	2,360	720.00	3,080
14' x 14' high	"	2,530	540.00	3,070
20' x 8' high	"	2,190	870.00	3,060
16' high	"	5,060	1,080	6,140
Sectional metal overhead doors, complete				
Residential grade, manual				
9' x 7'	EA.	510.00	170.00	680.00
16' x 7'	"	970.00	220.00	1,190
Commercial grade				
8' x 8'	EA.	650.00	360.00	1,010
10' x 10'	"	830.00	390.00	1,220
12' x 12'	"	1,470	430.00	1,900
20' x 14', with chain hoist	"	3,120	870.00	3,990
Sliding glass doors				
Tempered plate glass, 1/4" thick				
6' wide				
Economy grade	EA.	930.00	140.00	1,070
Premium grade	"	1,070	140.00	1,210
12' wide				
Economy grade	EA.	1,310	220.00	1,530
Premium grade	"	1,960	220.00	2,180
Insulating glass, 5/8" thick				
6' wide				
Economy grade	EA.	1,140	140.00	1,280
Premium grade	"	1,470	140.00	1,610
12' wide				
Economy grade	EA.	1,430	220.00	1,650
Premium grade	"	2,290	220.00	2,510
1" thick				
6' wide				
Economy grade	EA.	1,440	140.00	1,580
Premium grade	"	1,670	140.00	1,810
12' wide				
Economy grade	EA.	2,240	220.00	2,460
Premium grade	"	3,280	220.00	3,500
Added costs				
Custom quality, add to material, 30%				
Tempered glass, 6' wide, add	S.F.			3.68
Vertical lift doors, channel frame construction				
20' high x				
10' wide	EA.	24,960	730.00	25,690
15' wide	"	31,200	730.00	31,930
20' wide	"	34,320	1,300	35,620
25' wide	"	37,440	1,300	38,740
25' high x				
20' wide	EA.	39,000	1,300	40,300
25' wide	"	43,690	1,520	45,210
30' high x				
25' wide	EA.	46,810	1,520	48,330
30' wide	"	51,490	1,520	53,010

Wood And Plastic

08300.10 Special Doors

	UNIT	MAT.	INST.	TOTAL
35' wide	EA.	60,850	1,520	62,370
Residential storm door				
Minimum	EA.	140.00	72.00	212.00
Average	"	200.00	72.00	272.00
Maximum	"	440.00	110.00	550.00

Storefronts

08410.10 Storefronts

	UNIT	MAT.	INST.	TOTAL
Storefront, aluminum and glass				
Minimum	S.F.	15.55	6.00	21.55
Average	"	24.65	6.85	31.50
Maximum	"	42.80	8.00	50.80
Entrance doors, premium, incl. glass, closers, panic dev.,etc.				
1/2" thick glass				
3' x 7'	EA.	2,680	400.00	3,080
6' x 7'	"	4,620	600.00	5,220
3/4" thick glass				
3' x 7'	EA.	2,830	400.00	3,230
6' x 7'	"	4,760	600.00	5,360
1" thick glass				
3' x 7'	EA.	2,980	400.00	3,380
6' x 7'	"	5,360	600.00	5,960
Revolving doors				
7'diameter, 7' high				
Minimum	EA.	16,870	4,550	21,420
Average	"	22,940	7,270	30,210
Maximum	"	30,370	9,090	39,460

Metal Windows

08510.10 Steel Windows

	UNIT	MAT.	INST.	TOTAL
Steel windows, primed				
Casements				
Operable				
Minimum	S.F.	31.50	2.85	34.35
Maximum	"	47.30	3.20	50.50
Fixed sash	"	25.20	2.40	27.60
Double hung	"	47.30	2.65	49.95

08 DOORS AND WINDOWS

Metal Windows

Metal Windows	UNIT	MAT.	INST.	TOTAL
08510.10 Steel Windows				
Industrial windows				
Horizontally pivoted sash	S.F.	44.10	3.20	47.30
Fixed sash	"	34.70	2.65	37.35
Security sash				
Operable	S.F.	55.00	3.20	58.20
Fixed	"	48.90	2.65	51.55
Picture window	"	23.65	2.65	26.30
Projecting sash				
Minimum	S.F.	41.00	3.00	44.00
Maximum	"	50.50	3.00	53.50
Mullions	L.F.	10.80	2.40	13.20
08520.10 Aluminum Windows				
Jalousie				
3-0 x 4-0	EA.	260.00	60.00	320.00
3-0 x 5-0	"	310.00	60.00	370.00
Fixed window				
6 sf to 8 sf	S.F.	13.15	6.85	20.00
12 sf to 16 sf	"	11.70	5.35	17.05
Projecting window				
6 sf to 8 sf	S.F.	29.20	12.00	41.20
12 sf to 16 sf	"	26.30	8.00	34.30
Horizontal sliding				
6 sf to 8 sf	S.F.	19.00	6.00	25.00
12 sf to 16 sf	"	17.55	4.80	22.35
Double hung				
6 sf to 8 sf	S.F.	26.30	9.60	35.90
10 sf to 12 sf	"	23.40	8.00	31.40
Storm window, 0.5 cfm, up to				
60 u.i. (united inches)	EA.	61.50	24.05	85.55
70 u.i.	"	63.00	24.05	87.05
80 u.i.	"	70.00	24.05	94.05
90 u.i.	"	71.50	26.70	98.20
100 u.i.	"	73.00	26.70	99.70
2.0 cfm, up to				
60 u.i.	EA.	79.00	24.05	103.05
70 u.i.	"	80.50	24.05	104.55
80 u.i.	"	82.00	24.05	106.05
90 u.i.	"	87.50	26.70	114.20
100 u.i.	"	89.00	26.70	115.70

08 DOORS AND WINDOWS

Wood And Plastic

08600.10 Wood Windows

	UNIT	MAT.	INST.	TOTAL
Double hung				
24" x 36"				
Minimum	EA.	170.00	43.30	213.30
Average	"	260.00	54.00	314.00
Maximum	"	350.00	72.00	422.00
24" x 48"				
Minimum	EA.	200.00	43.30	243.30
Average	"	300.00	54.00	354.00
Maximum	"	430.00	72.00	502.00
30" x 48"				
Minimum	EA.	220.00	48.10	268.10
Average	"	310.00	62.00	372.00
Maximum	"	450.00	86.50	536.50
30" x 60"				
Minimum	EA.	230.00	48.10	278.10
Average	"	380.00	62.00	442.00
Maximum	"	480.00	86.50	566.50
Casement				
1 leaf, 22" x 38" high				
Minimum	EA.	260.00	43.30	303.30
Average	"	310.00	54.00	364.00
Maximum	"	370.00	72.00	442.00
2 leaf, 50" x 50" high				
Minimum	EA.	720.00	54.00	774.00
Average	"	810.00	72.00	882.00
Maximum	"	930.00	110.00	1,040
3 leaf, 71" x 62" high				
Minimum	EA.	1,070	54.00	1,124
Average	"	1,170	72.00	1,242
Maximum	"	1,380	110.00	1,490
4 leaf, 95" x 75" high				
Minimum	EA.	1,520	62.00	1,582
Average	"	1,670	86.50	1,757
Maximum	"	2,170	140.00	2,310
5 leaf, 119" x 75" high				
Minimum	EA.	1,920	62.00	1,982
Average	"	2,170	86.50	2,257
Maximum	"	2,750	140.00	2,890
Picture window, fixed glass, 54" x 54" high				
Minimum	EA.	400.00	54.00	454.00
Average	"	450.00	62.00	512.00
Maximum	"	830.00	72.00	902.00
68" x 55" high				
Minimum	EA.	770.00	54.00	824.00
Average	"	870.00	62.00	932.00
Maximum	"	1,120	72.00	1,192
Sliding, 40" x 31" high				
Minimum	EA.	220.00	43.30	263.30
Average	"	350.00	54.00	404.00
Maximum	"	430.00	72.00	502.00
52" x 39" high				
Minimum	EA.	280.00	54.00	334.00

08 DOORS AND WINDOWS

Wood And Plastic

	UNIT	MAT.	INST.	TOTAL
08600.10 Wood Windows				
Average	EA.	440.00	62.00	502.00
Maximum	"	500.00	72.00	572.00
64" x 72" high				
Minimum	EA.	460.00	54.00	514.00
Average	"	760.00	72.00	832.00
Maximum	"	830.00	86.50	916.50
Awning windows				
34" x 21" high				
Minimum	EA.	240.00	43.30	283.30
Average	"	290.00	54.00	344.00
Maximum	"	330.00	72.00	402.00
40" x 21" high				
Minimum	EA.	290.00	48.10	338.10
Average	"	320.00	62.00	382.00
Maximum	"	360.00	86.50	446.50
48" x 27" high				
Minimum	EA.	300.00	48.10	348.10
Average	"	370.00	62.00	432.00
Maximum	"	430.00	86.50	516.50
60" x 36" high				
Minimum	EA.	320.00	54.00	374.00
Average	"	590.00	72.00	662.00
Maximum	"	670.00	86.50	756.50
Window frame, milled				
Minimum	L.F.	3.20	8.65	11.85
Average	"	4.80	10.80	15.60
Maximum	"	7.25	14.45	21.70

Hardware

	UNIT	MAT.	INST.	TOTAL
08710.10 Hinges				
Hinges				
3 x 3 butts, steel, interior, plain bearing	PAIR			15.94
4 x 4 butts, steel, standard	"			23.92
5 x 4-1/2 butts, bronze/s. steel, heavy duty	"			63.74
Pivot hinges				
Top pivot	EA.			57.37
Intermediate pivot	"			62.16
Bottom pivot	"			127.50
BHMA specifications				
3-1/2 x 3-1/2, full mortise butts				
Plain bearing	PAIR			19.11
Ball bearing	"			23.92
Half surface butts	"			35.06
4 x 4				
Full mortise butts, plain bearing, standard duty	PAIR			20.73

08 DOORS AND WINDOWS

Hardware

Hardware	UNIT	MAT.	INST.	TOTAL
08710.10 Hinges				
Full mortise butts, ball bearing	PAIR			25.49
Half surface butts				
Standard duty	PAIR			35.06
Ball bearing	"			35.06
4-1/2 x 4-1/2				
Full mortise butts, plain bearing	PAIR			22.30
Ball bearing, heavy duty	"			55.78
Half mortise and half surface butts				
Plain bearing	PAIR			35.06
Full surface and half surface butts				
Standard duty	PAIR			54.19
Heavy duty	"			148.48
Full mortise and full slide-in butts, ball bearing	"			25.49
Half mortise butts, ball bearing				
Standard duty	PAIR			119.50
Heavy duty	"			140.23
5 x 5, ball bearing				
Full mortise butts	PAIR			55.78
Half mortise, full & half surface butts	"			124.30
Full mortise, full surface and half surface butts	"			167.73
4 x 4				
Full mortise butts, plain bearing, standard duty	PAIR			11.17
5 x 4-1/2				
Full mortise butts, ball bearing, heavy duty	PAIR			63.74
08710.20 Locksets				
Latchset, heavy duty				
Cylindrical	EA.	130.00	27.10	157.10
Mortise	"	130.00	43.30	173.30
Lockset, heavy duty				
Cylindrical	EA.	160.00	27.10	187.10
Mortise	"	190.00	43.30	233.30
Mortise locks and latchsets, chrome				
Latchset passage or closet latch	EA.	130.00	36.10	166.10
Privacy (bath or bedroom)	"	160.00	36.10	196.10
Entry lockset	"	200.00	36.10	236.10
Classroom lockset (outside key operated)	"	200.00	36.10	236.10
Storeroom lock	"	200.00	36.10	236.10
Front door lock	"	200.00	36.10	236.10
Dormitory or exit lock	"	200.00	36.10	236.10
Preassembled locks and latches, brass				
Latchset, passage or closet latch	EA.	210.00	36.10	246.10
Lockset				
Privacy (bath or bathroom)	EA.	260.00	36.10	296.10
Entry lock	"	310.00	36.10	346.10
Classroom lock (outside key, operated)	"	310.00	36.10	346.10
Storeroom lock	"	380.00	36.10	416.10
Bored locks and latches, satin chrome plated				
Latchset passage or closet latch	EA.	120.00	36.10	156.10
Lockset				
Privacy (bath or bedroom)	EA.	150.00	36.10	186.10
Entry lock	"	170.00	36.10	206.10

Hardware

	UNIT	MAT.	INST.	TOTAL
08710.20 Locksets				
Classroom lock	EA.	170.00	36.10	206.10
Corridor lock	"	170.00	36.10	206.10
Miscellaneous locks				
Exit lock with alarm, single door	EA.	550.00	170.00	720.00
Electric strike				
Rim mounted wrought steel	EA.	450.00	110.00	560.00
Mortised, wrought steel with bronze plating	"	150.00	170.00	320.00
Dead bolt				
Bored, wrought brass, keyed both sides	EA.	91.00	72.00	163.00
Mortised, cast brass	"	100.00	72.00	172.00
Lockset, cipher, mechanical	"	1,570	43.30	1,613
08710.30 Closers				
Door closers				
Surface mounted, traditional type, parallel arm				
Standard	EA.	150.00	54.00	204.00
Heavy duty	"	160.00	54.00	214.00
Modern type, parallel arm, standard duty	"	160.00	54.00	214.00
Overhead, concealed, pivot hung, single acting				
Interior	EA.	270.00	54.00	324.00
Exterior	"	520.00	54.00	574.00
Floor concealed, single acting, offset, pivoted				
Interior	EA.	490.00	140.00	630.00
Exterior	"	630.00	140.00	770.00
08710.40 Door Trim				
Door bumper, bronze, wall type	EA.	3.40	8.65	12.05
Wall type, 4" dia. with convex rubber pad, aluminum	"	27.50	8.65	36.15
Floor type				
Aluminum	EA.	4.30	8.65	12.95
Brass	"	4.60	8.65	13.25
Door holders				
Wall type, bronze	EA.	4.30	8.65	12.95
Overhead	"	43.10	21.65	64.75
Floor type	"	43.10	21.65	64.75
Plunger type	"	23.25	21.65	44.90
Wall type, aluminum	"	27.60	21.65	49.25
Surface bolt	"	20.65	8.65	29.30
Panic device				
Rim type with thumb piece	EA.	550.00	110.00	660.00
Mortise	"	700.00	110.00	810.00
Vertical rod	"	810.00	110.00	920.00
Labelled, rim type	"	720.00	110.00	830.00
Mortise	"	870.00	110.00	980.00
Vertical rod	"	980.00	110.00	1,090
Silencers, rubber type	"	0.16	0.87	1.03
Dust proof strike with plate, brass	"	15.50	14.45	29.95
Flush bolt, lever extension, brass, rated	"	27.60	8.65	36.25
Surface bolt with strike, brass, 6" long	"	20.65	8.65	29.30
Door coordinator, labeled, brass, satin chrome	"	100.00	30.90	130.90
Door plates				
Kick plate, aluminum, 3 beveled edges				

Hardware	UNIT	MAT.	INST.	TOTAL
08710.40 Door Trim				
10" x 28"	EA.	15.50	21.65	37.15
10" x 30"	"	17.20	21.65	38.85
10" x 34"	"	18.95	21.65	40.60
10" x 38"	"	21.45	21.65	43.10
Push plate, 4" x 16"				
Aluminum	EA.	15.30	8.65	23.95
Bronze	"	76.50	8.65	85.15
Stainless steel	"	61.50	8.65	70.15
Armor plate, 40" x 34"	"	69.00	17.30	86.30
Pull handle, 4" x 16"				
Aluminum	EA.	61.50	8.65	70.15
Bronze	"	120.00	8.65	128.65
Stainless steel	"	110.00	8.65	118.65
Hasp assembly				
3"	EA.	2.30	7.20	9.50
4-1/2"	"	3.05	9.60	12.65
6"	"	5.40	12.35	17.75
Electro-magnetic door holder				
Wall mounted	EA.	150.00	140.00	290.00
Floor mounted	"	240.00	140.00	380.00
Smoke detector door holder				
Photo electric type	EA.	250.00	140.00	390.00
Ionization type	"	240.00	140.00	380.00
Pneumatic operators, activated by rubber mats				
Swing				
Single	EA.	1,840	360.00	2,200
Double	"	3,060	540.00	3,600
Sliding				
Single	EA.	1,990	360.00	2,350
Double	"	3,520	540.00	4,060
08710.60 Weatherstripping				
Weatherstrip, head and jamb, metal strip, neoprene bulb				
Standard duty	L.F.	3.60	2.40	6.00
Heavy duty	"	4.55	2.70	7.25
Spring type				
Metal doors	EA.	46.90	110.00	156.90
Wood doors	"	46.90	140.00	186.90
Sponge type with adhesive backing	"	43.30	43.30	86.60
Astragal				
1-3/4" x 13 ga., aluminum	L.F.	4.60	3.60	8.20
1-3/8" x 5/8", oak	"	4.40	2.90	7.30
Thresholds				
Bronze	L.F.	45.00	10.80	55.80
Aluminum				
Plain	L.F.	16.90	10.80	27.70
Vinyl insert	"	25.70	10.80	36.50
Aluminum with grit	"	24.15	10.80	34.95
Steel				
Plain	L.F.	16.90	10.80	27.70
Interlocking	"	25.70	36.10	61.80

Glazing

08810.10 Glazing

Glazing	UNIT	MAT.	INST.	TOTAL
Sheet glass, 1/8" thick	S.F.	6.60	2.65	9.25
Plate glass, bronze or grey, 1/4" thick	"	9.80	4.35	14.15
Clear	"	7.55	4.35	11.90
Polished	"	8.35	4.35	12.70
Plexiglass				
1/8" thick	S.F.	2.85	4.35	7.20
1/4" thick	"	5.25	2.65	7.90
Float glass, clear				
3/16" thick	S.F.	4.80	4.00	8.80
1/4" thick	"	5.60	4.35	9.95
5/16" thick	"	10.45	4.80	15.25
3/8" thick	"	11.25	6.00	17.25
1/2" thick	"	18.95	8.00	26.95
5/8" thick	"	21.75	9.60	31.35
3/4" thick	"	23.75	12.00	35.75
1" thick	"	41.80	16.05	57.85
Tinted glass, polished plate, twin ground				
3/16" thick	S.F.	7.55	4.00	11.55
1/4" thick	"	7.55	4.35	11.90
3/8" thick	"	12.15	6.00	18.15
1/2" thick	"	19.65	8.00	27.65
Total, full vision, all glass window system				
To 10' high				
Minimum	S.F.	30.50	12.00	42.50
Average	"	40.10	12.00	52.10
Maximum	"	48.20	12.00	60.20
10' to 20' high				
Minimum	S.F.	36.90	12.00	48.90
Average	"	45.00	12.00	57.00
Maximum	"	56.00	12.00	68.00
Insulated glass, bronze or gray				
1/2" thick	S.F.	14.00	8.00	22.00
1" thick	"	16.70	12.00	28.70
Spandrel glass, polished bronze/grey, 1 side, 1/4" thick	"	11.25	4.35	15.60
Tempered glass (safety)				
Clear sheet glass				
1/8" thick	S.F.	7.70	2.65	10.35
3/16" thick	"	9.35	3.70	13.05
Clear float glass				
1/4" thick	S.F.	8.05	4.00	12.05
5/16" thick	"	14.45	4.80	19.25
3/8" thick	"	17.65	6.00	23.65
1/2" thick	"	24.10	8.00	32.10
5/8" thick	"	27.30	9.60	36.90
3/4" thick	"	33.70	16.05	49.75
Tinted float glass				
3/16" thick	S.F.	9.65	3.70	13.35
1/4" thick	"	10.60	4.00	14.60
3/8" thick	"	19.25	6.00	25.25
1/2" thick	"	25.70	8.00	33.70
Laminated glass				
Float safety glass with polyvinyl plastic interlayer				

Glazing

08810.10 Glazing	UNIT	MAT.	INST.	TOTAL
1/4", sheet or float				
Two lites, 1/8" thick, clear glass	S.F.	11.25	4.00	15.25
1/2" thick, float glass				
Two lites, 1/4" thick, clear glass	S.F.	16.70	8.00	24.70
Tinted glass	"	20.00	8.00	28.00
Insulating glass, two lites, clear float glass				
1/2" thick	S.F.	10.95	8.00	18.95
5/8" thick	"	12.65	9.60	22.25
3/4" thick	"	13.85	12.00	25.85
7/8" thick	"	14.70	13.75	28.45
1" thick	"	15.65	16.05	31.70
Glass seal edge				
3/8" thick	S.F.	9.20	8.00	17.20
Tinted glass				
1/2" thick	S.F.	18.80	8.00	26.80
1" thick	"	20.20	16.05	36.25
Tempered, clear				
1" thick	S.F.	36.80	16.05	52.85
Wire reinforced	"	46.80	16.05	62.85
Plate mirror glass				
1/4" thick				
15 sf	S.F.	9.65	4.80	14.45
Over 15 sf	"	8.85	4.35	13.20
Door type, 1/4" thick	"	10.10	4.80	14.90
Transparent, one way vision, 1/4" thick	"	21.40	4.80	26.20
Sheet mirror glass				
3/16" thick	S.F.	9.35	4.80	14.15
1/4" thick	"	9.80	4.00	13.80
Wall tiles, 12" x 12"				
Clear glass	S.F.	3.15	2.65	5.80
Veined glass	"	3.70	2.65	6.35
Wire glass, 1/4" thick				
Clear	S.F.	13.05	16.05	29.10
Hammered	"	13.20	16.05	29.25
Obscure	"	15.80	16.05	31.85
Bullet resistant glass, plate glass with inter-leaved vinyl				
1-3/16" thick				
To 15 sf	S.F.	35.10	24.05	59.15
Over 15 sf	"	40.10	24.05	64.15
2" thick				
To 15 sf	S.F.	67.00	40.10	107.10
Over 15 sf	"	83.50	40.10	123.60
Glazing accessories				
Neoprene glazing gaskets				
1/4" glass	L.F.	1.20	1.90	3.10
3/8" glass	"	1.45	2.00	3.45
1/2" glass	"	1.80	2.10	3.90
3/4" glass	"	2.45	2.20	4.65
1" glass	"	2.80	2.40	5.20
Mullion section				
1/4" glass	L.F.	0.52	0.96	1.48
3/8" glass	"	0.67	1.20	1.87

Glazing

Glazing	UNIT	MAT.	INST.	TOTAL
08810.10 Glazing				
1/2" glass	L.F.	0.99	1.35	2.34
3/4" glass	“	1.50	1.60	3.10
1" glass	“	2.00	1.90	3.90
Molded corners	EA.	2.15	32.10	34.25

Glazed Curtain Walls

Glazed Curtain Walls	UNIT	MAT.	INST.	TOTAL
08910.10 Glazed Curtain Walls				
Curtain wall, aluminum system, framing sections				
2" x 3"				
Jamb	L.F.	9.45	4.00	13.45
Horizontal	“	9.60	4.00	13.60
Mullion	“	12.80	4.00	16.80
2" x 4"				
Jamb	L.F.	12.80	6.00	18.80
Horizontal	“	13.15	6.00	19.15
Mullion	“	12.80	6.00	18.80
3" x 5-1/2"				
Jamb	L.F.	16.85	6.00	22.85
Horizontal	“	18.80	6.00	24.80
Mullion	“	17.05	6.00	23.05
4" corner mullion	“	22.55	8.00	30.55
Coping sections				
1/8" x 8"	L.F.	23.60	8.00	31.60
1/8" x 0"	“	23.75	8.00	31.75
1/8" x 12-1/2"	“	24.35	9.60	33.95
Sill section				
1/8" x 6"	L.F.	23.25	4.80	28.05
1/8" x 7"	“	23.55	4.80	28.35
1/8" x 8-1/2"	“	24.00	4.80	28.80
Column covers, aluminum				
1/8" x 26"	L.F.	23.25	12.00	35.25
1/8" x 34"	“	23.55	12.65	36.20
1/8" x 38"	“	23.75	12.65	36.40
Doors				
Aluminum framed, standard hardware				
Narrow stile				
2-6 x 7-0	EA.	520.00	240.00	760.00
3-0 x 7-0	“	530.00	240.00	770.00
3-6 x 7-0	“	550.00	240.00	790.00
Wide stile				
2-6 x 7-0	EA.	900.00	240.00	1,140
3-0 x 7-0	“	970.00	240.00	1,210
3-6 x 7-0	“	1,050	240.00	1,290
Flush panel doors, to match adjacent wall panels				
2-6 x 7-0	EA.	760.00	300.00	1,060

Glazed Curtain Walls

08910.10 Glazed Curtain Walls

Glazed Curtain Walls	UNIT	MAT.	INST.	TOTAL
3-0 x 7-0	EA.	800.00	300.00	1,100
3-6 x 7-0	"	830.00	300.00	1,130
Wall panel, insulated				
"U"=.08	S.F.	9.60	4.00	13.60
"U"=.10	"	9.05	4.00	13.05
"U"=.15	"	8.15	4.00	12.15
Window wall system, complete				
Minimum	S.F.	16.90	4.80	21.70
Average	"	27.10	5.35	32.45
Maximum	"	62.50	6.85	69.35
Added costs				
For bronze, add 20% to material				
For stainless steel, add 50% to material				

Support Systems

09110.10 Metal Studs

	UNIT	MAT.	INST.	TOTAL
Studs, non load bearing, galvanized				
2-1/2", 20 ga.				
12" o.c.	S.F.	0.85	0.90	1.75
16" o.c.	"	0.69	0.72	1.41
25 ga.				
12" o.c.	S.F.	0.58	0.90	1.48
16" o.c.	"	0.41	0.72	1.13
24" o.c.	"	0.37	0.60	0.97
3-5/8", 20 ga.				
12" o.c.	S.F.	1.00	1.10	2.10
16" o.c.	"	0.85	0.87	1.71
24" o.c.	"	0.67	0.72	1.39
25 ga.				
12" o.c.	S.F.	0.57	1.10	1.67
16" o.c.	"	0.49	0.87	1.35
24" o.c.	"	0.40	0.72	1.12
4", 20 ga.				
12" o.c.	S.F.	1.15	1.10	2.25
16" o.c.	"	0.93	0.87	1.79
24" o.c.	"	0.77	0.72	1.49
25 ga.				
12" o.c.	S.F.	0.67	1.10	1.77
16" o.c.	"	0.57	0.87	1.43
24" o.c.	"	0.48	0.72	1.20
6", 20 ga.				
12" o.c.	S.F.	1.40	1.35	2.75
16" o.c.	"	1.15	1.10	2.25
24" o.c.	"	0.91	0.90	1.81
25 ga.				
12" o.c.	S.F.	0.86	1.35	2.21
16" o.c.	"	0.69	1.10	1.79
24" o.c.	"	0.58	0.90	1.48
Load bearing studs, galvanized				
3-5/8", 16 ga.				
12" o.c.	S.F.	1.65	1.10	2.75
16" o.c.	"	1.35	0.87	2.22
18 ga.				
12" o.c.	S.F.	1.50	0.72	2.22
16" o.c.	"	1.20	0.87	2.07
4", 16 ga.				
12" o.c.	S.F.	1.80	1.10	2.90
16" o.c.	"	1.50	0.87	2.37
6", 16 ga.				
12" o.c.	S.F.	2.20	1.35	3.55
16" o.c.	"	1.80	1.10	2.90
Furring				
On beams and columns				
7/8" channel	L.F.	0.44	2.90	3.34
1-1/2" channel	"	0.52	3.35	3.87
On ceilings				
3/4" furring channels				
12" o.c.	S.F.	0.41	1.80	2.21

Support Systems

	UNIT	MAT.	INST.	TOTAL
09110.10 Metal Studs				
16" o.c.	S.F.	0.36	1.75	2.11
24" o.c.	"	0.26	1.55	1.81
1-1/2" furring channels				
12" o.c.	S.F.	0.59	1.95	2.54
16" o.c.	"	0.52	1.80	2.32
24" o.c.	"	0.44	1.65	2.09
On walls				
3/4" furring channels				
12" o.c.	S.F.	0.42	1.45	1.87
16" o.c.	"	0.34	1.35	1.69
24" o.c.	"	0.26	1.25	1.51
1-1/2" furring channels				
12" o.c.	S.F.	0.59	1.55	2.14
16" o.c.	"	0.52	1.45	1.97
24" o.c.	"	0.44	1.35	1.79

Lath And Plaster

	UNIT	MAT.	INST.	TOTAL
09205.10 Gypsum Lath				
Gypsum lath, 1/2" thick				
Clipped	S.Y.	6.30	2.40	8.70
Nailed	"	6.15	2.70	8.85
09205.20 Metal Lath				
Diamond expanded, galvanized				
2.5 lb., on walls				
Nailed	S.Y.	2.80	5.40	8.20
Wired	"	2.85	6.20	9.05
On ceilings				
Nailed	S.Y.	2.80	6.20	9.00
Wired	"	2.85	7.20	10.05
3.4 lb., on walls				
Nailed	S.Y.	3.25	5.40	8.65
Wired	"	3.30	6.20	9.50
On ceilings				
Nailed	S.Y.	3.25	6.20	9.45
Wired	"	3.30	7.20	10.50
Flat rib				
2.75 lb., on walls				
Nailed	S.Y.	2.80	5.40	8.20
Wired	"	2.85	6.20	9.05
On ceilings				
Nailed	S.Y.	2.80	6.20	9.00
Wired	"	2.85	7.20	10.05
3.4 lb., on walls				
Nailed	S.Y.	3.45	5.40	8.85

Lath And Plaster

	UNIT	MAT.	INST.	TOTAL
09205.20 Metal Lath				
Wired	S.Y.	3.55	6.20	9.75
On ceilings				
Nailed	S.Y.	3.45	6.20	9.65
Wired	"	3.55	7.20	10.75
Stucco lath				
1.8 lb.	S.Y.	4.50	5.40	9.90
3.6 lb.	"	5.05	5.40	10.45
Paper backed				
Minimum	S.Y.	3.50	4.35	7.85
Maximum	"	5.60	6.20	11.80
09205.60 Plaster Accessories				
Expansion joint, 3/4", 26 ga., galvanized, one piece	L.F.	0.87	1.10	1.97
Plaster corner beads, 3/4", galvanized	"	0.51	1.25	1.76
Casing bead, expanded flange, galvanized	"	0.51	1.10	1.61
Expanded wing, 1-1/4" wide, galvanized	"	0.43	1.10	1.53
Joint clips for lath	EA.	0.22	0.22	0.44
Metal base, galvanized, 2-1/2" high	L.F.	0.66	1.45	2.11
Stud clips for gypsum lath	EA.	0.22	0.22	0.44
Tie wire galvanized, 18 ga., 25 lb. hank	"			42.60
Sound deadening board, 1/4", nailed or clipped	S.F.	0.41	0.72	1.13
09210.10 Plaster				
Gypsum plaster, trowel finish, 2 coats				
Ceilings	S.Y.	5.30	13.30	18.60
Walls	"	5.30	12.55	17.85
3 coats				
Ceilings	S.Y.	7.35	18.50	25.85
Walls	"	7.35	16.40	23.75
Vermiculite plaster				
2 coats				
Ceilings	S.Y.	6.00	20.30	26.30
Walls	"	6.00	18.50	24.50
3 coats				
Ceilings	S.Y.	9.45	25.10	34.55
Walls	"	9.45	22.40	31.85
Keenes cement plaster				
2 coats				
Ceilings	S.Y.	8.40	16.40	24.80
Walls	"	8.40	14.20	22.60
3 coats				
Ceilings	S.Y.	10.25	18.50	28.75
Walls	"	10.25	16.40	26.65
On columns, add to installation, 50%	"			
Chases, fascia, and soffits, add to installation, 50%	"			
Beams, add to installation, 50%	"			
09220.10 Portland Cement Plaster				
Stucco, portland, gray, 3 coat, 1" thick				
Sand finish	S.Y.	6.90	18.50	25.40
Trowel finish	"	6.90	19.35	26.25
White cement				

Lath And Plaster

Lath And Plaster	UNIT	MAT.	INST.	TOTAL
09220.10 Portland Cement Plaster				
Sand finish	S.Y.	7.90	19.35	27.25
Trowel finish	"	7.90	21.30	29.20
Scratch coat				
For ceramic tile	S.Y.	2.50	4.25	6.75
For quarry tile	"	2.50	4.25	6.75
Portland cement plaster				
2 coats, 1/2"	S.Y.	5.00	8.50	13.50
3 coats, 7/8"	"	5.95	10.65	16.60
09250.10 Gypsum Board				
Drywall, plasterboard, 3/8" clipped to				
Metal furred ceiling	S.F.	0.31	0.48	0.79
Columns and beams	"	0.31	1.10	1.41
Walls	"	0.31	0.43	0.74
Nailed or screwed to				
Wood framed ceiling	S.F.	0.31	0.43	0.74
Columns and beams	"	0.31	0.96	1.27
Walls	"	0.31	0.39	0.70
1/2", clipped to				
Metal furred ceiling	S.F.	0.32	0.48	0.80
Columns and beams	"	0.32	1.10	1.42
Walls	"	0.32	0.43	0.75
Nailed or screwed to				
Wood framed ceiling	S.F.	0.32	0.43	0.75
Columns and beams	"	0.32	0.96	1.28
Walls	"	0.32	0.39	0.71
5/8", clipped to				
Metal furred ceiling	S.F.	0.35	0.54	0.89
Columns and beams	"	0.35	1.20	1.55
Walls	"	0.35	0.48	0.83
Nailed or screwed to				
Wood framed ceiling	S.F.	0.35	0.54	0.89
Columns and beams	"	0.35	1.20	1.55
Walls	"	0.35	0.48	0.83
Vinyl faced, clipped to metal studs				
1/2"	S.F.	1.55	0.54	2.09
5/8"	"	1.65	0.54	2.19
Add for				
Fire resistant	S.F.			0.10
Water resistant	"			0.15
Water and fire resistant	"			0.20
Taping and finishing joints				
Minimum	S.F.	0.04	0.29	0.33
Average	"	0.06	0.36	0.42
Maximum	"	0.08	0.43	0.52
Casing bead				
Minimum	L.F.	0.14	1.25	1.39
Average	"	0.15	1.45	1.60
Maximum	"	0.19	2.15	2.34
Corner bead				
Minimum	L.F.	0.15	1.25	1.40
Average	"	0.19	1.45	1.64

Lath And Plaster

09250.10 Gypsum Board	UNIT	MAT.	INST.	TOTAL
Maximum	L.F.	0.24	2.15	2.39

Tile

09310.10 Ceramic Tile	UNIT	MAT.	INST.	TOTAL
Glazed wall tile, 4-1/4" x 4-1/4"				
Minimum	S.F.	2.10	3.00	5.10
Average	"	3.35	3.50	6.85
Maximum	"	9.10	4.20	13.30
Base, 4-1/4" high				
Minimum	L.F.	3.25	5.25	8.50
Average	"	3.65	5.25	8.90
Maximum	"	5.80	5.25	11.05
Unglazed floor tile				
Portland cement bed, cushion edge, face mounted				
1" x 1"	S.F.	6.35	3.80	10.15
1" x 2"	"	10.20	3.65	13.85
2" x 2"	"	6.70	3.50	10.20
Adhesive bed, with white grout				
1" x 1"	S.F.	5.60	3.80	9.40
1" x 2"	"	5.85	3.65	9.50
2" x 2"	"	5.95	3.50	9.45
Organic adhesive bed, thin set, back mounted				
1" x 1"	S.F.	6.25	3.80	10.05
1" x 2"	"	6.40	3.65	10.05
2" x 2"	"	6.70	3.50	10.20
For group 2 colors, add to material, 10%				
For group 3 colors, add to material, 20%				
For abrasive surface, add to material, 25%				
Unglazed wall tile				
Organic adhesive, face mounted cushion edge				
1" x 1"				
Minimum	S.F.	2.95	3.50	6.45
Average	"	3.90	3.80	7.70
Maximum	"	5.80	4.20	10.00
1" x 2"				
Minimum	S.F.	3.25	3.35	6.60
Average	"	4.35	3.65	8.00
Maximum	"	6.40	4.00	10.40
2" x 2"				
Minimum	S.F.	3.45	3.20	6.65
Average	"	3.90	3.50	7.40
Maximum	"	4.50	3.80	8.30
Back mounted				
1" x 1"				
Minimum	S.F.	2.95	3.50	6.45

Tile

	UNIT	MAT.	INST.	TOTAL
09310.10 Ceramic Tile				
Average	S.F.	3.90	3.80	7.70
Maximum	"	5.80	4.20	10.00
1" x 2"				
Minimum	S.F.	3.25	3.35	6.60
Average	"	4.35	3.65	8.00
Maximum	"	6.30	4.00	10.30
2" x 2"				
Minimum	S.F.	3.45	3.20	6.65
Average	"	3.90	3.50	7.40
Maximum	"	6.30	3.80	10.10
For glazed finish, add to material, 25%				
For glazed mosaic, add to material, 100%				
For metallic colors, add to material, 125%				
For exterior wall use, add to total, 25%				
For exterior soffit, add to total, 25%				
For portland cement bed, add to total, 25%				
For dry set portland cement bed, add to total, 10%				
Conductive floor tile, unglazed square edged				
Portland cement bed				
1 x 1	S.F.	6.10	5.25	11.35
1-9/16 x 1-9/16	"	5.80	5.25	11.05
Dry set				
1 x 1	S.F.	6.40	5.25	11.65
1-9/16 x 1-9/16	"	6.00	5.25	11.25
Epoxy bed with epoxy joints				
1 x 1	S.F.	5.25	5.25	10.50
1-9/16 x 1-9/16	"	6.40	5.25	11.65
For WWF in bed add to total, 15%				
For abrasive surface, add to material, 40%				
Ceramic accessories				
Towel bar, 24" long				
Minimum	EA.	24.35	16.75	41.10
Average	"	31.90	20.95	52.85
Maximum	"	48.60	27.90	76.50
Soap dish				
Minimum	EA.	21.30	27.90	49.20
Average	"	25.90	34.90	60.80
Maximum	"	36.50	41.90	78.40
09330.10 Quarry Tile				
Floor				
4 x 4 x 1/2"	S.F.	4.85	5.60	10.45
6 x 6 x 1/2"	"	4.90	5.25	10.15
6 x 6 x 3/4"	"	5.80	5.25	11.05
Wall, applied to 3/4" portland cement bed				
4 x 4 x 1/2"	S.F.	4.45	8.40	12.85
6 x 6 x 3/4"	"	5.55	7.00	12.55
Cove base				
5 x 6 x 1/2" straight top	L.F.	4.35	7.00	11.35
6 x 6 x 3/4" round top	"	4.60	7.00	11.60
Stair treads 6 x 6 x 3/4"	"	7.70	10.45	18.15
Window sill 6 x 8 x 3/4"	"	6.20	8.40	14.60

Tile	UNIT	MAT.	INST.	TOTAL
09330.10 Quarry Tile				
For abrasive surface, add to material, 25%				
09410.10 Terrazzo				
Floors on concrete, 1-3/4" thick, 5/8" topping				
Gray cement	S.F.	4.20	6.10	10.30
White cement	"	4.55	6.10	10.65
Sand cushion, 3" thick, 5/8" top, 1/4"				
Gray cement	S.F.	4.95	7.10	12.05
White cement	"	5.50	7.10	12.60
Monolithic terrazzo, 3-1/2" base slab, 5/8" topping	"	3.55	5.30	8.85
Terrazzo wainscot, cast-in-place, 1/2" thick	"	7.40	10.65	18.05
Base, cast in place, terrazzo cove type, 6" high	L.F.	7.90	6.10	14.00
Curb, cast in place, 6" wide x 6" high, polished top	"	8.75	21.30	30.05
For venetian type terrazzo, add to material, 10%				
For abrasive heavy duty terrazzo, add to material, 15%				
Divider strips				
Zinc	L.F.			1.34
Brass	"			2.49
Stairs, cast-in-place, topping on concrete or metal				
1-1/2" thick treads, 12" wide	L.F.	5.25	21.30	26.55
Combined tread and riser	"	7.90	53.50	61.40
Precast terrazzo, thin set				
Terrazzo tiles, non-slip surface				
9" x 9" x 1" thick	S.F.	16.70	6.10	22.80
12" x 12"				
1" thick	S.F.	18.00	5.70	23.70
1-1/2" thick	"	18.80	6.10	24.90
18" x 18" x 1-1/2" thick	"	24.55	6.10	30.65
24" x 24" x 1-1/2" thick	"	31.60	5.00	36.60
For white cement, add to material, 10%				
For venetian type terrazzo, add to material, 25%				
Terrazzo wainscot				
12" x 12" x 1" thick	S.F.	18.45	10.65	29.10
18" x 18" x 1-1/2" thick	"	29.00	12.15	41.15
Base				
6" high				
Straight	L.F.	11.85	3.30	15.15
Coved	"	14.05	3.30	17.35
8" high				
Straight	L.F.	13.35	3.55	16.90
Coved	"	15.60	3.55	19.15
Terrazzo curbs				
8" wide x 8" high	L.F.	28.10	17.05	45.15
6" wide x 6" high	"	25.40	14.20	39.60
Precast terrazzo stair treads, 12" wide				
1-1/2" thick				
Diamond pattern	L.F.	29.80	7.75	37.55
Non-slip surface	"	33.40	7.75	41.15
2" thick				
Diamond pattern	L.F.	33.30	7.75	41.05
Non-slip surface	"	36.90	8.50	45.40
Stair risers, 1" thick to 6" high				

Tile

09410.10 Terrazzo	UNIT	MAT.	INST.	TOTAL
Straight sections	L.F.	12.55	4.25	16.80
Cove sections	"	14.75	4.25	19.00
Combined tread and riser				
Straight sections				
1-1/2" tread, 3/4" riser	L.F.	61.00	12.15	73.15
3" tread, 1" riser	"	64.50	12.15	76.65
Curved sections				
2" tread, 1" riser	L.F.	69.00	14.20	83.20
3" tread, 1" riser	"	71.00	14.20	85.20
Stair stringers, notched for treads and risers				
1" thick	L.F.	32.40	10.65	43.05
2" thick	"	33.70	14.20	47.90
Landings, structural, nonslip				
1-1/2" thick	S.F.	27.90	7.10	35.00
3" thick	"	27.20	8.50	35.70
Conductive terrazzo, spark proof industrial floor				
Epoxy terrazzo				
Floor	S.F.	11.75	2.65	14.40
Base	"	11.75	3.55	15.30
Polyacrylate				
Floor	S.F.	9.50	2.65	12.15
Base	"	9.50	3.55	13.05
Polyester				
Floor	S.F.	3.35	1.70	5.05
Base	"	3.35	2.15	5.50
Synthetic latex mastic				
Floor	S.F.	5.45	2.65	8.10
Base	"	5.45	3.55	9.00

Acoustical Treatment

09510.10 Ceilings And Walls	UNIT	MAT.	INST.	TOTAL
Acoustical panels, suspension system not included				
Fiberglass panels				
5/8" thick				
2' x 2'	S.F.	0.84	0.62	1.46
2' x 4'	"	0.84	0.48	1.32
3/4" thick				
2' x 2'	S.F.	1.00	0.62	1.62
2' x 4'	"	1.00	0.48	1.48
Glass cloth faced fiberglass panels				
3/4" thick	S.F.	2.30	0.72	3.02
1" thick	"	2.55	0.72	3.27
Mineral fiber panels				
5/8" thick				
2' x 2'	S.F.	1.00	0.62	1.62

Acoustical Treatment

09510.10 Ceilings And Walls	UNIT	MAT.	INST.	TOTAL
2' x 4'	S.F.	1.00	0.48	1.48
3/4" thick				
2' x 2'	S.F.	1.35	0.62	1.97
2' x 4'	"	1.35	0.48	1.83
For aluminum faced panels, add to material, 80%				
For vinyl faced panels, add to total, 125%				
For fire rated panels, add to material, 75%				
Wood fiber panels				
1/2" thick				
2' x 2'	S.F.	1.50	0.62	2.12
2' x 4'	"	1.50	0.48	1.98
5/8" thick				
2' x 2'	S.F.	1.65	0.62	2.27
2' x 4'	"	1.65	0.48	2.13
3/4" thick				
2' x 2'	S.F.	2.05	0.62	2.67
2' x 4'	"	2.05	0.48	2.53
2" thick				
2' x 2'	S.F.	2.40	0.72	3.12
2' x 4'	"	2.40	0.54	2.94
For flameproofing, add to material, 10%				
For sculptured finish, add to material, 15%				
Air distributing panels				
3/4" thick	S.F.	2.40	1.10	3.50
5/8" thick	"	2.05	0.87	2.92
Acoustical tiles, suspension system not included				
Fiberglass tile, 12" x 12"				
5/8" thick	S.F.	0.61	0.79	1.40
3/4" thick	"	0.65	0.96	1.62
Glass cloth faced fiberglass tile				
3/4" thick	S.F.	2.30	0.96	3.26
3" thick	"	2.35	1.10	3.45
Mineral fiber tile, 12" x 12"				
5/8" thick				
Standard	S.F.	1.30	0.87	2.17
Vinyl faced	"	1.45	0.87	2.32
3/4" thick				
Standard	S.F.	1.55	0.87	2.42
Vinyl faced	"	1.70	0.87	2.57
Fire rated	"	2.45	0.87	3.32
Aluminum or mylar faced	"	3.10	0.87	3.97
Wood fiber tile, 12" x 12"				
1/2" thick	S.F.	0.97	0.87	1.83
3/4" thick	"	1.40	0.87	2.27
For flameproofing, add to material, 10%				
For sculptured 3 dimensional, add to material, 50%				
Metal pan units, 24 ga. steel				
12" x 12"	S.F.	4.75	1.75	6.50
12" x 24"	"	4.90	1.45	6.35
Aluminum, .025" thick				
12" x 12"	S.F.	5.05	1.75	6.80
12" x 24"	"	5.20	1.45	6.65

09 FINISHES

Acoustical Treatment	UNIT	MAT.	INST.	TOTAL
09510.10 Ceilings And Walls				
Anodized aluminum, 0.25" thick				
12" x 12"	S.F.	5.80	1.75	7.55
12" x 24"	"	6.90	1.45	8.35
Stainless steel, 24 ga.				
12" x 12"	S.F.	11.60	1.75	13.35
12" x 24"	"	12.00	1.45	13.45
For flameproof sound absorbing pads, add to material	"			1.81
Metal ceiling systems				
.020" thick panels				
10', 12', and 16' lengths	S.F.	4.55	1.25	5.80
Custom lengths, 3' to 20'	"	4.60	1.25	5.85
.025" thick panels				
32 sf, 38 sf, and 52 sf pieces	S.F.	4.55	1.45	6.00
Custom lengths, 10 sf to 65 sf	"	5.30	1.45	6.75
Carriers, black, add	"			2.74
Recess filler strip, add	"			0.92
Custom lengths, add	"			1.37
Sound absorption walls, with fabric cover				
2-6" x 9' x 3/4"	S.F.	5.95	1.45	7.40
2' x 9' x 1"	"	6.50	1.45	7.95
Starter spline	L.F.	1.45	1.10	2.55
Internal spline	"	1.25	1.10	2.35
Acoustical treatment				
Barriers for plenums				
Leaded vinyl				
0.48 lb per sf	S.F.	2.70	2.05	4.75
0.87 lb per sf	"	2.70	2.15	4.85
Aluminum foil, fiberglass reinforcement				
Minimum	S.F.	1.00	1.45	2.45
Maximum	"	1.15	2.15	3.30
Aluminum mesh, paper backed	"	0.97	1.45	2.42
Fibered cement sheet, 3/16" thick	"	1.95	1.55	3.50
Sheet lead, 1/64" thick	"	2.80	1.10	3.90
Sound attenuation blanket				
1" thick	S.F.	0.49	4.35	4.84
1-1/2" thick	"	0.65	4.35	5.00
2" thick	"	0.82	4.35	5.17
3" thick	"	0.97	4.80	5.77
Ceiling suspension systems				
T bar system				
2' x 4'	S.F.	1.05	0.43	1.48
2' x 2'	"	1.15	0.48	1.63
Concealed Z bar suspension system, 12" module	"	0.98	0.72	1.70
For 1-1/2" carrier channels, 4' o.c., add	"			0.35
Carrier channel for recessed light fixtures	"			0.63

09 FINISHES

Flooring

	UNIT	MAT.	INST.	TOTAL
09550.10 Wood Flooring				
Wood strip flooring, unfinished				
Fir floor				
C and better				
Vertical grain	S.F.	3.35	1.45	4.80
Flat grain	"	3.15	1.45	4.60
Oak floor				
Minimum	S.F.	3.80	2.05	5.85
Average	"	4.50	2.05	6.55
Maximum	"	5.65	2.05	7.70
Maple floor				
25/32" x 2-1/4"				
Minimum	S.F.	4.10	2.05	6.15
Maximum	"	4.80	2.05	6.85
33/32" x 3-1/4"				
Minimum	S.F.	5.30	2.05	7.35
Maximum	"	6.00	2.05	8.05
Added costs				
For factory finish, add to material, 10%				
For random width floor, add to total, 20%				
For simulated pegs, add to total, 10%				
Wood block industrial flooring				
Creosoted				
2" thick	S.F.	3.95	1.15	5.10
2-1/2" thick	"	4.10	1.35	5.45
3" thick	"	4.30	1.45	5.75
Parquet, 5/16", white oak				
Finished	S.F.	8.70	2.15	10.85
Unfinished	"	3.10	2.15	5.25
Gym floor, 2 ply felt, 25/32" maple, finished, in mastic	"	7.40	2.40	9.80
Over wood sleepers	"	8.25	2.70	10.95
Finishing, sand, fill, finish, and wax	"	0.61	1.10	1.71
Refinish sand, seal, and 2 coats of polyurethane	"	1.10	1.45	2.55
Clean and wax floors	"	0.19	0.22	0.40
09630.10 Unit Masonry Flooring				
Clay brick				
9 x 4-1/2 x 3" thick				
Glazed	S.F.	7.40	3.60	11.00
Unglazed	"	7.05	3.60	10.65
8 x 4 x 3/4" thick				
Glazed	S.F.	6.65	3.75	10.40
Unglazed	"	6.55	3.75	10.30
For herringbone pattern, add to labor, 15%				
09660.10 Resilient Tile Flooring				
Solid vinyl tile, 1/8" thick, 12" x 12"				
Marble patterns	S.F.	2.40	1.10	3.50
Solid colors	"	3.10	1.10	4.20
Travertine patterns	"	3.50	1.10	4.60
Conductive resilient flooring, vinyl tile				
1/8" thick, 12" x 12"	S.F.	5.40	1.25	6.65

Flooring

Flooring	UNIT	MAT.	INST.	TOTAL
09665.10 Resilient Sheet Flooring				
Vinyl sheet flooring				
Minimum	S.F.	2.10	0.43	2.53
Average	"	3.00	0.52	3.52
Maximum	"	6.30	0.72	7.02
Cove, to 6"	L.F.	1.25	0.87	2.12
Fluid applied resilient flooring				
Polyurethane, poured in place, 3/8" thick	S.F.	8.55	3.60	12.15
Vinyl sheet goods, backed				
0.070" thick	S.F.	3.15	0.54	3.69
0.093" thick	"	4.90	0.54	5.44
0.125" thick	"	5.65	0.54	6.19
0.250" thick	"	6.45	0.54	6.99
09678.10 Resilient Base And Accessories				
Wall base, vinyl				
Group 1				
4" high	L.F.	0.84	1.45	2.29
6" high	"	1.15	1.45	2.60
Group 2				
4" high	L.F.	0.74	1.45	2.19
6" high	"	0.87	1.45	2.32
Group 3				
4" high	L.F.	0.81	1.45	2.26
6" high	"	0.95	1.45	2.40
Stair accessories				
Treads, 1/4" x 12", rubber diamond surface				
Marbled	L.F.	6.70	3.60	10.30
Plain	"	6.55	3.60	10.15
Grit strip safety tread, 12" wide, colors				
3/16" thick	L.F.	9.15	3.60	12.75
5/16" thick	"	12.85	3.60	16.45
Risers, 7" high, 1/8" thick, colors				
Flat	L.F.	1.90	2.15	4.05
Coved	"	3.20	2.15	5.35
Nosing, rubber				
3/16" thick, 3" wide				
Black	L.F.	3.20	2.15	5.35
Colors	"	3.40	2.15	5.55
6" wide				
Black	L.F.	4.00	3.60	7.60
Colors	"	4.20	3.60	7.80

Carpet	UNIT	MAT.	INST.	TOTAL
09680.10 Floor Leveling				
Repair and level floors to receive new flooring				
Minimum	S.Y.	0.80	1.45	2.25
Average	"	3.10	3.60	6.70
Maximum	"	4.55	4.35	8.90
09682.10 Carpet Padding				
Carpet padding				
Foam rubber, waffle type, 0.3" thick	S.Y.	5.65	2.15	7.80
Jute padding				
Minimum	S.Y.	3.85	1.95	5.80
Average	"	5.00	2.15	7.15
Maximum	"	7.50	2.40	9.90
Sponge rubber cushion				
Minimum	S.Y.	4.55	1.95	6.50
Average	"	6.00	2.15	8.15
Maximum	"	7.50	2.40	9.90
Urethane cushion, 3/8" thick				
Minimum	S.Y.	4.55	1.95	6.50
Average	"	5.30	2.15	7.45
Maximum	"	6.00	2.40	8.40
09685.10 Carpet				
Carpet, acrylic				
24 oz., light traffic	S.Y.	25.60	4.80	30.40
28 oz., medium traffic	"	32.50	4.80	37.30
Residential				
Nylon				
15 oz., light traffic	S.Y.	18.45	4.80	23.25
28 oz., medium traffic	"	22.10	4.80	26.90
Commercial				
Nylon				
28 oz., medium traffic	S.Y.	23.55	4.80	28.35
35 oz., heavy traffic	"	28.20	4.80	33.00
Wool				
30 oz., medium traffic	S.Y.	38.40	4.80	43.20
36 oz., medium traffic	"	40.60	4.80	45.40
42 oz., heavy traffic	"	53.50	4.80	58.30
Carpet tile				
Foam backed				
Minimum	S.F.	2.95	0.87	3.82
Average	"	3.40	0.96	4.36
Maximum	"	5.40	1.10	6.50
Tufted loop or shag				
Minimum	S.F.	3.20	0.87	4.07
Average	"	3.85	0.96	4.81
Maximum	"	6.20	1.10	7.30
Clean and vacuum carpet				
Minimum	S.Y.	0.27	0.17	0.44
Average	"	0.43	0.29	0.72
Maximum	"	0.59	0.43	1.02

09 FINISHES

Carpet

09700.10 Special Flooring	UNIT	MAT.	INST.	TOTAL
Epoxy flooring, marble chips				
Epoxy with colored quartz chips in 1/4" base	S.F.	9.95	2.40	12.35
Heavy duty epoxy topping, 3/16" thick	"	8.70	2.40	11.10
Epoxy terrazzo				
1/4" thick chemical resistant	S.F.	10.20	2.70	12.90

Painting

09905.10 Painting Preparation	UNIT	MAT.	INST.	TOTAL
Dropcloths				
Minimum	S.F.	0.02	0.03	0.05
Average	"	0.02	0.04	0.06
Maximum	"	0.02	0.05	0.07
Masking				
Paper and tape				
Minimum	L.F.	0.02	0.43	0.46
Average	"	0.02	0.54	0.56
Maximum	"	0.02	0.72	0.74
Doors				
Minimum	EA.	0.04	5.40	5.44
Average	"	0.04	7.20	7.24
Maximum	"	0.04	9.60	9.64
Windows				
Minimum	EA.	0.04	5.40	5.44
Average	"	0.04	7.20	7.24
Maximum	"	0.04	9.60	9.64
Sanding				
Walls and flat surfaces				
Minimum	S.F.		0.29	0.29
Average	"		0.36	0.36
Maximum	"		0.43	0.43
Doors and windows				
Minimum	EA.		7.20	7.20
Average	"		10.80	10.80
Maximum	"		14.45	14.45
Trim				
Minimum	L.F.		0.54	0.54
Average	"		0.72	0.72
Maximum	"		0.96	0.96
Puttying				
Minimum	S.F.	0.01	0.67	0.68
Average	"	0.01	0.87	0.88
Maximum	"	0.01	1.10	1.11

09 FINISHES

Painting	UNIT	MAT.	INST.	TOTAL
09910.05 Ext. Painting, Sitework				
Benches				
Brush				
First Coat				
Minimum	S.F.	0.16	0.43	0.60
Average	"	0.16	0.54	0.70
Maximum	"	0.16	0.72	0.88
Second Coat				
Minimum	S.F.	0.16	0.27	0.43
Average	"	0.16	0.31	0.47
Maximum	"	0.16	0.36	0.52
Roller				
First Coat				
Minimum	S.F.	0.16	0.22	0.38
Average	"	0.16	0.24	0.40
Maximum	"	0.16	0.27	0.43
Second Coat				
Minimum	S.F.	0.16	0.16	0.32
Average	"	0.16	0.18	0.34
Maximum	"	0.16	0.20	0.36
Brickwork				
Brush				
First Coat				
Minimum	S.F.	0.16	0.27	0.43
Average	"	0.16	0.36	0.52
Maximum	"	0.16	0.54	0.70
Second Coat				
Minimum	S.F.	0.16	0.24	0.40
Average	"	0.16	0.29	0.45
Maximum	"	0.16	0.36	0.52
Spray				
First Coat				
Minimum	S.F.	0.14	0.12	0.26
Average	"	0.14	0.16	0.29
Maximum	"	0.14	0.20	0.33
Second Coat				
Minimum	S.F.	0.14	0.11	0.25
Average	"	0.14	0.14	0.28
Maximum	"	0.14	0.18	0.32
Concrete Block				
Roller				
First Coat				
Minimum	S.F.	0.16	0.22	0.38
Average	"	0.16	0.29	0.45
Maximum	"	0.16	0.43	0.60
Second Coat				
Minimum	S.F.	0.16	0.18	0.34
Average	"	0.16	0.24	0.40
Maximum	"	0.16	0.36	0.52
Spray				
First Coat				
Minimum	S.F.	0.14	0.12	0.26
Average	"	0.14	0.14	0.28

Painting

09910.05 Ext. Painting, Sitework

	UNIT	MAT.	INST.	TOTAL
Maximum	S.F.	0.14	0.17	0.30
Second Coat				
Minimum	S.F.	0.14	0.08	0.21
Average	"	0.14	0.10	0.23
Maximum	"	0.14	0.14	0.27
Fences, Chain Link				
Brush				
First Coat				
Minimum	S.F.	0.11	0.43	0.54
Average	"	0.11	0.48	0.59
Maximum	"	0.11	0.54	0.65
Second Coat				
Minimum	S.F.	0.11	0.29	0.40
Average	"	0.11	0.33	0.44
Maximum	"	0.11	0.39	0.50
Roller				
First Coat				
Minimum	S.F.	0.11	0.31	0.42
Average	"	0.11	0.36	0.47
Maximum	"	0.11	0.41	0.52
Second Coat				
Minimum	S.F.	0.11	0.18	0.29
Average	"	0.11	0.22	0.33
Maximum	"	0.11	0.27	0.38
Spray				
First Coat				
Minimum	S.F.	0.08	0.14	0.22
Average	"	0.08	0.16	0.24
Maximum	"	0.08	0.18	0.26
Second Coat				
Minimum	S.F.	0.08	0.10	0.18
Average	"	0.08	0.12	0.20
Maximum	"	0.08	0.14	0.22
Fences, Wood or Masonry				
Brush				
First Coat				
Minimum	S.F.	0.16	0.46	0.62
Average	"	0.16	0.54	0.70
Maximum	"	0.16	0.72	0.88
Second Coat				
Minimum	S.F.	0.16	0.27	0.43
Average	"	0.16	0.33	0.50
Maximum	"	0.16	0.43	0.60
Roller				
First Coat				
Minimum	S.F.	0.16	0.24	0.40
Average	"	0.16	0.29	0.45
Maximum	"	0.16	0.33	0.50
Second Coat				
Minimum	S.F.	0.16	0.17	0.33
Average	"	0.16	0.21	0.37
Maximum	"	0.16	0.27	0.43

Painting	UNIT	MAT.	INST.	TOTAL
09910.05 Ext. Painting, Sitework				
Spray				
First Coat				
Minimum	S.F.	0.14	0.16	0.29
Average	"	0.14	0.20	0.33
Maximum	"	0.14	0.27	0.41
Second Coat				
Minimum	S.F.	0.14	0.11	0.24
Average	"	0.14	0.14	0.27
Maximum	"	0.14	0.18	0.32
Storage Tanks				
Roller				
First Coat				
Minimum	S.F.	0.14	0.18	0.32
Average	"	0.14	0.22	0.35
Maximum	"	0.14	0.27	0.41
Second Coat				
Minimum	S.F.	0.14	0.14	0.28
Average	"	0.14	0.17	0.31
Maximum	"	0.14	0.22	0.35
Spray				
First Coat				
Minimum	S.F.	0.11	0.11	0.22
Average	"	0.11	0.13	0.24
Maximum	"	0.11	0.16	0.26
Second Coat				
Minimum	S.F.	0.11	0.09	0.20
Average	"	0.11	0.10	0.21
Maximum	"	0.11	0.11	0.22
09910.15 Ext. Painting, Buildings				
Decks, Metal				
Spray				
First Coat				
Minimum	S.F.	0.11	0.20	0.31
Average	"	0.11	0.22	0.33
Maximum	"	0.11	0.24	0.35
Second Coat				
Minimum	S.F.	0.11	0.14	0.24
Average	"	0.11	0.16	0.26
Maximum	"	0.11	0.18	0.29
Decks, Wood, Stained				
Brush				
First Coat				
Minimum	S.F.	0.14	0.22	0.35
Average	"	0.14	0.24	0.38
Maximum	"	0.14	0.27	0.41
Second Coat				
Minimum	S.F.	0.14	0.16	0.29
Average	"	0.14	0.17	0.30
Maximum	"	0.14	0.18	0.32
Roller				
First Coat				

Painting	UNIT	MAT.	INST.	TOTAL
09910.15 Ext. Painting, Buildings				
Minimum	S.F.	0.14	0.16	0.29
Average	"	0.14	0.17	0.30
Maximum	"	0.14	0.18	0.32
Second Coat				
Minimum	S.F.	0.14	0.14	0.27
Average	"	0.14	0.14	0.28
Maximum	"	0.14	0.17	0.30
Spray				
First Coat				
Minimum	S.F.	0.11	0.14	0.24
Average	"	0.11	0.14	0.25
Maximum	"	0.11	0.17	0.28
Second Coat				
Minimum	S.F.	0.11	0.12	0.23
Average	"	0.11	0.13	0.24
Maximum	"	0.11	0.14	0.25
Doors, Metal				
Roller				
First Coat				
Minimum	S.F.	0.14	0.31	0.45
Average	"	0.14	0.36	0.50
Maximum	"	0.14	0.43	0.57
Second Coat				
Minimum	S.F.	0.14	0.22	0.35
Average	"	0.14	0.24	0.38
Maximum	"	0.14	0.27	0.41
Spray				
First Coat				
Minimum	S.F.	0.11	0.27	0.38
Average	"	0.11	0.31	0.42
Maximum	"	0.11	0.36	0.47
Second Coat				
Minimum	S.F.	0.11	0.20	0.31
Average	"	0.11	0.22	0.33
Maximum	"	0.11	0.24	0.35
Door Frames, Metal				
Brush				
First Coat				
Minimum	L.F.	0.16	0.54	0.70
Average	"	0.16	0.68	0.84
Maximum	"	0.16	0.79	0.95
Second Coat				
Minimum	L.F.	0.16	0.31	0.47
Average	"	0.16	0.36	0.52
Maximum	"	0.16	0.43	0.60
Spray				
First Coat				
Minimum	L.F.	0.11	0.24	0.35
Average	"	0.11	0.31	0.42
Maximum	"	0.11	0.43	0.54
Second Coat				
Minimum	L.F.	0.11	0.20	0.31

Painting

09910.15 Ext. Painting, Buildings

	UNIT	MAT.	INST.	TOTAL
Average	L.F.	0.11	0.22	0.33
Maximum	"	0.11	0.24	0.35
Doors, Wood				
Brush				
First Coat				
Minimum	S.F.	0.14	0.67	0.80
Average	"	0.14	0.87	1.00
Maximum	"	0.14	1.10	1.24
Second Coat				
Minimum	S.F.		0.54	0.54
Average	"	0.14	0.62	0.75
Maximum	"	0.14	0.72	0.86
Roller				
First Coat				
Minimum	S.F.	0.14	0.29	0.43
Average	"	0.14	0.36	0.50
Maximum	"	0.14	0.54	0.68
Second Coat				
Minimum	S.F.	0.14	0.22	0.35
Average	"	0.14	0.24	0.38
Maximum	"	0.14	0.36	0.50
Spray				
First Coat				
Minimum	S.F.	0.11	0.14	0.24
Average	"	0.11	0.17	0.28
Maximum	"	0.11	0.22	0.33
Second Coat				
Minimum	S.F.	0.11	0.11	0.22
Average	"	0.11	0.12	0.23
Maximum	"	0.11	0.14	0.25
Gutters and Downspouts				
Brush				
First Coat				
Minimum	L.F.	0.16	0.54	0.70
Average	"	0.16	0.62	0.78
Maximum	"	0.16	0.72	0.88
Second Coat				
Minimum	L.F.	0.16	0.36	0.52
Average	"	0.16	0.43	0.60
Maximum	"	0.16	0.54	0.70
Siding, Metal				
Roller				
First Coat				
Minimum	S.F.	0.14	0.18	0.32
Average	"	0.14	0.20	0.33
Maximum	"	0.14	0.22	0.35
Second Coat				
Minimum	S.F.	0.14	0.17	0.30
Average	"	0.14	0.18	0.32
Maximum	"	0.14	0.20	0.33
Spray				
First Coat				

Painting	UNIT	MAT.	INST.	TOTAL
09910.15 Ext. Painting, Buildings				
Minimum	S.F.	0.11	0.14	0.24
Average	"	0.11	0.16	0.26
Maximum	"	0.11	0.18	0.29
Second Coat				
Minimum	S.F.	0.11	0.09	0.20
Average	"	0.11	0.11	0.22
Maximum	"	0.11	0.14	0.25
Siding, Wood				
Roller				
First Coat				
Minimum	S.F.	0.14	0.16	0.29
Average	"	0.14	0.18	0.32
Maximum	"	0.14	0.20	0.33
Second Coat				
Minimum	S.F.	0.14	0.18	0.32
Average	"	0.14	0.20	0.33
Maximum	"	0.14	0.22	0.35
Spray				
First Coat				
Minimum	S.F.	0.11	0.14	0.25
Average	"	0.11	0.16	0.26
Maximum	"	0.11	0.17	0.28
Second Coat				
Minimum	S.F.	0.11	0.11	0.22
Average	"	0.11	0.14	0.25
Maximum	"	0.11	0.22	0.33
Stucco				
Roller				
First Coat				
Minimum	S.F.	0.16	0.20	0.36
Average	"	0.16	0.23	0.39
Maximum	"	0.16	0.27	0.43
Second Coat				
Minimum	S.F.	0.16	0.16	0.32
Average	"	0.16	0.18	0.35
Maximum	"	0.16	0.22	0.38
Spray				
First Coat				
Minimum	S.F.	0.12	0.14	0.26
Average	"	0.12	0.16	0.28
Maximum	"	0.12	0.18	0.30
Second Coat				
Minimum	S.F.	0.12	0.11	0.23
Average	"	0.12	0.12	0.25
Maximum	"	0.12	0.14	0.27
Trim				
Brush				
First Coat				
Minimum	L.F.	0.16	0.18	0.34
Average	"	0.16	0.22	0.38
Maximum	"	0.16	0.27	0.43
Second Coat				

Painting

09910.15 Ext. Painting, Buildings	UNIT	MAT.	INST.	TOTAL
Minimum	L.F.	0.16	0.14	0.30
Average	"	0.16	0.18	0.34
Maximum	"	0.16	0.27	0.43
Walls				
Roller				
First Coat				
Minimum	S.F.	0.14	0.16	0.29
Average	"	0.14	0.16	0.30
Maximum	"	0.14	0.17	0.31
Second Coat				
Minimum	S.F.	0.14	0.14	0.27
Average	"	0.14	0.14	0.28
Maximum	"	0.14	0.17	0.30
Spray				
First Coat				
Minimum	S.F.	0.11	0.07	0.18
Average	"	0.11	0.09	0.20
Maximum	"	0.11	0.11	0.22
Second Coat				
Minimum	S.F.	0.11	0.06	0.17
Average	"	0.11	0.07	0.18
Maximum	"	0.11	0.10	0.21
Windows				
Brush				
First Coat				
Minimum	S.F.	0.11	0.72	0.83
Average	"	0.11	0.87	0.98
Maximum	"	0.11	1.10	1.21
Second Coat				
Minimum	S.F.	0.11	0.62	0.73
Average	"	0.11	0.72	0.83
Maximum	"	0.11	0.87	0.98

09910.25 Ext. Painting, Misc.	UNIT	MAT.	INST.	TOTAL
Gratings, Metal				
Roller				
First Coat				
Minimum	S.F.	0.14	1.25	1.39
Average	"	0.14	1.45	1.59
Maximum	"	0.14	1.75	1.89
Second Coat				
Minimum	S.F.	0.14	0.87	1.00
Average	"	0.14	1.10	1.24
Maximum	"	0.14	1.45	1.59
Spray				
First Coat				
Minimum	S.F.	0.11	0.62	0.73
Average	"	0.11	0.72	0.83
Maximum	"	0.11	0.87	0.98
Second Coat				
Minimum	S.F.	0.11	0.48	0.59
Average	"	0.11	0.54	0.65

Painting

09910.25 Ext. Painting, Misc.

	UNIT	MAT.	INST.	TOTAL
Maximum	S.F.	0.11	0.62	0.73
Ladders				
Brush				
First Coat				
Minimum	L.F.	0.16	1.10	1.26
Average	"	0.16	1.25	1.41
Maximum	"	0.16	1.45	1.61
Second Coat				
Minimum	L.F.	0.16	0.87	1.03
Average	"	0.16	0.96	1.13
Maximum	"	0.16	1.10	1.26
Spray				
First Coat				
Minimum	L.F.	0.11	0.72	0.83
Average	"	0.11	0.79	0.90
Maximum	"	0.11	0.87	0.98
Second Coat				
Minimum	L.F.	0.11	0.62	0.73
Average	"	0.11	0.67	0.78
Maximum	"	0.11	0.72	0.83
Shakes				
Spray				
First Coat				
Minimum	S.F.	0.12	0.18	0.30
Average	"	0.12	0.20	0.32
Maximum	"	0.12	0.22	0.34
Second Coat				
Minimum	S.F.	0.12	0.17	0.29
Average	"	0.12	0.18	0.30
Maximum	"	0.12	0.20	0.32
Shingles, Wood				
Roller				
First Coat				
Minimum	S.F.	0.14	0.24	0.38
Average	"	0.14	0.27	0.41
Maximum	"	0.14	0.31	0.45
Second Coat				
Minimum	S.F.	0.14	0.17	0.30
Average	"	0.14	0.18	0.32
Maximum	"	0.14	0.20	0.33
Spray				
First Coat				
Minimum	L.F.	0.11	0.17	0.28
Average	"	0.11	0.18	0.29
Maximum	"	0.11	0.20	0.31
Second Coat				
Minimum	L.F.	0.11	0.13	0.24
Average	"	0.11	0.14	0.24
Maximum	"	0.11	0.14	0.25
Shutters and Louvres				
Brush				
First Coat				

Painting

09910.25 Ext. Painting, Misc.

Painting	UNIT	MAT.	INST.	TOTAL
Minimum	EA.	0.16	8.65	8.81
Average	"	0.16	10.80	10.96
Maximum	"	0.16	14.45	14.61
Second Coat				
Minimum	EA.	0.16	5.40	5.56
Average	"	0.16	6.65	6.81
Maximum	"	0.16	8.65	8.81
Spray				
First Coat				
Minimum	EA.	0.11	2.90	3.01
Average	"	0.11	3.45	3.56
Maximum	"	0.11	4.35	4.46
Second Coat				
Minimum	EA.	0.11	2.15	2.26
Average	"	0.11	2.90	3.01
Maximum	"	0.11	3.45	3.56
Stairs, metal				
Brush				
First Coat				
Minimum	S.F.	0.16	0.48	0.64
Average	"	0.16	0.54	0.70
Maximum	"	0.16	0.62	0.78
Second Coat				
Minimum	S.F.	0.16	0.27	0.43
Average	"	0.16	0.31	0.47
Maximum	"	0.16	0.36	0.52
Spray				
First Coat				
Minimum	S.F.	0.11	0.24	0.35
Average	"		0.31	0.31
Maximum	"	0.11	0.33	0.44
Second Coat				
Minimum	S.F.	0.11	0.18	0.29
Average	"	0.11	0.22	0.33
Maximum	"	0.11	0.27	0.38
Steel, Structural, Light				
Brush				
First Coat				
Minimum	S.F.	0.14	0.87	1.00
Average	"	0.14	1.10	1.24
Maximum	"	0.14	1.45	1.59
Second Coat				
Minimum	S.F.	0.11	0.62	0.73
Average	"	0.11	0.72	0.83
Maximum	"	0.11	0.87	0.98
Roller				
First Coat				
Minimum	S.F.	0.14	0.62	0.75
Average	"	0.14	0.72	0.86
Maximum	"	0.14	0.87	1.00
Second Coat				
Minimum	S.F.	0.11	0.36	0.47

Painting

09910.25 Ext. Painting, Misc.	UNIT	MAT.	INST.	TOTAL
Average	S.F.	0.11	0.43	0.54
Maximum	"	0.11	0.54	0.65
Spray				
First Coat				
Minimum	S.F.	0.11	0.36	0.47
Average	"	0.11	0.43	0.54
Maximum	"	0.11	0.54	0.65
Second Coat				
Minimum	S.F.	0.08	0.31	0.39
Average	"	0.08	0.36	0.44
Maximum	"	0.08	0.43	0.51
Steel, Medium to Heavy				
Brush				
First Coat				
Minimum	S.F.	0.14	0.43	0.57
Average	"	0.14	0.48	0.62
Maximum	"	0.14	0.54	0.68
Second Coat				
Minimum	S.F.	0.11	0.36	0.47
Average	"	0.11	0.43	0.54
Maximum	"	0.11	0.51	0.62
Roller				
First Coat				
Minimum	S.F.	0.14	0.36	0.50
Average	"	0.14	0.43	0.57
Maximum	"	0.14	0.51	0.65
Second Coat				
Minimum	S.F.	0.11	0.27	0.38
Average	"	0.11	0.31	0.42
Maximum	"	0.11	0.36	0.47
Spray				
First Coat				
Minimum	S.F.	0.11	0.24	0.35
Average	"	0.11	0.27	0.38
Maximum	"	0.11	0.31	0.42
Second Coat				
Minimum	S.F.	0.08	0.20	0.28
Average	"	0.08	0.22	0.30
Maximum	"	0.08	0.24	0.32
09910.35 Int. Painting, Buildings				
Acoustical Ceiling				
Roller				
First Coat				
Minimum	S.F.	0.16	0.27	0.43
Average	"	0.16	0.36	0.52
Maximum	"	0.16	0.54	0.70
Second Coat				
Minimum	S.F.	0.16	0.22	0.38
Average	"	0.16	0.27	0.43
Maximum	"	0.16	0.36	0.52
Spray				

Painting

09910.35 Int. Painting, Buildings	UNIT	MAT.	INST.	TOTAL
First Coat				
Minimum	S.F.	0.14	0.12	0.26
Average	"	0.14	0.14	0.28
Maximum	"	0.14	0.18	0.32
Second Coat				
Minimum	S.F.	0.14	0.10	0.23
Average	"	0.14	0.11	0.24
Maximum	"	0.14	0.12	0.26
Cabinets and Casework				
Brush				
First Coat				
Minimum	S.F.	0.16	0.43	0.60
Average	"	0.16	0.48	0.64
Maximum	"	0.16	0.54	0.70
Second Coat				
Minimum	S.F.	0.16	0.36	0.52
Average	"	0.16	0.39	0.56
Maximum	"	0.16	0.43	0.60
Spray				
First Coat				
Minimum	S.F.	0.14	0.22	0.35
Average	"	0.14	0.26	0.39
Maximum	"	0.14	0.31	0.45
Second Coat				
Minimum	S.F.	0.14	0.17	0.31
Average	"	0.14	0.19	0.32
Maximum	"	0.14	0.24	0.38
Ceilings				
Roller				
First Coat				
Minimum	S.F.	0.14	0.18	0.32
Average	"	0.14	0.20	0.33
Maximum	"	0.14	0.22	0.35
Second Coat				
Minimum	S.F.	0.14	0.14	0.28
Average	"	0.14	0.17	0.30
Maximum	"	0.14	0.18	0.32
Spray				
First Coat				
Minimum	S.F.	0.11	0.11	0.22
Average	"	0.11	0.12	0.23
Maximum	"	0.11	0.14	0.24
Second Coat				
Minimum	S.F.	0.11	0.08	0.19
Average	"	0.11	0.09	0.20
Maximum	"	0.11	0.11	0.22
Doors, Metal				
Roller				
First Coat				
Minimum	L.F.	0.16	0.29	0.45
Average	"	0.16	0.33	0.50
Maximum	"	0.16	0.39	0.56

Painting

09910.35 Int. Painting, Buildings

	UNIT	MAT.	INST.	TOTAL
Second Coat				
Minimum	L.F.	0.16	0.21	0.37
Average	"	0.16	0.23	0.39
Maximum	"	0.16	0.26	0.42
Spray				
First Coat				
Minimum	L.F.	0.11	0.24	0.35
Average	"	0.11	0.27	0.38
Maximum	"	0.11	0.31	0.42
Second Coat				
Minimum	L.F.	0.11	0.18	0.29
Average	"	0.11	0.20	0.31
Maximum	"	0.11	0.22	0.33
Doors, Wood				
Brush				
First Coat				
Minimum	S.F.	0.16	0.62	0.78
Average	"	0.16	0.79	0.95
Maximum	"	0.16	0.96	1.13
Second Coat				
Minimum	S.F.	0.16	0.48	0.64
Average	"	0.16	0.54	0.70
Maximum	"	0.16	0.62	0.78
Spray				
First Coat				
Minimum	S.F.	0.11	0.13	0.24
Average	"	0.11	0.16	0.26
Maximum	"	0.11	0.20	0.31
Second Coat				
Minimum	S.F.	0.11	0.10	0.21
Average	"	0.11	0.12	0.23
Maximum	"	0.11	0.14	0.24
Ductwork				
Brush				
Minimum	L.F.	0.14	0.54	0.68
Average	"	0.14	0.62	0.75
Maximum	"	0.14	0.72	0.86
Roller				
Minimum	S.F.	0.14	0.36	0.50
Average	"	0.14	0.39	0.53
Maximum	"	0.14	0.43	0.57
Spray				
Minimum	S.F.	0.11	0.16	0.26
Average	"	0.11	0.17	0.28
Maximum	"	0.11	0.18	0.29
Floors				
Roller				
First Coat				
Minimum	S.F.	0.14	0.14	0.27
Average	"	0.14	0.16	0.30
Maximum	"	0.14	0.18	0.32
Second Coat				

Painting

09910.35 Int. Painting, Buildings

	UNIT	MAT.	INST.	TOTAL
Minimum	S.F.	0.14	0.11	0.24
Average	"	0.14	0.12	0.26
Maximum	"	0.14	0.13	0.26
Spray				
First Coat				
Minimum	S.F.	0.11	0.10	0.21
Average	"	0.11	0.10	0.21
Maximum	"	0.11	0.11	0.22
Second Coat				
Minimum	S.F.	0.11	0.08	0.19
Average	"	0.11	0.09	0.20
Maximum	"	0.11	0.10	0.21
Pipes to 6" diameter				
Brush				
Minimum	L.F.	0.16	0.54	0.70
Average	"	0.16	0.62	0.78
Maximum	"	0.16	0.72	0.88
Spray				
Minimum	L.F.	0.14	0.18	0.32
Average	"	0.14	0.22	0.35
Maximum	"	0.14	0.29	0.43
Pipes to 12" diameter				
Brush				
Minimum	L.F.	0.27	1.10	1.37
Average	"	0.27	1.25	1.52
Maximum	"	0.27	1.45	1.72
Spray				
Minimum	L.F.	0.25	0.36	0.61
Average	"	0.25	0.43	0.68
Maximum	"	0.25	0.54	0.79
Trim				
Brush				
First Coat				
Minimum	L.F.	0.16	0.17	0.34
Average	"	0.16	0.20	0.36
Maximum	"	0.16	0.24	0.40
Second Coat				
Minimum	L.F.	0.16	0.13	0.29
Average	"	0.16	0.17	0.33
Maximum	"	0.16	0.24	0.40
Walls				
Roller				
First Coat				
Minimum	S.F.	0.14	0.16	0.29
Average	"	0.14	0.16	0.30
Maximum	"	0.14	0.18	0.32
Second Coat				
Minimum	S.F.	0.14	0.14	0.27
Average	"	0.14	0.14	0.28
Maximum	"	0.14	0.17	0.30
Spray				
First Coat				

Painting	UNIT	MAT.	INST.	TOTAL
09910.35 Int. Painting, Buildings				
Minimum	S.F.	0.11	0.07	0.18
Average	"	0.11	0.08	0.19
Maximum	"	0.11	0.11	0.22
Second Coat				
Minimum	S.F.	0.11	0.06	0.17
Average	"	0.11	0.08	0.18
Maximum	"	0.11	0.10	0.21
09955.10 Wall Covering				
Vinyl wall covering				
Medium duty	S.F.	1.35	0.62	1.97
Heavy duty	"	1.95	0.72	2.67
Over pipes and irregular shapes				
Lightweight, 13 oz.	S.F.	1.35	0.87	2.22
Medium weight, 25 oz.	"	1.55	0.96	2.51
Heavy weight, 34 oz.	"	1.95	1.10	3.05
Cork wall covering				
1' x 1' squares				
1/4" thick	S.F.	2.05	1.10	3.15
1/2" thick	"	2.60	1.10	3.70
3/4" thick	"	2.95	1.10	4.05
Wall fabrics				
Natural fabrics, grass cloths				
Minimum	S.F.	1.30	0.67	1.97
Average	"	1.50	0.72	2.22
Maximum	"	4.90	0.87	5.77
Flexible gypsum coated wall fabric, fire resistant	"	1.50	0.43	1.93
Vinyl corner guards				
3/4" x 3/4" x 8'	EA.	6.95	5.40	12.35
2-3/4" x 2-3/4" x 4'	"	4.10	5.40	9.50
09980.15 Paint				
Paint, enamel				
600 sf per gal.	GAL			37.41
550 sf per gal.	"			33.73
500 sf per gal.	"			24.53
450 sf per gal.	"			22.99
350 sf per gal.	"			21.46
Filler, 60 sf per gal.	"			26.07
Latex, 400 sf per gal.	"			21.46
Aluminum				
400 sf per gal.	GAL			30.66
500 sf per gal.	"			55.17
Red lead, 350 sf per gal.	"			45.99
Primer				
400 sf per gal.	GAL			28.34
300 sf per gal.	"			28.45
Latex base, interior, white	"			23.81
Sealer and varnish				
400 sf per gal.	GAL			21.98
425 sf per gal.	"			32.16
600 sf per gal.	"			43.09

10 SPECIALTIES

Specialties	UNIT	MAT.	INST.	TOTAL
10110.10 Chalkboards				
Chalkboard, metal frame, 1/4" thick				
48"x60"	EA.	260.00	43.30	303.30
48"x96"	"	420.00	48.10	468.10
48"x144"	"	640.00	54.00	694.00
48"x192"	"	830.00	62.00	892.00
Liquid chalkboard				
48"x60"	EA.	310.00	43.30	353.30
48"x96"	"	470.00	48.10	518.10
48"x144"	"	720.00	54.00	774.00
48"x192"	"	920.00	62.00	982.00
Map rail, deluxe	L.F.	7.85	2.15	10.00
Average	PCT.			4.00
10165.10 Toilet Partitions				
Toilet partition, plastic laminate				
Ceiling mounted	EA.	800.00	140.00	940.00
Floor mounted	"	750.00	110.00	860.00
Metal				
Ceiling mounted	EA.	580.00	140.00	720.00
Floor mounted	"	550.00	110.00	660.00
Wheel chair partition, plastic laminate				
Ceiling mounted	EA.	1,070	140.00	1,210
Floor mounted	"	1,000	110.00	1,110
Painted metal				
Ceiling mounted	EA.	840.00	140.00	980.00
Floor mounted	"	760.00	110.00	870.00
Urinal screen, plastic laminate				
Wall hung	EA.	380.00	54.00	434.00
Floor mounted	"	320.00	54.00	374.00
Porcelain enameled steel, floor mounted	"	440.00	54.00	494.00
Painted metal, floor mounted	"	280.00	54.00	334.00
Stainless steel, floor mounted	"	580.00	54.00	634.00
Metal toilet partitions				
Toilet partitions, front door and side divider, floor mounted				
Porcelain enameled steel	EA.	990.00	110.00	1,100
Painted steel	"	550.00	110.00	660.00
Stainless steel	"	1,310	110.00	1,420
10185.10 Shower Stalls				
Shower receptors				
Precast, terrazzo				
32" x 32"	EA.	440.00	39.80	479.80
32" x 48"	"	560.00	47.70	607.70
Concrete				
32" x 32"	EA.	220.00	39.80	259.80
48" x 48"	"	260.00	53.00	313.00
Shower door, trim and hardware				
Economy, 24" wide, chrome frame, tempered glass	EA.	230.00	47.70	277.70
Porcelain enameled steel, flush	"	430.00	47.70	477.70
Baked enameled steel, flush	"	250.00	47.70	297.70
Aluminum frame, tempered glass, 48" wide, sliding	"	540.00	59.50	599.50
Folding	"	520.00	59.50	579.50

Specialties	UNIT	MAT.	INST.	TOTAL
10185.10 Shower Stalls				
Aluminum frame and tempered glass, molded plastic				
Complete with receptor and door				
32" x 32"	EA.	680.00	120.00	800.00
36" x 36"	"	770.00	120.00	890.00
40" x 40"	"	860.00	140.00	1,000
Shower compartment, precast concrete receptor				
Single entry type				
Porcelain enameled steel	EA.	2,010	480.00	2,490
Baked enameled steel	"	1,160	480.00	1,640
Stainless steel	"	2,320	480.00	2,800
Double entry type				
Porcelain enameled steel	EA.	5,730	600.00	6,330
Baked enameled steel	"	3,720	600.00	4,320
Stainless steel	"	5,810	600.00	6,410
10190.10 Cubicles				
Hospital track				
Ceiling hung	L.F.	5.00	4.80	9.80
Suspended	"	5.65	6.20	11.85
Hospital metal dividers, galvanized steel				
Baked enamel finish				
54" high				
10" glass light	L.F.	110.00	21.65	131.65
14" glass light	"	110.00	21.65	131.65
24" glass light	"	100.00	21.65	121.65
60" high				
10" glass light	L.F.	120.00	24.05	144.05
14" glass light	"	120.00	24.05	144.05
24" glass light	"	110.00	24.05	134.05
Stainless steel				
54" high				
10" glass light	L.F.	190.00	24.05	214.05
14" glass light	"	210.00	24.05	234.05
24" glass light	"	240.00	24.05	264.05
60" high				
14" glass light	L.F.	210.00	27.10	237.10
14" glass light	"	210.00	27.10	237.10
24" glass light	"	260.00	27.10	287.10
10210.10 Vents And Wall Louvres				
Block vent, 8"x16"x4" aluminum, w/screen, mill finish	EA.	120.00	16.05	136.05
Standard	"	67.50	15.00	82.50
Vents w/screen, 4" deep, 8" wide, 5" high				
Modular	EA.	79.00	15.00	94.00
Aluminum gable louvers	S.F.	14.10	8.00	22.10
Vent screen aluminum, 4" wide, continuous	L.F.	3.90	1.60	5.50
Louvers, aluminum, anodized, fixed blade				
Horizontal line	S.F.	45.40	12.00	57.40
Vertical line	"	45.40	12.00	57.40
Wall louvre, aluminum mill finish				
Under, 2 sf	S.F.	29.80	6.00	35.80
2 to 4 sf	"	25.00	5.35	30.35

Specialties	UNIT	MAT.	INST.	TOTAL
10210.10 Vents And Wall Louvres				
5 to 10 sf	S.F.	24.25	5.35	29.60
Galvanized steel				
Under 2 sf	S.F.	28.20	6.00	34.20
2 to 4 sf	"	20.35	5.35	25.70
5 to 10 sf	"	18.80	5.35	24.15
10225.10 Door Louvres				
Fixed, 1" thick, enameled steel				
8"x8"	EA.	46.30	5.40	51.70
12"x8"	"	52.50	5.40	57.90
12"x12"	"	59.00	6.20	65.20
16"x12"	"	84.50	6.65	91.15
18"x12"	"	84.50	10.80	95.30
20"x8"	"	94.00	6.20	100.20
20"x12"	"	110.00	12.35	122.35
20"x16"	"	110.00	14.45	124.45
20"x20"	"	110.00	17.30	127.30
24"x12"	"	95.00	14.45	109.45
24"x16"	"	100.00	15.45	115.45
24"x18"	"	120.00	16.65	136.65
24"x20"	"	120.00	18.05	138.05
24"x24"	"	120.00	19.65	139.65
26"x26"	"	140.00	27.10	167.10
10270.40 Access & Pedestal Floor				
Panels, no covering, 2'x2'				
Plain	S.F.	8.70	0.54	9.24
Perforated	"	11.85	21.65	33.50
Pedestals				
For 6" to 12" clearance	EA.	6.65	4.35	11.00
Stringers				
2'	L.F.	1.15	2.05	3.20
6'	"	1.15	1.45	2.60
Accessories				
Ramp assembly	S.F.	41.10	1.75	42.85
Elevated floor assembly	"	64.00	1.60	65.60
Handrail	L.F.	49.80	21.65	71.45
Fascia plate	"	22.90	10.80	33.70
For carpet tiles, add	S.F.			5.52
For vinyl flooring, add	"			7.11
RF shielding components, floor liner				
Hot rolled steel sheet				
14 ga.	S.F.	2.65	1.10	3.75
11 ga.	"	2.85	3.35	6.20
10290.10 Pest Control				
Termite control				
Under slab spraying				
Minimum	S.F.	0.17	0.09	0.26
Average	"	0.23	0.17	0.39
Maximum	"	0.31	0.34	0.64

10 SPECIALTIES

Specialties	UNIT	MAT.	INST.	TOTAL
10350.10 Flagpoles				
Installed in concrete base				
Fiberglass				
25' high	EA.	1,160	290.00	1,450
50' high	"	5,210	720.00	5,930
Aluminum				
25' high	EA.	1,450	290.00	1,740
50' high	"	4,770	720.00	5,490
Bonderized steel				
25' high	EA.	2,270	330.00	2,600
50' high	"	5,300	870.00	6,170
Freestanding tapered, fiberglass				
30' high	EA.	1,300	310.00	1,610
40' high	"	2,290	390.00	2,680
50' high	"	4,630	430.00	5,060
60' high	"	6,080	510.00	6,590
Wall mounted, with collar, brushed aluminum finish				
15' long	EA.	1,010	220.00	1,230
18' long	"	1,300	220.00	1,520
20' long	"	1,450	230.00	1,680
24' long	"	1,590	250.00	1,840
Outrigger, wall, including base				
10' long	EA.	870.00	290.00	1,160
20' long	"	1,300	360.00	1,660
10400.10 Identifying Devices				
Directory and bulletin boards				
Open face boards				
Chrome plated steel frame	S.F.	26.80	21.65	48.45
Aluminum framed	"	23.60	21.65	45.25
Bronze framed	"	25.40	21.65	47.05
Stainless steel framed	"	50.50	21.65	72.15
Tack board, aluminum framed	"	15.15	21.65	36.80
Visual aid board, aluminum framed	"	15.15	21.65	36.80
Glass encased boards, hinged and keyed				
Aluminum framed	S.F.	58.50	54.00	112.50
Bronze framed	"	63.50	54.00	117.50
Stainless steel framed	"	150.00	54.00	204.00
Chrome plated steel framed	"	140.00	54.00	194.00
Metal plaque				
Cast bronze	S.F.	380.00	36.10	416.10
Aluminum	"	360.00	36.10	396.10
Metal engraved plaque				
Porcelain steel	S.F.	480.00	36.10	516.10
Stainless steel	"	440.00	36.10	476.10
Brass	"	480.00	36.10	516.10
Aluminum	"	220.00	36.10	256.10
Metal built-up plaque				
Bronze	S.F.	370.00	43.30	413.30
Copper and bronze	"	380.00	43.30	423.30
Copper and aluminum	"	420.00	43.30	463.30
Metal nameplate plaques				
Cast bronze	S.F.	370.00	27.10	397.10

Specialties	UNIT	MAT.	INST.	TOTAL
10400.10 Identifying Devices				
Cast aluminum	S.F.	340.00	27.10	367.10
Engraved, 1-1/2" x 6"				
Bronze	EA.	120.00	27.10	147.10
Aluminum	"	120.00	27.10	147.10
Letters, on masonry or concrete, aluminum, satin finish				
1/2" thick				
2" high	EA.	13.85	17.30	31.15
4" high	"	15.45	21.65	37.10
6" high	"	20.35	24.05	44.40
3/4" thick				
8" high	EA.	26.80	27.10	53.90
10" high	"	34.20	30.90	65.10
1" thick				
12" high	EA.	44.70	36.10	80.80
14" high	"	50.50	43.30	93.80
16" high	"	70.00	54.00	124.00
For polished aluminum add, 15%				
For clear anodized aluminum add, 15%				
For colored anodic aluminum add, 30%				
For profiled and color enameled letters add, 50%				
Cast bronze, satin finish letters				
3/8" thick				
2" high	EA.	25.20	17.30	42.50
4" high	"	27.30	21.65	48.95
1/2" thick, 6" high	"	35.50	24.05	59.55
5/8" thick, 8" high	"	45.90	27.10	73.00
1" thick				
10" high	EA.	69.50	30.90	100.40
12" high	"	88.50	36.10	124.60
14" high	"	100.00	43.30	143.30
16" high	"	130.00	54.00	184.00
Interior door signs, adhesive, flexible				
2" x 8"	EA.	16.55	8.45	25.00
4" x 4"	"	17.05	8.45	25.50
6" x 7"	"	22.30	8.45	30.75
6" x 9"	"	29.30	8.45	37.75
10" x 9"	"	39.10	8.45	47.55
10" x 12"	"	50.50	8.45	58.95
Hard plastic type, no frame				
3" x 8"	EA.	35.80	8.45	44.25
4" x 4"	"	34.20	8.45	42.65
4" x 12"	"	42.30	8.45	50.75
Hard plastic type, with frame				
3" x 8"	EA.	120.00	8.45	128.45
4" x 4"	"	92.50	8.45	100.95
4" x 12"	"	150.00	8.45	158.45
10450.10 Control				
Access control, 7' high, indoor or outdoor impenetrability				
Remote or card control, type B	EA.	3,910	590.00	4,500
Free passage, type B	"	3,610	590.00	4,200
Remote or card control, type AA	"	5,110	590.00	5,700

10 SPECIALTIES

Specialties	UNIT	MAT.	INST.	TOTAL
10450.10 Control				
Free passage, type AA	EA.	4,660	590.00	5,250
10500.10 Lockers				
Locker bench, floor mounted, laminated maple				
4'	EA.	120.00	36.10	156.10
6'	"	160.00	36.10	196.10
Wardrobe locker, 12" x 60" x 15", baked on enamel				
1-tier	EA.	160.00	21.65	181.65
2-tier	"	170.00	21.65	191.65
3-tier	"	180.00	22.80	202.80
4-tier	"	200.00	22.80	222.80
12" x 72" x 15", baked on enamel				
1-tier	EA.	160.00	21.65	181.65
2-tier	"	190.00	21.65	211.65
4-tier	"	230.00	22.80	252.80
5-tier	"	230.00	22.80	252.80
15" x 60" x 15", baked on enamel				
1-tier	EA.	200.00	21.65	221.65
4-tier	"	220.00	22.80	242.80
Wardrobe locker, single tier type				
12" x 15" x 72"	EA.	160.00	43.30	203.30
18" x 15" x 72"	"	190.00	45.60	235.60
12" x 18" x 72"	"	180.00	48.10	228.10
18" x 18" x 72"	"	210.00	51.00	261.00
Double tier type				
12" x 15" x 36"	EA.	110.00	21.65	131.65
18" x 15" x 36"	"	110.00	21.65	131.65
12" x 18" x 36"	"	100.00	21.65	121.65
18" x 18" x 36"	"	120.00	21.65	141.65
Two person unit				
18" x 15" x 72"	EA.	310.00	72.00	382.00
18" x 18" x 72"	"	180.00	86.50	266.50
Duplex unit				
15" x 15" x 72"	EA.	300.00	43.30	343.30
15" x 21" x 72"	"	310.00	43.30	353.30
Basket lockers, basket sets with baskets				
24 basket set	SET	640.00	220.00	860.00
30 basket set	"	770.00	270.00	1,040
36 basket set	"	860.00	360.00	1,220
42 basket set	"	960.00	430.00	1,390
10520.10 Fire Protection				
Portable fire extinguishers				
Water pump tank type				
2.5 gal.				
Red enameled galvanized	EA.	110.00	22.55	132.55
Red enameled copper	"	120.00	22.55	142.55
Polished copper	"	190.00	22.55	212.55
Carbon dioxide type, red enamel steel				
Squeeze grip with hose and horn				
2.5 lb	EA.	62.00	22.55	84.55
5 lb	"	140.00	26.00	166.00

10 SPECIALTIES

Specialties

Specialties	UNIT	MAT.	INST.	TOTAL
10520.10 Fire Protection				
10 lb	EA.	210.00	33.80	243.80
15 lb	"	250.00	42.30	292.30
20 lb	"	290.00	42.30	332.30
Wheeled type				
125 lb	EA.	1,560	67.50	1,628
250 lb	"	1,610	67.50	1,678
500 lb	"	3,010	67.50	3,078
Dry chemical, pressurized type				
Red enameled steel				
2.5 lb	EA.	37.20	22.55	59.75
5 lb	"	62.00	26.00	88.00
10 lb	"	77.50	33.80	111.30
20 lb	"	120.00	42.30	162.30
30 lb	"	320.00	42.30	362.30
Chrome plated steel, 2.5 lb	"	77.50	22.55	100.05
Other type extinguishers				
2.5 gal, stainless steel, pressurized water tanks	EA.	77.50	22.55	100.05
Soda and acid type	"	140.00	22.55	162.55
Cartridge operated, water type	"	230.00	22.55	252.55
Loaded stream, water type	"	110.00	22.55	132.55
Foam type	"	90.00	22.55	112.55
40 gal, wheeled foam type	"	4,700	67.50	4,768
Fire extinguisher cabinets				
Enameled steel				
8" x 12" x 27"	EA.	170.00	67.50	237.50
8" x 16" x 38"	"	230.00	67.50	297.50
Aluminum				
8" x 12" x 27"	EA.	230.00	67.50	297.50
8" x 16" x 38"	"	270.00	67.50	337.50
8" x 12" x 27"	"	380.00	67.50	447.50
Stainless steel				
8" x 16" x 38"	EA.	490.00	67.50	557.50
10550.10 Postal Specialties				
Mail chutes				
Single mail chute				
Finished aluminum	L.F.	610.00	110.00	720.00
Bronze	"	890.00	110.00	1,000
Single mail chute receiving box				
Finished aluminum	EA.	960.00	220.00	1,180
Bronze	"	1,790	220.00	2,010
Twin mail chute, double parallel				
Finished aluminum	FLR	1,140	220.00	1,360
Bronze	"	1,790	220.00	2,010
Receiving box, 36" x 20" x 12"				
Finished aluminum	EA.	1,990	360.00	2,350
Bronze	"	2,850	360.00	3,210
Locked receiving mail box				
Finished aluminum	EA.	960.00	220.00	1,180
Bronze	"	1,620	220.00	1,840
Commercial postal accessories for mail chutes				
Letter slot, brass	EA.	320.00	72.00	392.00

Specialties	UNIT	MAT.	INST.	TOTAL
10550.10 Postal Specialties				
Bulk mail slot, brass	EA.	910.00	72.00	982.00
Mail boxes				
Residential postal accessories				
Letter slot	EA.	94.50	21.65	116.15
Rural letter box	"	66.50	54.00	120.50
Apartment house, keyed, 3.5" x 4.5" x 16"	"	97.50	14.45	111.95
Ranch style	"	130.00	21.65	151.65
Commercial postal accessories				
Letter box, with combination lock	EA.	79.50	15.45	94.95
Key lock	"	72.50	15.45	87.95
Mail box, aluminum w/glass front, 4x5				
Horizontal rear load	EA.	100.00	12.35	112.35
Vertical front load	"	77.00	12.35	89.35
10600.10 Movable Partitions				
Partition, movable office type, 2-1/2" thick, vinyl-gypsum	S.F.	12.50	2.15	14.65
Enameled steel frame, with 1/4" thick clear glass	"	17.20	2.15	19.35
Door frame and hardware for movable partitions	EA.	500.00	140.00	640.00
Add for acoustic movable partition	S.F.			1.08
Accordion partition, 12' high				
Vinyl	S.F.	7.70	7.20	14.90
Acoustical	"	9.10	7.20	16.30
Standard office cubicles, 8' high, steel framed				
Baked enamel finish				
100% flush	L.F.	120.00	10.80	130.80
75% flush and 25% glass	"	140.00	12.00	152.00
50% flush and 50% glass	"	210.00	12.00	222.00
100% glass	"	250.00	14.45	264.45
Natural hardwood panels				
100% flush	L.F.	130.00	14.45	144.45
50% flush and 50% glass	"	170.00	15.45	185.45
Plastic laminated panels				
100% flush	L.F.	150.00	14.45	164.45
75% flush and 25% glass	"	190.00	15.45	205.45
50% flush and 50% glass	"	220.00	15.45	235.45
Vinyl covered panels				
100% flush	L.F.	130.00	14.90	144.90
75% flush and 25% glass	"	170.00	16.05	186.05
50% and 50% glass	"	200.00	16.05	216.05
Aluminum framed				
Enameled or anodized aluminum panels				
100% flush	L.F.	100.00	10.80	110.80
75% flush and 25% glass	"	98.50	12.00	110.50
50% flush and 50% glass	"	91.50	12.00	103.50
Vinyl covered panels				
100% flush	L.F.	110.00	14.90	124.90
75% flush and 25% glass	"	140.00	16.05	156.05
50% flush and 50% glass	"	160.00	16.05	176.05
60" high partitions, steel framed				
Enameled panels	L.F.	54.50	9.60	64.10
Natural hardwood panels, two sides	"	230.00	10.05	240.05
Plastic laminated panels	"	120.00	10.05	130.05

Specialties	UNIT	MAT.	INST.	TOTAL
10600.10 Movable Partitions				
Vinyl covered panels	L.F.	160.00	9.60	169.60
Aluminum framed				
Anodized or baked enamel panels	L.F.	79.00	9.60	88.60
Natural hardwood panels	"	240.00	10.05	250.05
Plastic laminated panels	"	120.00	10.05	130.05
Vinyl covered panels	"	150.00	9.60	159.60
Wire mesh partitions				
Wall panels				
4' x 7'	EA.	120.00	27.10	147.10
4' x 8'	"	130.00	28.90	158.90
4' x 10'	"	160.00	33.30	193.30
Wall filler panels				
1' x 7'	EA.	69.00	27.10	96.10
1' x 8'	"	70.50	28.90	99.40
1' x 10'	"	94.00	30.90	124.90
2' x 7'	"	67.00	27.10	94.10
2' x 8'	"	72.00	28.90	100.90
2' x 10'	"	84.50	30.90	115.40
3' x 7'	"	89.00	27.10	116.10
3' x 8'	"	97.00	28.90	125.90
3' x 10'	"	120.00	30.90	150.90
Ceiling panels				
10' x 2'	EA.	81.50	62.00	143.50
10' x 4'	"	150.00	86.50	236.50
Wall panel with service window				
5' wide				
7' high	EA.	380.00	27.10	407.10
8' high	"	400.00	28.90	428.90
10' high	"	410.00	30.90	440.90
Doors				
Sliding				
3' x 7'	EA.	470.00	110.00	580.00
3' x 8'	"	430.00	120.00	550.00
3' x 10'	"	440.00	170.00	610.00
4' x 7'	"	440.00	120.00	560.00
4' x 8'	"	450.00	170.00	620.00
4' x 10'	"	490.00	220.00	710.00
5' x 7'	"	480.00	170.00	650.00
5' x 8'	"	580.00	220.00	800.00
5' x 10'	"	670.00	220.00	890.00
Swing door				
3' x 7'	EA.	340.00	110.00	450.00
4' x 7'	"	380.00	120.00	500.00
Swing door, with 1' transom				
3' x 7'	EA.	370.00	120.00	490.00
4' x 7'	"	420.00	170.00	590.00
Swing door, with 3' transom				
3' x 7'	EA.	320.00	170.00	490.00
4' x 7'	"	350.00	220.00	570.00

Specialties	UNIT	MAT.	INST.	TOTAL
10670.10 Shelving				
Shelving, enamel, closed side and back, 12" x 36"				
5 shelves	EA.	160.00	72.00	232.00
8 shelves	"	200.00	96.00	296.00
Open				
5 shelves	EA.	100.00	72.00	172.00
8 shelves	"	100.00	96.00	196.00
Metal storage shelving, baked enamel				
7 shelf unit, 72" or 84" high				
10" shelf	L.F.	43.50	43.30	86.80
12" shelf	"	45.10	45.60	90.70
15" shelf	"	51.50	48.10	99.60
18" shelf	"	51.50	51.00	102.50
24" shelf	"	59.50	54.00	113.50
30" shelf	"	64.50	57.50	122.00
36" shelf	"	77.50	62.00	139.50
4 shelf unit, 40" high				
10" shelf	L.F.	37.00	36.10	73.10
12" shelf	"	38.60	39.30	77.90
15" shelf	"	48.30	43.30	91.60
18" shelf	"	54.50	45.60	100.10
24" shelf	"	54.50	48.10	102.60
3 shelf unit, 32" high				
10" shelf	L.F.	30.60	21.65	52.25
12" shelf	"	30.60	22.80	53.40
15" shelf	"	33.80	24.05	57.85
18" shelf	"	35.40	25.50	60.90
24" shelf	"	40.30	27.10	67.40
Single shelf unit, attached to masonry				
10" shelf	L.F.	10.95	7.20	18.15
12" shelf	"	12.10	7.85	19.95
15" shelf	"	12.40	8.30	20.70
18" shelf	"	14.00	8.85	22.85
24" shelf	"	16.90	9.40	26.30
For stainless steel, add to material, 120%				
For attachment to gypsum board, add to labor, 50%				
Built-in wood shelves				
Posts and trimmed plywood	L.F.	2.65	6.20	8.85
Solid clear pine	"	4.45	6.65	11.10
Closet shelf, pine with rod	"	3.00	6.65	9.65
For lumber edge band, add to material	"			1.57
For prefinished shelves, add to material, 225%				
10750.10 Telephone Enclosures				
Telephone enclosure, wall mounted, shelf, 28" x 30" x 15"	EA.	1,030	110.00	1,140
Directory shelf, stainless steel, 3 binders	"	800.00	72.00	872.00
10800.10 Bath Accessories				
Ash receiver, wall mounted, aluminum	EA.	170.00	21.65	191.65
Grab bar, 1-1/2" dia., stainless steel, wall mounted				
24" long	EA.	47.60	21.65	69.25
36" long	"	69.00	22.80	91.80
42" long	"	79.00	24.05	103.05

Specialties

10800.10 Bath Accessories	UNIT	MAT.	INST.	TOTAL
48" long	EA.	95.00	25.50	120.50
52" long	"	100.00	27.10	127.10
1" dia., stainless steel				
12" long	EA.	37.80	18.80	56.60
18" long	"	41.60	19.65	61.25
24" long	"	44.30	21.65	65.95
30" long	"	49.30	22.80	72.10
36" long	"	56.00	24.05	80.05
48" long	"	57.50	25.50	83.00
Hand dryer, surface mounted, 110 volt	"	530.00	54.00	584.00
Medicine cabinet, 16 x 22, baked enamel, steel, lighted	"	85.50	17.30	102.80
With mirror, lighted	"	150.00	28.90	178.90
Mirror, 1/4" plate glass, up to 10 sf	S.F.	8.55	4.35	12.90
Mirror, stainless steel frame				
18"x24"	EA.	130.00	14.45	144.45
18"x32"	"	140.00	17.30	157.30
18"x36"	"	150.00	21.65	171.65
24"x30"	"	170.00	21.65	191.65
24"x36"	"	190.00	24.05	214.05
24"x48"	"	340.00	36.10	376.10
24"x60"	"	350.00	43.30	393.30
30"x30"	"	190.00	43.30	233.30
30"x72"	"	540.00	54.00	594.00
48"x72"	"	560.00	72.00	632.00
With shelf, 18"x24"	"	190.00	17.30	207.30
Sanitary napkin dispenser, stainless steel, wall mounted	"	460.00	28.90	488.90
Shower rod, 1" diameter				
Chrome finish over brass	EA.	77.00	21.65	98.65
Stainless steel	"	74.00	21.65	95.65
Soap dish, stainless steel, wall mounted	"	93.50	28.90	122.40
Toilet tissue dispenser, stainless, wall mounted				
Single roll	EA.	46.00	10.80	56.80
Double roll	"	72.00	12.35	84.35
Towel dispenser, stainless steel				
Flush mounted	EA.	130.00	24.05	154.05
Surface mounted	"	120.00	21.65	141.65
Combination towel dispenser and waste receptacle	"	410.00	28.90	438.90
Towel bar, stainless steel				
18" long	EA.	39.40	17.30	56.70
24" long	"	46.00	19.65	65.65
30" long	"	49.30	21.65	70.95
36" long	"	54.00	24.05	78.05
Toothbrush and tumbler holder	"	34.50	14.45	48.95
Waste receptacle, stainless steel, wall mounted	"	280.00	36.10	316.10

10900.10 Wardrobe Specialties	UNIT	MAT.	INST.	TOTAL
Hospital wardrobe units, 24" x 24" x 76", with door				
Baked enameled steel	EA.	2,070	240.00	2,310
Hardwood	"	1,110	240.00	1,350
Stainless steel	"	3,480	240.00	3,720
Plastic laminated	"	1,240	240.00	1,480
Dormitory wardrobe units, 24" x 76", with door				

Specialties	UNIT	MAT.	INST.	TOTAL
10900.10 Wardrobe Specialties				
Hardwood	EA.	1,110	240.00	1,350
Plastic laminated	"	980.00	240.00	1,220
Hat and coat rack				
Single tier				
Baked enameled steel	L.F.	130.00	10.80	140.80
Stainless steel	"	150.00	10.80	160.80
Aluminum	"	140.00	10.80	150.80
Double tier				
Baked enameled steel	L.F.	220.00	12.35	232.35
Stainless steel	"	310.00	12.35	322.35
Aluminum	"	270.00	12.35	282.35

11 EQUIPMENT

Architectural Equipment

Architectural Equipment	UNIT	MAT.	INST.	TOTAL
11010.10 Maintenance Equipment				
Vacuum cleaning system				
3 valves				
1.5 hp	EA.	1,490	480.00	1,970
2.5 hp	"	1,790	620.00	2,410
5 valves	"	2,380	870.00	3,250
7 valves	"	2,680	1,080	3,760
11020.10 Security Equipment				
Bulletproof teller window				
4' x 4'	EA.	10,520	720.00	11,240
5' x 4'	"	12,020	870.00	12,890
Bulletproof partitions				
Up to 12' high, 2.5" thick	S.F.	57.00	2.90	59.90
Counter for banks				
Minimum	L.F.	650.00	86.50	736.50
Maximum	"	2,920	140.00	3,060
Drive-up window				
Minimum	EA.	7,740	620.00	8,360
Maximum	"	37,580	1,440	39,020
Night depository				
Minimum	EA.	10,520	620.00	11,140
Maximum	"	21,040	1,440	22,480
Office safes, 30" x 20" x 20", 1 hr rating	"	2,400	110.00	2,510
30" x 16" x 15", 2 hr rating	"	2,250	86.50	2,337
30" x 28" x 20", H&G rating	"	10,520	54.00	10,574
Service windows, pass through painted steel				
24" x 36"	EA.	4,510	430.00	4,940
48" x 40"	"	6,760	540.00	7,300
72" x 40"	"	7,520	870.00	8,390
Special doors and windows				
3' x 7' bulletproof door with frame	EA.	5,260	620.00	5,880
12" x 12" vision panel	"	3,010	310.00	3,320
Surveillance system				
Minimum	EA.	5,600	870.00	6,470
Maximum	"	43,430	4,330	47,760
Vault door, 3' wide, 6'6" high				
3-1/2" thick	EA.	20,530	5,410	25,940
7" thick	"	24,480	7,210	31,690
10" thick	"	29,210	8,660	37,870
Insulated vault door				
2 hr rating				
32" wide	EA.	3,310	430.00	3,740
40" wide	"	3,630	460.00	4,090
4 hr rating				
32" wide	EA.	3,550	480.00	4,030
40" wide	"	3,710	540.00	4,250
6 hr rating				
32" wide	EA.	3,790	480.00	4,270
40" wide	"	4,180	540.00	4,720
Insulated file room door				
1 hr rating				
32" wide	EA.	3,790	430.00	4,220

Architectural Equipment	UNIT	MAT.	INST.	TOTAL
11020.10 Security Equipment				
40" wide	EA.	4,030	480.00	4,510
11060.10 Theater Equipment				
Roll out stage, steel frame, wood floor				
Manual	S.F.	33.50	2.70	36.20
Electric	"	36.70	4.35	41.05
Portable stages				
8" high	S.F.	10.65	2.15	12.80
18" high	"	12.30	2.40	14.70
36" high	"	13.55	2.55	16.10
48" high	"	14.35	2.70	17.05
Band risers				
Minimum	S.F.	29.50	2.15	31.65
Maximum	"	47.80	2.15	49.95
Chairs for risers				
Minimum	EA.	81.50	1.55	83.05
Maximum	"	150.00	1.55	151.55
11080.10 Police Equipment				
Firing range equipment, rifle				
3 position	EA.	12,280	1,440	13,720
4 position	"	16,110	2,160	18,270
5 position	"	19,950	2,400	22,350
6 position	"	24,560	2,550	27,110
11090.10 Checkroom Equipment				
Motorized checkroom equipment				
No shelf system, 6'4" height				
7'6" length	EA.	2,450	430.00	2,880
14'6" length	"	2,760	430.00	3,190
28' length	"	3,810	430.00	4,240
One shelf, 6'8" height				
7'6" length	EA.	3,000	430.00	3,430
14'6" length	"	3,630	430.00	4,060
28' length	"	5,370	430.00	5,800
Two shelves, 7'5" height				
7'6" length	EA.	3,630	430.00	4,060
14'6" length	"	4,420	430.00	4,850
28' length	"	6,310	430.00	6,740
Three shelves, 8' height				
7'6" length	EA.	3,790	870.00	4,660
14'6" length	"	4,580	870.00	5,450
28' length	"	6,470	870.00	7,340
Four shelves, 8'7" height				
7'6" length	EA.	3,790	870.00	4,660
14'6" length	"	4,730	870.00	5,600
28' length	"	6,470	870.00	7,340
11110.10 Laundry Equipment				
High capacity, heavy duty				
Washer extractors				
135 lb				
Standard	EA.	55,010	360.00	55,370

11 EQUIPMENT

Architectural Equipment

Architectural Equipment	UNIT	MAT.	INST.	TOTAL
11110.10 Laundry Equipment				
Pass through	EA.	63,530	360.00	63,890
200 lb				
Standard	EA.	61,210	360.00	61,570
Pass through	"	68,960	360.00	69,320
110 lb dryer	"	10,840	360.00	11,200
Hand operated presser	"	7,440	480.00	7,920
Mushroom press	"	4,800	480.00	5,280
Spreader feeders				
2 station	EA.	71,280	480.00	71,760
4 station	"	78,250	870.00	79,120
Delivery carts				
12 bushel	EA.	470.00	5.40	475.40
16 bushel	"	560.00	5.75	565.75
18 bushel	"	570.00	6.20	576.20
30 bushel	"	820.00	7.20	827.20
40 bushel	"	1,010	8.65	1,019
Low capacity				
Pressers				
Air operated	EA.	5,890	170.00	6,060
Hand operated	"	4,800	170.00	4,970
Extractor, low capacity	"	6,890	170.00	7,060
Ironer, 48"	"	3,250	86.50	3,337
Coin washers				
10 lb capacity	EA.	1,460	86.50	1,547
20 lb capacity	"	3,740	86.50	3,827
Coin dryer	"	600.00	54.00	654.00
Coin dry cleaner, 20 lb	"	29,440	170.00	29,610
11161.10 Loading Dock Equipment				
Dock leveler, 10 ton capacity				
6' x 8'	EA.	4,340	430.00	4,770
7' x 8'	"	4,480	430.00	4,910
Bumpers, laminated rubber				
4-1/2" thick				
6" x 14"	EA.	73.50	8.65	82.15
6" x 36"	"	75.00	9.60	84.60
10" x 14"	"	79.00	10.80	89.80
10" x 24"	"	89.50	12.35	101.85
10" x 36"	"	100.00	14.45	114.45
12" x 14"	"	86.50	11.40	97.90
12" x 24"	"	92.50	13.50	106.00
12" x 36"	"	120.00	16.05	136.05
6" thick				
10" x 14"	EA.	92.50	12.35	104.85
10" x 24"	"	120.00	14.90	134.90
10" x 36"	"	130.00	21.65	151.65
Extruded rubber bumpers				
T-section, 22" x 22" x 3"	EA.	120.00	8.65	128.65
Molded rubber bumpers				
24" x 12" x 3" thick	EA.	71.50	21.65	93.15
Door seal, 12" x 12", vinyl covered	L.F.	41.10	10.80	51.90
Dock boards, heavy duty, 5' x 5'				

11 EQUIPMENT

Architectural Equipment

	UNIT	MAT.	INST.	TOTAL
11161.10 Loading Dock Equipment				
5000 lb				
Minimum	EA.	1,150	360.00	1,510
Maximum	"	1,970	360.00	2,330
9000 lb				
Minimum	EA.	1,320	360.00	1,680
Maximum	"	2,200	390.00	2,590
15,000 lb	"	1,740	390.00	2,130
Truck shelters				
Minimum	EA.	1,380	330.00	1,710
Maximum	"	2,280	620.00	2,900
11170.10 Waste Handling				
Incinerator, electric				
100 lb/hr				
Minimum	EA.	21,180	440.00	21,620
Maximum	"	36,520	440.00	36,960
400 lb/hr				
Minimum	EA.	37,600	880.00	38,480
Maximum	"	73,030	880.00	73,910
1000 lb/hr				
Minimum	EA.	73,040	1,340	74,380
Maximum	"	116,860	1,340	118,200
Incinerator, medical-waste				
25 lb/hr, 2-7 x 4-0	EA.	21,910	880.00	22,790
50 lb/hr, 2-11 x 4-11	"	23,370	880.00	24,250
75 lb/hr, 3-8 x 5-0	"	32,140	1,770	33,910
100 lb/hr, 3-8 x 6-0	"	36,520	1,770	38,290
Industrial compactor				
1 cy	EA.	10,660	490.00	11,150
3 cy	"	21,480	630.00	22,110
5 cy	"	27,750	880.00	28,630
Trash chutes steel, including sprinklers				
18" dia.	L.F.	72.00	220.00	292.00
24" dia.	"	93.50	230.00	323.50
30" dia.	"	120.00	240.00	360.00
36" dia.	"	160.00	250.00	410.00
Refuse bottom hopper	EA.	2,300	240.00	2,540
11400.10 Food Service Equipment				
Unit kitchens				
30" compact kitchen				
Refrigerator, with range, sink	EA.	3,530	220.00	3,750
Sink only	"	3,220	150.00	3,370
Range only	"	2,990	110.00	3,100
Cabinet for upper wall section	"	410.00	63.00	473.00
Stainless shield, for rear wall	"	130.00	17.70	147.70
Side wall	"	70.50	17.70	88.20
42" compact kitchen				
Refrigerator with range, sink	EA.	4,370	250.00	4,620
Sink only	"	3,760	220.00	3,980
Cabinet for upper wall section	"	570.00	73.50	643.50
Stainless shield, for rear wall	"	180.00	18.45	198.45

Architectual Equipment

11400.10 Food Service Equipment

Architectural Equipment	UNIT	MAT.	INST.	TOTAL
Side wall	EA.	70.50	18.45	88.95
54" compact kitchen				
Refrigerator, oven, range, sink	EA.	5,220	320.00	5,540
Cabinet for upper wall section	"	750.00	88.50	838.50
Stainless shield, for				
Rear wall	EA.	260.00	20.10	280.10
Side wall	"	70.50	20.10	90.60
60" compact kitchen				
Refrigerator, oven, range, sink	EA.	6,060	320.00	6,380
Cabinet for upper wall section	"	860.00	88.50	948.50
Stainless shield, for				
Rear wall	EA.	260.00	20.10	280.10
Side wall	"	70.50	20.10	90.60
72" compact kitchen				
Refrigerator, oven, range, sink	EA.	5,910	370.00	6,280
Cabinet for upper wall section	"	1,690	88.50	1,779
Stainless shield for				
Rear wall	EA.	310.00	22.10	332.10
Side wall	"	70.50	22.10	92.60
Bake oven				
Single deck				
Minimum	EA.	1,990	55.50	2,046
Maximum	"	4,600	110.00	4,710
Double deck				
Minimum	EA.	2,300	73.50	2,374
Maximum	"	8,770	110.00	8,880
Triple deck				
Minimum	EA.	3,220	73.50	3,294
Maximum	"	10,960	150.00	11,110
Convection type oven, electric, 40" x 45" x 57"				
Minimum	EA.	3,220	55.50	3,276
Maximum	"	5,840	110.00	5,950
Broiler, without oven, 69" x 26" x 39"				
Minimum	EA.	3,210	55.50	3,266
Maximum	"	5,520	73.50	5,594
Coffee urns, 10 gallons				
Minimum	EA.	6,290	150.00	6,440
Maximum	"	10,130	220.00	10,350
Fryer, with submerger				
Single				
Minimum	EA.	1,460	88.50	1,549
Maximum	"	36,530	150.00	36,680
Double				
Minimum	EA.	2,780	110.00	2,890
Maximum	"	5,840	150.00	5,990
Griddle, counter				
3' long				
Minimum	EA.	1,900	73.50	1,974
Maximum	"	4,760	88.50	4,849
5' long				
Minimum	EA.	2,780	110.00	2,890
Maximum	"	6,140	150.00	6,290

Architectural Equipment

11400.10 Food Service Equipment

Architectural Equipment	UNIT	MAT.	INST.	TOTAL
Kettles, steam, jacketed				
20 gallons				
Minimum	EA.	2,920	110.00	3,030
Maximum	"	5,980	220.00	6,200
40 gallons				
Minimum	EA.	3,510	110.00	3,620
Maximum	"	6,600	220.00	6,820
60 gallons				
Minimum	EA.	3,990	110.00	4,100
Maximum	"	8,820	220.00	9,040
Range				
Heavy duty, single oven, open top				
Minimum	EA.	2,220	55.50	2,276
Maximum	"	6,140	150.00	6,290
Fry top				
Minimum	EA.	3,210	55.50	3,266
Maximum	"	6,720	150.00	6,870
Steamers, electric				
27 kw				
Minimum	EA.	8,440	110.00	8,550
Maximum	"	16,880	150.00	17,030
18 kw				
Minimum	EA.	4,380	110.00	4,490
Maximum	"	10,670	150.00	10,820
Dishwasher, rack type				
Single tank, 190 racks/hr	EA.	13,200	220.00	13,420
Double tank				
234 racks/hr	EA.	19,190	250.00	19,440
265 racks/hr	"	21,180	290.00	21,470
Dishwasher, automatic 100 meals/hr	"	62,010	150.00	62,160
Disposals				
100 gal/hr	EA.	1,460	150.00	1,610
120 gal/hr	"	2,070	150.00	2,220
250 gal/hr	"	5,300	160.00	5,460
Exhaust hood for dishwasher, gutter 4 sides, s-steel				
4'x4'x2'	EA.	2,340	160.00	2,500
4'x7'x2'	"	4,090	180.00	4,270
Food preparation machines				
Vertical cutter mixers				
25 quart	EA.	8,210	150.00	8,360
40 quart	"	9,890	150.00	10,040
80 quart	"	11,280	220.00	11,500
130 quart	"	12,430	370.00	12,800
Choppers				
5 lb	EA.	2,340	110.00	2,450
16 lb	"	2,780	150.00	2,930
40 lb	"	4,970	220.00	5,190
Mixers, floor models				
20 quart	EA.	3,650	55.50	3,706
60 quart	"	10,230	55.50	10,286
80 quart	"	12,430	63.00	12,493
140 quart	"	17,530	88.50	17,619

Architectural Equipment

Architectural Equipment	UNIT	MAT.	INST.	TOTAL
11400.10 Food Service Equipment				
Ice cube maker				
50 lb per day				
Minimum	EA.	1,460	440.00	1,900
Maximum	"	2,610	440.00	3,050
500 lb per day				
Minimum	EA.	4,990	740.00	5,730
Maximum	"	9,050	740.00	9,790
Ice flakers				
300 lb per day	EA.	4,090	440.00	4,530
600 lb per day	"	5,550	740.00	6,290
1000 lb per day	"	7,310	980.00	8,290
2000 lb per day	"	13,150	1,110	14,260
Refrigerated cases				
Dairy products				
Multi deck type	L.F.	700.00	29.50	729.50
For rear sliding doors, add	"			79.83
Delicatessen case, service deli				
Single deck	L.F.	660.00	220.00	880.00
Multi deck	"	720.00	280.00	1,000
Meat case				
Single deck	L.F.	470.00	260.00	730.00
Multi deck	"	520.00	280.00	800.00
Produce case				
Single deck	L.F.	500.00	260.00	760.00
Multi deck	"	570.00	280.00	850.00
Bottle coolers				
6' long				
Minimum	EA.	1,920	880.00	2,800
Maximum	"	3,910	880.00	4,790
10' long				
Minimum	EA.	2,610	1,470	4,080
Maximum	"	5,060	1,470	6,530
Frozen food cases				
Chest type	L.F.	510.00	260.00	770.00
Reach-in, glass door	"	1,060	280.00	1,340
Island case, single	"	520.00	260.00	780.00
Multi deck	"	1,150	280.00	1,430
Ice storage bins				
500 lb capacity	EA.	1,460	630.00	2,090
1000 lb capacity	"	2,840	1,260	4,100
11450.10 Residential Equipment				
Compactor, 4 to 1 compaction	EA.	1,420	110.00	1,530
Dishwasher, built-in				
2 cycles	EA.	1,210	220.00	1,430
4 or more cycles	"	1,910	220.00	2,130
Disposal				
Garbage disposer	EA.	490.00	150.00	640.00
Heaters, electric, built-in				
Ceiling type	EA.	250.00	150.00	400.00
Wall type				
Minimum	EA.	170.00	110.00	280.00

Architectural Equipment

Architectural Equipment	UNIT	MAT.	INST.	TOTAL
11450.10 Residential Equipment				
Maximum	EA.	330.00	150.00	480.00
Hood for range, 2-speed, vented				
30" wide	EA.	170.00	150.00	320.00
42" wide	"	680.00	150.00	830.00
Ice maker, automatic				
30 lb per day	EA.	840.00	63.00	903.00
50 lb per day	"	1,910	220.00	2,130
Folding access stairs, disappearing metal stair				
8' long	EA.	1,990	63.00	2,053
11' long	"	2,070	63.00	2,133
12' long	"	2,230	63.00	2,293
Wood frame, wood stair				
22" x 54" x 8'9" long	EA.	110.00	44.20	154.20
25" x 54" x 10' long	"	140.00	44.20	184.20
Ranges electric				
Built-in, 30", 1 oven	EA.	1,200	150.00	1,350
2 oven	"	1,830	150.00	1,980
Counter top, 4 burner, standard	"	670.00	110.00	780.00
With grill	"	1,990	110.00	2,100
Free standing, 21", 1 oven	"	760.00	150.00	910.00
30", 1 oven	"	990.00	88.50	1,079
2 oven	"	2,950	88.50	3,039
Water softener				
30 grains per gallon	EA.	1,000	150.00	1,150
70 grains per gallon	"	1,910	220.00	2,130
11470.10 Darkroom Equipment				
Dryers				
36" x 25" x 68"	EA.	8,850	240.00	9,090
48" x 25" x 68"	"	9,140	240.00	9,380
Processors, film				
Black and white	EA.	14,020	240.00	14,260
Color negatives	"	15,490	240.00	15,730
Prints	"	16,730	240.00	16,970
Transparencies	"	18,590	240.00	18,830
Sinks with cabinet and/or stand				
5" sink with stand				
24" x 48"	EA.	1,470	120.00	1,590
32" x 64"	"	1,620	160.00	1,780
38" x 52"	"	1,770	160.00	1,930
42" x 132"	"	2,950	240.00	3,190
48" x 52"	"	2,650	240.00	2,890
5" sink with cabinet				
24" x 48"	EA.	2,360	120.00	2,480
32" x 64"	"	2,510	160.00	2,670
38" x 52"	"	2,510	160.00	2,670
42" x 132"	"	4,720	240.00	4,960
48" x 52"	"	3,100	240.00	3,340
10" sink with stand				
24" x 48"	EA.	1,470	120.00	1,590
32" x 64"	"	1,770	160.00	1,930
38" x 52"	"	2,060	160.00	2,220

11 EQUIPMENT

Architectural Equipment

Architectural Equipment	UNIT	MAT.	INST.	TOTAL
11470.10 Darkroom Equipment				
10" sink with cabinet				
24" x 48"	EA.	2,510	120.00	2,630
38" x 52"	"	2,880	160.00	3,040
11480.10 Athletic Equipment				
Basketball backboard				
Fixed	EA.	830.00	540.00	1,370
Swing-up	"	1,910	870.00	2,780
Portable, hydraulic	"	10,600	220.00	10,820
Suspended type, standard	"	3,740	870.00	4,610
For glass backboard, add	"			1,146
For electrically operated, add	"			1,512
Bleacher, telescoping, manual				
15 tier, minimum	SEAT	110.00	8.65	118.65
Maximum	"	120.00	8.65	128.65
20 tier, minimum	"	110.00	9.60	119.60
Maximum	"	120.00	9.60	129.60
30 tier, minimum	"	110.00	14.45	124.45
Maximum	"	120.00	14.45	134.45
Boxing ring elevated, complete, 22' x 22'	EA.	10,580	6,180	16,760
Gym divider curtain				
Minimum	S.F.	4.30	0.58	4.88
Maximum	"	5.90	0.58	6.48
Scoreboards, single face				
Minimum	EA.	3,900	430.00	4,330
Maximum	"	25,180	2,160	27,340
Parallel bars				
Minimum	EA.	1,300	430.00	1,730
Maximum	"	3,660	720.00	4,380
11500.10 Industrial Equipment				
Vehicular paint spray booth, solid back, 14'4" x 9'6"				
24' deep	EA.	13,210	430.00	13,640
26'6" deep	"	13,370	430.00	13,800
28'6" deep	"	13,620	430.00	14,050
Drive through, 14'9" x 9'6"				
24' deep	EA.	13,620	430.00	14,050
26'6" deep	"	13,940	430.00	14,370
28'6" deep	"	14,100	430.00	14,530
Water wash, paint spray booth				
5' x 11'2" x 10'8"	EA.	6,850	430.00	7,280
6' x 11'2" x 10'8"	"	7,010	430.00	7,440
8' x 11'2" x 10'8"	"	8,860	430.00	9,290
10' x 11'2" x 11'2"	"	9,830	430.00	10,260
12' x 12'2" x 11'2"	"	12,010	430.00	12,440
14' x 12'2" x 11'2"	"	14,100	430.00	14,530
16' x 12'2" x 11'2"	"	15,870	430.00	16,300
20' x 12'2" x 11'2"	"	19,020	430.00	19,450
Dry type spray booth, with paint arrestors				
5'4" x 7'2" x 6'8"	EA.	2,980	430.00	3,410
6'4" x 7'2" x 6'8"	"	3,630	430.00	4,060
8'4" x 7'2" x 9'2"	"	4,110	430.00	4,540

11 EQUIPMENT

Architectural Equipment

11500.10 Industrial Equipment	UNIT	MAT.	INST.	TOTAL
10'4" x 7'2" x 9'2"	EA.	4,750	430.00	5,180
12'4" x 7'6" x 9'2"	"	4,670	430.00	5,100
14'4" x 7'6" x 9'8"	"	6,360	430.00	6,790
16'4" x 7'7" x 9'8"	"	7,170	430.00	7,600
20'4" x 7'7" x 10'8"	"	8,060	430.00	8,490
Air compressor, electric				
1 hp				
115 volt	EA.	3,630	290.00	3,920
5 hp				
115 volt	EA.	7,490	430.00	7,920
230 volt	"	6,360	430.00	6,790
Hydraulic lifts				
8,000 lb capacity	EA.	5,560	1,080	6,640
11,000 lb capacity	"	7,810	1,730	9,540
24,000 lb capacity	"	11,040	2,890	13,930
Power tools				
Band saws				
10"	EA.	760.00	36.10	796.10
14"	"	1,320	43.30	1,363
Motorized shaper	"	710.00	33.30	743.30
Motorized lathe	"	840.00	36.10	876.10
Bench saws				
9" saw	EA.	1,510	28.90	1,539
10" saw	"	2,340	30.90	2,371
12" saw	"	2,980	36.10	3,016
Electric grinders				
1/3 hp	EA.	290.00	17.30	307.30
1/2 hp	"	420.00	18.80	438.80
3/4 hp	"	500.00	18.80	518.80

11600.10 Laboratory Equipment	UNIT	MAT.	INST.	TOTAL
Cabinets, base				
Minimum	L.F.	130.00	36.10	166.10
Maximum	"	240.00	36.10	276.10
Full storage, 7' high				
Minimum	L.F.	200.00	36.10	236.10
Maximum	"	310.00	36.10	346.10
Wall				
Minimum	L.F.	88.50	43.30	131.80
Maximum	"	150.00	43.30	193.30
Counter tops				
Minimum	S.F.	18.90	5.40	24.30
Average	"	44.40	6.20	50.60
Maximum	"	72.50	7.20	79.70
Tables				
Open underneath	S.F.	100.00	21.65	121.65
Doors underneath	"	190.00	27.10	217.10
Medical laboratory equipment				
Analyzer				
Chloride	EA.	590.00	22.10	612.10
Blood	"	26,280	36.90	26,317
Bath, water, utility, countertop unit	"	430.00	44.20	474.20

Architectural Equipment

Architectural Equipment	UNIT	MAT.	INST.	TOTAL
11600.10 Laboratory Equipment				
Hot plate, lab, countertop	EA.	140.00	40.20	180.20
Stirrer	"	240.00	40.20	280.20
Incubator, anaerobic, 23x23x36"	"	6,900	220.00	7,120
Dry heat bath	"	510.00	73.50	583.50
Incinerator, for sterilizing	"	210.00	4.40	214.40
Meter, serum protein	"	920.00	5.55	925.55
Ph analog, general purpose	"	690.00	6.30	696.30
Refrigerator, blood bank, undercounter type 153 litres	"	5,090	73.50	5,164
5.4 cf, undercounter type	"	4,270	73.50	4,344
Refrigerator/freezer, 4.4 cf, undercounter type	"	480.00	73.50	553.50
Sealer, impulse, free standing, 20x12x4"	"	260.00	14.75	274.75
Timer, electric, 1-60 minutes, bench or wall mounted	"	69.00	24.60	93.60
Glassware washer - dryer, undercounter	"	7,800	550.00	8,350
Balance, torsion suspension, tabletop, 4.5 lb capacity	"	550.00	24.60	574.60
Binocular microscope, with in base illuminator	"	2,210	17.00	2,227
Centrifuge, table model, 19x16x13"	"	1,280	17.70	1,298
Clinical model, with four place head	"	670.00	9.85	679.85
11700.10 Medical Equipment				
Hospital equipment, lights				
Examination, portable	EA.	1,100	36.90	1,137
Meters				
Air flow meter	EA.	64.00	24.60	88.60
Oxygen flow meters	"	43.90	18.45	62.35
Racks				
40 chart, revolving open frame; mobile caddy	EA.	1,270	36.90	1,307
Scales.				
Clinical, metric with measure rod, 350 lb	EA.	390.00	40.20	430.20
Physical therapy				
Chair, hydrotherapy	EA.	320.00	7.20	327.20
Diathermy, shortwave, portable, on casters	"	2,870	17.30	2,887
Exercise bicycle, floor standing, 35" x 15"	"	790.00	14.45	804.45
Hydrocollator, 4 pack, portable, 129 x 90 x 160"	"	320.00	6.20	326.20
Lamp, infrared, mobile with variable heat control	"	420.00	33.30	453.30
Ultra violet, base mounted	"	390.00	33.30	423.30
Mirror, posture training, 27" wide and 72" high	"	440.00	10.80	450.80
Parallel bars, adjustable	"	1,940	54.00	1,994
Platform mat 10'x6', 1" thick	"	520.00	10.80	530.80
Pulley, duplex, wall mounted	"	370.00	140.00	510.00
Rack, crutch, wall mounted, 66 x 16 x 13"	"	810.00	43.30	853.30
Stimulator, galvanic-faradic, hand held	"	83.50	2.90	86.40
Ultrasound, muscle stimulator, portable, 13x13x8"	"	2,110	3.60	2,114
Sandbag set, velcro straps, saddle bag type	"	110.00	6.20	116.20
Whirlpool, 85 gallon	"	2,870	220.00	3,090
65 gallon capacity	"	2,450	220.00	2,670
Radiology				
Radiographic table, motor driven tilting table	EA.	31,260	4,330	35,590
Fluoroscope image/tv system	"	87,880	8,660	96,540
Processor for washing and drying radiographs				
Water filter unit, 30" x 48-1/2" x 37-1/2"	EA.	28.20	740.00	768.20
Cassette transfer cabinet	"	880.00	36.90	916.90
Base storage cabinets, sectional design				

Architectural Equipment

11700.10 Medical Equipment

Architectural Equipment	UNIT	MAT.	INST.	TOTAL
With back splash, 24" deep and 35" high	L.F.	2,160	36.90	2,197
Wall storage cabinets	"	160.00	55.50	215.50
Steam sterilizers				
For heat and moisture stable materials	EA.	3,210	44.20	3,254
For fast drying after sterilization	"	2,200	55.50	2,256
Compact unit	"	1,550	55.50	1,606
Semi-automatic	"	6,420	220.00	6,640
Floor loading				
Single door	EA.	116,600	370.00	116,970
Double door	"	147,020	440.00	147,460
Utensil washer, sanitizer	"	6,590	340.00	6,930
Automatic washer/sterilizer	"	25,350	880.00	26,230
16 x 16 x 26", including generator & accessories	"	24,500	1,470	25,970
Steam generator, elec., 10 kw to 180 kw	"	22,810	880.00	23,690
Surgical scrub				
Minimum	EA.	2,500	150.00	2,650
Maximum	"	5,240	150.00	5,390
Gas sterilizers				
Automatic, free standing, 21x19x29"	EA.	4,480	440.00	4,920
Surgical tables				
Minimum	EA.	11,610	630.00	12,240
Maximum	"	52,390	880.00	53,270
Surgical lights, ceiling mounted				
Minimum	EA.	4,390	740.00	5,130
Maximum	"	7,100	880.00	7,980
Water stills				
4 liters/hr	EA.	2,280	150.00	2,430
8 liters/hr	"	3,290	150.00	3,440
19 liters/hr	"	9,120	370.00	9,490
X-ray equipment				
Mobile unit				
Minimum	EA.	7,770	220.00	7,990
Maximum	"	38,870	440.00	39,310
Film viewers				
Minimum	EA.	160.00	73.50	233.50
Maximum	"	690.00	150.00	840.00
Autopsy table				
Minimum	EA.	4,390	440.00	4,830
Maximum	"	8,700	440.00	9,140
Incubators				
15 cf	EA.	4,050	220.00	4,270
29 cf	"	4,900	370.00	5,270
Infant transport, portable	"	10,310	230.00	10,540
Beds				
Stretcher, with pad, 30" x 78"	EA.	1,570	110.00	1,680
Transfer, for patient transport	"	2,700	110.00	2,810
Headwall				
Aluminum, with back frame and console	EA.	3,630	220.00	3,850
Hospital ground detection system				
Power ground module	EA.	780.00	130.00	910.00
Ground slave module	"	540.00	96.00	636.00
Master ground module	"	340.00	83.50	423.50

11 EQUIPMENT

Architectural Equipment

11700.10 Medical Equipment

Architectural Equipment	UNIT	MAT.	INST.	TOTAL
Remote indicator	EA.	370.00	88.50	458.50
X-ray indicator	"	1,130	96.00	1,226
Micro ammeter	"	1,380	110.00	1,490
Supervisory module	"	1,070	96.00	1,166
Ground cords	"	89.50	16.40	105.90
Hospital isolation monitors, 5 ma				
120v	EA.	2,200	190.00	2,390
208v	"	2,110	190.00	2,300
240v	"	2,110	190.00	2,300
Digital clock-timers separate display	"	860.00	88.50	948.50
One display	"	830.00	88.50	918.50
Remote control	"	490.00	69.00	559.00
Battery pack	"	110.00	69.00	179.00
Surgical chronometer clock and 3 timers	"	1,820	140.00	1,960
Auxilary control	"	660.00	64.00	724.00

11700.20 Dental Equipment

Architectural Equipment	UNIT	MAT.	INST.	TOTAL
Dental care equipment				
Drill console with accessories	EA.	12,970	740.00	13,710
Amalgamator	"	320.00	22.10	342.10
Lathe	"	340.00	14.75	354.75
Finish polisher	"	250.00	29.50	279.50
Model trimmer	"	410.00	20.10	430.10
Motor, wall mounted	"	450.00	20.10	470.10
Cleaner, ultrasonic	"	1,690	44.20	1,734
Curing unit, bench mounted	"	2,310	73.50	2,384
Oral evacuation system, dual pump	"	12,540	55.50	12,596
Sterilizer, table top, self contained	"	880.00	24.60	904.60
Dental lights				
Light, floor or ceiling mounted	EA.	2,080	220.00	2,300
X-ray unit				
Portable	EA.	[illegible],220	110.00	[illegible]
Wall mounted with remote control	"	6,650	370.00	7,020
Illuminator, single panel	"	9,470	630.00	10,100
X-ray film processor	"	1,550	370.00	1,920
Shield, portable x-ray, lead lined	"	340.00	29.50	369.50

12 FURNISHINGS

Interior

	UNIT	MAT.	INST.	TOTAL
12302.10 Casework				
Kitchen base cabinet, prefinished, 24" deep, 35" high				
12"wide	EA.	180.00	43.30	223.30
18" wide	"	220.00	43.30	263.30
24" wide	"	240.00	48.10	288.10
27" wide	"	280.00	48.10	328.10
36" wide	"	320.00	54.00	374.00
48" wide	"	360.00	54.00	414.00
Corner cabinet, 36" wide	"	250.00	54.00	304.00
Wall cabinet, 12" deep, 12" high				
30" wide	EA.	140.00	43.30	183.30
36" wide	"	140.00	43.30	183.30
15" high				
30" wide	EA.	140.00	48.10	188.10
36" wide	"	160.00	48.10	208.10
24" high				
30" wide	EA.	170.00	48.10	218.10
36" wide	"	210.00	48.10	258.10
30" high				
12" wide	EA.	130.00	54.00	184.00
18" wide	"	150.00	54.00	204.00
24" wide	"	160.00	54.00	214.00
27" wide	"	190.00	54.00	244.00
30" wide	"	220.00	62.00	282.00
36" wide	"	220.00	62.00	282.00
Corner cabinet, 30" high				
24" wide	EA.	210.00	72.00	282.00
30" wide	"	260.00	72.00	332.00
36" wide	"	280.00	72.00	352.00
Wardrobe	"	660.00	110.00	770.00
Vanity with top, laminated plastic				
24" wide	EA.	410.00	110.00	520.00
30" wide	"	500.00	110.00	610.00
36" wide	"	520.00	140.00	660.00
48" wide	"	580.00	170.00	750.00
12390.10 Counter Tops				
Stainless steel, counter top, with backsplash	S.F.	110.00	10.80	120.80
Acid-proof, kemrock surface	"	32.30	7.20	39.50
12500.10 Window Treatment				
Drapery tracks, wall or ceiling mounted				
Basic traverse rod				
50 to 90"	EA.	41.10	21.65	62.75
84 to 156"	"	61.50	24.05	85.55
136 to 250"	"	79.00	24.05	103.05
165 to 312"	"	110.00	27.10	137.10
Traverse rod with stationary curtain rod				
30 to 50"	EA.	49.30	21.65	70.95
50 to 90"	"	61.00	21.65	82.65
84 to 156"	"	88.50	24.05	112.55
136 to 250"	"	100.00	27.10	127.10
Double traverse rod				

Interior	UNIT	MAT.	INST.	TOTAL
12500.10 Window Treatment				
30 to 50"	EA.	59.00	21.65	80.65
50 to 84"	"	79.00	21.65	100.65
84 to 156"	"	97.00	24.05	121.05
136 to 250"	"	120.00	27.10	147.10
12510.10 Blinds				
Venetian blinds				
2" slats	S.F.	4.65	1.10	5.75
1" slats	"	6.30	1.10	7.40
12690.40 Floor Mats				
Recessed entrance mat, 3/8" thick, aluminum link	S.F.	23.00	21.65	44.65
Steel, flexible	"	10.15	21.65	31.80

13 SPECIAL

Construction

	UNIT	MAT.	INST.	TOTAL
13056.10 Vaults				
Floor safes				
Class C				
1.0 cf	EA.	680.00	36.10	716.10
1.3 cf	"	770.00	54.00	824.00
1.9 cf	"	850.00	72.00	922.00
5.2 cf	"	2,040	72.00	2,112
13121.10 Pre-engineered Buildings				
Pre-engineered metal building, 40'x100'				
14' eave height	S.F.	10.95	3.60	14.55
16' eave height	"	11.55	4.15	15.70
20' eave height	"	13.15	5.40	18.55
60'x100'				
14' eave height	S.F.	9.45	3.60	13.05
16' eave height	"	9.90	4.15	14.05
20' eave height	"	11.05	5.40	16.45
80'x100'				
14' eave height	S.F.	9.45	3.60	13.05
16' eave height	"	9.75	4.15	13.90
20' eave height	"	11.20	5.40	16.60
100'x100'				
14' eave height	S.F.	13.85	3.60	17.45
16' eave height	"	9.90	4.15	14.05
20' eave height	"	10.85	5.40	16.25
100'x150'				
14' eave height	S.F.	8.60	3.60	12.20
16' eave height	"	9.25	4.15	13.40
20' eave height	"	10.25	5.40	15.65
120'x150'				
14' eave height	S.F.	8.60	3.60	12.20
16' eave height	"	9.10	4.15	13.25
20' eave height	"	9.90	5.40	15.30
140'x150'				
14' eave height	S.F.	8.10	3.60	11.70
16' eave height	"	8.80	4.15	12.95
20' eave height	"	9.75	5.40	15.15
160'x200'				
14' eave height	S.F.	7.95	3.60	11.55
16' eave height	"	8.05	4.15	12.20
20' eave height	"	9.10	5.40	14.50
200'x200'				
14' eave height	S.F.	7.95	3.60	11.55
16' eave height	"	8.30	4.15	12.45
20' eave height	"	8.95	5.40	14.35
Hollow metal door and frame, 6' x 7'	EA.			841.92
Sectional steel overhead door, manually operated				
8' x 8'	EA.			909.84
12' x 12'	"			1,367
Roll-up steel door, manually operated				
10' x 10'	EA.			1,687
12' x 12'	"			2,026
For gravity ridge ventilator with birdscreen	"			622.60

13 SPECIAL

Construction

	UNIT	MAT.	INST.	TOTAL
13121.10 Pre-engineered Buildings				
9" throat x 10'	EA.			673.54
12" throat x 10'	"			707.50
For 20" rotary vent with damper	"			285.83
For 4' x 3' fixed louver	"			302.81
For 4' x 3' aluminum sliding window	"			311.30
For 3' x 9' fiberglass panels	"			124.95
Liner panel, 26 ga, painted steel	S.F.	1.95	1.20	3.15
Wall panel insulated, 26 ga. steel, foam core	"	6.25	1.20	7.45
Roof panel, 26 ga. painted steel	"	1.60	0.69	2.29
Plastic (sky light)	"	4.60	0.69	5.29
Insulation, 3-1/2" thick blanket, R11	"	1.45	0.32	1.77
13152.10 Swimming Pool Equipment				
Diving boards				
14' long				
Aluminum	EA.	1,550	190.00	1,740
Fiberglass	"	810.00	190.00	1,000
Ladders, heavy duty				
2 steps				
Minimum	EA.	530.00	67.50	597.50
Maximum	"	840.00	67.50	907.50
4 steps				
Minimum	EA.	590.00	84.50	674.50
Maximum	"	1,110	84.50	1,195
Lifeguard chair				
Minimum	EA.	1,390	340.00	1,730
Maximum	"	1,950	340.00	2,290
Lights, underwater				
12 volt, with transformer	EA.	390.00	84.50	474.50
110 volt				
Minimum	EA.	360.00	84.50	444.50
Maximum	"	1,110	84.50	1,195
Ground fault interrupter for 110 volt, each light	"	140.00	28.20	168.20
Pool covers				
Reinforced polyethylene	S.F.	0.66	2.60	3.26
Vinyl water tube				
Minimum	S.F.	0.74	2.60	3.34
Maximum	"	1.15	2.60	3.75
Slides with water tube				
Minimum	EA.	460.00	280.00	740.00
Maximum	"	1,110	280.00	1,390

Elevators

14210.10 Elevators

Elevators	UNIT	MAT.	INST.	TOTAL
Passenger elevators, electric, geared				
Based on a shaft of 6 stops and 6 openings				
50 fpm, 2000 lb	EA.	95,640	1,820	97,460
100 fpm, 2000 lb	"	99,190	2,020	101,210
150 fpm				
2000 lb	EA.	109,810	2,270	112,080
3000 lb	"	138,150	2,600	140,750
4000 lb	"	143,470	3,030	146,500
200 fpm				
2500 lb	EA.	133,720	2,600	136,320
3000 lb	"	137,270	2,800	140,070
4000 lb	"	145,240	3,030	148,270
250 fpm				
2500 lb	EA.	138,150	2,600	140,750
3000 lb	"	145,240	2,800	148,040
4000 lb	"	148,780	3,030	151,810
300 fpm				
2500 lb	EA.	138,150	2,600	140,750
3000 lb	"	143,470	2,800	146,270
4000 lb	"	148,780	1,820	150,600
For each additional; 50 fpm, add per stop, $3000				
500 lb, add per stop, $4000				
Opening, add per stop, $4500				
Stop, add per stop, $4000				
Bonderized steel door, add per opening, $150				
Colored aluminum door, add per opening, $850				
Stainless steel door, add per opening, $600				
Cast bronze door, add per opening, $1100				
Two speed door, add per opening, $360				
Bi-parting door, add per opening, $850				
Custom cab interior add, $4800				
Based on a shaft of 8 stops and 8 openings				
300 fpm				
3000 lb	EA.	164,720	3,640	168,360
3500 lb	"	168,400	3,640	172,040
4000 lb	"	177,120	4,040	181,160
5000 lb	"	198,380	4,330	202,710
400 fpm				
3000 lb	EA.	173,580	3,640	177,220
3500 lb	"	176,230	3,640	179,870
4000 lb	"	194,840	4,040	198,880
5000 lb	"	212,550	4,330	216,880
600 fpm				
3000 lb	EA.	247,970	4,040	252,010
3500 lb	"	251,510	4,330	255,840
4000 lb	"	253,280	4,430	257,710
5000 lb	"	264,800	4,550	269,350
800 fpm				
3000 lb	EA.	292,250	4,040	296,290
3500 lb	"	295,790	4,330	300,120
4000 lb	"	299,330	4,430	303,760
5000 lb	"	300,930	4,550	305,480

Elevators	UNIT	MAT.	INST.	TOTAL
14210.10 Elevators				
For each additional; 100 fpm add per stop, $13,000				
500 lb, add per stop, $6500				
Opening add per stop, $12,000				
Stop add per stop, $4800				
Bypass floor, add per each, $2000				
Bonderized steel door, add per opening, $150				
Colored aluminum door, add per opening, $900				
Stainless steel door, add per opening, $600				
Cast bronze door, add per opening, $600				
Two speed bi-parting door, add per opening, $1000				
Custom cab interior, add $5000				
Hydraulic, based on a shaft of 3 stops, 3 openings				
50 fpm				
2000 lb	EA.	62,600	1,520	64,120
2500 lb	"	66,990	1,520	68,510
3000 lb	"	70,680	1,580	72,260
100 fpm				
2000 lb	EA.	68,750	1,520	70,270
2500 lb	"	72,440	1,580	74,020
3000 lb	"	77,010	1,650	78,660
150 fpm				
2000 lb	EA.	74,200	1,520	75,720
2500 lb	"	81,230	1,580	82,810
3000 lb	"	86,860	1,730	88,590
For each additional; 50 fpm add per stop, $3500				
500 lb, add per stop, $3500				
Opening, add, $4200				
Stop, add per stop, $5300				
Bonderized steel door, add per opening, $400				
Colored aluminum door, add per opening, $1500				
Stainless steel door, add per opening, $650				
Cast bronze door, add per opening, $1200				
Two speed door, add per opening, $400				
Bi-parting door, add per opening, $900				
Custom cab interior, add per cab, $5000				
Small elevators, 4 to 6 passenger capacity				
Electric, push				
2 stops	EA.	22,860	1,520	24,380
3 stops	"	28,130	1,650	29,780
4 stops	"	32,530	1,820	34,350
Freight elevators, electric				
Based on a shaft of 6 stops and 6 openings				
50 fpm				
3500 lb	EA.	172,450	2,020	174,470
4000 lb	"	173,110	2,020	175,130
5000 lb	"	174,750	2,270	177,020
100 fpm				
3500 lb	EA.	181,480	2,270	183,750
4000 lb	"	182,310	2,270	184,580
5000 lb	"	177,870	2,600	180,470
200 fpm				
3500 lb	EA.	178,690	2,600	181,290

Elevators

	UNIT	MAT.	INST.	TOTAL
14210.10 Elevators				
4000 lb	EA.	179,510	2,600	182,110
5000 lb	"	181,160	3,030	184,190
For elevator with manual door, deduct 15%				
For variable voltage control, add 20%				
Based on shaft of 8 stops and 8 openings				
100 fpm				
4000 lb	EA.	181,980	2,270	184,250
6000 lb	"	175,740	2,330	178,070
8000 lb	"	179,020	2,460	181,480
150 fpm				
4000 lb	EA.	175,740	2,600	178,340
6000 lb	"	179,020	2,670	181,690
8000 lb	"	180,660	2,840	183,500
200 fpm				
4000 lb	EA.	177,380	3,030	180,410
6000 lb	"	180,660	3,130	183,790
8000 lb	"	188,870	3,310	192,180
For each additional; 50 fpm, add per stop, $2000				
500 lb, add per stop, $600				
Opening, add per stop, $7000				
Stop, add per stop, $5500				
For variable voltage, add 20%				
Hydraulic, based on 3 stops and 3 openings				
50 fpm				
3000 lb	EA.	73,680	1,300	74,980
4000 lb	"	80,380	1,350	81,730
6000 lb	"	93,780	1,400	95,180
100 fpm				
3000 lb	EA.	83,060	1,300	84,360
4000 lb	"	88,750	1,350	90,100
6000 lb	"	103,660	1,400	105,060
150 fpm				
3000 lb	EA.	91,270	1,300	92,570
4000 lb	"	97,960	1,350	99,310
6000 lb	"	112,370	1,400	113,770
For each additional; 50 fpm, add per stop, $2000				
500 lb, add per stop, $600				
Opening, add per stop, $5500				
Stop, add per stop, $5500				
For elevator with manual door deduct from total, 15%				
14300.10 Escalators				
Escalators				
32" wide, floor to floor				
12' high	EA.	125,760	3,030	128,790
15' high	"	136,380	3,640	140,020
18' high	"	147,010	4,550	151,560
22' high	"	159,410	6,060	165,470
25' high	"	177,120	7,270	184,390
48" wide				
12' high	EA.	139,930	3,130	143,060
15' high	"	152,320	3,790	156,110

14 CONVEYING

Elevators

	UNIT	MAT.	INST.	TOTAL
14300.10 Escalators				
18' high	EA.	162,950	4,780	167,730
22' high	"	182,430	6,490	188,920
25' high	"	194,740	7,270	202,010

Lifts

	UNIT	MAT.	INST.	TOTAL
14410.10 Personnel Lifts				
Electrically operated, 1 or 2 person lift				
With attached foot platforms				
3 stops	EA.			20,127
5 stops	"			25,762
7 stops	"			30,592
For each additional stop, add $1250				
Residential stair climber, per story	EA.	5,960	370.00	6,330
14450.10 Vehicle Lifts				
Automotive hoist, one post, semi-hydraulic, 8,000 lb	EA.	4,860	1,820	6,680
Full hydraulic, 8,000 lb	"	4,600	1,820	6,420
2 post, semi-hydraulic, 10,000 lb	"	5,020	2,600	7,620
Full hydraulic				
10,000 lb	EA.	5,360	2,600	7,960
13,000 lb	"	6,610	4,550	11,160
18,500 lb	"	8,620	4,550	13,170
24,000 lb	"	11,640	4,550	16,190
26,000 lb	"	13,480	4,550	18,030
Pneumatic hoist, fully hydraulic				
11,000 lb	EA.	6,450	6,060	12,510
24,000 lb	"	11,300	6,060	17,360

Material Handling

	UNIT	MAT.	INST.	TOTAL
14560.10 Chutes				
Linen chutes, stainless steel, with supports				
18" dia.	L.F.	120.00	3.45	123.45
24" dia.	"	160.00	3.70	163.70
30" dia.	"	170.00	4.00	174.00
Hopper	EA.	1,960	32.10	1,992
Skylight	"	1,180	48.10	1,228
Sprinkler unit at top	"	430.00	53.50	483.50
For galvanized metal, deduct from material cost, 35%				

14 CONVEYING

Material Handling

	UNIT	MAT.	INST.	TOTAL
14560.10 Chutes				
For aluminum, deduct from material cost, 25%				
14580.10 Pneumatic Systems				
Pneumatic message tube system				
Average, 20 station job				
3" round system	E.A.	310,470	4,020	314,490
4" round system	"	388,090	4,420	392,510
6" round system	"	586,450	4,920	591,370
4" x 7" oval system	"	612,320	8,850	621,170
Trash and linen tube system				
10 stations	EA.	198,360	9,090	207,450
15 stations	"	250,100	12,120	262,220
20 stations	"	335,510	13,990	349,500
30 stations	"	439,830	16,530	456,360

Hoists And Cranes

	UNIT	MAT.	INST.	TOTAL
14600.10 Industrial Hoists				
Industrial hoists, electric, light to medium duty				
500 lb	EA.	6,600	220.00	6,820
1000 lb	"	6,930	230.00	7,160
2000 lb	"	7,100	250.00	7,350
3000 lb	"	7,230	260.00	7,490
4000 lb	"	8,180	280.00	8,460
5000 lb	"	11,560	290.00	11,850
6000 lb	"	11,950	310.00	12,260
7500 lb	"	12,780	320.00	13,100
10,000 lb	"	21,570	330.00	21,900
15,000 lb	"	32,030	340.00	32,370
20,000 lb	"	37,830	370.00	38,200
25,000 lb	"	41,480	400.00	41,880
30,000 lb	"	40,670	440.00	41,110
Heavy duty				
500 lb	EA.	9,740	220.00	9,960
1000 lb	"	10,040	230.00	10,270
2000 lb	"	11,170	250.00	11,420
3000 lb	"	11,780	260.00	12,040
4000 lb	"	12,440	280.00	12,720
5000 lb	"	12,940	290.00	13,230
6000 lb	"	13,270	310.00	13,580
7500 lb	"	18,250	320.00	18,570
10,000 lb	"	19,080	330.00	19,410
15,000 lb	"	21,070	340.00	21,410
20,000 lb	"	26,050	370.00	26,420
25,000 lb	"	37,840	400.00	38,240
30,000 lb	"	39,490	440.00	39,930

Hoists And Cranes

Hoists And Cranes	UNIT	MAT.	INST.	TOTAL
14600.10 Industrial Hoists				
Air powered hoists				
500 lb	EA.	3,650	220.00	3,870
1000 lb	"	4,060	220.00	4,280
2000 lb	"	4,440	230.00	4,670
4000 lb	"	5,180	260.00	5,440
6000 lb	"	6,180	340.00	6,520
Overhead traveling bridge crane				
Single girder, 20' span				
3 ton	EA.	20,740	910.00	21,650
5 ton	"	22,820	910.00	23,730
7.5 ton	"	28,870	910.00	29,780
10 ton	"	30,900	1,140	32,040
15 ton	"	35,680	1,140	36,820
30' span				
3 ton	EA.	22,070	910.00	22,980
5 ton	"	24,390	910.00	25,300
10 ton	"	31,200	1,140	32,340
15 ton	"	37,340	1,140	38,480
Double girder, 40' span				
3 ton	EA.	42,810	2,020	44,830
5 ton	"	43,970	2,020	45,990
7.5 ton	"	45,470	2,020	47,490
10 ton	"	47,790	2,600	50,390
15 ton	"	53,760	2,600	56,360
25 ton	"	79,650	2,600	82,250
50' span				
3 ton	EA.	46,710	2,020	48,730
5 ton	"	48,370	2,020	50,390
7.5 ton	"	50,120	2,020	52,140
10 ton	"	54,430	2,600	57,030
15 ton	"	63,060	2,600	65,660
25 ton	"	87,010	2,600	89,610
14650.10 Jib Cranes				
Self supporting, swinging 8' boom, 200 deg rotation				
1000 lb	EA.	1,910	400.00	2,310
2000 lb	"	1,990	400.00	2,390
3000 lb	"	2,320	800.00	3,120
4000 lb	"	2,470	800.00	3,270
6000 lb	"	2,990	800.00	3,790
10,000 lb	"	3,930	800.00	4,730
Wall mounted, 180 deg rotation				
2000 lb	EA.	1,060	400.00	1,460
3000 lb	"	1,260	400.00	1,660
4000 lb	"	1,410	800.00	2,210
6000 lb	"	1,590	800.00	2,390
10,000 lb	"	3,150	800.00	3,950

Basic Materials

Basic Materials	UNIT	MAT.	INST.	TOTAL
15100.10 Specialties				
Wall penetration				
Concrete wall, 6" thick				
2" dia.	EA.		11.30	11.30
4" dia.	"		16.90	16.90
8" dia.	"		24.15	24.15
12" thick				
2" dia.	EA.		15.40	15.40
4" dia.	"		24.15	24.15
8" dia.	"		37.60	37.60
Non-destructive testing, piping systems				
X-ray of welds				
3" dia. pipe	EA.	59.00	47.70	106.70
4" dia. pipe	"	50.50	47.70	98.20
6" dia. pipe	"	43.20	47.70	90.90
8" dia. pipe	"	46.20	59.50	105.70
10" dia. pipe	"	49.20	59.50	108.70
Liquid penetration of welds				
2" dia. pipe	EA.	29.80	29.80	59.60
3" dia. pipe	"	31.30	29.80	61.10
4" dia. pipe	"	34.30	29.80	64.10
6" dia. pipe	"	34.30	29.80	64.10
8" dia. pipe	"	41.70	29.80	71.50
10" dia. pipe	"	41.70	29.80	71.50
15120.10 Backflow Preventers				
Backflow preventer, flanged, cast iron, with valves				
3" pipe	EA.	3,470	240.00	3,710
4" pipe	"	4,590	270.00	4,860
6" pipe	"	7,090	400.00	7,490
8" pipe	"	12,610	480.00	13,090
Threaded				
3/4" pipe	EA.	790.00	29.80	819.80
2" pipe	"	1,420	47.70	1,468
Reduced pressure assembly, bronze, threaded				
3/4"	EA.	910.00	29.80	939.80
1"	"	960.00	34.10	994.10
1-1/4"	"	1,390	39.80	1,430
1-1/2"	"	1,420	47.70	1,468
15140.10 Pipe Hangers, Heavy				
Hangers				
1/2" pipe, clevis pipe hanger				
Black steel	EA.	2.55	15.90	18.45
Galvanized	"	2.90	15.90	18.80
U bolt	"	1.40	4.75	6.15
3/4" pipe, clevis pipe hanger				
Black steel	EA.	2.55	15.90	18.45
Galvanized	"	3.00	15.90	18.90
1" pipe, clevis pipe hanger				
Black steel	EA.	2.70	15.90	18.60
Galvanized	"	3.10	15.90	19.00
U bolt	"	1.70	4.75	6.45

Basic Materials

	UNIT	MAT.	INST.	TOTAL
15140.10 Pipe Hangers, Heavy				
2" pipe, clevis pipe hanger				
Black steel	EA.	4.00	15.90	19.90
Galvanized	"	4.50	15.90	20.40
3" pipe, clevis pipe hanger				
Black steel	EA.	6.75	15.90	22.65
Galvanized	"	11.95	15.90	27.85
4" pipe, clevis pipe hanger				
Black steel	EA.	9.35	15.90	25.25
Galvanized	"	11.80	15.90	27.70
6" pipe, clevis pipe hanger				
Black steel	EA.	14.60	19.10	33.70
Galvanized	"	21.65	19.10	40.75
12" pipe, clevis pipe hanger				
Black steel	EA.	23.30	19.10	42.40
Galvanized	"	33.00	19.10	52.10
Threaded rod, galvanized				
3/8"	L.F.			0.45
1/2"	"			0.76
Hex nuts, galvanized				
3/8"	EA.			0.05
1/2"	"			0.10
1"	"			0.80
C-clamp, steel, with lock nut				
3/8"	EA.	2.25	5.95	8.20
1/2"	"	2.70	5.95	8.65
15140.11 Pipe Hangers, Light				
A band, black iron				
1/2"	EA.	0.63	3.40	4.02
1"	"	0.78	3.55	4.33
1-1/4"	"	0.85	3.65	4.50
1-1/2"	"	0.97	4.00	4.97
2"	"	1.10	4.35	5.45
2-1/2"	"	1.45	4.75	6.20
3"	"	1.55	5.30	6.85
4"	"	2.25	5.95	8.20
5"	"	2.90	6.35	9.25
6"	"	4.05	6.80	10.85
8"	"	6.95	7.95	14.90
Copper				
1/2"	EA.	0.79	3.40	4.19
3/4"	"	0.79	3.55	4.34
1"	"	0.85	3.55	4.40
1-1/4"	"	0.86	3.65	4.51
1-1/2"	"	0.99	4.00	4.99
2"	"	1.15	4.35	5.50
2-1/2"	"	1.55	4.75	6.30
3"	"	1.70	5.30	7.00
4"	"	2.45	5.95	8.40
Black riser friction hangers				
3/4"	EA.	3.35	4.00	7.35
1"	"	3.35	4.15	7.50

Basic Materials

15140.11 Pipe Hangers, Light

	UNIT	MAT.	INST.	TOTAL
1-1/4"	EA.	4.25	4.35	8.60
1-1/2"	"	4.55	4.55	9.10
2"	"	4.70	4.75	9.45
2-1/2"	"	5.05	5.30	10.35
3"	"	5.20	5.95	11.15
4"	"	6.60	6.80	13.40
Short pattern black riser clamps				
1-1/2"	EA.	2.85	4.35	7.20
2"	"	2.95	4.55	7.50
3"	"	3.20	4.75	7.95
4"	"	3.60	5.30	8.90
Copper riser friction hanger				
1/2"	EA.	3.60	3.65	7.25
3/4"	"	3.60	3.80	7.40
1"	"	3.95	4.00	7.95
1-1/4"	"	4.00	4.15	8.15
1-1/2"	"	4.15	4.35	8.50
2"	"	4.85	4.55	9.40
2-1/2"	"	7.45	4.75	12.20
3"	"	8.20	4.75	12.95
4"	"	9.35	5.30	14.65
Auto grip hangers, galvanized				
1/2"	EA.	0.81	3.40	4.21
1"	"	0.81	3.80	4.61
2"	"	0.94	4.35	5.29
3"	"	4.10	4.75	8.85
4"	"	2.65	5.30	7.95
Copper				
1/2"	EA.	1.15	3.40	4.55
1"	"	1.25	3.80	5.05
2"	"	1.40	4.35	5.75
3"	"	3.40	4.75	8.15
4"	"	4.70	5.30	10.00
Split rings (F&M), galvanized				
1/2"	EA.	2.10	3.65	5.75
1"	"	2.85	4.00	6.85
2"	"	4.10	4.55	8.65
3"	"	12.05	5.00	17.05
4"	"	14.60	5.30	19.90
Copper				
1/2"	EA.	2.05	3.65	5.70
1"	"	2.45	4.00	6.45
2"	"	3.70	4.55	8.25
3"	"	8.75	5.00	13.75
4"	"	21.90	5.30	27.20
2 hole clips, galvanized				
3/4"	EA.	0.21	3.20	3.41
1"	"	0.24	3.30	3.54
1-1/4"	"	0.31	3.40	3.71
1-1/2"	"	0.38	3.55	3.92
2"	"	0.49	3.65	4.14
2-1/2"	"	0.89	3.80	4.69

Basic Materials

15140.11 Pipe Hangers, Light	UNIT	MAT.	INST.	TOTAL
3"	EA.	1.30	4.00	5.30
4"	"	2.75	4.35	7.10
Perforated strap				
3/4"				
Galvanized, 20 ga.	L.F.	0.32	2.40	2.72
Copper, 22 ga.	"	0.38	2.40	2.77
Threaded rod-couplings				
1/4"	EA.	1.05	3.00	4.05
3/4"	"	1.05	3.20	4.25
1/2"	"	1.20	3.40	4.60
5/8"	"	1.85	3.65	5.50
Reducing rod coupling, 1/2" x 3/8"	"	1.65	3.40	5.05
C-clamps				
3/4"	EA.	1.55	4.75	6.30
Top beam clamp				
3/8"	EA.	2.45	4.00	6.45
1/2"	"	3.00	4.35	7.35
Side beam connector				
3/8"	EA.	1.10	4.00	5.10
1/2"	"	2.35	4.35	6.70
Hex nuts, heavy				
1"	EA.			2.35
Heavy washers				
3/8"	EA.			0.10
1/2"	"			0.21
5/8"	"			0.42
3/4"	"			0.60
Lag rod, 3/8" x				
4"	EA.			0.81
4-1/2"	"			0.81
6"	"			0.94
8"	"			1.19
10"	"			1.62
12"	"			2.00
18"	"			2.71
J-Hooks				
1/2"	EA.	0.35	2.15	2.50
3/4"	"	0.38	2.15	2.52
1"	"	0.39	2.25	2.64
1-1/4"	"	0.40	2.35	2.75
1-1/2"	"	0.42	2.40	2.82
2"	"	0.43	2.40	2.83
3"	"	0.50	2.50	3.00
4"	"	0.50	2.50	3.00
PVC coated hangers, galvanized, 28 ga.				
1-1/2" x 12"	EA.	1.05	3.20	4.25
2" x 12"	"	1.10	3.40	4.50
3" x 12"	"	1.25	3.65	4.90
4" x 12"	"	1.35	4.00	5.35
Copper, 30 ga.				
1-1/2" x 12"	EA.	1.25	3.20	4.45
2" x 12"	"	1.45	3.40	4.85

Basic Materials

Basic Materials	UNIT	MAT.	INST.	TOTAL
15140.11 Pipe Hangers, Light				
3" x 12"	EA.	1.60	3.65	5.25
4" x 12"	"	1.80	4.00	5.80
Wire hook hangers				
Black wire, 1/2" x				
4"	EA.	0.36	2.40	2.76
6"	"	0.42	2.50	2.92
8"	"	0.46	2.65	3.11
10"	"	0.58	2.65	3.23
12"	"	0.69	2.80	3.49
3/4" x				
4"	EA.	0.44	2.50	2.94
6"	"	0.47	2.65	3.12
8"	"	0.49	2.80	3.29
10"	"	0.67	3.00	3.67
12"	"	0.68	3.20	3.88
1" x				
4"	EA.	0.44	2.65	3.09
6"	"	0.46	2.80	3.26
8"	"	0.49	3.00	3.49
10"	"	0.67	3.20	3.87
12"	"	0.72	3.40	4.12
Copper wire hooks				
1/2" x				
4"	EA.	0.42	2.40	2.82
6"	"	0.47	2.50	2.97
8"	"	0.53	2.65	3.18
10"	"	0.67	2.80	3.47
12"	"	0.76	3.00	3.76
3/4" x				
4"	EA.	0.42	2.50	2.92
6"	"	0.51	2.65	3.16
8"	"	0.60	2.80	3.40
10"	"	0.68	3.00	3.68
12"	"	0.82	3.20	4.02
15175.60 Expansion Tanks				
Expansion tank, 125 psi, steel				
20 gallon	EA.	500.00	59.50	559.50
80 gallon	"	760.00	140.00	900.00
15240.10 Vibration Control				
Vibration isolator, in-line, stainless connector, screwed				
1/2"	EA.	77.50	26.50	104.00
1"	"	96.00	29.80	125.80
2"	"	150.00	36.70	186.70
3"	"	400.00	43.40	443.40
6"	"	800.00	53.00	853.00

Insulation

	UNIT	MAT.	INST.	TOTAL
15260.10 Fiberglass Pipe Insulation				
Fiberglass insulation on 1/2" pipe				
1" thick	L.F.	1.55	1.60	3.15
1-1/2" thick	"	3.75	2.00	5.75
3/4" pipe				
1" thick	L.F.	3.30	1.60	4.90
1-1/2" thick	"	4.00	2.00	6.00
1" pipe				
1" thick	L.F.	3.40	1.60	5.00
1-1/2" thick	"	4.30	2.00	6.30
2" pipe				
1" thick	L.F.	4.80	2.00	6.80
1-1/2" thick	"	5.65	2.15	7.80
2" thick	"	8.30	2.40	10.70
2-1/2" pipe				
1" thick	L.F.	5.30	2.00	7.30
1-1/2" thick	"	6.10	2.15	8.25
2" thick	"	8.95	2.40	11.35
3" pipe				
1" thick	L.F.	5.50	2.25	7.75
1-1/2" thick	"	6.65	2.40	9.05
2" thick	"	9.15	2.65	11.80
4" pipe				
1" thick	L.F.	6.80	2.25	9.05
1-1/2" thick	"	8.45	2.40	10.85
2" thick	"	11.75	2.65	14.40
5" pipe				
1" thick	L.F.	6.80	2.25	9.05
2" thick	"	12.60	2.40	15.00
6" pipe				
1" thick	L.F.	8.45	2.50	10.95
2" thick	"	13.35	2.65	16.00
8" pipe				
2" thick	L.F.	17.80	2.50	20.30
3" thick	"	23.70	2.65	26.35
10" pipe				
2" thick	L.F.	19.70	2.50	22.20
3" thick	"	27.30	2.65	29.95
12" pipe				
2" thick	L.F.	21.45	2.50	23.95
3" thick	"	30.60	2.65	33.25
15260.60 Exterior Pipe Insulation				
Fiberglass insulation, aluminum jacket				
1/2" pipe				
1" thick	L.F.	4.15	3.65	7.80
1-1/2" thick	"	5.10	4.00	9.10
3/4" pipe				
1" thick	L.F.	4.50	3.65	8.15
1-1/2" thick	"	5.25	4.00	9.25
1" pipe				
1" thick	L.F.	4.65	3.65	8.30
1-1/2" thick	"	5.70	4.00	9.70

Insulation

	UNIT	MAT.	INST.	TOTAL
15260.60 Exterior Pipe Insulation				
1-1/4" pipe				
1" thick	L.F.	5.35	4.35	9.70
1-1/2" thick	"	6.05	4.55	10.60
1-1/2" pipe				
1" thick	L.F.	5.50	4.35	9.85
1-1/2" thick	"	6.55	4.55	11.10
2" pipe				
1" thick	L.F.	5.90	4.35	10.25
1-1/2" thick	"	7.05	4.55	11.60
3" pipe				
1" thick	L.F.	6.55	4.75	11.30
1-1/2" thick	"	7.80	5.00	12.80
6" pipe				
1" thick	L.F.	8.55	5.30	13.85
2" thick	"	14.15	5.60	19.75
10" pipe				
2" thick	L.F.	21.60	5.30	26.90
3" thick	"	31.10	5.60	36.70
15260.90 Pipe Insulation Fittings				
Insulation protection saddle				
1" thick covering				
1/2" pipe	EA.	4.75	19.10	23.85
3/4" pipe	"	4.75	19.10	23.85
1" pipe	"	4.75	19.10	23.85
2" pipe	"	6.85	19.10	25.95
3" pipe	"	7.75	21.70	29.45
6" pipe	"	9.70	29.80	39.50
1-1/2" thick covering				
3/4" pipe	EA.	7.80	19.10	26.90
1" pipe	"	7.80	19.10	26.90
2" pipe	"	8.75	19.10	27.85
3" pipe	"	9.85	19.10	28.95
6" pipe	"	14.55	29.80	44.35
10" pipe	"	18.30	39.80	58.10
15280.10 Equipment Insulation				
Equipment insulation, 2" thick, cellular glass	S.F.	2.80	3.00	5.80
Urethane, rigid, field applied jacket, plastered finish	"	3.00	5.95	8.95
Fiberglass, rigid, with vapor barrier	"	2.80	2.65	5.45
15290.10 Ductwork Insulation				
Fiberglass duct insulation, plain blanket				
1-1/2" thick	S.F.	0.87	0.60	1.47
2" thick	"	1.05	0.80	1.85
With vapor barrier				
1-1/2" thick	S.F.	0.54	0.60	1.14
2" thick	"	0.61	0.80	1.41
Rigid with vapor barrier				
2" thick	S.F.	1.80	1.60	3.40
3" thick	"	2.90	1.90	4.80
4" thick	"	3.75	2.40	6.15

Insulation

15290.10 Ductwork Insulation	UNIT	MAT.	INST.	TOTAL
6" thick	S.F.	5.20	3.20	8.40
Weatherproof, polystyrene, 3" thick, w/vapor barrier	"	3.85	4.75	8.60
Urethane board with vapor barrier	"	5.45	5.95	11.40

Fire Protection

15330.10 Wet Sprinkler System	UNIT	MAT.	INST.	TOTAL
Sprinkler head, 212 deg, brass, exposed piping	EA.	7.40	19.10	26.50
Chrome, concealed piping	"	7.10	26.50	33.60
Water motor alarm	"	99.00	79.50	178.50
Fire department inlet connection	"	150.00	95.50	245.50
Wall plate for fire dept connection	"	72.50	39.80	112.30
Swing check valve flanged iron body, 4"	"	150.00	160.00	310.00
Check valve, 6"	"	690.00	240.00	930.00
Wet pipe valve, flange to groove, 4"	"	600.00	53.00	653.00
Flange to flange				
6"	EA.	770.00	79.50	849.50
8"	"	1,080	160.00	1,240
Alarm valve, flange to flange, (wet valve)				
4"	EA.	720.00	53.00	773.00
8"	"	1,110	400.00	1,510
Inspector's test connection	"	46.40	39.80	86.20
Wall hydrant, polished brass, 2-1/2" x 2-1/2", single	"	370.00	34.10	404.10
2-way	"	830.00	34.10	864.10
3 way	"	1,700	34.10	1,734
Wet valve trim, includes retard chamber & gauges, 4"-6"	"	590.00	39.80	629.80
Retard pressure switch for wet systems	"	1,020	95.50	1,116
Air maintenance device	"	310.00	39.80	349.80
Wall hydrant non-freeze, 8" thick wall, vacuum breaker	"	34.00	23.85	57.85
12" thick wall	"	37.10	23.85	60.95

15330.50 Dry Sprinkler System	UNIT	MAT.	INST.	TOTAL
Dry pipe valve, flange to flange				
4"	EA.	1,170	95.50	1,266
6"	"	1,480	120.00	1,600
Trim, 4" and 6", includes gauges	"	550.00	39.80	589.80
Field testing and flushing	"		400.00	400.00
Disinfection	"		400.00	400.00
Pressure switch double circuit, open/close contacts	"	280.00	120.00	400.00
Low air				
Supervisory unit	EA.	910.00	79.50	989.50
Pressure switch	"	290.00	39.80	329.80

Plumbing	UNIT	MAT.	INST.	TOTAL
15410.05 C.i. Pipe, Above Ground				
No hub pipe				
1-1/2" pipe	L.F.	5.75	3.40	9.15
2" pipe	"	6.60	4.00	10.60
3" pipe	"	9.30	4.75	14.05
4" pipe	"	12.75	7.95	20.70
6" pipe	"	19.10	9.55	28.65
8" pipe	"	28.60	15.90	44.50
10" pipe	"	44.60	19.10	63.70
No hub fittings, 1-1/2" pipe				
1/4 bend	EA.	7.95	15.90	23.85
1/8 bend	"	5.65	15.90	21.55
Sanitary tee	"	9.70	23.85	33.55
Coupling	"			10.59
Wye	"	10.60	23.85	34.45
2" pipe				
1/4 bend	EA.	8.10	19.10	27.20
1/8 bend	"	5.85	19.10	24.95
Sanitary tee	"	10.10	31.80	41.90
Coupling	"			10.59
Wye	"	11.30	39.80	51.10
3" pipe				
1/4 bend	EA.	10.25	23.85	34.10
1/8 bend	"	9.20	23.85	33.05
Sanitary tee	"	12.00	29.80	41.80
Coupling	"			12.36
Wye	"	12.35	39.80	52.15
4" pipe				
1/4 bend	EA.	14.15	23.85	38.00
1/8 bend	"	12.00	23.85	35.85
Sanitary tee	"	17.65	39.80	57.45
Coupling	"			13.81
Wye	"	17.65	39.80	57.45
8" deep	"	24.75	23.85	48.60
6" pipe				
1/4 bend	EA.	33.60	39.80	73.40
1/8 bend	"	24.75	39.80	64.55
Sanitary tee	"	53.00	47.70	100.70
Coupling	"			28.28
Wye	"	42.40	47.70	90.10
8" pipe				
1/4 bend	EA.	69.00	39.80	108.80
1/8 bend	"	62.00	39.80	101.80
Sanitary tee	"	120.00	59.50	179.50
Coupling	"			31.81
Wye	"	85.00	47.70	132.70
10" pipe				
1/4 bend	EA.	130.00	39.80	169.80
1/8 bend	"	120.00	39.80	159.80
Coupling	"			44.18
Wye	"	170.00	79.50	249.50
10x6" wye	"	140.00	79.50	219.50

Plumbing	UNIT	MAT.	INST.	TOTAL
15410.06 C.i. Pipe, Below Ground				
No hub pipe				
1-1/2" pipe	L.F.	6.05	2.40	8.45
2" pipe	"	6.60	2.65	9.25
3" pipe	"	9.30	3.00	12.30
4" pipe	"	12.75	4.00	16.75
6" pipe	"	19.10	4.35	23.45
8" pipe	"	28.60	5.30	33.90
10" pipe	"	44.60	5.95	50.55
Fittings, 1-1/2"				
1/4 bend	EA.	7.95	13.65	21.60
1/8 bend	"	6.20	13.65	19.85
Wye	"	10.60	19.10	29.70
2"				
1/4 bend	EA.	8.10	15.90	24.00
1/8 bend	"	5.85	15.90	21.75
3"				
1/4 bend	EA.	10.25	19.10	29.35
1/8 bend	"	9.20	19.10	28.30
Wye	"	12.35	29.80	42.15
4"				
1/4 bend	EA.	14.15	19.10	33.25
1/8 bend	"	12.00	19.10	31.10
Wye	"	17.65	29.80	47.45
6"				
1/4 bend	EA.	33.60	29.80	63.40
1/8 bend	"	24.75	29.80	54.55
8"				
1/4 bend	EA.	69.00	29.80	98.80
1/8 bend	"	62.00	29.80	91.80
Wye	"	85.00	39.80	124.80
10"				
1/4 bend	EA.	130.00	29.80	159.80
1/8 bend	"	120.00	29.80	149.80
Plug	"			31.81
Wye	"	170.00	59.50	229.50
15410.08 Extra Heavy Soil Pipe				
Extra heavy soil pipe, single hub				
2" x 5'	EA.	47.30	9.55	56.85
3" x 5'	"	56.50	10.15	66.65
4" x 5'	"	71.00	11.10	82.10
6" x 5'	"	100.00	11.95	111.95
Double hub				
2" x 5'	EA.	48.70	11.95	60.65
4" x 5'	"	71.00	12.55	83.55
5" x 5'	"	100.00	13.25	113.25
6" x 5'	"	110.00	13.65	123.65
Single hub				
3" x 10'	EA.	71.00	7.95	78.95
4" x 10'	"	80.50	8.25	88.75
6" x 10'	"	130.00	8.70	138.70

15 MECHANICAL

Plumbing

	UNIT	MAT.	INST.	TOTAL
15410.09 Service Weight Pipe				
Service weight pipe, single hub				
3" x 5'	EA.	38.20	10.15	48.35
4" x 5'	"	46.10	10.60	56.70
6" x 5'	"	73.00	11.95	84.95
Double hub				
3" x 5'	EA.	41.30	12.90	54.20
4" x 5'	"	49.30	13.65	62.95
6" x 5'	"	84.50	15.90	100.40
Single hub				
3" x 10'	EA.	47.70	12.90	60.60
4" x 10'	"	59.00	13.65	72.65
6" x 10'	"	95.50	15.90	111.40
1/8 bend				
3"	EA.	8.75	19.10	27.85
4"	"	12.15	21.70	33.85
6"	"	19.40	23.85	43.25
1/4 bend				
3"	EA.	11.00	19.10	30.10
4"	"	15.35	21.70	37.05
6"	"	26.10	23.85	49.95
Sweep				
3"	EA.	16.85	19.10	35.95
4"	"	23.60	21.70	45.30
6"	"	42.20	23.85	66.05
Sanitary T				
3"	EA.	17.75	34.10	51.85
4"	"	21.95	39.80	61.75
6"	"	43.90	43.40	87.30
Wye				
3"	EA.	19.40	26.50	45.90
4"	"	24.45	28.10	52.55
6"	"	50.50	34.10	84.60
15410.10 Copper Pipe				
Type "K" copper				
1/2"	L.F.	2.70	1.50	4.20
3/4"	"	4.95	1.60	6.55
1"	"	6.25	1.70	7.95
1-1/4"	"	7.85	1.85	9.70
1-1/2"	"	9.90	2.00	11.90
2"	"	13.60	2.15	15.75
2-1/2"	"	17.60	2.40	20.00
3"	"	23.80	2.50	26.30
4"	"	38.40	2.65	41.05
DWV, copper				
1-1/4"	L.F.	4.95	2.00	6.95
1-1/2"	"	6.40	2.15	8.55
2"	"	7.20	2.40	9.60
3"	"	13.60	2.65	16.25
4"	"	22.40	3.00	25.40
6"	"	72.00	3.40	75.40
Refrigeration tubing, copper, sealed				

Plumbing	UNIT	MAT.	INST.	TOTAL
15410.10 Copper Pipe				
1/8"	L.F.	0.72	1.90	2.62
3/16"	"	0.78	2.00	2.78
1/4"	"	0.89	2.05	2.94
5/16"	"	1.15	2.15	3.30
3/8"	"	1.30	2.25	3.55
1/2"	"	1.70	2.40	4.10
7/8"	"	2.90	2.75	5.65
1-1/8"	"	4.65	3.20	7.85
1-3/8"	"	5.60	3.65	9.25
Type "L" copper				
1/4"	L.F.	1.60	1.40	3.00
3/8"	"	1.75	1.40	3.15
1/2"	"	2.70	1.50	4.20
3/4"	"	3.05	1.60	4.65
1"	"	4.15	1.70	5.85
1-1/4"	"	5.60	1.85	7.45
1-1/2"	"	6.70	2.00	8.70
2"	"	10.00	2.15	12.15
2-1/2"	"	15.20	2.40	17.60
3"	"	20.85	2.50	23.35
3-1/2"	"	25.60	2.60	28.20
4"	"	32.00	2.65	34.65
Type "M" copper				
1/2"	L.F.	1.95	1.50	3.45
3/4"	"	2.70	1.60	4.30
1"	"	3.30	1.70	5.00
1-1/4"	"	4.80	1.85	6.65
2"	"	9.25	2.15	11.40
2-1/2"	"	14.40	2.40	16.80
3"	"	17.60	2.50	20.10
4"	"	28.80	2.65	31.45
15410.11 Copper Fittings				
Coupling, with stop				
1/4"	EA.	0.58	15.90	16.48
3/8"	"	0.97	19.10	20.06
1/2"	"	1.15	20.75	21.90
5/8"	"	1.50	23.85	25.35
3/4"	"	1.65	26.50	28.15
1"	"	3.30	28.10	31.40
3"	"	29.60	47.70	77.30
4"	"	59.00	59.50	118.50
Reducing coupling				
1/4" x 1/8"	EA.	1.90	19.10	21.00
3/8" x 1/4"	"	2.10	20.75	22.85
1/2" x				
3/8"	EA.	2.00	23.85	25.85
1/4"	"	2.00	23.85	25.85
1/8"	"	2.00	23.85	25.85
3/4" x				
3/8"	EA.	2.70	26.50	29.20
1/2"	"	2.80	26.50	29.30

Plumbing	UNIT	MAT.	INST.	TOTAL
15410.11 Copper Fittings				
1" x				
3/8"	EA.	5.60	29.80	35.40
1" x 1/2"	"	5.75	29.80	35.55
1" x 3/4"	"	5.95	29.80	35.75
1-1/4" x				
1/2"	EA.	6.75	31.80	38.55
3/4"	"	6.90	31.80	38.70
1"	"	7.25	31.80	39.05
1-1/2" x				
1/2"	EA.	11.50	34.10	45.60
3/4"	"	11.70	34.10	45.80
1"	"	11.85	34.10	45.95
1-1/4"	"	12.05	34.10	46.15
2" x				
1/2"	EA.	17.30	39.80	57.10
3/4"	"	17.60	39.80	57.40
1"	"	18.10	39.80	57.90
1-1/4"	"	18.45	39.80	58.25
1-1/2"	"	18.60	39.80	58.40
2-1/2" x				
1"	EA.	37.90	47.70	85.60
1-1/4"	"	38.70	47.70	86.40
1-1/2"	"	38.70	47.70	86.40
2"	"	39.50	47.70	87.20
3" x				
1-1/2"	EA.	47.70	59.50	107.20
2"	"	49.40	59.50	108.90
2-1/2"	"	51.00	59.50	110.50
4" x				
2"	EA.	100.00	68.00	168.00
2-1/2"	"	100.00	68.00	168.00
3"	"	100.00	68.00	168.00
Slip coupling				
1/4"	EA.	0.51	15.90	16.40
1/2"	"	0.77	19.10	19.87
3/4"	"	1.45	23.85	25.30
1"	"	3.10	26.50	29.60
1-1/4"	"	5.75	29.80	35.55
1-1/2"	"	7.75	31.80	39.55
2"	"	11.20	39.80	51.00
2-1/2"	"	15.30	39.80	55.10
3"	"	29.60	47.70	77.30
4"	"	57.50	59.50	117.00
Coupling with drain				
1/2"	EA.	6.75	23.85	30.60
3/4"	"	9.85	26.50	36.35
1"	"	12.35	29.80	42.15
Reducer				
3/8" x 1/4"	EA.	2.15	19.10	21.25
1/2" x 3/8"	"	2.30	19.10	21.40
3/4" x				
1/4"	EA.	3.45	21.70	25.15

Plumbing	UNIT	MAT.	INST.	TOTAL
15410.11 Copper Fittings				
3/8"	EA.	3.60	21.70	25.30
1/2"	"	3.80	21.70	25.50
1" x				
1/2"	EA.	5.75	23.85	29.60
3/4"	"	5.95	23.85	29.80
1-1/4" x				
1/2"	EA.	7.75	26.50	34.25
3/4"	"	9.70	26.50	36.20
1"	"	9.85	26.50	36.35
1-1/2" x				
1/2"	EA.	11.50	29.80	41.30
3/4"	"	11.60	29.80	41.40
1"	"	11.85	29.80	41.65
1-1/4"	"	12.05	29.80	41.85
2" x				
1/2"	EA.	18.10	34.10	52.20
3/4"	"	18.45	34.10	52.55
1"	"	18.45	34.10	52.55
1-1/4"	"	18.90	34.10	53.00
1-1/2"	"	18.90	34.10	53.00
2-1/2" x				
1"	EA.	39.50	39.80	79.30
1-1/4"	"	41.10	39.80	80.90
1-1/2"	"	41.10	39.80	80.90
2"	"	42.80	39.80	82.60
3" x				
1-1/4"	EA.	52.50	47.70	100.20
1-1/2"	"	52.50	47.70	100.20
2"	"	52.50	47.70	100.20
2-1/2"	"	52.50	47.70	100.20
4" x				
2"	EA.	89.00	59.50	148.50
3"	"	92.00	59.50	151.50
Female adapters				
1/4"	EA.	7.60	19.10	26.70
3/8"	"	7.95	21.70	29.65
1/2"	"	3.95	23.85	27.80
3/4"	"	5.75	26.50	32.25
1"	"	9.70	26.50	36.20
1-1/4"	"	15.65	29.80	45.45
1-1/2"	"	23.05	29.80	52.85
2"	"	29.60	31.80	61.40
2-1/2"	"	95.50	34.10	129.60
3"	"	130.00	39.80	169.80
4"	"	210.00	47.70	257.70
Increasing female adapters				
1/8" x				
3/8"	EA.	5.75	19.10	24.85
1/2"	"	5.75	19.10	24.85
1/4" x 1/2"	"	5.75	20.75	26.50
3/8" x 1/2"	"	5.75	21.70	27.45
1/2" X				

Plumbing	UNIT	MAT.	INST.	TOTAL
15410.11 Copper Fittings				
3/4"	EA.	7.75	23.85	31.60
1"	"	14.30	23.85	38.15
3/4" X				
1"	EA.	13.20	26.50	39.70
1-1/4"	"	23.85	26.50	50.35
1" x				
1-1/4"	EA.	23.85	26.50	50.35
1-1/2"	"	27.70	26.50	54.20
1-1/4" x				
1-1/2"	EA.	27.80	29.80	57.60
2"	"	39.50	29.80	69.30
1-1/2" x 2"	"	41.10	31.80	72.90
Reducing female adapters				
3/8" x 1/4"	EA.	6.30	21.70	28.00
1/2" x				
1/4"	EA.	6.30	23.85	30.15
3/8"	"	6.60	23.85	30.45
3/4" x 1/2"	"	8.10	26.50	34.60
1" x				
1/2"	EA.	13.00	26.50	39.50
3/4"	"	13.00	26.50	39.50
1-1/4" x				
1/2"	EA.	24.70	29.80	54.50
3/4"	"	24.70	29.80	54.50
1"	"	24.70	29.80	54.50
1-1/2" x				
1"	EA.	25.40	31.80	57.20
1-1/4"	"	25.40	31.80	57.20
2" x				
1"	EA.	37.90	34.10	72.00
1-1/4"	"	39.20	34.10	73.30
1-1/2"	"	41.10	34.10	75.20
Female fitting adapters				
1/2"	EA.	5.95	23.85	29.80
3/4"	"	7.95	23.85	31.80
3/4" x 1/2"	"	13.20	25.10	38.30
1"	"	12.35	26.50	38.85
1-1/4"	"	18.90	28.10	47.00
1-1/2"	"	24.70	29.80	54.50
2"	"	26.30	31.80	58.10
Male adapters				
1/4"	EA.	7.25	21.70	28.95
3/8"	"	4.10	21.70	25.80
3"	"	85.50	39.80	125.30
4"	"	130.00	47.70	177.70
Increasing male adapters				
3/8" x 1/2"	EA.	4.30	21.70	26.00
1/2" x				
3/4"	EA.	6.30	23.85	30.15
1"	"	11.50	23.85	35.35
3/4" x				
1"	EA.	11.50	25.10	36.60

Plumbing	UNIT	MAT.	INST.	TOTAL
15410.11 Copper Fittings				
1-1/4"	EA.	16.45	25.10	41.55
1" x 1-1/4"	"	16.45	26.50	42.95
1-1/2" x				
3/4"	EA.	21.40	28.10	49.50
1"	"	21.75	28.10	49.85
1-1/4"	"	22.05	28.10	50.15
2" x				
1"	EA.	36.20	29.80	66.00
1-1/4"	"	36.20	29.80	66.00
1-1/2"	"	37.90	29.80	67.70
2" x 2-1/2"	"	120.00	31.80	151.80
Copper pipe fittings				
1/2"				
90 deg ell	EA.	1.15	10.60	11.75
45 deg ell	"	1.50	10.60	12.10
Tee	"	1.95	13.65	15.60
Cap	"	0.80	5.30	6.10
Coupling	"	0.86	10.60	11.46
Union	"	5.95	11.95	17.90
3/4"				
90 deg ell	EA.	2.55	11.95	14.50
45 deg ell	"	3.00	11.95	14.95
Tee	"	4.30	15.90	20.20
Cap	"	1.55	5.60	7.15
Coupling	"	1.75	11.95	13.70
Union	"	8.75	13.65	22.40
1"				
90 deg ell	EA.	5.95	15.90	21.85
45 deg ell	"	7.75	15.90	23.65
Tee	"	9.70	19.10	28.80
Cap	"	2.90	7.95	10.86
Coupling	"	4.30	15.90	20.20
Union	"	11.45	15.90	27.35
1-1/4"				
90 deg ell	EA.	8.10	13.65	21.75
45 deg ell	"	10.00	13.65	23.65
Tee	"	13.20	23.85	37.05
Cap	"	2.30	7.95	10.25
Union	"	18.90	17.05	35.95
1-1/2"				
90 deg ell	EA.	10.55	17.05	27.60
45 deg ell	"	12.50	17.05	29.55
Tee	"	17.30	26.50	43.80
Cap	"	2.30	7.95	10.25
Coupling	"	7.75	15.90	23.65
Union	"	28.80	21.70	50.50
2"				
90 deg ell	EA.	20.55	19.10	39.65
45 deg ell	"	18.90	29.80	48.70
Tee	"	29.60	29.80	59.40
Cap	"	4.80	9.55	14.35
Coupling	"	12.50	19.10	31.60

Plumbing

	UNIT	MAT.	INST.	TOTAL
15410.11 Copper Fittings				
Union	EA.	31.30	23.85	55.15
2-1/2"				
90 deg ell	EA.	39.50	23.85	63.35
45 deg ell	"	34.60	23.85	58.45
Tee	"	39.50	34.10	73.60
Cap	"	9.70	11.95	21.65
Coupling	"	18.90	23.85	42.75
Union	"	57.50	26.50	84.00
15410.15 Brass Fittings				
Compression fittings, union				
3/8"	EA.	3.95	7.95	11.90
1/2"	"	5.85	7.95	13.80
5/8"	"	7.25	7.95	15.20
Union elbow				
3/8"	EA.	6.55	7.95	14.50
1/2"	"	10.35	7.95	18.30
5/8"	"	12.95	7.95	20.90
Union tee				
3/8"	EA.	8.60	7.95	16.55
1/2"	"	13.80	7.95	21.75
5/8"	"	19.80	7.95	27.75
90 deg elbow union				
3/8"	EA.	6.90	7.70	14.60
1/2"	"	9.65	7.70	17.35
5/8"	"	13.80	7.70	21.50
15410.17 Chrome Plated Fittings				
Fittings				
90 ell				
3/8"	EA.	19.30	11.95	31.25
1/2"	"	24.80	11.95	36.75
45 ell				
3/8"	EA.	24.80	11.95	36.75
1/2"	"	32.70	11.95	44.65
Tee				
3/8"	EA.	23.90	15.90	39.80
1/2"	"	28.30	15.90	44.20
Coupling				
3/8"	EA.	15.05	11.95	27.00
1/2"	"	15.05	11.95	27.00
Union				
3/8"	EA.	24.80	11.95	36.75
1/2"	"	25.70	11.95	37.65
Tee				
1/2" x 3/8" x 3/8"	EA.	28.30	15.90	44.20
1/2" x 3/8" x 1/2"	"	28.80	15.90	44.70
15410.18 Glass Pipe				
Glass pipe				
1-1/2" dia.	L.F.	12.00	9.55	21.55
2" dia.	"	16.85	10.60	27.45

Plumbing	UNIT	MAT.	INST.	TOTAL
15410.18 Glass Pipe				
3" dia.	L.F.	22.50	11.95	34.45
4" dia.	"	39.00	13.65	52.65
6" dia.	"	75.00	15.90	90.90
15410.30 Pvc/cpvc Pipe				
PVC schedule 40				
1/2" pipe	L.F.	2.35	2.00	4.35
3/4" pipe	"	2.70	2.15	4.85
1" pipe	"	3.10	2.40	5.50
1-1/4" pipe	"	3.50	2.65	6.15
1-1/2" pipe	"	3.70	3.00	6.70
2" pipe	"	4.25	3.40	7.65
2-1/2" pipe	"	5.85	4.00	9.85
3" pipe	"	6.20	4.75	10.95
4" pipe	"	9.10	5.95	15.05
6" pipe	"	15.35	11.95	27.30
8" pipe	"	22.40	15.90	38.30
PVC schedule 80 pipe				
1-1/2" pipe	L.F.	2.95	3.00	5.95
2" pipe	"	3.75	3.40	7.15
3" pipe	"	5.75	4.75	10.50
4" pipe	"	7.90	5.95	13.85
Polypropylene, acid resistant, DWV pipe				
Schedule 40				
1-1/2" pipe	L.F.	6.00	3.40	9.40
2" pipe	"	8.05	4.00	12.05
3" pipe	"	14.85	4.75	19.60
4" pipe	"	21.00	5.95	26.95
6" pipe	"	37.70	11.95	49.65
Polyethylene pipe and fittings				
SDR-21				
3" pipe	L.F.	3.95	5.95	9.90
4" pipe	"	5.95	7.95	13.90
6" pipe	"	9.90	11.95	21.85
8" pipe	"	15.85	13.65	29.50
10" pipe	"	17.85	15.90	33.75
15410.33 Abs Dwv Pipe				
Schedule 40 ABS				
1-1/2" pipe	L.F.	3.05	2.40	5.45
2" pipe	"	4.05	2.65	6.70
3" pipe	"	6.10	3.40	9.50
4" pipe	"	9.10	4.75	13.85
6" pipe	"	14.20	5.95	20.15
15410.35 Plastic Pipe				
Fiberglass reinforced pipe				
2" pipe	L.F.	15.85	3.65	19.50
3" pipe	"	21.40	4.00	25.40
4" pipe	"	27.20	4.35	31.55
6" pipe	"	42.70	4.75	47.45
8" pipe	"	62.00	7.95	69.95

Plumbing

	UNIT	MAT.	INST.	TOTAL
15410.35 Plastic Pipe				
10" pipe	L.F.	77.50	9.55	87.05
12" pipe	"	97.00	11.95	108.95
15410.70 Stainless Steel Pipe				
Stainless steel, schedule 40, threaded				
1/2" pipe	L.F.	7.85	6.80	14.65
3/4" pipe	"	10.45	7.00	17.45
1" pipe	"	12.20	7.35	19.55
1-1/2" pipe	"	16.60	7.95	24.55
2" pipe	"	24.95	8.70	33.65
2-1/2" pipe	"	34.90	9.55	44.45
3" pipe	"	48.90	10.60	59.50
4" pipe	"	63.00	11.95	74.95
15410.80 Steel Pipe				
Black steel, extra heavy pipe, threaded				
1/2" pipe	L.F.	1.85	1.90	3.75
3/4" pipe	"	2.20	1.90	4.10
1" pipe	"	3.10	2.40	5.50
1-1/2" pipe	"	4.65	2.65	7.30
2-1/2" pipe	"	9.25	5.95	15.20
3" pipe	"	12.35	7.95	20.30
4" pipe	"	18.70	9.55	28.25
5" pipe	"	25.20	11.95	37.15
6" pipe	"	31.50	11.95	43.45
8" pipe	"	46.40	15.90	62.30
10" pipe	"	64.50	19.10	83.60
12" pipe	"	82.00	23.85	105.85
Fittings, malleable iron, threaded, 1/2" pipe				
90 deg ell	EA.	1.90	15.90	17.80
45 deg ell	"	3.00	15.90	18.90
Tee	"	2.50	23.85	26.35
3/4" pipe				
90 deg ell	EA.	2.50	15.90	18.40
45 deg ell	"	4.10	23.85	27.95
Tee	"	4.15	23.85	28.00
1" pipe				
90 deg ell	EA.	4.10	19.10	23.20
45 deg ell	"	4.55	19.10	23.65
Tee	"	6.15	26.50	32.65
1-1/2" pipe				
90 deg ell	EA.	7.80	23.85	31.65
45 deg ell	"	9.10	23.85	32.95
Tee	"	11.25	34.10	45.35
2-1/2" pipe				
90 deg ell	EA.	31.50	59.50	91.00
45 deg ell	"	39.30	59.50	98.80
Tee	"	42.50	79.50	122.00
3" pipe				
90 deg ell	EA.	45.60	79.50	125.10
45 deg ell	"	52.00	79.50	131.50
Tee	"	58.00	120.00	178.00

Plumbing	UNIT	MAT.	INST.	TOTAL
15410.80 Steel Pipe				
4" pipe				
90 deg ell	EA.	91.00	95.50	186.50
45 deg ell	"	93.00	95.50	188.50
Tee	"	130.00	160.00	290.00
6" pipe				
90 deg ell	EA.	260.00	95.50	355.50
45 deg ell	"	310.00	95.50	405.50
Tee	"	360.00	160.00	520.00
8" pipe				
90 deg ell	EA.	270.00	190.00	460.00
45 deg ell	"	330.00	190.00	520.00
Tee	"	390.00	300.00	690.00
10" pipe				
90 deg ell	EA.	300.00	240.00	540.00
45 deg ell	"	350.00	240.00	590.00
Tee	"	450.00	300.00	750.00
12" pipe				
90 deg ell	EA.	350.00	300.00	650.00
45 deg ell	"	410.00	300.00	710.00
Tee	"	520.00	400.00	920.00
15410.82 Galvanized Steel Pipe				
Galvanized pipe				
1/2" pipe	L.F.	2.10	4.75	6.85
3/4" pipe	"	2.75	5.95	8.70
1" pipe	"	4.20	6.80	11.00
1-1/4" pipe	"	4.85	7.95	12.80
1-1/2" pipe	"	5.20	9.55	14.75
2" pipe	"	7.45	11.95	19.40
2-1/2" pipe	"	10.65	15.90	26.55
3" pipe	"	14.85	17.05	31.90
4" pipe	"	20.20	19.90	40.10
6" pipe	"	40.40	39.80	80.20
15430.23 Cleanouts				
Cleanout, wall				
2"	EA.	77.50	31.80	109.30
3"	"	85.00	31.80	116.80
4"	"	110.00	39.80	149.80
6"	"	190.00	47.70	237.70
8"	"	230.00	59.50	289.50
Floor				
2"	EA.	80.00	39.80	119.80
3"	"	100.00	39.80	139.80
4"	"	110.00	47.70	157.70
6"	"	150.00	59.50	209.50
8"	"	280.00	68.00	348.00
15430.24 Grease Traps				
Grease traps, cast iron, 3" pipe				
35 gpm, 70 lb capacity	EA.	2,600	480.00	3,080
50 gpm, 100 lb capacity	"	3,310	600.00	3,910

Plumbing

	UNIT	MAT.	INST.	TOTAL
15430.25 Hose Bibbs				
Hose bibb				
1/2"	EA.	7.35	15.90	23.25
3/4"	"	7.80	15.90	23.70
15430.60 Valves				
Gate valve, 125 lb, bronze, soldered				
1/2"	EA.	19.80	11.95	31.75
3/4"	"	23.40	11.95	35.35
1"	"	28.80	15.90	44.70
1-1/2"	"	50.50	19.10	69.60
2"	"	70.50	23.85	94.35
2-1/2"	"	140.00	29.80	169.80
Threaded				
1/4", 125 lb	EA.	19.45	19.10	38.55
1/2"				
125 lb	EA.	21.60	19.10	40.70
300 lb	"	50.50	19.10	69.60
3/4"				
125 lb	EA.	25.20	19.10	44.30
300 lb	"	65.00	19.10	84.10
1"				
125 lb	EA.	32.40	19.10	51.50
300 lb	"	89.00	23.85	112.85
1-1/2"				
125 lb	EA.	100.00	23.85	123.85
300 lb	"	170.00	26.50	196.50
2"				
125 lb	EA.	68.50	34.10	102.60
300 lb	"	250.00	39.80	289.80
Cast iron, flanged				
2", 150 lb	EA.	290.00	39.80	329.80
2-1/2"				
125 lb	EA.	280.00	39.80	319.80
250 lb	"	560.00	39.80	599.80
3"				
125 lb	EA.	330.00	47.70	377.70
250 lb	"	710.00	47.70	757.70
4"				
125 lb	EA.	410.00	68.00	478.00
250 lb	"	950.00	68.00	1,018
6"				
125 lb	EA.	620.00	95.50	715.50
250 lb	"	1,590	95.50	1,686
8"				
125 lb	EA.	1,240	120.00	1,360
250 lb	"	3,010	120.00	3,130
OS&Y, flanged				
2"				
125 lb	EA.	210.00	39.80	249.80
250 lb	"	490.00	39.80	529.80
2-1/2"				
125 lb	EA.	230.00	39.80	269.80

Plumbing

15430.60 Valves

Plumbing	UNIT	MAT.	INST.	TOTAL
250 lb	EA.	560.00	47.70	607.70
3"				
125 lb	EA.	230.00	47.70	277.70
250 lb	"	620.00	47.70	667.70
4"				
125 lb	EA.	330.00	79.50	409.50
250 lb	"	950.00	79.50	1,030
6"				
125 lb	EA.	520.00	95.50	615.50
250 lb	"	1,540	95.50	1,636
Ball valve, bronze, 250 lb, threaded				
1/2"	EA.	10.75	19.10	29.85
3/4"	"	16.10	19.10	35.20
1"	"	18.40	23.85	42.25
1-1/4"	"	29.90	26.50	56.40
1-1/2"	"	47.50	29.80	77.30
2"	"	53.50	34.10	87.60
Angle valve, bronze, 150 lb, threaded				
1/2"	EA.	64.50	17.05	81.55
3/4"	"	84.00	19.10	103.10
1"	"	120.00	19.10	139.10
1-1/4"	"	140.00	23.85	163.85
1-1/2"	"	180.00	26.50	206.50
Balancing valve, with meter connections, circuit setter				
1/2"	EA.	62.00	19.10	81.10
3/4"	"	65.50	21.70	87.20
1"	"	85.00	23.85	108.85
1-1/4"	"	120.00	26.50	146.50
1-1/2"	"	140.00	31.80	171.80
2"	"	200.00	39.80	239.80
2-1/2"	"	390.00	47.70	437.70
3"	"	570.00	59.50	629.50
4"	"	810.00	79.50	889.50
Pressure reducing valve, bronze, threaded, 250 lb				
1/2"	EA.	370.00	29.80	399.80
3/4"	"	370.00	29.80	399.80
1"	"	400.00	29.80	429.80
1-1/4"	"	470.00	34.10	504.10
1-1/2"	"	540.00	39.80	579.80
Pressure regulating valve, bronze, class 300				
1"	EA.	770.00	29.80	799.80
1-1/2"	"	1,040	36.70	1,077
2"	"	1,180	47.70	1,228
3"	"	1,330	68.00	1,398
4"	"	1,670	95.50	1,766
5"	"	2,530	120.00	2,650
6"	"	2,560	160.00	2,720
Solar water temperature regulating valve				
3/4"	EA.	480.00	39.80	519.80
1"	"	490.00	47.70	537.70
1-1/4"	"	540.00	53.00	593.00
1-1/2"	"	590.00	59.50	649.50

Plumbing

Plumbing	UNIT	MAT.	INST.	TOTAL
15430.60 Valves				
2"	EA.	740.00	68.00	808.00
2-1/2"	"	1,390	120.00	1,510
Tempering valve, threaded				
3/4"	EA.	74.50	15.90	90.40
1"	"	190.00	19.10	209.10
1-1/4"	"	230.00	23.85	253.85
1-1/2"	"	310.00	23.85	333.85
2"	"	380.00	29.80	409.80
2-1/2"	"	480.00	39.80	519.80
3"	"	590.00	47.70	637.70
4"	"	1,300	68.00	1,368
Thermostatic mixing valve, threaded				
1/2"	EA.	88.50	17.05	105.55
3/4"	"	93.50	19.10	112.60
1"	"	330.00	20.75	350.75
1-1/2"	"	370.00	23.85	393.85
2"	"	470.00	29.80	499.80
Sweat connection				
1/2"	EA.	100.00	17.05	117.05
3/4"	"	120.00	19.10	139.10
Mixing valve, sweat connection				
1/2"	EA.	41.60	17.05	58.65
3/4"	"	46.80	19.10	65.90
Liquid level gauge, aluminum body				
3/4"	EA.	330.00	19.10	349.10
4125 psi, pvc body				
3/4"	EA.	330.00	19.10	349.10
150 psi, crs body				
3/4"	EA.	260.00	19.10	279.10
1"	"	290.00	19.10	309.10
175 psi, bronze body, 1/2"	"	540.00	17.05	557.05
15430.65 Vacuum Breakers				
Vacuum breaker, atmospheric, threaded connection				
3/4"	EA.	34.70	19.10	53.80
1"	"	50.50	19.10	69.60
Anti-siphon, brass				
3/4"	EA.	39.40	19.10	58.50
1"	"	55.00	19.10	74.10
1-1/4"	"	86.50	23.85	110.35
1-1/2"	"	100.00	26.50	126.50
2"	"	160.00	29.80	189.80
15430.68 Strainers				
Strainer, Y pattern, 125 psi, cast iron body, threaded				
3/4"	EA.	14.90	17.05	31.95
1"	"	17.85	19.10	36.95
1-1/4"	"	23.80	23.85	47.65
1-1/2"	"	29.80	23.85	53.65
2"	"	44.70	29.80	74.50
250 psi, brass body, threaded				
3/4"	EA.	37.70	19.10	56.80

15 MECHANICAL

Plumbing

	UNIT	MAT.	INST.	TOTAL
15430.68 Strainers				
1"	EA.	52.50	19.10	71.60
1-1/4"	"	65.50	23.85	89.35
1-1/2"	"	95.00	23.85	118.85
2"	"	160.00	29.80	189.80
Cast iron body, threaded				
3/4"	EA.	23.90	19.10	43.00
1"	"	30.40	19.10	49.50
1-1/4"	"	40.60	23.85	64.45
1-1/2"	"	54.00	23.85	77.85
2"	"	68.00	29.80	97.80
15430.70 Drains, Roof & Floor				
Floor drain, cast iron, with cast iron top				
2"	EA.	68.50	39.80	108.30
3"	"	100.00	39.80	139.80
4"	"	140.00	39.80	179.80
6"	"	250.00	47.70	297.70
Roof drain, cast iron				
2"	EA.	170.00	39.80	209.80
3"	"	180.00	39.80	219.80
4"	"	200.00	39.80	239.80
5"	"	230.00	47.70	277.70
6"	"	270.00	47.70	317.70
15430.80 Traps				
Bucket trap, threaded				
3/4"	EA.	150.00	29.80	179.80
1"	"	440.00	31.80	471.80
1-1/4"	"	520.00	36.70	556.70
1-1/2"	"	780.00	43.40	823.40
Inverted bucket steam trap, threaded				
3/4"	EA.	190.00	29.80	219.80
1"	"	370.00	29.80	399.80
1-1/4"	"	560.00	26.50	586.50
1-1/2"	"	590.00	39.80	629.80
Float trap, 15 psi				
3/4"	EA.	130.00	29.80	159.80
1"	"	200.00	31.80	231.80
1-1/4"	"	270.00	34.10	304.10
1-1/2"	"	340.00	39.80	379.80
2"	"	590.00	47.70	637.70
Float and thermostatic trap, 15 psi				
3/4"	EA.	140.00	29.80	169.80
1"	"	160.00	31.80	191.80
1-1/4"	"	250.00	34.10	284.10
1-1/2"	"	320.00	39.80	359.80
2"	"	590.00	47.70	637.70
Steam trap, cast iron body, threaded, 125 psi				
3/4"	EA.	170.00	29.80	199.80
1"	"	200.00	31.80	231.80
1-1/4"	"	280.00	34.10	314.10
1-1/2"	"	460.00	39.80	499.80

Plumbing Fixtures

	UNIT	MAT.	INST.	TOTAL
15440.10 Baths				
Bath tub, 5' long				
Minimum	EA.	460.00	160.00	620.00
Average	"	1,010	240.00	1,250
Maximum	"	2,310	480.00	2,790
6' long				
Minimum	EA.	520.00	160.00	680.00
Average	"	1,050	240.00	1,290
Maximum	"	2,990	480.00	3,470
Square tub, whirlpool, 4'x4'				
Minimum	EA.	1,590	240.00	1,830
Average	"	2,250	480.00	2,730
Maximum	"	6,860	600.00	7,460
5'x5'				
Minimum	EA.	1,590	240.00	1,830
Average	"	2,250	480.00	2,730
Maximum	"	6,990	600.00	7,590
6'x6'				
Minimum	EA.	1,930	240.00	2,170
Average	"	2,820	480.00	3,300
Maximum	"	8,110	600.00	8,710
For trim and rough-in				
Minimum	EA.	160.00	160.00	320.00
Average	"	240.00	240.00	480.00
Maximum	"	690.00	480.00	1,170
15440.12 Disposals & Accessories				
Continuous feed				
Minimum	EA.	96.50	95.50	192.00
Maximum	"	350.00	160.00	510.00
Batch feed, 1/2 hp				
Minimum	EA.	350.00	95.50	445.50
Maximum	"	860.00	160.00	1,020
Hot water dispenser				
Minimum	EA.	140.00	95.50	235.50
Maximum	"	460.00	160.00	620.00
15440.15 Faucets				
Kitchen				
Minimum	EA.	120.00	79.50	199.50
Average	"	210.00	95.50	305.50
Maximum	"	260.00	120.00	380.00
Bath				
Minimum	EA.	190.00	79.50	269.50
Average	"	300.00	95.50	395.50
Maximum	"	340.00	120.00	460.00
Lavatory, domestic				
Minimum	EA.	130.00	79.50	209.50
Average	"	270.00	95.50	365.50
Maximum	"	420.00	120.00	540.00
Hospital, patient rooms				
Minimum	EA.	310.00	120.00	430.00
Average	"	460.00	160.00	620.00

Plumbing Fixtures

Plumbing Fixtures	UNIT	MAT.	INST.	TOTAL
15440.15 Faucets				
Maximum	EA.	610.00	240.00	850.00
Operating room				
Minimum	EA.	360.00	120.00	480.00
Average	"	500.00	160.00	660.00
Maximum	"	730.00	240.00	970.00
Washroom				
Minimum	EA.	180.00	79.50	259.50
Average	"	310.00	95.50	405.50
Maximum	"	460.00	120.00	580.00
Handicapped				
Minimum	EA.	230.00	95.50	325.50
Average	"	330.00	120.00	450.00
Maximum	"	510.00	160.00	670.00
Shower				
Minimum	EA.	150.00	79.50	229.50
Average	"	290.00	95.50	385.50
Maximum	"	460.00	120.00	580.00
For trim and rough-in				
Minimum	EA.	71.50	95.50	167.00
Average	"	110.00	120.00	230.00
Maximum	"	180.00	240.00	420.00
15440.18 Hydrants				
Wall hydrant				
8" thick	EA.	150.00	79.50	229.50
12" thick	"	180.00	95.50	275.50
18" thick	"	200.00	110.00	310.00
24" thick	"	230.00	120.00	350.00
Ground hydrant				
2' deep	EA.	300.00	59.50	359.50
4' deep	"	320.00	68.00	388.00
6' deep	"	350.00	79.50	429.50
8' deep	"	440.00	120.00	560.00
15440.20 Lavatories				
Lavatory, counter top, porcelain enamel on cast iron				
Minimum	EA.	160.00	95.50	255.50
Average	"	250.00	120.00	370.00
Maximum	"	470.00	160.00	630.00
Wall hung, china				
Minimum	EA.	240.00	95.50	335.50
Average	"	280.00	120.00	400.00
Maximum	"	700.00	160.00	860.00
Handicapped				
Minimum	EA.	390.00	120.00	510.00
Average	"	450.00	160.00	610.00
Maximum	"	750.00	240.00	990.00
For trim and rough-in				
Minimum	EA.	170.00	120.00	290.00
Average	"	210.00	160.00	370.00
Maximum	"	380.00	240.00	620.00

Plumbing Fixtures

Plumbing Fixtures	UNIT	MAT.	INST.	TOTAL
15440.30 Showers				
Shower, fiberglass, 36"x34"x84"				
Minimum	EA.	650.00	340.00	990.00
Average	"	910.00	480.00	1,390
Maximum	"	1,310	480.00	1,790
Steel, 1 piece, 36"x36"				
Minimum	EA.	600.00	340.00	940.00
Average	"	910.00	480.00	1,390
Maximum	"	1,070	480.00	1,550
Receptor, molded stone, 36"x36"				
Minimum	EA.	200.00	160.00	360.00
Average	"	340.00	240.00	580.00
Maximum	"	520.00	400.00	920.00
For trim and rough-in				
Minimum	EA.	170.00	220.00	390.00
Average	"	220.00	270.00	490.00
Maximum	"	290.00	480.00	770.00
15440.40 Sinks				
Service sink, 24"x29"				
Minimum	EA.	580.00	120.00	700.00
Average	"	720.00	160.00	880.00
Maximum	"	1,060	240.00	1,300
Kitchen sink, single, stainless steel, single bowl				
Minimum	EA.	220.00	95.50	315.50
Average	"	250.00	120.00	370.00
Maximum	"	460.00	160.00	620.00
Double bowl				
Minimum	EA.	250.00	120.00	370.00
Average	"	280.00	160.00	440.00
Maximum	"	490.00	240.00	730.00
Porcelain enamel, cast iron, single bowl				
Minimum	EA.	180.00	95.50	275.50
Average	"	210.00	120.00	330.00
Maximum	"	370.00	160.00	530.00
Double bowl				
Minimum	EA.	250.00	120.00	370.00
Average	"	310.00	160.00	470.00
Maximum	"	500.00	240.00	740.00
Mop sink, 24"x36"x10"				
Minimum	EA.	440.00	95.50	535.50
Average	"	530.00	120.00	650.00
Maximum	"	710.00	160.00	870.00
Washing machine box				
Minimum	EA.	160.00	120.00	280.00
Average	"	230.00	160.00	390.00
Maximum	"	280.00	240.00	520.00
For trim and rough-in				
Minimum	EA.	260.00	160.00	420.00
Average	"	400.00	240.00	640.00
Maximum	"	510.00	320.00	830.00

Plumbing Fixtures

Plumbing Fixtures	UNIT	MAT.	INST.	TOTAL
15440.50 Urinals				
Urinal, flush valve, floor mounted				
Minimum	EA.	740.00	120.00	860.00
Average	"	880.00	160.00	1,040
Maximum	"	1,030	240.00	1,270
Wall mounted				
Minimum	EA.	370.00	120.00	490.00
Average	"	510.00	160.00	670.00
Maximum	"	660.00	240.00	900.00
For trim and rough-in				
Minimum	EA.	140.00	120.00	260.00
Average	"	210.00	240.00	450.00
Maximum	"	270.00	320.00	590.00
15440.60 Water Closets				
Water closet flush tank, floor mounted				
Minimum	EA.	290.00	120.00	410.00
Average	"	590.00	160.00	750.00
Maximum	"	1,160	240.00	1,400
Handicapped				
Minimum	EA.	340.00	160.00	500.00
Average	"	610.00	240.00	850.00
Maximum	"	930.00	480.00	1,410
Bowl, with flush valve, floor mounted				
Minimum	EA.	420.00	120.00	540.00
Average	"	460.00	160.00	620.00
Maximum	"	900.00	240.00	1,140
Wall mounted				
Minimum	EA.	420.00	120.00	540.00
Average	"	490.00	160.00	650.00
Maximum	"	940.00	240.00	1,180
For trim and rough-in				
Minimum	EA.	160.00	120.00	280.00
Average	"	190.00	160.00	350.00
Maximum	"	250.00	240.00	490.00
15440.70 Water Heaters				
Water heater, electric				
6 gal	EA.	270.00	79.50	349.50
10 gal	"	290.00	79.50	369.50
15 gal	"	300.00	79.50	379.50
20 gal	"	350.00	95.50	445.50
30 gal	"	360.00	95.50	455.50
40 gal	"	380.00	95.50	475.50
52 gal	"	440.00	120.00	560.00
66 gal	"	640.00	120.00	760.00
80 gal	"	690.00	120.00	810.00
100 gal	"	860.00	160.00	1,020
120 gal	"	1,100	160.00	1,260
Oil fired				
20 gal	EA.	1,130	240.00	1,370
50 gal	"	1,770	340.00	2,110

Plumbing Fixtures

	UNIT	MAT.	INST.	TOTAL
15440.90 Miscellaneous Fixtures				
Electric water cooler				
Floor mounted	EA.	690.00	160.00	850.00
Wall mounted	"	860.00	160.00	1,020
Wash fountain				
Wall mounted	EA.	1,200	240.00	1,440
Circular, floor supported	"	2,400	480.00	2,880
Deluge shower and eye wash	"	870.00	240.00	1,110
15440.95 Fixture Carriers				
Water fountain, wall carrier				
Minimum	EA.	96.50	47.70	144.20
Average	"	130.00	59.50	189.50
Maximum	"	160.00	79.50	239.50
Lavatory, wall carrier				
Minimum	EA.	130.00	47.70	177.70
Average	"	190.00	59.50	249.50
Maximum	"	240.00	79.50	319.50
Sink, industrial, wall carrier				
Minimum	EA.	140.00	47.70	187.70
Average	"	170.00	59.50	229.50
Maximum	"	200.00	79.50	279.50
Toilets, water closets, wall carrier				
Minimum	EA.	170.00	47.70	217.70
Average	"	200.00	59.50	259.50
Maximum	"	260.00	79.50	339.50
Floor support				
Minimum	EA.	82.00	39.80	121.80
Average	"	98.50	47.70	146.20
Maximum	"	110.00	59.50	169.50
Urinals, wall carrier				
Minimum	EA.	110.00	47.70	157.70
Average	"	140.00	59.50	199.50
Maximum	"	170.00	79.50	249.50
Floor support				
Minimum	EA.	90.50	39.80	130.30
Average	"	130.00	47.70	177.70
Maximum	"	150.00	59.50	209.50
15450.30 Pumps				
In-line pump, bronze, centrifugal				
5 gpm, 20' head	EA.	670.00	29.80	699.80
20 gpm, 40' head	"	1,250	29.80	1,280
50 gpm				
50' head	EA.	1,640	59.50	1,700
Cast iron, centrifugal				
50 gpm, 200' head	EA.	3,460	59.50	3,520
100 gpm				
100' head	EA.	3,190	79.50	3,270
Centrifugal, close coupled, c.i., single stage				
50 gpm, 100' head	EA.	1,400	59.50	1,460
100 gpm, 100' head	"	1,700	79.50	1,780
Base mounted				

Plumbing Fixtures

Plumbing Fixtures	UNIT	MAT.	INST.	TOTAL
15450.30 Pumps				
50 gpm, 100' head	EA.	2,290	59.50	2,350
100 gpm, 50' head	"	2,590	79.50	2,670
200 gpm, 100' head	"	3,320	120.00	3,440
300 gpm, 175' head	"	4,310	120.00	4,430
Condensate pump, simplex				
1000 sf EDR, 2 gpm	EA.	1,170	400.00	1,570
2000 sf EDR, 3 gpm	"	1,190	400.00	1,590
4000 sf EDR, 6 gpm	"	1,200	430.00	1,630
6000 sf EDR, 9 gpm	"	1,220	430.00	1,650
Duplex, bronze				
8000 sf EDR, 12 gpm	EA.	1,670	430.00	2,100
10,000 sf EDR, 15 gpm	"	1,740	600.00	2,340
15,000 sf EDR, 23 gpm	"	2,090	680.00	2,770
20,000 sf EDR, 30 gpm	"	2,430	950.00	3,380
25,000 sf EDR, 38 gpm	"	2,510	950.00	3,460
15450.40 Storage Tanks				
Hot water storage tank, cement lined				
10 gallon	EA.	4,140	160.00	4,300
70 gallon	"	4,140	240.00	4,380
200 gallon	"	4,290	340.00	4,630
900 gallon	"	4,840	600.00	5,440
1100 gallon	"	5,570	600.00	6,170
2000 gallon	"	10,770	600.00	11,370
15480.10 Special Systems				
Air compressor, air cooled, two stage				
5.0 cfm, 175 psi	EA.	3,560	950.00	4,510
10 cfm, 175 psi	"	4,350	1,060	5,410
20 cfm, 175 psi	"	6,000	1,140	7,140
50 cfm, 125 psi	"	8,740	1,260	10,000
80 cfm, 125 psi	"	12,520	1,360	13,880
Single stage, 125 psi				
1.0 cfm	EA.	3,440	680.00	4,120
1.5 cfm	"	3,500	680.00	4,180
2.0 cfm	"	3,570	680.00	4,250
Automotive compressor, hose reel, air and water, 50' hose	"	1,600	400.00	2,000
Lube equipment, 3 reel, with pumps	"	8,380	1,910	10,290
Tire changer				
Truck	EA.	13,120	680.00	13,800
Passenger car	"	3,100	370.00	3,470
Air hose reel, includes, 50' hose	"	820.00	370.00	1,190
Hose reel, 5 reel, motor oil, gear oil, lube, air & water	"	7,650	1,910	9,560
Water hose reel, 50' hose	"	820.00	370.00	1,190
Pump, air operated, for motor or gear oil, fits 55 gal drum	"	1,070	47.70	1,118
For chassis lube	"	1,750	47.70	1,798
Fuel dispensing pump, lighted dial, one product				
One hose	EA.	3,820	400.00	4,220
Two hose	"	6,740	400.00	7,140
Two products, two hose	"	7,100	400.00	7,500

Heating & Ventilating

15555.10 Boilers

Heating & Ventilating	UNIT	MAT.	INST.	TOTAL
Cast iron, gas fired, hot water				
115 mbh	EA.	1,890	1,520	3,410
175 mbh	"	2,210	1,650	3,860
235 mbh	"	3,350	1,820	5,170
940 mbh	"	9,790	3,640	13,430
1600 mbh	"	13,420	4,550	17,970
3000 mbh	"	28,410	6,060	34,470
6000 mbh	"	66,300	9,090	75,390
Steam				
115 mbh	EA.	2,730	1,520	4,250
175 mbh	"	3,480	1,650	5,130
235 mbh	"	3,810	1,820	5,630
940 mbh	"	9,630	3,640	13,270
1600 mbh	"	17,360	4,550	21,910
3000 mbh	"	33,150	6,060	39,210
6000 mbh	"	71,040	9,090	80,130
Electric, hot water				
115 mbh	EA.	5,360	910.00	6,270
175 mbh	"	6,440	910.00	7,350
235 mbh	"	8,730	910.00	9,640
940 mbh	"	15,330	1,820	17,150
1600 mbh	"	22,990	3,640	26,630
3000 mbh	"	39,850	4,550	44,400
6000 mbh	"	59,770	6,060	65,830
Steam				
115 mbh	EA.	8,680	910.00	9,590
175 mbh	"	9,940	910.00	10,850
235 mbh	"	11,680	910.00	12,590
940 mbh	"	23,680	1,820	25,500
1600 mbh	"	39,470	3,640	43,110
3000 mbh	"	45,780	4,550	50,330
6000 mbh	"	71,040	6,060	77,100
Oil fired, hot water				
115 mbh	EA.	2,600	1,210	3,810
175 mbh	"	3,220	1,400	4,620
235 mbh	"	4,210	1,650	5,860
940 mbh	"	12,410	3,030	15,440
1600 mbh	"	21,530	3,640	25,170
3000 mbh	"	36,780	4,550	41,330
6000 mbh	"	68,970	9,090	78,060
Steam				
115 mbh	EA.	2,730	1,210	3,940
175 mbh	"	3,300	1,400	4,700
235 mbh	"	4,340	1,650	5,990
940 mbh	"	14,000	3,030	17,030
1600 mbh	"	22,850	3,640	26,490
3000 mbh	"	39,590	4,550	44,140
6000 mbh	"	70,500	9,090	79,590

15 MECHANICAL

Heading & Ventilating

15610.10 Furnaces

	UNIT	MAT.	INST.	TOTAL
Electric, hot air				
40 mbh	EA.	720.00	240.00	960.00
60 mbh	"	830.00	250.00	1,080
80 mbh	"	770.00	270.00	1,040
100 mbh	"	1,240	280.00	1,520
125 mbh	"	1,380	290.00	1,670
160 mbh	"	1,990	300.00	2,290
200 mbh	"	3,670	310.00	3,980
400 mbh	"	6,530	320.00	6,850
Gas fired hot air				
40 mbh	EA.	720.00	240.00	960.00
60 mbh	"	760.00	250.00	1,010
80 mbh	"	870.00	270.00	1,140
100 mbh	"	920.00	280.00	1,200
125 mbh	"	1,000	290.00	1,290
160 mbh	"	1,200	300.00	1,500
200 mbh	"	2,120	310.00	2,430
400 mbh	"	7,590	320.00	7,910
Oil fired hot air				
40 mbh	EA.	960.00	240.00	1,200
60 mbh	"	1,150	250.00	1,400
80 mbh	"	1,250	270.00	1,520
100 mbh	"	1,460	280.00	1,740
125 mbh	"	1,630	290.00	1,920
160 mbh	"	1,760	300.00	2,060
200 mbh	"	3,760	310.00	4,070
400 mbh	"	7,970	320.00	8,290

Refrigeration

15670.10 Condensing Units

	UNIT	MAT.	INST.	TOTAL
Air cooled condenser, single circuit				
3 ton	EA.	2,040	79.50	2,120
5 ton	"	3,400	79.50	3,480
7.5 ton	"	5,440	230.00	5,670
20 ton	"	11,480	240.00	11,720
25 ton	"	13,390	240.00	13,630
30 ton	"	15,430	240.00	15,670
40 ton	"	21,260	340.00	21,600
50 ton	"	25,860	340.00	26,200
60 ton	"	29,770	300.00	30,070
With low ambient dampers				
3 ton	EA.	2,380	120.00	2,500
5 ton	"	3,740	120.00	3,860
7.5 ton	"	5,780	240.00	6,020
20 ton	"	11,910	320.00	12,230

15 MECHANICAL

Refrigeration

Refrigeration	UNIT	MAT.	INST.	TOTAL
15670.10 Condensing Units				
25 ton	EA.	13,780	320.00	14,100
30 ton	"	15,820	320.00	16,140
40 ton	"	22,280	400.00	22,680
50 ton	"	26,870	430.00	27,300
60 ton	"	30,790	430.00	31,220
Dual circuit				
10 ton	EA.	5,100	240.00	5,340
15 ton	"	7,480	340.00	7,820
20 ton	"	12,760	340.00	13,100
25 ton	"	14,630	340.00	14,970
30 ton	"	17,010	340.00	17,350
40 ton	"	24,320	400.00	24,720
50 ton	"	26,870	400.00	27,270
60 ton	"	27,900	400.00	28,300
80 ton	"	34,870	530.00	35,400
100 ton	"	40,820	530.00	41,350
120 ton	"	48,480	530.00	49,010
With low ambient dampers				
15 ton	EA.	8,330	340.00	8,670
20 ton	"	13,610	340.00	13,950
25 ton	"	15,480	340.00	15,820
30 ton	"	16,670	340.00	17,010
40 ton	"	25,170	400.00	25,570
50 ton	"	27,730	400.00	28,130
60 ton	"	28,750	400.00	29,150
80 ton	"	36,570	530.00	37,100
100 ton	"	42,520	530.00	43,050
120 ton	"	50,690	530.00	51,220
15680.10 Chillers				
Chiller, reciprocal				
Air cooled, remote condenser, starter				
20 ton	EA.	26,190	610.00	26,800
25 ton	"	29,560	610.00	30,170
30 ton	"	31,250	610.00	31,860
40 ton	"	46,220	910.00	47,130
50 ton	"	50,330	1,010	51,340
100 ton	"	84,200	1,820	86,020
200 ton	"	153,430	3,030	156,460
Water cooled, with starter				
20 ton	EA.	22,450	610.00	23,060
25 ton	"	24,880	610.00	25,490
30 ton	"	29,940	910.00	30,850
40 ton	"	41,160	910.00	42,070
50 ton	"	44,910	1,010	45,920
100 ton	"	65,490	1,820	67,310
200 ton	"	104,780	3,030	107,810
Packaged, air cooled, with starter				
20 ton	EA.	25,070	450.00	25,520
25 ton	"	27,130	450.00	27,580
30 ton	"	31,620	450.00	32,070
40 ton	"	36,300	450.00	36,750

Refrigeration

	UNIT	MAT.	INST.	TOTAL
15680.10 Chillers				
Heat recovery, air cooled, with starter				
50 ton	EA.	48,650	910.00	49,560
100 ton	"	69,610	1,820	71,430
Water cooled, with starter				
40 ton	EA.	41,540	910.00	42,450
100 ton	"	75,030	2,020	77,050
Centrifugal, single bundle condenser, with starter				
80 ton	EA.	98,230	2,600	100,830
230 ton	"	117,690	4,040	121,730
460 ton	"	178,320	6,060	184,380
670 ton	"	23,780	7,270	31,050
15710.10 Cooling Towers				
Cooling tower, propeller type				
100 ton	EA.	10,110	610.00	10,720
400 ton	"	33,710	1,820	35,530
1000 ton	"	80,920	4,550	85,470
Centrifugal				
100 ton	EA.	10,110	610.00	10,720
400 ton	"	40,460	1,820	42,280
1000 ton	"	94,400	4,550	98,950

Heat Transfer

	UNIT	MAT.	INST.	TOTAL
15780.10 Computer Room A/c				
Air cooled, alarm, high efficiency filter, elec. heat				
3 ton	EA.	12,580	370.00	12,950
5 ton	"	13,490	400.00	13,890
7.5 ton	"	22,960	480.00	23,440
10 ton	"	23,690	600.00	24,290
15 ton	"	26,250	680.00	26,930
Steam heat				
3 ton	EA.	14,580	370.00	14,950
5 ton	"	15,490	400.00	15,890
7.5 ton	"	24,600	480.00	25,080
10 ton	"	25,330	600.00	25,930
15 ton	"	28,070	680.00	28,750
Hot water heat				
3 ton	EA.	14,580	370.00	14,950
5 ton	"	15,490	400.00	15,890
7.5 ton	"	24,600	480.00	25,080
10 ton	"	25,330	600.00	25,930
15 ton	"	28,070	680.00	28,750
Air cooled condenser, low ambient damper				
3 ton	EA.	3,280	95.50	3,376
5 ton	"	3,640	120.00	3,760

Heat Transfer

	UNIT	MAT.	INST.	TOTAL
15780.10 Computer Room A/c				
7.5 ton	EA.	5,280	240.00	5,520
10 ton	"	5,280	340.00	5,620
15 ton	"	5,830	280.00	6,110
Water cooled, high efficiency filter, alarm, elec. heat				
3 ton	EA.	14,400	340.00	14,740
5 ton	"	15,490	400.00	15,890
7.5 ton	"	24,600	600.00	25,200
10 ton	"	25,520	680.00	26,200
15 ton	"	29,890	800.00	30,690
Steam heat				
3 ton	EA.	16,400	340.00	16,740
5 ton	"	18,750	400.00	19,150
7.5 ton	"	26,430	600.00	27,030
10 ton	"	27,340	680.00	28,020
15 ton	"	31,710	800.00	32,510
Hot water heat				
3 ton	EA.	16,400	340.00	16,740
5 ton	"	17,500	400.00	17,900
7.5 ton	"	26,430	600.00	27,030
10 ton	"	27,340	680.00	28,020
15 ton	"	31,710	800.00	32,510
15780.20 Rooftop Units				
Packaged, single zone rooftop unit, with roof curb				
2 ton	EA.	4,420	480.00	4,900
3 ton	"	5,100	480.00	5,580
4 ton	"	6,120	600.00	6,720
5 ton	"	6,190	800.00	6,990
7.5 ton	"	10,040	950.00	10,990
15820.10 Dehumidifiers				
Dessicant dehumidifier, 1125 cfm	EA.			17,490
15830.20 Fan Coil Units				
Fan coil unit, 2 pipe, complete				
200 cfm ceiling hung	EA.	790.00	160.00	950.00
Floor mounted	"	740.00	120.00	860.00
300 cfm, ceiling hung	"	840.00	190.00	1,030
Floor mounted	"	800.00	160.00	960.00
400 cfm, ceiling hung	"	880.00	230.00	1,110
Floor mounted	"	850.00	160.00	1,010
500 cfm, ceiling hung	"	1,030	240.00	1,270
Floor mounted	"	990.00	180.00	1,170
600 cfm, ceiling hung	"	1,310	260.00	1,570
Floor mounted	"	1,210	220.00	1,430
15830.70 Unit Heaters				
Steam unit heater, horizontal				
12,500 btuh, 200 cfm	EA.	530.00	79.50	609.50
17,000 btuh, 300 cfm	"	610.00	79.50	689.50
40,000 btuh, 500 cfm	"	760.00	79.50	839.50
60,000 btuh, 700 cfm	"	850.00	79.50	929.50
70,000 btuh, 1000 cfm	"	1,070	120.00	1,190

Heat Transfer

Heat Transfer	UNIT	MAT.	INST.	TOTAL
15830.70 Unit Heaters				
Vertical				
12,500 btuh, 200 cfm	EA.	460.00	79.50	539.50
17,000 btuh, 300 cfm	"	760.00	79.50	839.50
40,000 btuh, 500 cfm	"	920.00	79.50	999.50
60,000 btuh, 700 cfm	"	1,070	79.50	1,150
70,000 btuh, 1000 cfm	"	1,220	79.50	1,300
Gas unit heater, horizontal				
27,400 btuh	EA.	760.00	190.00	950.00
38,000 btuh	"	840.00	190.00	1,030
56,000 btuh	"	920.00	190.00	1,110
82,200 btuh	"	1,010	190.00	1,200
103,900 btuh	"	1,050	300.00	1,350
125,700 btuh	"	1,080	300.00	1,380
133,200 btuh	"	1,150	300.00	1,450
149,000 btuh	"	1,340	300.00	1,640
172,000 btuh	"	1,450	300.00	1,750
190,000 btuh	"	1,530	300.00	1,830
225,000 btuh	"	1,680	300.00	1,980
Hot water unit heater, horizontal				
12,500 btuh, 200 cfm	EA.	460.00	79.50	539.50
17,000 btuh, 300 cfm	"	530.00	79.50	609.50
25,000 btuh, 500 cfm	"	570.00	79.50	649.50
30,000 btuh, 700 cfm	"	690.00	79.50	769.50
50,000 btuh, 1000 cfm	"	760.00	120.00	880.00
60,000 btuh, 1300 cfm	"	1,070	120.00	1,190
Vertical				
12,500 btuh, 200 cfm	EA.	570.00	79.50	649.50
17,000 btuh, 300 cfm	"	610.00	79.50	689.50
25,000 btuh, 500 cfm	"	650.00	79.50	729.50
30,000 btuh, 700 cfm	"	700.00	79.50	779.50
50,000 btuh, 1000 cfm	"	760.00	79.50	839.50
60,000 btuh, 1300 cfm	"	840.00	79.50	919.50
Cabinet unit heaters, ceiling, exposed, hot water				
200 cfm	EA.	1,070	160.00	1,230
300 cfm	"	1,150	190.00	1,340
400 cfm	"	1,190	230.00	1,420
600 cfm	"	1,220	250.00	1,470
800 cfm	"	1,530	300.00	1,830
1000 cfm	"	1,990	340.00	2,330
1200 cfm	"	2,140	400.00	2,540
2000 cfm	"	3,670	530.00	4,200

Air Handling

	UNIT	MAT.	INST.	TOTAL
15855.10 Air Handling Units				
Air handling unit, medium pressure, single zone				
1500 cfm	EA.	3,440	300.00	3,740
3000 cfm	"	4,290	530.00	4,820
4000 cfm	"	4,810	600.00	5,410
5000 cfm	"	5,500	640.00	6,140
6000 cfm	"	6,190	680.00	6,870
7000 cfm	"	6,870	730.00	7,600
8500 cfm	"	8,080	800.00	8,880
10,500 cfm	"	9,110	950.00	10,060
12,500 cfm	"	11,680	1,060	12,740
Rooftop air handling units				
4950 cfm	EA.	10,330	530.00	10,860
7370 cfm	"	11,190	680.00	11,870
9790 cfm	"	13,940	800.00	14,740
14,300 cfm	"	19,680	680.00	20,360
21,725 cfm	"	27,890	680.00	28,570
33,000 cfm	"	39,370	800.00	40,170
15870.20 Exhaust Fans				
Belt drive roof exhaust fans				
640 cfm, 2618 fpm	EA.	890.00	59.50	949.50
940 cfm, 2604 fpm	"	1,150	59.50	1,210
1050 cfm, 3325 fpm	"	1,020	59.50	1,080
1170 cfm, 2373 fpm	"	1,490	59.50	1,550
2440 cfm, 4501 fpm	"	1,170	59.50	1,230
2760 cfm, 4950 fpm	"	1,290	59.50	1,350
3890 cfm, 6769 fpm	"	1,470	59.50	1,530
2380 cfm, 3382 fpm	"	1,640	59.50	1,700
2880 cfm, 3859 fpm	"	1,710	59.50	1,770
3200 cfm, 4173 fpm	"	1,720	79.50	1,800
3660 cfm, 3437 fpm	"	1,750	79.50	1,830
Direct drive fans				
60 to 390 cfm	EA.	730.00	59.50	789.50
145 to 590 cfm	"	880.00	59.50	939.50
295 to 860 cfm	"	1,060	59.50	1,120
235 to 1300 cfm	"	1,140	59.50	1,200
415 to 1630 cfm	"	1,290	59.50	1,350
590 to 2045 cfm	"	1,490	59.50	1,550

Air Distribution

	UNIT	MAT.	INST.	TOTAL
15890.10 Metal Ductwork				
Rectangular duct				
Galvanized steel				
Minimum	Lb.	1.50	4.35	5.85
Average	"	1.90	5.30	7.20

Air Distribution	UNIT	MAT.	INST.	TOTAL
15890.10 Metal Ductwork				
Maximum	Lb.	2.85	7.95	10.80
Aluminum				
Minimum	Lb.	3.55	9.55	13.10
Average	"	4.75	11.95	16.70
Maximum	"	5.95	15.90	21.85
Fittings				
Minimum	EA.	12.50	15.90	28.40
Average	"	18.75	23.85	42.60
Maximum	"	27.50	47.70	75.20
For work				
10-20' high, add per pound, $.30				
30-50', add per pound, $.50				
15890.30 Flexible Ductwork				
Flexible duct, 1.25" fiberglass				
5" dia.	L.F.	2.25	2.40	4.65
6" dia.	"	2.65	2.65	5.30
7" dia.	"	2.95	2.80	5.75
8" dia.	"	3.40	3.00	6.40
10" dia.	"	3.85	3.40	7.25
12" dia.	"	4.75	3.65	8.40
14" dia.	"	5.75	4.00	9.75
16" dia.	"	7.10	4.35	11.45
Flexible duct connector, 3" wide fabric	"	2.00	7.95	9.95
15895.10 Roof Curbs				
8" high, insulated, with liner and raised can				
15" x 15"	EA.	83.00	23.85	106.85
21" x 21"	"	100.00	23.85	123.85
36" x 36"	"	150.00	34.10	184.10
48" x 48"	"	420.00	36.70	456.70
60" x 60"	"	780.00	47.70	827.70
72" x 72"	"	1,240	59.50	1,300
15910.10 Dampers				
Horizontal parallel aluminum backdraft damper				
12" x 12"	EA.	59.00	11.95	70.95
16" x 16"	"	61.00	13.65	74.65
24" x 24"	"	95.50	23.85	119.35
36" x 36"	"	220.00	34.10	254.10
48" x 48"	"	390.00	47.70	437.70
"Up", parallel dampers				
12" x 12"	EA.	76.50	11.95	88.45
16" x 16"	"	110.00	13.65	123.65
24" x 24"	"	130.00	23.85	153.85
36" x 36"	"	220.00	34.10	254.10
48" x 48"	"	440.00	47.70	487.70
"Down", parallel dampers				
12" x 12"	EA.	76.50	11.95	88.45
16" x 16"	"	110.00	13.65	123.65
24" x 24"	"	130.00	23.85	153.85
36" x 36"	"	220.00	34.10	254.10

Air Distribution	UNIT	MAT.	INST.	TOTAL
15910.10 Dampers				
48" x 48"	EA.	440.00	47.70	487.70
Fire damper, 1.5 hr rating				
12" x 12"	EA.	26.10	23.85	49.95
16" x 16"	"	41.70	23.85	65.55
24" x 24"	"	52.00	23.85	75.85
36" x 36"	"	87.00	47.70	134.70
48" x 48"	"	170.00	68.00	238.00
15940.10 Diffusers				
Ceiling diffusers, round, baked enamel finish				
6" dia.	EA.	42.70	15.90	58.60
12" dia.	"	77.00	19.90	96.90
16" dia.	"	110.00	21.70	131.70
20" dia.	"	160.00	23.85	183.85
Rectangular				
6x6"	EA.	47.10	15.90	63.00
12x12"	"	83.00	23.85	106.85
18x18"	"	130.00	23.85	153.85
24x24"	"	230.00	29.80	259.80
Lay in, flush mounted, perforated face, with grid				
6x6/24x24	EA.	78.00	19.10	97.10
12x12/24x24	"	87.50	19.10	106.60
18x18/24x24	"	91.00	19.10	110.10
Two-way slot diffuser with balancing damper, 4'	"	97.50	47.70	145.20
15940.20 Relief Ventilators				
Intake ventilator, aluminum, with screen, no curbs				
12" x 12"	EA.	160.00	39.80	199.80
16" x 16"	"	220.00	47.70	267.70
36" x 36"	"	860.00	79.50	939.50
48" x 48"	"	1,410	95.50	1,506
15940.40 Registers And Grilles				
Lay in flush mounted, perforated face, return				
6x6/24x24	EA.	39.70	19.10	58.80
8x8/24x24	"	39.70	19.10	58.80
9x9/24x24	"	43.10	19.10	62.20
10x10/24x24	"	46.40	19.10	65.50
12x12/24x24	"	46.40	19.10	65.50
Rectangular, ceiling return, single deflection				
10x10	EA.	19.85	23.85	43.70
12x12	"	23.20	23.85	47.05
14x14	"	28.20	23.85	52.05
16x8	"	23.20	23.85	47.05
16x16	"	23.20	23.85	47.05
24x12	"	63.00	23.85	86.85
24x18	"	81.00	23.85	104.85
36x24	"	160.00	26.50	186.50
36x30	"	230.00	26.50	256.50
Wall, return air register				
12x12	EA.	33.00	11.95	44.95
16x16	"	48.70	11.95	60.65

Air Distribution

	UNIT	MAT.	INST.	TOTAL
15940.40 Registers And Grilles				
18x18	EA.	57.50	11.95	69.45
20x20	"	68.50	11.95	80.45
24x24	"	94.00	11.95	105.95
Ceiling, return air grille				
6x6	EA.	12.20	15.90	28.10
8x8	"	13.90	19.10	33.00
10x10	"	15.65	19.10	34.75
Ceiling, exhaust grille, aluminum egg crate				
6x6	EA.	15.65	15.90	31.55
8x8	"	15.65	19.10	34.75
10x10	"	17.40	19.10	36.50
12x12	"	22.60	23.85	46.45
14x14	"	29.60	23.85	53.45
16x16	"	34.80	23.85	58.65
18x18	"	41.70	23.85	65.55
15940.80 Penthouse Louvers				
Penthouse louvers				
12" high, extruded aluminum, 4" louver				
6' perimeter	EA.	390.00	120.00	510.00
12' perimeter	"	940.00	120.00	1,060
20' perimeter	"	1,950	320.00	2,270
16" high x 4' perimeter	"	320.00	120.00	440.00
6' perimeter	"	450.00	120.00	570.00
12' perimeter	"	1,060	120.00	1,180
20' perimeter	"	2,270	320.00	2,590
20" high x 4' perimeter	"	420.00	120.00	540.00
6' perimeter	"	480.00	120.00	600.00
12' perimeter	"	1,190	120.00	1,310
20' perimeter	"	2,440	320.00	2,760
24" high x 4' perimeter	"	450.00	120.00	570.00
6' perimeter	"	540.00	120.00	660.00
12' perimeter	"	1,320	120.00	1,440
20' perimeter	"	2,740	320.00	3,060

Controls

	UNIT	MAT.	INST.	TOTAL
15950.10 Hvac Controls				
Pressure gauge, direct reading gage cock and siphon	EA.	78.00	29.80	107.80
Control valve, 1", modulating				
2-way	EA.	780.00	39.80	819.80
3-way	"	880.00	59.50	939.50
Self contained control valve w/ sensing elmnt, 3/4"	"	140.00	29.80	169.80
Inst air syst 2-1/2 hp comp, rcvr refrg dryer	"			6,601
Thermostat primary control device	"			142.47
Humidistat primary control device	"			107.72

Controls	UNIT	MAT.	INST.	TOTAL
15950.10 Hvac Controls				
Timers primary control device, indoor/outdoor, 24 hour	EA.			220.71
Thermometer, dir. reading, 3 dial	"			107.72
Control dampers, round				
6" dia.	EA.	80.00	19.10	99.10
8" dia	"	120.00	19.10	139.10
10" dia	"	150.00	19.10	169.10
12" dia	"	200.00	19.10	219.10
12" dia	"	290.00	23.85	313.85
18" dia	"	310.00	23.85	333.85
20" dia	"	410.00	23.85	433.85
Rectangular, parallel blade standard leakage				
12" x 12"	EA.	59.00	23.85	82.85
16" x 16"	"	87.00	23.85	110.85
20" x 20"	"	100.00	23.85	123.85
48" x 48"	"	310.00	68.00	378.00
48" x 60"	"	400.00	79.50	479.50
48" x 72"	"	480.00	79.50	559.50
Low leakage				
12" x 12"	EA.	110.00	23.85	133.85
16" x 16"	"	150.00	23.85	173.85
36" x 36"	"	370.00	39.80	409.80
48" x 48"	"	660.00	68.00	728.00
48" x 72"	"	1,020	79.50	1,100
Rectangular, opposed horizontal blade				
12" x 12"	EA.	76.50	23.85	100.35
16" x 16"	"	110.00	23.85	133.85
24" x 24"	"	140.00	23.85	163.85
36" x 36"	"	230.00	39.80	269.80
48" x 72"	"	660.00	79.50	739.50

Basic Materials

Basic Materials	UNIT	MAT.	INST.	TOTAL
16110.12 Cable Tray				
Cable tray, 6"	L.F.	16.00	3.30	19.30
Ventilated cover	"	6.50	1.65	8.15
Solid cover	"	5.05	1.65	6.70
16110.20 Conduit Specialties				
Rod beam clamp, 1/2"	EA.	4.95	2.75	7.70
Hanger rod				
3/8"	L.F.	1.05	2.20	3.25
1/2"	"	1.70	2.75	4.45
All thread rod				
1/4"	L.F.	0.64	1.65	2.29
3/8"	"	1.05	2.20	3.25
1/2"	"	1.70	2.75	4.45
5/8"	"	2.70	4.40	7.10
Hanger channel, 1-1/2"				
No holes	EA.	4.05	1.65	5.70
Holes	"	4.50	1.65	6.15
Channel strap				
1/2"	EA.	1.05	2.75	3.80
3/4"	"	1.25	2.75	4.00
1"	"	1.40	2.75	4.15
1-1/4"	"	1.50	4.40	5.90
1-1/2"	"	1.70	4.40	6.10
2"	"	2.00	4.40	6.40
2-1/2"	"	2.15	6.80	8.95
3"	"	2.35	6.80	9.15
3-1/2"	"	2.90	6.80	9.70
4"	"	3.30	8.05	11.35
5"	"	4.60	8.05	12.65
6"	"	5.50	8.05	13.55
Conduit penetrations, roof and wall, 8" thick				
1/2"	EA.		34.00	34.00
3/4"	"		34.00	34.00
1"	"		44.20	44.20
1-1/4"	"		44.20	44.20
1-1/2"	"		44.20	44.20
2"	"		88.50	88.50
2-1/2"	"		88.50	88.50
3"	"		88.50	88.50
3-1/2"	"		110.00	110.00
4"	"		110.00	110.00
Fireproofing, for conduit penetrations				
1/2"	EA.	2.45	27.70	30.15
3/4"	"	2.50	27.70	30.20
1"	"	2.55	27.70	30.25
1-1/4"	"	3.50	43.40	46.90
1-1/2"	"	3.60	40.20	43.80
2"	"	3.65	40.20	43.85
2-1/2"	"	7.00	49.70	56.70
3"	"	7.25	53.50	60.75
3-1/2"	"	8.45	69.00	77.45
4"	"	10.20	83.50	93.70

Basic Materials

	UNIT	MAT.	INST.	TOTAL
16110.21 Aluminum Conduit				
Aluminum conduit				
1/2"	L.F.	1.10	1.65	2.75
3/4"	"	1.40	2.20	3.60
1"	"	1.45	2.75	4.20
1-1/4"	"	2.70	3.30	6.00
1-1/2"	"	3.40	4.40	7.80
2"	"	4.50	4.90	9.40
2-1/2"	"	7.15	5.55	12.70
3"	"	9.40	5.90	15.30
3-1/2"	"	11.25	6.80	18.05
4"	"	13.35	8.05	21.40
5"	"	19.10	10.05	29.15
6"	"	25.20	11.05	36.25
16110.22 Emt Conduit				
EMT conduit				
1/2"	L.F.	0.38	1.65	2.03
3/4"	"	0.53	2.20	2.73
1"	"	0.73	2.75	3.48
1-1/4"	"	1.05	3.30	4.35
1-1/2"	"	1.35	4.40	5.75
2"	"	1.50	4.90	6.40
2-1/2"	"	3.40	5.55	8.95
3"	"	4.80	6.80	11.60
3-1/2"	"	6.30	8.05	14.35
4"	"	7.05	10.05	17.10
16110.23 Flexible Conduit				
Flexible conduit, steel				
3/8"	L.F.	0.30	1.65	1.95
1/2	"	0.31	1.65	1.96
3/4"	"	0.45	2.20	2.64
1"	"	0.86	2.20	3.06
1-1/4"	"	1.10	2.75	3.85
1-1/2"	"	1.45	3.30	4.75
2"	"	1.85	4.40	6.25
2-1/2"	"	2.20	4.90	7.10
3"	"	2.70	5.90	8.60
16110.24 Galvanized Conduit				
Galvanized rigid steel conduit				
1/2"	L.F.	1.30	2.20	3.50
3/4"	"	1.65	2.75	4.40
1"	"	2.10	3.30	5.40
1-1/4"	"	3.10	4.40	7.50
1-1/2"	"	3.55	4.90	8.45
2"	"	4.45	5.55	10.00
2-1/2"	"	7.30	8.05	15.35
3"	"	9.60	10.05	19.65
3-1/2"	"	9.85	10.55	20.40
4"	"	14.25	11.65	25.90
5"	"	29.20	15.80	45.00

16 ELECTRICAL

Basic Materials

	UNIT	MAT.	INST.	TOTAL
16110.24 Galvanized Conduit				
6"	L.F.	42.10	21.05	63.15
16110.27 Plastic Coated Conduit				
Rigid steel conduit, plastic coated				
1/2"	L.F.	3.40	2.75	6.15
3/4"	"	4.00	3.30	7.30
1"	"	5.20	4.40	9.60
1-1/4"	"	6.45	5.55	12.00
1-1/2"	"	7.90	6.80	14.70
2"	"	10.40	8.05	18.45
2-1/2"	"	15.80	10.55	26.35
3"	"	19.90	12.30	32.20
3-1/2"	"	24.15	13.80	37.95
4"	"	29.50	17.00	46.50
5"	"	50.50	21.05	71.55
16110.28 Steel Conduit				
Intermediate metal conduit (IMC)				
1/2"	L.F.	1.20	1.65	2.85
3/4"	"	1.40	2.20	3.60
1"	"	1.90	2.75	4.65
1-1/4"	"	2.45	3.30	5.75
1-1/2"	"	2.95	4.40	7.35
2"	"	3.95	4.90	8.85
2-1/2"	"	6.00	6.60	12.60
3"	"	8.50	8.05	16.55
3-1/2"	"	11.05	10.05	21.10
4"	"	12.90	10.55	23.45
16110.35 Wiremold				
Wiremold raceway with fittings, surface mounted				
#200	L.F.	0.78	1.65	2.43
#500	"	0.96	1.65	2.61
#700	"	1.05	2.20	3.25
#800	"	1.70	2.20	3.90
Fittings, #200, 90 degree flat elbow	EA.	1.50	2.75	4.25
Internal elbow	"	3.00	2.75	5.75
Extension adapter	"	5.35	3.30	8.65
#200, #500, #700				
Single pole switch and box	EA.	10.10	22.10	32.20
Duplex receptacle with box	"	11.70	17.00	28.70
#500, #700				
90 deg. flat elbow	EA.	1.30	4.40	5.70
Internal elbow	"	1.70	4.40	6.10
Junction box	"	8.55	7.35	15.90
Fixture box	"	8.65	7.35	16.00
Shallow switch and receptacle box	"	7.05	4.40	11.45
#800				
90 deg. flat elbow	EA.	1.65	4.40	6.05
Internal elbow	"	1.65	4.40	6.05
Junction box	"	5.05	6.80	11.85

Basic Materials

	UNIT	MAT.	INST.	TOTAL
16110.80 Wireways				
Wireway, hinge cover type				
2-1/2" x 2-1/2"				
1' section	EA.	12.00	8.50	20.50
2'	"	17.00	10.55	27.55
3'	"	23.10	13.80	36.90
5'	"	38.10	21.05	59.15
10'	"	79.00	36.90	115.90
4" x 4"				
1'	EA.	11.60	13.80	25.40
2'	"	19.20	13.80	33.00
3'	"	28.50	17.00	45.50
4'	"	38.60	17.00	55.60
10'	"	41.40	44.20	85.60
16120.43 Copper Conductors				
Copper conductors, type THW, solid				
#14	L.F.	0.11	0.22	0.33
#12	"	0.12	0.28	0.40
#10	"	0.14	0.33	0.47
Stranded				
#14	L.F.	0.14	0.22	0.36
#12	"	0.15	0.28	0.43
#10	"	0.18	0.33	0.52
#8	"	0.29	0.44	0.73
#6	"	0.38	0.50	0.88
#4	"	0.55	0.55	1.10
#3	"	0.78	0.55	1.33
#2	"	0.96	0.67	1.63
#1	"	1.25	0.78	2.03
1/0	"	1.50	0.89	2.39
2/0	"	1.75	1.10	2.85
3/0	"	2.20	1.40	3.60
4/0	"	2.70	1.55	4.25
250 MCM	"	3.35	1.65	5.00
300 MCM	"	3.95	1.85	5.80
350 MCM	"	4.85	2.20	7.05
400 MCM	"	5.45	2.45	7.90
500 MCM	"	6.55	2.85	9.40
600 MCM	"	9.35	3.30	12.65
750 MCM	"	11.90	3.70	15.60
1000 MCM	"	13.20	4.20	17.40
THHN-THWN, solid				
#14	L.F.	0.11	0.22	0.33
#12	"	0.14	0.28	0.41
#10	"	0.17	0.33	0.50
Stranded				
#14	L.F.	0.14	0.22	0.36
#12	"	0.17	0.28	0.45
#10	"	0.18	0.33	0.52
#8	"	0.29	0.44	0.73
#6	"	0.38	0.50	0.88
#4	"	0.55	0.55	1.10

Basic Materials

Basic Materials	UNIT	MAT.	INST.	TOTAL
16120.43 Copper Conductors				
#2	L.F.	0.96	0.67	1.63
#1	"	1.20	0.78	1.98
1/0	"	1.50	0.89	2.39
2/0	"	1.75	1.10	2.85
3/0	"	2.20	1.40	3.60
4/0	"	2.70	1.55	4.25
250 MCM	"	3.35	1.65	5.00
350 MCM	"	4.85	2.20	7.05
16120.47 Sheathed Cable				
Non-metallic sheathed cable				
Type NM cable with ground				
#14/2	L.F.	0.26	0.83	1.09
#12/2	"	0.34	0.89	1.22
#10/2	"	0.54	0.98	1.52
#8/2	"	1.05	1.10	2.15
#6/2	"	1.55	1.40	2.95
#14/3	"	0.40	1.45	1.85
#12/3	"	0.55	1.45	2.00
#10/3	"	0.81	1.50	2.31
#8/3	"	1.90	1.55	3.45
#6/3	"	2.25	1.55	3.80
#4/3	"	3.35	1.75	5.10
#2/3	"	4.80	1.90	6.70
Type U.F. cable with ground				
#14/2	L.F.	0.32	0.89	1.21
#12/2	"	0.40	1.05	1.45
#10/2	"	0.50	1.10	1.60
#8/2	"	1.60	1.25	2.85
#6/2	"	1.90	1.50	3.40
#14/3	"	0.43	1.10	1.53
#12/3	"	0.57	1.20	1.77
#10/3	"	0.81	1.40	2.21
#8/3	"	2.10	1.55	3.65
#6/3	"	3.05	1.75	4.80
16130.60 Pull And Junction Boxes				
4"				
Octagon box	EA.	1.80	6.30	8.10
Box extension	"	2.60	3.30	5.90
Plaster ring	"	1.40	3.30	4.70
Cover blank	"	0.58	3.30	3.88
Square box	"	2.15	6.30	8.45
Box extension	"	2.35	3.30	5.65
Plaster ring	"	1.05	3.30	4.35
Cover blank	"	0.65	3.30	3.95
4-11/16"				
Square box	EA.	3.85	6.30	10.15
Box extension	"	5.15	3.30	8.45
Plaster ring	"	2.95	3.30	6.25
Cover blank	"	1.15	3.30	4.45
Switch and device boxes				

Basic Materials

Basic Materials	UNIT	MAT.	INST.	TOTAL
16130.60 Pull And Junction Boxes				
2 gang	EA.	11.15	6.30	17.45
3 gang	"	13.75	6.30	20.05
4 gang	"	19.45	8.85	28.30
Device covers				
2 gang	EA.	5.45	3.30	8.75
3 gang	"	6.45	3.30	9.75
4 gang	"	9.35	3.30	12.65
Handy box	"	1.65	6.30	7.95
Extension	"	2.25	3.30	5.55
Switch cover	"	0.65	3.30	3.95
Switch box with knockout	"	1.85	8.05	9.90
Weatherproof cover, spring type	"	6.45	4.40	10.85
Cover plate, dryer receptacle 1 gang plastic	"	2.95	5.55	8.50
For 4" receptacle, 2 gang	"	5.15	5.55	10.70
Duplex receptacle cover plate, plastic	"	0.58	3.30	3.88
4", vertical bracket box, 1-1/2" with				
RMX clamps	EA.	2.70	8.05	10.75
BX clamps	"	4.10	8.05	12.15
4", octagon device cover				
1 switch	EA.	1.75	3.30	5.05
1 duplex recept	"	1.65	3.30	4.95
4", square face bracket boxes, 1-1/2"				
RMX	EA.	4.50	8.05	12.55
BX	"	4.75	8.05	12.80
16130.65 Pull Boxes And Cabinets				
Galvanized pull boxes, screw cover				
4x4x4	EA.	6.50	10.55	17.05
4x6x4	"	8.10	10.55	18.65
6x6x4	"	10.55	10.55	21.10
6x8x4	"	12.70	10.55	23.25
8x8x4	"	15.75	13.80	29.55
8x10x4	"	18.15	13.40	31.55
8x12x4	"	19.95	13.80	33.75
Screw cover				
10x10x4	EA.	19.90	17.00	36.90
12x12x6	"	30.40	24.60	55.00
12x15x6	"	34.40	24.60	59.00
12x18x6	"	40.30	27.70	68.00
15x18x6	"	45.60	31.60	77.20
18x24x6	"	82.50	34.00	116.50
18x30x6	"	91.00	40.20	131.20
24x36x6	"	140.00	40.20	180.20
16130.80 Receptacles				
125 volt, 20a, duplex, grounding type, standard grade	EA.	7.35	11.05	18.40
Ground fault interrupter type	"	54.50	16.40	70.90
250 volt, 20a, 2 pole, single receptacle, ground type	"	8.25	11.05	19.30
120/208v, 4 pole, single receptacle, twist lock				
20a	EA.	17.80	19.25	37.05
50a	"	33.90	19.25	53.15
125/250v, 3 pole, flush receptacle				

16 ELECTRICAL

Basic Materials

	UNIT	MAT.	INST.	TOTAL
16130.80 Receptacles				
30a	EA.	11.30	16.40	27.70
50a	"	13.05	16.40	29.45
60a	"	52.50	19.25	71.75
277v, 20a, 2 pole, grounding type, twist lock	"	13.05	11.05	24.10
Dryer receptacle, 250v, 30a/50a, 3 wire	"	21.85	16.40	38.25
Clock receptacle, 2 pole, grounding type	"	9.10	11.05	20.15
125v, 20a single recept. grounding type				
Standard grade	EA.	6.65	11.05	17.70
Specification	"	10.30	11.05	21.35
Hospital	"	10.60	11.05	21.65
Isolated ground orange	"	18.50	13.80	32.30
Duplex				
Specification grade	EA.	10.65	11.05	21.70
Hospital	"	13.80	11.05	24.85
Isolated ground orange	"	20.75	13.80	34.55
GFI hospital grade recepts, 20a, 125v, duplex	"	63.00	16.40	79.40
16198.10 Electric Manholes				
Precast, handhole, 4' deep				
2'x2'	EA.	450.00	190.00	640.00
3'x3'	"	610.00	310.00	920.00
4'x4'	"	1,310	570.00	1,880
Power manhole, complete, precast, 8' deep				
4'x4'	EA.	1,880	780.00	2,660
6'x6'	"	2,510	1,110	3,620
8'x8'	"	2,970	1,160	4,130
6' deep, 9' x 12'	"	3,280	1,380	4,660
Cast in place, power manhole, 8' deep				
4'x4'	EA.	2,230	780.00	3,010
6'x6'	"	2,870	1,110	3,980
8'x8'	"	3,190	1,160	4,350
16199.10 Utility Poles & Fittings				
Wood pole, creosoted				
25'	EA.	400.00	130.00	530.00
40'	"	770.00	210.00	980.00
55'	"	1,200	420.00	1,620
Treated, wood preservative, 6"x6"				
8'	EA.	40.40	27.70	68.10
16'	"	94.50	88.50	183.00
20'	"	140.00	110.00	250.00
Aluminum, brushed, no base				
8'	EA.	470.00	110.00	580.00
20'	"	800.00	180.00	980.00
40'	"	2,410	350.00	2,760
Steel, no base				
10'	EA.	600.00	140.00	740.00
20'	"	730.00	210.00	940.00
35'	"	1,750	350.00	2,100
Concrete, no base				
13'	EA.	720.00	310.00	1,030
30'	"	1,970	670.00	2,640

Basic Materials

16199.10 Utility Poles & Fittings	UNIT	MAT.	INST.	TOTAL
50'	EA.	4,360	1,010	5,370
60'	"	5,550	1,110	6,660

Power Generation

16210.10 Generators	UNIT	MAT.	INST.	TOTAL
Diesel generator, with auto transfer switch				
50kw	EA.	29,800	1,700	31,500
125kw	"	44,870	2,770	47,640
300kw	"	73,500	5,530	79,030
750kw	"	214,320	11,060	225,380

16320.10 Transformers	UNIT	MAT.	INST.	TOTAL
Floor mounted, single phase, int. dry, 480v-120/240v				
3 kva	EA.	330.00	100.00	430.00
5 kva	"	450.00	170.00	620.00
7.5 kva	"	670.00	190.00	860.00
10 kva	"	770.00	210.00	980.00
15 kva	"	1,040	240.00	1,280
100 kva	"	3,640	640.00	4,280
Three phase, 480v-120/208v				
15 kva	EA.	1,250	330.00	1,580
30 kva	"	1,610	520.00	2,130
45 kva	"	2,150	600.00	2,750
225 kva	"	7,070	850.00	7,920

16350.10 Circuit Breakers	UNIT	MAT.	INST.	TOTAL
Molded case, 240v, 15-60a, bolt-on				
1 pole	EA.	13.95	13.80	27.75
2 pole	"	27.90	19.25	47.15
70-100a, 2 pole	"	77.50	29.50	107.00
15-60a, 3 pole	"	93.00	22.10	115.10
70-100a, 3 pole	"	120.00	34.00	154.00
480v, 2 pole				
15-60a	EA.	200.00	16.40	216.40
70-100a	"	250.00	22.10	272.10
3 pole				
15-60a	EA.	250.00	22.10	272.10
70-100a	"	290.00	24.60	314.60
70-225a	"	620.00	34.00	654.00
Load center circuit breakers, 240v				
1 pole, 10-60a	EA.	13.95	13.80	27.75
2 pole				
10-60a	EA.	27.90	22.10	50.00
70-100a	"	46.40	36.90	83.30
110-150a	"	170.00	40.20	210.20

Power Generation

	UNIT	MAT.	INST.	TOTAL
16350.10 Circuit Breakers				
3 pole				
10-60a	EA.	93.00	27.70	120.70
70-100a	"	140.00	40.20	180.20
Load center, G.F.I. breakers, 240v				
1 pole, 15-30a	EA.	93.00	16.40	109.40
2 pole, 15-30a	"	160.00	22.10	182.10
Key operated breakers, 240v, 1 pole, 10-30a	"	15.50	16.40	31.90
Tandem breakers, 240v				
1 pole, 15-30a	EA.	26.30	22.10	48.40
2 pole, 15-30a	"	48.00	29.50	77.50
Bolt-on, G.F.I. breakers, 240v, 1 pole, 15-30a	"	110.00	19.25	129.25
16360.10 Safety Switches				
Fused, 3 phase, 30 amp, 600v, heavy duty				
NEMA 1	EA.	170.00	63.00	233.00
NEMA 3r	"	290.00	63.00	353.00
NEMA 4	"	770.00	88.50	858.50
NEMA 12	"	280.00	96.00	376.00
60a				
NEMA 1	EA.	230.00	63.00	293.00
NEMA 3r	"	350.00	63.00	413.00
NEMA 4	"	920.00	88.50	1,009
NEMA 12	"	290.00	96.00	386.00
100a				
NEMA 1	EA.	430.00	96.00	526.00
NEMA 3r	"	600.00	96.00	696.00
NEMA 4	"	1,840	110.00	1,950
NEMA 12	"	500.00	140.00	640.00
200a				
NEMA 1	EA.	610.00	140.00	750.00
NEMA 3r	"	800.00	140.00	940.00
NEMA 4	"	2,600	150.00	2,750
NEMA 12	"	780.00	190.00	970.00
Non-fused, 240-600v, heavy duty, 3 phase, 30 amp				
NEMA 1	EA.	120.00	63.00	183.00
NEMA 3r	"	200.00	63.00	263.00
NEMA 4	"	770.00	96.00	866.00
NEMA 12	"	230.00	96.00	326.00
60a				
NEMA1	EA.	200.00	63.00	263.00
NEMA 3r	"	320.00	63.00	383.00
NEMA 4	"	920.00	96.00	1,016
NEMA 12	"	290.00	96.00	386.00
100a				
NEMA 1	EA.	310.00	96.00	406.00
NEMA 3r	"	460.00	96.00	556.00
NEMA 4	"	1,840	140.00	1,980
NEMA 12	"	400.00	140.00	540.00
200a, NEMA 1	"	470.00	140.00	610.00
600a, NEMA 12	"	2,070	680.00	2,750

16 ELECTRICAL

Power Generation

	UNIT	MAT.	INST.	TOTAL
16365.10 Fuses				
Fuse, one-time, 250v				
30a	EA.	1.50	2.75	4.25
60a	"	2.25	2.75	5.00
100a	"	9.10	2.75	11.85
200a	"	22.75	2.75	25.50
400a	"	39.40	2.75	42.15
600a	"	76.00	2.75	78.75
600v				
30a	EA.	6.85	2.75	9.60
60a	"	7.60	2.75	10.35
100a	"	21.25	2.75	24.00
200a	"	45.50	2.75	48.25
400a	"	91.00	2.75	93.75
16395.10 Grounding				
Ground rods, copper clad, 1/2" x				
6'	EA.	12.75	36.90	49.65
8'	"	19.10	40.20	59.30
10'	"	23.90	55.50	79.40
5/8" x				
5'	EA.	17.50	34.00	51.50
6'	"	19.10	40.20	59.30
8'	"	23.90	55.50	79.40
10'	"	27.10	69.00	96.10
3/4" x				
8'	EA.	35.00	40.20	75.20
10'	"	47.80	44.20	92.00
Ground rod clamp				
5/8"	EA.	5.75	6.80	12.55
3/4"	"	7.40	6.80	14.20

Service And Distribution

	UNIT	MAT.	INST.	TOTAL
16425.10 Switchboards				
Switchboard, 90" high, no main disconnect, 208/120v				
400a	EA.	2,240	440.00	2,680
600a	"	3,470	440.00	3,910
1000a	"	4,360	440.00	4,800
1200a	"	4,620	550.00	5,170
1600a	"	5,070	660.00	5,730
2000a	"	5,450	780.00	6,230
2500a	"	4,470	880.00	5,350
277/480v				
600a	EA.	3,980	450.00	4,430
800a	"	4,360	450.00	4,810
1600a	"	5,500	660.00	6,160

Service And Distribution

	UNIT	MAT.	INST.	TOTAL
16425.10 Switchboards				
2000a	EA.	5,880	780.00	6,660
2500a	"	6,260	880.00	7,140
3000a	"	7,210	1,530	8,740
4000a	"	8,730	1,640	10,370
16470.10 Panelboards				
3 phase, 480/277v, main lugs only, 120a, 30 circuits	EA.	890.00	190.00	1,080
277/480v, 4 wire, flush surface				
225a, 30 circuits	EA.	950.00	220.00	1,170
400a, 30 circuits	"	1,090	280.00	1,370
600a, 42 circuits	"	1,360	330.00	1,690
208/120v, main circuit breaker, 3 phase, 4 wire				
100a				
12 circuits	EA.	770.00	280.00	1,050
20 circuits	"	960.00	350.00	1,310
30 circuits	"	1,420	390.00	1,810
400a				
30 circuits	EA.	2,980	820.00	3,800
42 circuits	"	3,580	880.00	4,460
600a, 42 circuits	"	6,950	1,010	7,960
120/208v, flush, 3 ph., 4 wire, main only				
100a				
12 circuits	EA.	540.00	280.00	820.00
20 circuits	"	750.00	350.00	1,100
30 circuits	"	1,130	390.00	1,520
400a				
30 circuits	EA.	2,190	820.00	3,010
42 circuits	"	3,190	880.00	4,070
600a, 42 circuits	"	4,970	1,010	5,980
16480.10 Motor Controls				
Motor generator set, 3 phase, 480/277v, w/controls				
10kw	EA.	10,020	1,530	11,550
15kw	"	13,060	1,700	14,760
20kw	"	14,500	1,770	16,270
40kw	"	20,450	2,110	22,560
100kw	"	33,450	3,400	36,850
200kw	"	76,470	4,020	80,490
300kw	"	95,590	4,420	100,010
2 pole, 230 volt starter, w/NEMA-1				
1 hp, 9 amp, size 00	EA.	150.00	55.50	205.50
2 hp, 18amp, size 0	"	190.00	55.50	245.50
3 hp, 27amp, size 1	"	200.00	55.50	255.50
5 hp, 45amp, size 1p	"	260.00	55.50	315.50
7-1/2 hp, 45a, size 2	"	450.00	55.50	505.50
15 hp, 90a, size 3	"	680.00	55.50	735.50
16490.10 Switches				
Fused interrupter load, 35kv				
20A				
1 pole	EA.	20,270	880.00	21,150
2 pole	"	21,960	940.00	22,900

Service And Distribution

	UNIT	MAT.	INST.	TOTAL
16490.10 Switches				
3 way	EA.	23,650	940.00	24,590
4 way	"	25,340	1,010	26,350
30a, 1 pole	"	20,270	880.00	21,150
3 way	"	23,650	940.00	24,590
4 way	"	25,340	1,010	26,350
Weatherproof switch, including box & cover, 20a				
1 pole	EA.	21,960	880.00	22,840
2 pole	"	23,650	940.00	24,590
3 way	"	25,340	1,010	26,350
4 way	"	27,030	1,010	28,040
Specification grade toggle switches, 20a, 120-277v				
Single pole	EA.	14.50	11.05	25.55
Double pole	"	16.65	16.40	33.05
3 way	"	15.35	13.80	29.15
4 way	"	39.10	16.40	55.50
Switch plates, plastic ivory				
1 gang	EA.	0.56	4.40	4.96
2 gang	"	1.10	5.55	6.65
3 gang	"	1.65	6.60	8.25
4 gang	"	2.75	8.05	10.80
5 gang	"	6.30	8.85	15.15
6 gang	"	7.45	10.05	17.50
Stainless steel				
1 gang	EA.	1.80	4.40	6.20
2 gang	"	3.50	5.55	9.05
3 gang	"	6.10	6.80	12.90
4 gang	"	8.45	8.05	16.50
5 gang	"	11.65	8.85	20.50
6 gang	"	14.45	10.05	24.50
Brass				
1 gang	EA.	3.95	4.40	8.35
2 gang	"	7.90	5.55	13.45
3 gang	"	14.40	6.80	21.20
4 gang	"	19.35	8.05	27.40
5 gang	"	23.95	8.85	32.80
6 gang	"	28.90	10.05	38.95
16490.20 Transfer Switches				
Automatic transfer switch 600v, 3 pole				
30a	EA.	2,050	190.00	2,240
100a	"	4,260	260.00	4,520
400a	"	9,380	550.00	9,930
800a	"	15,700	1,010	16,710
1200a	"	25,600	1,260	26,860
2600a	"	59,730	2,330	62,060
16490.80 Safety Switches				
Safety switch, 600v, 3 pole, heavy duty, NEMA-1				
30a	EA.	170.00	55.50	225.50
60a	"	230.00	63.00	293.00
100a	"	430.00	88.50	518.50
200a	"	610.00	140.00	750.00

Service And Distribution

16490.80 Safety Switches	UNIT	MAT.	INST.	TOTAL
400a	EA.	1,530	310.00	1,840
600a	"	2,760	440.00	3,200
800a	"	4,240	580.00	4,820
1200a	"	5,250	790.00	6,040

Lighting

16510.05 Interior Lighting	UNIT	MAT.	INST.	TOTAL
Recessed fluorescent fixtures, 2'x2'				
2 lamp	EA.	69.00	40.20	109.20
4 lamp	"	92.50	40.20	132.70
2 lamp w/flange	"	86.50	55.50	142.00
4 lamp w/flange	"	110.00	55.50	165.50
1'x4'				
2 lamp	EA.	70.00	36.90	106.90
3 lamp	"	97.00	36.90	133.90
2 lamp w/flange	"	86.50	40.20	126.70
3 lamp w/flange	"	110.00	40.20	150.20
2'x4'				
2 lamp	EA.	86.50	40.20	126.70
3 lamp	"	100.00	40.20	140.20
4 lamp	"	99.00	40.20	139.20
2 lamp w/flange	"	100.00	55.50	155.50
3 lamp w/flange	"	120.00	55.50	175.50
4 lamp w/flange	"	120.00	55.50	175.50
4'x4'				
4 lamp	EA.	210.00	55.50	265.50
6 lamp	"	240.00	55.50	295.50
8 lamp	"	250.00	55.50	305.50
4 lamp w/flange	"	250.00	83.50	333.50
6 lamp w/flange	"	310.00	83.50	393.50
8 lamp, w/flange	"	350.00	83.50	433.50
Surface mounted incandescent fixtures				
40w	EA.	62.50	36.90	99.40
75w	"	68.00	36.90	104.90
100w	"	80.50	36.90	117.40
150w	"	95.00	36.90	131.90
Pendant				
40w	EA.	68.00	44.20	112.20
75w	"	75.50	44.20	119.70
100w	"	86.00	44.20	130.20
150w	"	98.50	44.20	142.70
Recessed incandescent fixtures				
40w	EA.	120.00	83.50	203.50
75w	"	120.00	83.50	203.50
100w	"	120.00	83.50	203.50

Lighting	UNIT	MAT.	INST.	TOTAL
16510.05 Interior Lighting				
150w	EA.	140.00	83.50	223.50
Exit lights, 120v				
Recessed	EA.	89.50	69.00	158.50
Back mount	"	62.50	40.20	102.70
Universal mount	"	71.50	40.20	111.70
Emergency battery units, 6v-120v, 50 unit	"	130.00	83.50	213.50
With 1 head	"	160.00	83.50	243.50
With 2 heads	"	180.00	83.50	263.50
Light track single circuit				
2'	EA.	32.70	27.70	60.40
4'	"	59.50	27.70	87.20
8'	"	110.00	55.50	165.50
12'	"	160.00	83.50	243.50
Fixtures, square				
R-20	EA.	85.00	8.05	93.05
R-30	"	88.50	8.05	96.55
Mini spot	"	120.00	8.05	128.05
16510.10 Lighting Industrial				
Surface mounted fluorescent, wrap around lens				
1 lamp	EA.	64.50	44.20	108.70
2 lamps	"	70.00	49.20	119.20
4 lamps	"	110.00	55.50	165.50
Wall mounted fluorescent				
2-20w lamps	EA.	51.50	27.70	79.20
2-30w lamps	"	55.50	27.70	83.20
2-40w lamps	"	64.50	36.90	101.40
Indirect, with wood shielding, 2049w lamps				
4'	EA.	110.00	55.50	165.50
8'	"	160.00	88.50	248.50
Industrial fluorescent, 2 lamp				
4'	EA.	110.00	40.20	150.20
8'	"	170.00	73.50	243.50
Strip fluorescent				
4'				
1 lamp	EA.	36.90	36.90	73.80
2 lamps	"	46.10	36.90	83.00
8'				
1 lamp	EA.	74.00	40.20	114.20
2 lamps	"	83.00	49.20	132.20
Wire guard for strip fixture, 4' long	"	18.45	19.25	37.70
Strip fluorescent, 8' long, two 4' lamps	"	110.00	73.50	183.50
With four 4' lamps	"	150.00	88.50	238.50
Wet location fluorescent, plastic housing				
4' long				
1 lamp	EA.	130.00	55.50	185.50
2 lamps	"	140.00	73.50	213.50
8' long				
2 lamps	EA.	230.00	88.50	318.50
4 lamps	"	310.00	96.00	406.00
Parabolic troffer, 2'x2'				
With 2 "U" lamps	EA.	160.00	55.50	215.50

Lighting

16510.10 Lighting Industrial

Description	UNIT	MAT.	INST.	TOTAL
With 3 "U" lamps	EA.	180.00	63.00	243.00
2'x4'				
With 2 40w lamps	EA.	210.00	63.00	273.00
With 3 40w lamps	"	230.00	73.50	303.50
With 4 40w lamps	"	260.00	73.50	333.50
1'x4'				
With 1 T-12 lamp, 9 cell	EA.	160.00	40.20	200.20
With 2 T-12 lamps	"	170.00	49.20	219.20
With 1 T-12 lamp, 20 cell	"	180.00	40.20	220.20
With 2 T-12 lamps	"	190.00	49.20	239.20
Steel sided surface fluorescent, 2'x4'				
3 lamps	EA.	140.00	73.50	213.50
4 lamps	"	150.00	73.50	223.50
Outdoor sign fluor., 1 lamp, remote ballast				
4' long	EA.	2,640	330.00	2,970
6' long	"	3,160	440.00	3,600
Recess mounted, commercial, 2'x2', 13" high				
100w	EA.	880.00	220.00	1,100
250w	"	960.00	250.00	1,210
High pressure sodium, hi-bay open				
400w	EA.	470.00	96.00	566.00
1000w	"	810.00	130.00	940.00
Enclosed				
400w	EA.	730.00	130.00	860.00
1000w	"	1,070	160.00	1,230
Metal halide hi-bay, open				
400w	EA.	430.00	96.00	526.00
1000w	"	720.00	130.00	850.00
Enclosed				
400w	EA.	680.00	130.00	810.00
1000w	"	1,020	160.00	1,180
High pressure sodium, low bay, surface mounted				
100w	EA.	300.00	55.50	355.50
150w	"	320.00	63.00	383.00
250w	"	360.00	73.50	433.50
400w	"	470.00	88.50	558.50
Metal halide, low bay, pendant mounted				
175w	EA.	320.00	73.50	393.50
250w	"	430.00	88.50	518.50
400w	"	520.00	120.00	640.00
Indirect luminare, square, metal halide, freestanding				
175w	EA.	850.00	55.50	905.50
250w	"	870.00	55.50	925.50
400w	"	940.00	55.50	995.50
High pressure sodium				
150w	EA.	870.00	55.50	925.50
250w	"	910.00	55.50	965.50
400w	"	990.00	55.50	1,046
Round, metal halide				
175w	EA.	800.00	55.50	855.50
250w	"	850.00	55.50	905.50
400w	"	900.00	55.50	955.50

Lighting

16510.10 Lighting Industrial	UNIT	MAT.	INST.	TOTAL
High pressure sodium				
150w	EA.	820.00	55.50	875.50
250w	"	880.00	55.50	935.50
400w	"	920.00	55.50	975.50
Wall mounted, metal halide				
175w	EA.	660.00	140.00	800.00
250w	"	710.00	140.00	850.00
400w	"	1,180	180.00	1,360
High pressure sodium				
150w	EA.	760.00	140.00	900.00
250w	"	940.00	140.00	1,080
400w	"	1,210	180.00	1,390
Wall pack lithonia, high pressure sodium				
35w	EA.	200.00	49.20	249.20
55w	"	230.00	55.50	285.50
150w	"	250.00	88.50	338.50
250w	"	320.00	96.00	416.00
Low pressure sodium				
35w	EA.	290.00	96.00	386.00
55w	"	440.00	110.00	550.00
Wall pack hubbell, high pressure sodium				
35w	EA.	220.00	49.20	269.20
150w	"	280.00	88.50	368.50
250w	"	360.00	96.00	456.00
Compact fluorescent				
2-7w	EA.	140.00	55.50	195.50
2-13w	"	150.00	73.50	223.50
1-18w	"	190.00	73.50	263.50
Handball & racquet ball court, 2'x2', metal halide				
250w	EA.	480.00	140.00	620.00
400w	"	590.00	150.00	740.00
High pressure sodium				
250w	EA.	520.00	140.00	660.00
400w	"	580.00	150.00	730.00
Bollard light, 42" w/found., high pressure sodium				
70w	EA.	810.00	140.00	950.00
100w	"	830.00	140.00	970.00
150w	"	830.00	140.00	970.00
Light fixture lamps				
Lamp				
20w med. bipin base, cool white, 24"	EA.	6.00	8.05	14.05
30w cool white, rapid start, 36"	"	7.75	8.05	15.80
40w cool white "U", 3"	"	14.85	8.05	22.90
40w cool white, rapid start, 48"	"	3.60	8.05	11.65
70w high pressure sodium, mogul base	"	77.00	11.05	88.05
75w slimline, 96"	"	9.90	11.05	20.95
100w				
Incandescent, 100a, inside frost	EA.	3.05	5.55	8.60
Mercury vapor, clear, mogul base	"	37.60	11.05	48.65
High pressure sodium, mogul base	"	82.00	11.05	93.05
150w				
Par 38 flood or spot, incandescent	EA.	9.20	5.55	14.75

Lighting

Lighting	UNIT	MAT.	INST.	TOTAL
16510.10 Lighting Industrial				
High pressure sodium, 1/2 mogul base	EA.	85.50	11.05	96.55
175w				
Mercury vapor, clear, mogul base	EA.	65.00	11.05	76.05
Metal halide, clear, mogul base	"	29.00	11.05	40.05
High pressure sodium, mogul base	"	39.30	11.05	50.35
250w				
Mercury vapor, clear, mogul base	EA.	51.00	11.05	62.05
Metal halide, clear, mogul base	"	82.00	11.05	93.05
High pressure sodium, mogul base	"	90.50	11.05	101.55
400w				
Mercury vapor, clear, mogul base	EA.	40.90	11.05	51.95
Metal halide, clear, mogul base	"	77.00	11.05	88.05
High pressure sodium, mogul base	"	97.50	11.05	108.55
1000w				
Mercury vapor, clear, mogul base	EA.	89.00	13.80	102.80
High pressure sodium, mogul base	"	230.00	13.80	243.80
16510.30 Exterior Lighting				
Exterior light fixtures				
Rectangle, high pressure sodium				
70w	EA.	520.00	140.00	660.00
100w	"	540.00	140.00	680.00
150w	"	580.00	140.00	720.00
250w	"	780.00	150.00	930.00
400w	"	880.00	190.00	1,070
Flood, rectangular, high pressure sodium				
70w	EA.	540.00	140.00	680.00
100w	"	560.00	140.00	700.00
150w	"	620.00	140.00	760.00
400w	"	930.00	190.00	1,120
1000w	"	1,410	250.00	1,660
Round				
400w	EA.	960.00	190.00	1,150
1000w	"	1,500	250.00	1,750
Round, metal halide				
400w	EA.	1,060	190.00	1,250
1000w	"	1,570	250.00	1,820
Light fixture arms, cobra head, 6', high press. sodium				
100w	EA.	580.00	110.00	690.00
150w	"	910.00	140.00	1,050
250w	"	950.00	140.00	1,090
400w	"	980.00	160.00	1,140
Flood, metal halide				
400w	EA.	1,010	190.00	1,200
1000w	"	1,530	250.00	1,780
1500w	"	2,070	330.00	2,400
Mercury vapor				
250w	EA.	590.00	150.00	740.00
400w	"	1,030	190.00	1,220
Incandescent				
300w	EA.	110.00	96.00	206.00
500w	"	180.00	110.00	290.00

16 ELECTRICAL

Lighting

Lighting	UNIT	MAT.	INST.	TOTAL
16510.30 Exterior Lighting				
1000w	EA.	200.00	180.00	380.00
16610.30 Uninterruptible Power				
Uninterruptible power systems, (U.P.S.), 3kva	EA.	13,650	440.00	14,090
5 kva	"	15,320	610.00	15,930
7.5 kva	"	18,380	880.00	19,260
10 kva	"	22,970	1,220	24,190
15 kva	"	27,570	1,260	28,830
20 kva	"	38,290	1,330	39,620
25 kva	"	49,010	1,380	50,390
30 kva	"	50,540	1,440	51,980
35 kva	"	53,600	1,490	55,090
40 kva	"	58,200	1,550	59,750
45 kva	"	61,260	1,600	62,860
50 kva	"	65,860	1,660	67,520
62.5 kva	"	78,110	1,770	79,880
75 kva	"	90,360	1,930	92,290
100 kva	"	120,990	1,990	122,980
150 kva	"	183,780	2,770	186,550
200 kva	"	245,050	3,050	248,100
300 kva	"	367,570	4,130	371,700
400 kva	"	549,880	4,970	554,850
500 kva	"	687,360	6,060	693,420
16670.10 Lightning Protection				
Lightning protection				
Copper point, nickel plated, 12'				
1/2" dia.	EA.	50.00	55.50	105.50
5/8" dia.	"	60.00	55.50	115.50

Communications

Communications	UNIT	MAT.	INST.	TOTAL
16720.10 Fire Alarm Systems				
Master fire alarm box, pedestal mounted	EA.	6,090	880.00	6,970
Master fire alarm box	"	3,130	330.00	3,460
Box light	"	170.00	27.70	197.70
Ground assembly for box	"	87.00	36.90	123.90
Bracket for pole type box	"	110.00	40.20	150.20
Pull station				
Waterproof	EA.	87.00	27.70	114.70
Manual	"	52.50	22.10	74.60
Horn, waterproof	"	78.50	55.50	134.00
Interior alarm	"	52.50	40.20	92.70
Coded transmitter, automatic	"	920.00	110.00	1,030
Control panel, 8 zone	"	1,840	440.00	2,280
Battery charger and cabinet	"	610.00	110.00	720.00

Communications

	UNIT	MAT.	INST.	TOTAL
16720.10 Fire Alarm Systems				
Batteries, nickel cadmium or lead calcium	EA.	460.00	280.00	740.00
CO2 pressure switch connection	"	87.00	40.20	127.20
Annunciator panels				
Fire detection annunciator, remote type, 8 zone	EA.	310.00	100.00	410.00
12 zone	"	400.00	110.00	510.00
16 zone	"	500.00	140.00	640.00
Fire alarm systems				
Bell	EA.	96.00	34.00	130.00
Weatherproof bell	"	120.00	36.90	156.90
Horn	"	55.50	40.20	95.70
Siren	"	570.00	110.00	680.00
Chime	"	69.50	34.00	103.50
Audio/visual	"	100.00	40.20	140.20
Strobe light	"	93.00	40.20	133.20
Smoke detector	"	150.00	36.90	186.90
Heat detection	"	26.10	27.70	53.80
Thermal detector	"	24.40	27.70	52.10
Ionization detector	"	120.00	29.50	149.50
Duct detector	"	410.00	150.00	560.00
Test switch	"	69.50	27.70	97.20
Remote indicator	"	35.20	31.60	66.80
Door holder	"	150.00	40.20	190.20
Telephone jack	"	26.10	16.40	42.50
Fireman phone	"	370.00	55.50	425.50
Speaker	"	74.50	44.20	118.70
Remote fire alarm annunciator panel				
24 zone	EA.	1,920	370.00	2,290
48 zone	"	3,830	720.00	4,550
Control panel				
12 zone	EA.	1,300	160.00	1,460
16 zone	"	1,700	250.00	1,950
24 zone	"	2,610	370.00	2,980
48 zone	"	4,870	880.00	5,750
Power supply	"	300.00	83.50	383.50
Status command	"	8,270	280.00	8,550
Printer	"	3,220	83.50	3,304
Transponder	"	260.00	49.70	309.70
Transformer	"	190.00	36.90	226.90
Transceiver	"	260.00	40.20	300.20
Relays	"	100.00	27.70	127.70
Flow switch	"	340.00	110.00	450.00
Tamper switch	"	200.00	160.00	360.00
End of line resistor	"	14.75	19.25	34.00
Printed ckt. card	"	130.00	27.70	157.70
Central processing unit	"	17,050	340.00	17,390
UPS backup to c.p.u.	"	16,550	500.00	17,050
Smoke detector, fixed temp. & rate of rise comb.	"	260.00	88.50	348.50
16720.50 Security Systems				
Sensors				
Balanced magnetic door switch, surface mounted	EA.	130.00	27.70	157.70
With remote test	"	170.00	55.50	225.50

Communications

16720.50 Security Systems

Communications	UNIT	MAT.	INST.	TOTAL
Flush mounted	EA.	120.00	100.00	220.00
Mounted bracket	"	9.35	19.25	28.60
Mounted bracket spacer	"	8.45	19.25	27.70
Photoelectric sensor, for fence				
6 beam	EA.	13,780	150.00	13,930
9 beam	"	16,840	240.00	17,080
Photoelectric sensor, 12 volt dc				
500' range	EA.	410.00	88.50	498.50
800' range	"	450.00	110.00	560.00
Capacitance wire grid kit				
Surface	EA.	110.00	55.50	165.50
Duct	"	84.50	88.50	173.00
Tube grid kit	"	140.00	27.70	167.70
Vibration sensor, 30 max per zone	"	170.00	27.70	197.70
Audio sensor, 30 max per zone	"	180.00	27.70	207.70
Inertia sensor				
Outdoor	EA.	130.00	40.20	170.20
Indoor	"	84.50	27.70	112.20
Ultrasonic transmitter, 20 max per zone				
Omni-directional	EA.	94.00	88.50	182.50
Directional	"	100.00	73.50	173.50
Transceiver				
Omni-directional	EA.	100.00	55.50	155.50
Directional	"	110.00	55.50	165.50
Passive infra-red sensor, 20 max per zone	"	730.00	88.50	818.50
Access/secure control unit, balanced magnetic switch	"	450.00	88.50	538.50
Photoelectric sensor	"	750.00	88.50	838.50
Photoelectric fence sensor	"	770.00	88.50	858.50
Capacitance sensor	"	890.00	96.00	986.00
Audio and vibration sensor	"	790.00	88.50	878.50
Inertia sensor	"	1,050	88.50	1,139
Ultrasonic sensor	"	1,200	96.00	1,296
Infra-red sensor	"	750.00	110.00	860.00
Monitor panel, with access/secure tone, standard	"	510.00	96.00	606.00
High security	"	750.00	110.00	860.00
Emergency power indicator	"	310.00	27.70	337.70
Monitor rack with 115v power supply				
1 zone	EA.	420.00	55.50	475.50
10 zone	"	2,160	140.00	2,300
Monitor cabinet, wall mounted				
1 zone	EA.	660.00	55.50	715.50
5 zone	"	2,350	88.50	2,439
10 zone	"	1,070	96.00	1,166
20 zone	"	3,270	110.00	3,380
Floor mounted, 50 zone	"	3,400	220.00	3,620
Security system accessories				
Tamper assembly for monitor cabinet	EA.	84.50	24.60	109.10
Monitor panel blank	"	11.25	19.25	30.50
Audible alarm	"	96.00	27.70	123.70
Audible alarm control	"	390.00	19.25	409.25
Termination screw, terminal cabinet				
25 pair	EA.	280.00	88.50	368.50

Communications

	UNIT	MAT.	INST.	TOTAL
16720.50 Security Systems				
50 pair	EA.	450.00	140.00	590.00
150 pair	"	740.00	280.00	1,020
Universal termination, cabinets & panel				
Remote test	EA.	66.00	96.00	162.00
No remote test	"	47.00	40.20	87.20
High security line supervision termination	"	330.00	55.50	385.50
Door cord for capacitance sensor, 12"	"	11.25	27.70	38.95
Insulation block kit for capacitance sensor	"	52.50	19.25	71.75
Termination block for capacitance sensor	"	11.45	19.25	30.70
Guard alert display	"	1,200	34.00	1,234
Uninterrupted power supply	"	1,220	440.00	1,660
Plug-in 40kva transformer				
12 volt	EA.	48.00	19.25	67.25
18 volt	"	32.00	19.25	51.25
24 volt	"	22.60	19.25	41.85
Test relay	"	72.50	19.25	91.75
Coaxial cable, 50 ohm	L.F.	0.33	0.33	0.66
Door openers	EA.	84.50	27.70	112.20
Push buttons				
Standard	EA.	18.80	19.25	38.05
Weatherproof	"	28.20	24.60	52.80
Bells	"	71.50	40.20	111.70
Horns				
Standard	EA.	75.00	55.50	130.50
Weatherproof	"	140.00	69.00	209.00
Chimes	"	110.00	36.90	146.90
Flasher	"	79.00	34.00	113.00
Motion detectors	"	330.00	83.50	413.50
Intercom units	"	75.00	40.20	115.20
Remote annunciator	"	3,380	280.00	3,660
16740.10 Telephone Systems				
Communication cable				
25 pair	L.F.	0.66	1.45	2.10
100 pair	"	3.10	1.60	4.70
400 pair	"	11.05	2.45	13.50
Cable tap in manhole or junction box				
25 pair cable	EA.	5.15	210.00	215.15
100 pair cable	"	20.70	830.00	850.70
400 pair cable	"	83.00	3,400	3,483
Cable terminations, manhole or junction box				
25 pair cable	EA.	5.15	210.00	215.15
100 pair cable	"	20.70	830.00	850.70
400 pair cable	"	64.50	3,400	3,465
16780.50 Television Systems				
TV outlet, self terminating, w/cover plate	EA.	10.20	17.00	27.20
Thru splitter	"	22.30	88.50	110.80
End of line	"	18.60	73.50	92.10
In line splitter multitap				
4 way	EA.	37.20	100.00	137.20
2 way	"	27.90	94.00	121.90

Communications

	UNIT	MAT.	INST.	TOTAL
16780.50 Television Systems				
Equipment cabinet	EA.	93.00	88.50	181.50
Antenna				
Broad band uhf	EA.	190.00	190.00	380.00
Lightning arrester	"	50.50	40.20	90.70
TV cable	L.F.	0.42	0.28	0.70
16850.10 Electric Heating				
Baseboard heater				
2', 375w	EA.	57.50	55.50	113.00
3', 500w	"	67.00	55.50	122.50
4', 750w	"	73.00	63.00	136.00
5', 935w	"	100.00	73.50	173.50
6', 1125w	"	110.00	88.50	198.50
7', 1310w	"	130.00	100.00	230.00
8', 1500w	"	140.00	110.00	250.00
9', 1680w	"	200.00	120.00	320.00
10', 1875w	"	210.00	130.00	340.00
Unit heater, wall mounted				
1500w	EA.	260.00	92.00	352.00
2500w	"	280.00	100.00	380.00
4000w	"	380.00	130.00	510.00
Thermostat				
Integral	EA.	44.10	27.70	71.80
Line voltage	"	44.60	27.70	72.30
Electric heater connection	"	1.90	13.80	15.70

Man-Hour Tables

The man-hour productivities used to develop the labor costs are listed in the following section of this book. These productivities represent typical installation labor for thousands of construction items. The data takes into account all activities involved in normal construction under commonly experienced working conditions. As with the Costbook pages, these items are listed according to the CSI MASTERFORMAT. In order to best use the information in this book, please review this sample page and read the "Features in this Book" section.

CSI MASTERFORMAT Division

CSI Broadscope Category

Unit of Measurement

Each item is defined in terms of the common estimating unit. Quantities listed are defined as man-hour per unit.

CSI Mediumscope Category (First 5 Digits)

Man-Hours

Man-hour quantities represent typical installation times and take into account all activities involved in normal construction under commonly experienced working conditions.

Detailed Descriptions

Complete descriptions of items may include information listed above a particular line. Review of the whole category is recommended for a complete description.

10 SPECIALTIES

SPECIALTIES	UNIT	MAN/ HOURS
10110.10 CHALKBOARDS		
Chalkboard, metal frame, 1/4" thick		
48"x60"	EA.	0.800
48"x96"	"	0.889
48"x144"	"	1.000
48"x192"	"	1.143
Liquid chalkboard		
48"x60"	EA.	0.800
48"x96"	"	0.889
48"x144"	"	1.000
48"x192"	"	1.143
Map rail, deluxe	L.F.	0.040
Average	PCT.	
10165.10 TOILET PARTITIONS		
Toilet partition, plastic laminate		
Ceiling mounted	EA.	2.667
Floor mounted	"	2.000
Metal		
Ceiling mounted	EA.	2.667
Floor mounted	"	2.000
Wheel chair partition, plastic laminate		
Ceiling mounted	EA.	2.667
Floor mounted	"	2.000
Painted metal		
Ceiling mounted	EA.	2.667
Floor mounted	"	2.000
Urinal screen, plastic laminate		
Wall hung	[illegible]	[illegible]
Floor mounted	"	1.000
Porcelain enameled steel, floor mounted	"	1.000
Painted metal, floor mounted	"	1.000
Stainless steel, floor mounted	"	1.000
Metal toilet partitions		
Toilet partitions, front door and side divider, floor mounted		
Porcelain enameled steel	EA.	2.000
Painted steel	"	2.000
Stainless steel	"	2.000
10185.10 SHOWER STALLS		
Shower receptors		
Precast, terrazzo		
32" x 32"	EA.	0.667
32" x 48"	"	0.800
Concrete		
32" x 32"	EA.	0.667
48" x 48"	"	0.889
Shower door, trim and hardware		
Economy, 24" wide, chrome frame, tempered glass	EA.	0.800
Porcelain enameled steel, flush	"	0.800
Baked enameled steel, flush	"	0.800
Aluminum frame, tempered glass, 48" wide, sliding	"	1.000
Folding	"	1.000
Aluminum frame and tempered glass, molded plastic		

SPECIALTIES	UNIT	MAN/ HOURS
10185.10 SHOWER STALLS		
Complete with receptor and door		
32" x 32"	EA.	2.000
36" x 36"	"	2.000
40" x 40"	"	2.286
Shower compartment, precast concrete receptor		
Single entry type		
Porcelain enameled steel	EA.	8.000
Baked enameled steel	"	8.000
Stainless steel	"	8.000
Double entry type		
Porcelain enameled steel	EA.	10.000
Baked enameled steel	"	10.000
Stainless steel	"	10.000
10190.10 CUBICLES		
Hospital track		
Ceiling hung	L.F.	0.089
Suspended	"	0.114
Hospital metal dividers, galvanized steel		
Baked enamel finish		
54" high		
10" glass light	L.F.	0.400
14" glass light	"	0.400
24" glass light	"	0.400
60" high		
10" glass light	L.F.	0.444
14" glass light	"	0.444
24" glass light	"	0.444
Stainless steel		
54" high		
10" glass light	L.F.	0.444
14" glass light	"	0.444
24" glass light	"	0.444
60" high		
14" glass light	L.F.	0.500
14" glass light	"	0.500
24" glass light	"	0.500
10210.10 VENTS AND WALL LOUVERS		
Block vent, 8"x16"x4" aluminum, w/screen, mill finish	EA.	0.267
Standard	"	0.250
Vents w/screen, 4" deep, 8" wide, 5" high		
Modular	EA.	0.250
Aluminum gable louvers	S.F.	0.133
Vent screen aluminum, 4" wide, continuous	L.F.	0.027
Louvers, aluminum, anodized, fixed blade		
Horizontal line	S.F.	0.200
Vertical line	"	0.200
Wall louver, aluminum mill finish		
Under, 2 sf	S.F.	0.100
2 to 4 sf	"	0.089
5 to 10 sf	"	0.089
Galvanized steel		

Soil Tests

02010.10 Soil Boring

	UNIT	MAN/ HOURS
Borings, uncased, stable earth		
2-1/2" dia.		
Minimum	L.F.	0.200
Average	L.F.	0.300
Maximum	L.F.	0.480
4" dia.		
Minimum	L.F.	0.218
Average	L.F.	0.343
Maximum	L.F.	0.600
Cased, including samples		
2-1/2" dia.		
Minimum	L.F.	0.240
Average	L.F.	0.400
Maximum	L.F.	0.800
4" dia.		
Minumum	L.F.	0.480
Average	L.F.	0.686
Maximum	L.F.	0.960
Drilling in rock		
No sampling		
Minimum	L.F.	0.436
Average	L.F.	0.632
Maximum	L.F.	0.857
With casing and sampling		
Minimum	L.F.	0.600
Average	L.F.	0.800
Maximum	L.F.	1.200
Test pits		
Light soil		
Minimum	EA.	3.000
Average	EA.	4.000
Maximum	EA.	8.000
Heavy soil		
Minimum	EA.	4.800
Average	EA.	6.000
Maximum	EA.	12.000

Demolition

02060.10 Building Demolition

	UNIT	MAN/ HOURS
Building, complete with disposal		
Wood frame	C.F.	0.003
Concrete	C.F.	0.004
Steel frame	C.F.	0.005
Partition removal		
Concrete block partitions		
4" thick	S.F.	0.040
8" thick	S.F.	0.053
12" thick	S.F.	0.073
Brick masonry partitions		
4" thick	S.F.	0.040
8" thick	S.F.	0.050
12" thick	S.F.	0.067
16" thick	S.F.	0.100
Cast in place concrete partitions		
Unreinforced		
6" thick	S.F.	0.160
8" thick	S.F.	0.171
10" thick	S.F.	0.200
12" thick	S.F.	0.240
Reinforced		
6" thick	S.F.	0.185
8" thick	S.F.	0.240
10" thick	S.F.	0.267
12" thick	S.F.	0.320
Terra cotta		
To 6" thick	S.F.	0.040
Stud partitions		
Metal or wood, with drywall both sides	S.F.	0.040
Metal studs, both sides, lath and plaster	S.F.	0.053
Door and frame removal		
Hollow metal in masonry wall		
Single		
2'6"x6'8"	EA.	1.000
3'x7'	EA.	1.333
Double		
3'x7'	EA.	1.600
4'x8'	EA.	1.600
Wood in framed wall		
Single		
2'6"x6'8"	EA.	0.571
3'x6'8"	EA.	0.667
Double		
2'6"x6'8"	EA.	0.800
3'x6'8"	EA.	0.889
Remove for re-use		
Hollow metal	EA.	2.000
Wood	EA.	1.333
Floor removal		
Brick flooring	S.F.	0.032
Ceramic or quarry tile	S.F.	0.018
Terrazzo	S.F.	0.036
Heavy wood	S.F.	0.021
Residential wood	S.F.	0.023
Resilient tile or linoleum	S.F.	0.008
Ceiling removal		
Acoustical tile ceiling		
Adhesive fastened	S.F.	0.008
Furred and glued	S.F.	0.007
Suspended grid	S.F.	0.005
Drywall ceiling		

Demolition

02060.10 Building Demolition

Description	UNIT	MAN/HOURS
Furred and nailed	S.F.	0.009
Nailed to framing	S.F.	0.008
Plastered ceiling		
Furred on framing	S.F.	0.020
Suspended system	S.F.	0.027
Roofing removal		
Steel frame		
Corrugated metal roofing	S.F.	0.016
Built-up roof on metal deck	S.F.	0.027
Wood frame		
Built up roof on wood deck	S.F.	0.025
Roof shingles	S.F.	0.013
Roof tiles	S.F.	0.027
Concrete frame	C.F.	0.053
Concrete plank	S.F.	0.040
Built-up roof on concrete	S.F.	0.023
Cut-outs		
Concrete, elevated slabs, mesh reinforcing		
Under 5 cf	C.F.	0.800
Over 5 cf	C.F.	0.667
Bar reinforcing		
Under 5 cf	C.F.	1.333
Over 5 cf	C.F.	1.000
Window removal		
Metal windows, trim included		
2'x3'	EA.	0.800
2'x4'	EA.	0.889
2'x6'	EA.	1.000
3'x4'	EA.	1.000
3'x6'	EA.	1.143
3'x8'	EA.	1.333
4'x4'	EA.	1.333
4'x6'	EA.	1.600
4'x8'	EA.	2.000
Wood windows, trim included		
2'x3'	EA.	0.444
2'x4'	EA.	0.471
2'x6'	EA.	0.500
3'x4'	EA.	0.533
3'x6'	EA.	0.571
3'x8'	EA.	0.615
6'x4'	EA.	0.667
6'x6'	EA.	0.727
6'x8'	EA.	0.800
Walls, concrete, bar reinforcing		
Small jobs	C.F.	0.533
Large jobs	C.F.	0.444
Brick walls, not including toothing		
4" thick	S.F.	0.040
8" thick	S.F.	0.050
12" thick	S.F.	0.067
16" thick	S.F.	0.100
Concrete block walls, not including toothing		
4" thick	S.F.	0.044

Demolition

02060.10 Building Demolition

Description	UNIT	MAN/HOURS
6" thick	S.F.	0.047
8" thick	S.F.	0.050
10" thick	S.F.	0.057
12" thick	S.F.	0.067
Rubbish handling		
Load in dumpster or truck		
Minimum	C.F.	0.018
Maximum	C.F.	0.027
For use of elevators, add		
Minimum	C.F.	0.004
Maximum	C.F.	0.008
Rubbish hauling		
Hand loaded on trucks, 2 mile trip	C.Y.	0.320
Machine loaded on trucks, 2 mile trip	C.Y.	0.240

Highway Demolition

02065.10 Pavement Demolition

Description	UNIT	MAN/HOURS
Bituminous pavement, up to 3" thick		
On streets		
Minimum	S.Y.	0.069
Average	S.Y.	0.096
Maximum	S.Y.	0.160
On pipe trench		
Minimum	S.Y.	0.096
Average	S.Y.	0.120
Maximum	S.Y.	0.240
Concrete pavement, 6" thick		
No reinforcement		
Minimum	S.Y.	0.120
Average	S.Y.	0.160
Maximum	S.Y.	0.240
With wire mesh		
Minimum	S.Y.	0.185
Average	S.Y.	0.240
Maximum	S.Y.	0.300
With rebars		
Minimum	S.Y.	0.240
Average	S.Y.	0.300
Maximum	S.Y.	0.400
9" thick		
No reinforcement		
Minimum	S.Y.	0.160
Average	S.Y.	0.200
Maximum	S.Y.	0.240
With wire mesh		
Minimum	S.Y.	0.253

Highway Demolition	UNIT	MAN/ HOURS
02065.10 Pavement Demolition		
Average	S.Y.	0.300
Maximum	S.Y.	0.369
With rebars		
Minimum	S.Y.	0.320
Average	S.Y.	0.400
Maximum	S.Y.	0.533
12" thick		
No reinforcement		
Minimum	S.Y.	0.200
Average	S.Y.	0.240
Maximum	S.Y.	0.300
With wire mesh		
Minimum	S.Y.	0.282
Average	S.Y.	0.343
Maximum	S.Y.	0.436
With rebars		
Minimum	S.Y.	0.400
Average	S.Y.	0.480
Maximum	S.Y.	0.600
Sidewalk, 4" thick, with disposal		
Minimum	S.Y.	0.057
Average	S.Y.	0.080
Maximum	S.Y.	0.114
Removal of pavement markings by waterblasting		
Minimum	S.F.	0.003
Average	S.F.	0.004
Maximum	S.F.	0.008
02065.15 Saw Cutting Pavement		
Pavement, bituminous		
2" thick	L.F.	0.016
3" thick	L.F.	0.020
4" thick	L.F.	0.025
5" thick	L.F.	0.027
6" thick	L.F.	0.029
Concrete pavement, with wire mesh		
4" thick	L.F.	0.031
5" thick	L.F.	0.033
6" thick	L.F.	0.036
8" thick	L.F.	0.040
10" thick	L.F.	0.044
Plain concrete, unreinforced		
4" thick	L.F.	0.027
5" thick	L.F.	0.031
6" thick	L.F.	0.033
8" thick	L.F.	0.036
10" thick	L.F.	0.040
02065.80 Curb & Gutter		
Curb removal		
Concrete, unreinforced		
Minimum	L.F.	0.048
Average	L.F.	0.060
Maximum	L.F.	0.075

Highway Demolition	UNIT	MAN/ HOURS
02065.80 Curb & Gutter		
Reinforced		
Minimum	L.F.	0.077
Average	L.F.	0.086
Maximum	L.F.	0.096
Combination curb and 2'gutter		
Unreinforced		
Minimum	L.F.	0.063
Average	L.F.	0.083
Maximum	L.F.	0.120
Reinforced		
Minimum	L.F.	0.100
Average	L.F.	0.133
Maximum	L.F.	0.240
Granite curb		
Minimum	L.F.	0.069
Average	L.F.	0.080
Maximum	L.F.	0.092
Asphalt curb		
Minimum	L.F.	0.040
Average	L.F.	0.048
Maximum	L.F.	0.057
02065.85 Guardrails		
Remove standard guardrail		
Steel		
Minimum	L.F.	0.060
Average	L.F.	0.080
Maximum	L.F.	0.120
Wood		
Minimum	L.F.	0.052
Average	L.F.	0.062
Maximum	L.F.	0.100
02075.80 Core Drilling		
Concrete		
6" thick		
3" dia.	EA.	0.571
4" dia.	EA.	0.667
6" dia.	EA.	0.800
8" dia.	EA.	1.333
8" thick		
3" dia.	EA.	0.800
4" dia.	EA.	1.000
6" dia.	EA.	1.143
8" dia.	EA.	1.600
10" thick		
3" dia.	EA.	1.000
4" dia.	EA.	1.143
6" dia.	EA.	1.333
8" dia.	EA.	2.000
12" thick		
3" dia.	EA.	1.333
4" dia.	EA.	1.600
6" dia.	EA.	2.000

Highway Demolition

02075.80 Core Drilling	UNIT	MAN/ HOURS
8" dia.	EA.	2.667

Hazardous Waste

02080.10 Asbestos Removal	UNIT	MAN/ HOURS
Enclosure using wood studs & poly, install & remove	S.F.	0.020

02080.12 Duct Insulation Removal	UNIT	MAN/ HOURS
Remove duct insulation, duct size		
6" x 12"	L.F.	0.044
x 18"	L.F.	0.062
x 24"	L.F.	0.089
8" x 12"	L.F.	0.067
x 18"	L.F.	0.073
x 24"	L.F.	0.100
12" x 12"	L.F.	0.067
x 18"	L.F.	0.089
x 24"	L.F.	0.114

02080.15 Pipe Insulation Removal	UNIT	MAN/ HOURS
Removal, asbestos insulation		
2" thick, pipe		
1" to 3" dia.	L.F.	0.067
4" to 6" dia.	L.F.	0.076
3" thick		
7" to 8" dia.	L.F.	0.080
9" to 10" dia.	L.F.	0.084
11" to 12" dia.	L.F.	0.089
13" to 14" dia.	L.F.	0.094
15" to 18" dia.	L.F.	0.100

Site Demolition

02105.10 Catch Basins/manholes	UNIT	MAN/ HOURS
Abandon catch basin or manhole (fill with sand)		
Minimum	EA.	3.000
Average	EA.	4.800
Maximum	EA.	8.000
Remove and reset frame and cover		
Minimum	EA.	1.600
Average	EA.	2.400
Maximum	EA.	4.000
Remove catch basin, to 10' deep		
Masonry		
Minumum	EA.	4.800
Average	EA.	6.000
Maximum	EA.	8.000
Concrete		
Minimum	EA.	6.000
Average	EA.	8.000
Maximum	EA.	9.600

02105.20 Fences	UNIT	MAN/ HOURS
Remove fencing		
Chain link, 8' high		
For disposal	L.F.	0.040
For reuse	L.F.	0.100
Wood		
4' high	S.F.	0.027
6' high	S.F.	0.032
8' high	S.F.	0.040
Masonry		
8" thick		
4' high	S.F.	0.080
6' high	S.F.	0.100
8' high	S.F.	0.114
12" thick		
4' high	S.F.	0.133
6' high	S.F.	0.160
8' high	S.F.	0.200
12' high	S.F.	0.267

02105.30 Hydrants	UNIT	MAN/ HOURS
Remove fire hydrant		
Minimum	EA.	3.000
Average	EA.	4.000
Maximum	EA.	6.000
Remove and reset fire hydrant		
Minimum	EA.	8.000
Average	EA.	12.000
Maximum	EA.	24.000

02105.42 Drainage Piping	UNIT	MAN/ HOURS
Remove drainage pipe, not including excavation		
12" dia.		
Minimum	L.F.	0.080
Average	L.F.	0.100
Maximum	L.F.	0.126
18" dia.		
Minimum	L.F.	0.109
Average	L.F.	0.126
Maximum	L.F.	0.160
24" dia.		
Minimum	L.F.	0.133
Average	L.F.	0.160

Site Demolition

02105.42 Drainage Piping

	UNIT	MAN/ HOURS
Maximum	L.F.	0.200
36" dia.		
Minimum	L.F.	0.160
Average	L.F.	0.200
Maximum	L.F.	0.253

02105.43 Gas Piping

	UNIT	MAN/ HOURS
Remove welded steel pipe, not including excavation		
4" dia.		
Minimum	L.F.	0.120
Average	L.F.	0.150
Maximum	L.F.	0.200
5" dia.		
Minimum	L.F.	0.200
Average	L.F.	0.240
Maximum	L.F.	0.300
6" dia.		
Minimum	L.F.	0.253
Average	L.F.	0.300
Maximum	L.F.	0.400
8" dia.		
Minimum	L.F.	0.369
Average	L.F.	0.480
Maximum	L.F.	0.632
10" dia.		
Minimum	L.F.	0.480
Average	L.F.	0.600
Maximum	L.F.	0.800

02105.45 Sanitary Piping

	UNIT	MAN/ HOURS
Remove sewer pipe, not including excavation		
4" dia.		
Minimum	L.F.	0.067
Average	L.F.	0.096
Maximum	L.F.	0.160
6" dia.		
Minimum	L.F.	0.075
Average	L.F.	0.109
Maximum	L.F.	0.200
8" dia.		
Minimum	L.F.	0.080
Average	L.F.	0.120
Maximum	L.F.	0.240
10" dia.		
Minimum	L.F.	0.086
Average	L.F.	0.126
Maximum	L.F.	0.267
12" dia.		
Minimum	L.F.	0.092
Average	L.F.	0.133
Maximum	L.F.	0.300
15" dia.		
Minimum	L.F.	0.100
Average	L.F.	0.141

Site Demolition

02105.45 Sanitary Piping

	UNIT	MAN/ HOURS
Maximum	L.F.	0.343
18" dia.		
Minimum	L.F.	0.109
Average	L.F.	0.160
Maximum	L.F.	0.400
24" dia.		
Minimum	L.F.	0.120
Average	L.F.	0.200
Maximum	L.F.	0.480
30" dia.		
Minimum	L.F.	0.133
Average	L.F.	0.240
Maximum	L.F.	0.600
36" dia.		
Minimum	L.F.	0.160
Average	L.F.	0.300
Maximum	L.F.	0.800

02105.48 Water Piping

	UNIT	MAN/ HOURS
Remove water pipe, not including excavation		
4" dia.		
Minimum	L.F.	0.096
Average	L.F.	0.109
Maximum	L.F.	0.126
6" dia.		
Minimum	L.F.	0.100
Average	L.F.	0.114
Maximum	L.F.	0.133
8" dia.		
Minimum	L.F.	0.109
Average	L.F.	0.126
Maximum	L.F.	0.150
10" dia.		
Minimum	L.F.	0.114
Average	L.F.	0.133
Maximum	L.F.	0.160
12" dia.		
Minimum	L.F.	0.120
Average	L.F.	0.141
Maximum	L.F.	0.171
14" dia.		
Minimum	L.F.	0.126
Average	L.F.	0.150
Maximum	L.F.	0.185
16" dia.		
Minimum	L.F.	0.133
Average	L.F.	0.160
Maximum	L.F.	0.200
18" dia.		
Minimum	L.F.	0.141
Average	L.F.	0.171
Maximum	L.F.	0.218
20" dia.		
Minimum	L.F.	0.150

Site Demolition

02105.48 Water Piping	UNIT	MAN/HOURS
Average	L.F.	0.185
Maximum	L.F.	0.240
Remove valves		
6"	EA.	1.200
10"	EA.	1.333
14"	EA.	1.500
18"	EA.	2.000

02105.60 Underground Tanks	UNIT	MAN/HOURS
Remove underground storage tank, and backfill		
50 to 250 gals	EA.	8.000
600 gals	EA.	8.000
1000 gals	EA.	12.000
4000 gals	EA.	19.200
5000 gals	EA.	19.200
10,000 gals	EA.	32.000
12,000 gals	EA.	40.000
15,000 gals	EA.	48.000
20,000 gals	EA.	60.000

02105.66 Septic Tanks	UNIT	MAN/HOURS
Remove septic tank		
1000 gals	EA.	2.000
2000 gals	EA.	2.400
5000 gals	EA.	3.000
15,000 gals	EA.	24.000
25,000 gals	EA.	32.000
40,000 gals	EA.	48.000

02105.80 Walls, Exterior	UNIT	MAN/HOURS
Concrete wall		
Light reinforcing		
6" thick	S.F.	0.120
8" thick	S.F.	0.126
10" thick	S.F.	0.133
12" thick	S.F.	0.150
Medium reinforcing		
6" thick	S.F.	0.126
8" thick	S.F.	0.133
10" thick	S.F.	0.150
12" thick	S.F.	0.171
Heavy reinforcing		
6" thick	S.F.	0.141
8" thick	S.F.	0.150
10" thick	S.F.	0.171
12" thick	S.F.	0.200
Masonry		
No reinforcing		
8" thick	S.F.	0.053
12" thick	S.F.	0.060
16" thick	S.F.	0.069
Horizontal reinforcing		
8" thick	S.F.	0.060
12" thick	S.F.	0.065

Site Demolition

02105.80 Walls, Exterior	UNIT	MAN/HOURS
16" thick	S.F.	0.077
Vertical reinforcing		
8" thick	S.F.	0.077
12" thick	S.F.	0.089
16" thick	S.F.	0.109
Remove concrete headwall		
15" pipe	EA.	1.714
18" pipe	EA.	2.000
24" pipe	EA.	2.182
30" pipe	EA.	2.400
36" pipe	EA.	2.667
48" pipe	EA.	3.429
60" pipe	EA.	4.800

02110.10 Clearing And Grubbing	UNIT	MAN/HOURS
Clear wooded area		
Light density	ACRE	60.000
Medium density	ACRE	80.000
Heavy density	ACRE	96.000

02110.50 Tree Cutting & Clearing	UNIT	MAN/HOURS
Cut trees and clear out stumps		
9" to 12" dia.	EA.	4.800
To 24" dia.	EA.	6.000
24" dia. and up	EA.	8.000
Loading and trucking		
For machine load, per load, round trip		
1 mile	EA.	0.960
3 mile	EA.	1.091
5 mile	EA.	1.200
10 mile	EA.	1.600
20 mile	EA.	2.400
Hand loaded, round trip		
1 mile	EA.	2.000
3 mile	EA.	2.286
5 mile	EA.	2.667
10 mile	EA.	3.200
20 mile	EA.	4.000
Tree trimming for pole line construction		
Light cutting	L.F.	0.012
Medium cutting	L.F.	0.016
Heavy cutting	L.F.	0.024

Dewatering

Dewatering	UNIT	MAN/ HOURS
02144.10 Wellpoint Systems		
Pumping, gas driven, 50' hose		
3" header pipe	DAY	8.000
6" header pipe	DAY	10.000
Wellpoint system per job		
6" header pipe	L.F.	0.032
8" header pipe	L.F.	0.040
10" header pipe	L.F.	0.053
Jetting wellpoint system		
14' long	EA.	0.533
18' long	EA.	0.667
Sand filter for wellpoints	L.F.	0.013
Replacement of wellpoint components	EA.	0.160

Shoring And Underpinning

Shoring And Underpinning	UNIT	MAN/ HOURS
02162.10 Trench Sheeting		
Closed timber, including pull and salvage, excavation		
8' deep	S.F.	0.064
10' deep	S.F.	0.067
12' deep	S.F.	0.071
14' deep	S.F.	0.075
16' deep	S.F.	0.080
18' deep	S.F.	0.091
20' deep	S.F.	0.098
02170.10 Cofferdams		
Cofferdam, steel, driven from shore		
15' deep	S.F.	0.137
20' deep	S.F.	0.128
25' deep	S.F.	0.120
30' deep	S.F.	0.113
40' deep	S.F.	0.107
Driven from barge		
20' deep	S.F.	0.148
30' deep	S.F.	0.137
40' deep	S.F.	0.128
50' deep	S.F.	0.120

Earthwork

Earthwork	UNIT	MAN/ HOURS
02210.10 Hauling Material		
Haul material by 10 cy dump truck, round trip distance		
1 mile	C.Y.	0.044
2 mile	C.Y.	0.053
5 mile	C.Y.	0.073
10 mile	C.Y.	0.080
20 mile	C.Y.	0.089
30 mile	C.Y.	0.107
Site grading, cut & fill, sandy clay, 200' haul, 75 hp dozer	C.Y.	0.032
Spread topsoil by equipment on site	C.Y.	0.036
Site grading (cut and fill to 6") less than 1 acre		
75 hp dozer	C.Y.	0.053
1.5 cy backhoe/loader	C.Y.	0.080
02210.30 Bulk Excavation		
Excavation, by small dozer		
Large areas	C.Y.	0.016
Small areas	C.Y.	0.027
Trim banks	C.Y.	0.040
Drag line		
1-1/2 cy bucket		
Sand or gravel	C.Y.	0.040
Light clay	C.Y.	0.053
Heavy clay	C.Y.	0.060
Unclassified	C.Y.	0.064
2 cy bucket		
Sand or gravel	C.Y.	0.037
Light clay	C.Y.	0.048
Heavy clay	C.Y.	0.053
Unclassified	C.Y.	0.056
2-1/2 cy bucket		
Sand or gravel	C.Y.	0.034
Light clay	C.Y.	0.044
Heavy clay	C.Y.	0.048
Unclassified	C.Y.	0.051
3 cy bucket		
Sand or gravel	C.Y.	0.030
Light clay	C.Y.	0.040
Heavy clay	C.Y.	0.044
Unclassified	C.Y.	0.046
Hydraulic excavator		
1 cy capacity		
Light material	C.Y.	0.040
Medium material	C.Y.	0.048
Wet material	C.Y.	0.060
Blasted rock	C.Y.	0.069
1-1/2 cy capacity		
Light material	C.Y.	0.010
Medium material	C.Y.	0.013
Wet material	C.Y.	0.016
Blasted rock	C.Y.	0.020
2 cy capacity		
Light material	C.Y.	0.009
Medium material	C.Y.	0.011
Wet material	C.Y.	0.013

02 SITEWORK

Earthwork

02210.30 Bulk Excavation

Earthwork	UNIT	MAN/HOURS
Blasted rock	C.Y.	0.016
Wheel mounted front-end loader		
7/8 cy capacity		
Light material	C.Y.	0.020
Medium material	C.Y.	0.023
Wet material	C.Y.	0.027
Blasted rock	C.Y.	0.032
1-1/2 cy capacity		
Light material	C.Y.	0.011
Medium material	C.Y.	0.012
Wet material	C.Y.	0.013
Blasted rock	C.Y.	0.015
2-1/2 cy capacity		
Light material	C.Y.	0.009
Medium material	C.Y.	0.010
Wet material	C.Y.	0.011
Blasted rock	C.Y.	0.011
3-1/2 cy capacity		
Light material	C.Y.	0.009
Medium material	C.Y.	0.009
Wet material	C.Y.	0.010
Blasted rock	C.Y.	0.011
6 cy capacity		
Light material	C.Y.	0.005
Medium material	C.Y.	0.006
Wet material	C.Y.	0.006
Blasted rock	C.Y.	0.007
Track mounted front-end loader		
1-1/2 cy capacity		
Light material	C.Y.	0.013
Medium material	C.Y.	0.015
Wet material	C.Y.	0.016
Blasted rock	C.Y.	0.018
2-3/4 cy capacity		
Light material	C.Y.	0.008
Medium material	C.Y.	0.009
Wet material	C.Y.	0.010
Blasted rock	C.Y.	0.011

02220.10 Borrow

Earthwork	UNIT	MAN/HOURS
Borrow fill, F.O.B. at pit		
Sand, haul to site, round trip		
10 mile	C.Y.	0.080
20 mile	C.Y.	0.133
30 mile	C.Y.	0.200
Place borrow fill and compact		
Less than 1 in 4 slope	C.Y.	0.040
Greater than 1 in 4 slope	C.Y.	0.053

02220.40 Building Excavation

Earthwork	UNIT	MAN/HOURS
Structural excavation, unclassified earth		
3/8 cy backhoe	C.Y.	0.107
3/4 cy backhoe	C.Y.	0.080
1 cy backhoe	C.Y.	0.067
Foundation backfill and compaction by machine	C.Y.	0.160

02220.50 Utility Excavation

Earthwork	UNIT	MAN/HOURS
Trencher, sandy clay, 8" wide trench		
18" deep	L.F.	0.018
24" deep	L.F.	0.020
36" deep	L.F.	0.023
Trench backfill, 95% compaction		
Tamp by hand	C.Y.	0.500
Vibratory compaction	C.Y.	0.400
Trench backfilling, with borrow sand, place & compact	C.Y.	0.400

02220.60 Trenching

Earthwork	UNIT	MAN/HOURS
Trenching and continuous footing excavation		
By gradall		
1 cy capacity		
Light soil	C.Y.	0.023
Medium soil	C.Y.	0.025
Heavy/wet soil	C.Y.	0.027
Loose rock	C.Y.	0.029
Blasted rock	C.Y.	0.031
By hydraulic excavator		
1/2 cy capacity		
Light soil	C.Y.	0.027
Medium soil	C.Y.	0.029
Heavy/wet soil	C.Y.	0.032
Loose rock	C.Y.	0.036
Blasted rock	C.Y.	0.040
1 cy capacity		
Light soil	C.Y.	0.019
Medium soil	C.Y.	0.020
Heavy/wet soil	C.Y.	0.021
Loose rock	C.Y.	0.023
Blasted rock	C.Y.	0.025
1-1/2 cy capacity		
Light soil	C.Y.	0.017
Medium soil	C.Y.	0.018
Heavy/wet soil	C.Y.	0.019
Loose rock	C.Y.	0.020
Blasted rock	C.Y.	0.021
2 cy capacity		
Light soil	C.Y.	0.016
Medium soil	C.Y.	0.017
Heavy/wet soil	C.Y.	0.018
Loose rock	C.Y.	0.019
Blasted rock	C.Y.	0.020
2-1/2 cy capacity		
Light soil	C.Y.	0.015
Medium soil	C.Y.	0.015
Heavy/wet soil	C.Y.	0.016
Loose rock	C.Y.	0.017
Blasted rock	C.Y.	0.018
Trencher, chain, 1' wide to 4' deep		
Light soil	C.Y.	0.020

02 SITEWORK

Earthwork

02220.60 Trenching	UNIT	MAN/ HOURS
Medium soil	C.Y.	0.023
Heavy soil	C.Y.	0.027
Hand excavation		
Bulk, wheeled 100'		
Normal soil	C.Y.	0.889
Sand or gravel	C.Y.	0.800
Medium clay	C.Y.	1.143
Heavy clay	C.Y.	1.600
Loose rock	C.Y.	2.000
Trenches, up to 2' deep		
Normal soil	C.Y.	1.000
Sand or gravel	C.Y.	0.889
Medium clay	C.Y.	1.333
Heavy clay	C.Y.	2.000
Loose rock	C.Y.	2.667
Trenches, to 6' deep		
Normal soil	C.Y.	1.143
Sand or gravel	C.Y.	1.000
Medium clay	C.Y.	1.600
Heavy clay	C.Y.	2.667
Loose rock	C.Y.	4.000
Backfill trenches		
With compaction		
By hand	C.Y.	0.667
By 60 hp tracked dozer	C.Y.	0.020
By 200 hp tracked dozer	C.Y.	0.009
By small front-end loader	C.Y.	0.023
Spread dumped fill or gravel, no compaction		
6" layers	S.Y.	0.013
12" layers	S.Y.	0.016
Compaction in 6" layers		
By hand with air tamper	S.Y.	0.016
Backfill trenches, sand bedding, no compaction		
By hand	C.Y.	0.667
By small front-end loader	C.Y.	0.023

02220.70 Roadway Excavation	UNIT	MAN/ HOURS
Roadway excavation		
1/4 mile haul	C.Y.	0.016
2 mile haul	C.Y.	0.027
5 mile haul	C.Y.	0.040
Excavation of open ditches	C.Y.	0.011
Trim banks, swales or ditches	S.Y.	0.013
Bulk swale excavation by dragline		
Small jobs	C.Y.	0.060
Large jobs	C.Y.	0.034
Spread base course	C.Y.	0.020
Roll and compact	C.Y.	0.027

02220.71 Base Course	UNIT	MAN/ HOURS
Base course, crushed stone		
3" thick	S.Y.	0.004
4" thick	S.Y.	0.004
6" thick	S.Y.	0.005

Earthwork

02220.71 Base Course	UNIT	MAN/ HOURS
8" thick	S.Y.	0.005
10" thick	S.Y.	0.006
12" thick	S.Y.	0.007
Base course, bank run gravel		
4" deep	S.Y.	0.004
6" deep	S.Y.	0.005
8" deep	S.Y.	0.005
10" deep	S.Y.	0.005
12" deep	S.Y.	0.006
Prepare and roll sub base		
Minimum	S.Y.	0.004
Average	S.Y.	0.005
Maximum	S.Y.	0.007

02220.90 Hand Excavation	UNIT	MAN/ HOURS
Excavation		
To 2' deep		
Normal soil	C.Y.	0.889
Sand and gravel	C.Y.	0.800
Medium clay	C.Y.	1.000
Heavy clay	C.Y.	1.143
Loose rock	C.Y.	1.333
To 6' deep		
Normal soil	C.Y.	1.143
Sand and gravel	C.Y.	1.000
Medium clay	C.Y.	1.333
Heavy clay	C.Y.	1.600
Loose rock	C.Y.	2.000
Backfilling foundation without compaction, 6" lifts	C.Y.	0.500
Compaction of backfill around structures or in trench		
By hand with air tamper	C.Y.	0.571
By hand with vibrating plate tamper	C.Y.	0.533
1 ton roller	C.Y.	0.400
Miscellaneous hand labor		
Trim slopes, sides of excavation	S.F.	0.001
Trim bottom of excavation	S.F.	0.002
Excavation around obstructions and services	C.Y.	2.667

02240.05 Soil Stabilization	UNIT	MAN/ HOURS
Straw bale secured with rebar	L.F.	0.027
Filter barrier, 18" high filter fabric	L.F.	0.080
Sediment fence, 36" fabric with 6" mesh	L.F.	0.100
Soil stabilization with tar paper, burlap, straw and stakes	S.F.	0.001

02240.30 Geotextile	UNIT	MAN/ HOURS
Filter cloth, light reinforcement		
Woven		
12'-6" wide x 50' long	S.F.	0.001
Various lengths	S.F.	0.001
Non-woven		
14'-8" wide x 430' long	S.F.	0.001
Various lengths	S.F.	0.001

Earthwork

02270.10 Slope Protection

Earthwork	UNIT	MAN/HOURS
Gabions, stone filled		
6" deep	S.Y.	0.200
9" deep	S.Y.	0.229
12" deep	S.Y.	0.267
18" deep	S.Y.	0.320
36" deep	S.Y.	0.533

02270.40 Riprap

Earthwork	UNIT	MAN/HOURS
Riprap		
Crushed stone blanket, max size 2-1/2"	TON	0.533
Stone, quarry run, 300 lb. stones	TON	0.492
400 lb. stones	TON	0.457
500 lb. stones	TON	0.427
750 lb. stones	TON	0.400
Dry concrete riprap in bags 3" thick, 80 lb. per bag	BAG	0.027

02280.20 Soil Treatment

Earthwork	UNIT	MAN/HOURS
Soil treatment, termite control pretreatment		
Under slabs	S.F.	0.004
By walls	S.F.	0.005

02290.30 Weed Control

Earthwork	UNIT	MAN/HOURS
Weed control, bromicil, 15 lb./acre, wettable powder	ACRE	4.000
Vegetation control, by application of plant killer	S.Y.	0.003
Weed killer, lawns and fields	S.Y.	0.002

Tunneling

02300.10 Pipe Jacking

Tunneling	UNIT	MAN/HOURS
Pipe casing, horizontal jacking		
18" dia.	L.F.	0.711
21" dia.	L.F.	0.762
24" dia.	L.F.	0.800
27" dia.	L.F.	0.800
30" dia.	L.F.	0.842
36" dia.	L.F.	0.914
42" dia.	L.F.	1.000
48" dia.	L.F.	1.067

Piles And Caissons

02360.50 Prestressed Piling

Piles And Caissons	UNIT	MAN/HOURS
Prestressed concrete piling, less than 60' long		
10" sq.	L.F.	0.040
12" sq.	L.F.	0.042
14" sq.	L.F.	0.043
16" sq.	L.F.	0.044
18" sq.	L.F.	0.047
20" sq.	L.F.	0.048
24" sq.	L.F.	0.049
More than 60' long		
12" sq.	L.F.	0.034
14" sq.	L.F.	0.035
16" sq.	L.F.	0.036
18" sq.	L.F.	0.036
20" sq.	L.F.	0.037
24" sq.	L.F.	0.038
Straight cylinder, less than 60' long		
12" dia.	L.F.	0.044
14" dia.	L.F.	0.045
16" dia.	L.F.	0.046
18" dia.	L.F.	0.047
20" dia.	L.F.	0.048
24" dia.	L.F.	0.049
More than 60' long		
12" dia.	L.F.	0.035
14" dia.	L.F.	0.036
16" dia.	L.F.	0.036
18" dia.	L.F.	0.037
20" dia.	L.F.	0.038
24" dia.	L.F.	0.038
Concrete sheet piling		
12" thick x 20' long	S.F.	0.096
25' long	S.F.	0.087
30' long	S.F.	0.080
35' long	S.F.	0.074
40' long	S.F.	0.069
16" thick x 40' long	S.F.	0.053
45' long	S.F.	0.051
50' long	S.F.	0.048
55' long	S.F.	0.046
60' long	S.F.	0.044

02360.60 Steel Piles

Piles And Caissons	UNIT	MAN/HOURS
H-section piles		
8x8		
36 lb/ft		
30' long	L.F.	0.080
40' long	L.F.	0.064
50' long	L.F.	0.053
10x10		
42 lb/ft		
30' long	L.F.	0.080
40' long	L.F.	0.064
50' long	L.F.	0.053
57 lb/ft		

Piles And Caissons

02360.60 Steel Piles	UNIT	MAN/ HOURS
30' long	L.F.	0.080
40' long	L.F.	0.064
50' long	L.F.	0.053
12x12		
53 lb/ft		
30' long	L.F.	0.087
40' long	L.F.	0.069
50' long	L.F.	0.053
74 lb/ft		
30' long	L.F.	0.087
40' long	L.F.	0.069
50' long	L.F.	0.053
14x14		
73 lb/ft		
40' long	L.F.	0.087
50' long	L.F.	0.069
60' long	L.F.	0.053
89 lb/ft		
40' long	L.F.	0.087
50' long	L.F.	0.069
60' long	L.F.	0.053
102 lb/ft		
40' long	L.F.	0.087
50' long	L.F.	0.069
60' long	L.F.	0.053
117 lb/ft		
40' long	L.F.	0.091
50' long	L.F.	0.071
60' long	L.F.	0.055
Splice		
8"	EA.	1.333
10"	EA.	1.600
12"	EA.	1.600
14"	EA.	2.000
Driving cap		
8"	EA.	0.800
10"	EA.	1.000
12"	EA.	1.000
14"	EA.	1.143
Standard point		
8"	EA.	0.800
10"	EA.	1.000
12"	EA.	1.143
14"	EA.	1.333
Heavy duty point		
8"	EA.	0.889
10"	EA.	1.143
12"	EA.	1.333
14"	EA.	1.600
Tapered friction piles, with fluted steel casing, up to 50'		
With 4000 psi concrete no reinforcing		
12" dia.	L.F.	0.048
14" dia.	L.F.	0.049
16" dia.	L.F.	0.051

Piles And Caissons

02360.60 Steel Piles	UNIT	MAN/ HOURS
18" dia.	L.F.	0.056

02360.65 Steel Pipe Piles	UNIT	MAN/ HOURS
Concrete filled, 3000# concrete, up to 40'		
8" dia.	L.F.	0.069
10" dia.	L.F.	0.071
12" dia.	L.F.	0.074
14" dia.	L.F.	0.077
16" dia.	L.F.	0.080
18" dia.	L.F.	0.083
Pipe piles, non-filled		
8" dia.	L.F.	0.053
10" dia.	L.F.	0.055
12" dia.	L.F.	0.056
14" dia.	L.F.	0.060
16" dia.	L.F.	0.062
18" dia.	L.F.	0.064
Splice		
8" dia.	EA.	1.600
10" dia.	EA.	1.600
12" dia.	EA.	2.000
14" dia.	EA.	2.000
16" dia.	EA.	2.667
18" dia.	EA.	2.667
Standard point		
8" dia.	EA.	1.600
10" dia.	EA.	1.600
12" dia.	EA.	2.000
14" dia.	EA.	2.000
16" dia.	EA.	2.667
18" dia.	EA.	2.667
Heavy duty point		
8" dia.	EA.	2.000
10" dia.	EA.	2.000
12" dia.	EA.	2.667
14" dia.	EA.	2.667
16" dia.	EA.	3.200
18" dia.	EA.	3.200

02360.70 Steel Sheet Piling	UNIT	MAN/ HOURS
Steel sheet piling,12" wide		
20' long	S.F.	0.096
35' long	S.F.	0.069
50' long	S.F.	0.048
Over 50' long	S.F.	0.044

02360.80 Wood And Timber Piles	UNIT	MAN/ HOURS
Treated wood piles, 12" butt, 8" tip		
25' long	L.F.	0.096
30' long	L.F.	0.080
35' long	L.F.	0.069
40' long	L.F.	0.060
12" butt, 7" tip		
40' long	L.F.	0.060

Piles And Caissons

02360.80 Wood And Timber Piles	UNIT	MAN/HOURS
45' long	L.F.	0.053
50' long	L.F.	0.048
55' long	L.F.	0.044
60' long	L.F.	0.040

02380.10 Caissons	UNIT	MAN/HOURS
Caisson, including 3000# concrete, in stable ground		
18" dia.	L.F.	0.192
24" dia.	L.F.	0.200
30" dia.	L.F.	0.240
36" dia.	L.F.	0.274
48" dia.	L.F.	0.320
60" dia.	L.F.	0.436
72" dia.	L.F.	0.533
84" dia.	L.F.	0.686
Wet ground, casing required but pulled		
18" dia.	L.F.	0.240
24" dia.	L.F.	0.267
30" dia.	L.F.	0.300
36" dia.	L.F.	0.320
48" dia.	L.F.	0.400
60" dia.	L.F.	0.533
72" dia.	L.F.	0.800
84" dia.	L.F.	1.200
Soft rock		
18" dia.	L.F.	0.686
24" dia.	L.F.	1.200
30" dia.	L.F.	1.600
36" dia.	L.F.	2.400
48" dia.	L.F.	3.200
60" dia.	L.F.	4.800
72" dia.	L.F.	5.333
84" dia.	L.F.	6.000

Railroad Work

02450.10 Railroad Work	UNIT	MAN/HOURS
Rail		
90 lb	L.F.	0.010
100 lb	L.F.	0.010
115 lb	L.F.	0.010
132 lb	L.F.	0.010
Rail relay		
90 lb	L.F.	0.010
100 lb	L.F.	0.010
115 lb	L.F.	0.010
132 lb	L.F.	0.010

Railroad Work

02450.10 Railroad Work	UNIT	MAN/HOURS
New angle bars, per pair		
90 lb	EA.	0.012
100 lb	EA.	0.012
115 lb	EA.	0.012
132 lb	EA.	0.012
Angle bar relay		
90 lb	EA.	0.012
100 lb	EA.	0.012
115 lb	EA.	0.012
132 lb	EA.	0.012
New tie plates		
90 lb	EA.	0.009
100 lb	EA.	0.009
115 lb	EA.	0.009
132 lb	EA.	0.009
Tie plate relay		
90 lb	EA.	0.009
100 lb	EA.	0.009
115 lb	EA.	0.009
132 lb	EA.	0.009
Track accessories		
Wooden cross ties, 8'	EA.	0.060
Concrete cross ties, 8'	EA.	0.120
Tie plugs, 5"	EA.	0.006
Track bolts and nuts, 1"	EA.	0.006
Lockwashers, 1"	EA.	0.004
Track spikes, 6"	EA.	0.024
Wooden switch ties	B.F.	0.006
Rail anchors	EA.	0.022
Ballast	TON	0.120
Gauge rods	EA.	0.096
Compromise splice bars	EA.	0.160
Turnout		
90 lb	EA.	24.000
100 lb	EA.	24.000
110 lb	EA.	24.000
115 lb	EA.	24.000
132 lb	EA.	24.000
Turnout relay		
90 lb	EA.	24.000
100 lb	EA.	24.000
110 lb	EA.	24.000
115 lb	EA.	24.000
132 lb	EA.	24.000
Railroad track in place, complete		
New rail		
90 lb	L.F.	0.240
100 lb	L.F.	0.240
110 lb	L.F.	0.240
115 lb	L.F.	0.240
132 lb	L.F.	0.240
Rail relay		
90 lb	L.F.	0.240
100 lb	L.F.	0.240

Railroad Work

02450.10 Railroad Work

Railroad Work	UNIT	MAN/ HOURS
110 lb	L.F.	0.240
115 lb	L.F.	0.240
132 lb	L.F.	0.240
No. 8 turnout		
90 lb	EA.	32.000
100 lb	EA.	32.000
110 lb	EA.	32.000
115 lb	EA.	32.000
132 lb	EA.	32.000
No. 8 turnout relay		
90 lb	EA.	32.000
100 lb	EA.	32.000
110 lb	EA.	32.000
115 lb	EA.	32.000
132 lb	EA.	32.000
Railroad crossings, asphalt, based on 8" thick x 20'		
Including track and approach		
12' roadway	EA.	6.000
15' roadway	EA.	6.857
18' roadway	EA.	8.000
21' roadway	EA.	9.600
24' roadway	EA.	12.000
Precast concrete inserts		
12' roadway	EA.	2.400
15' roadway	EA.	3.000
18' roadway	EA.	4.000
21' roadway	EA.	4.800
24' roadway	EA.	5.333
Molded rubber, with headers		
12' roadway	EA.	2.400
15' roadway	EA.	3.000
18' roadway	EA.	4.000
21' roadway	EA.	4.800
24' roadway	EA.	5.333

Paving And Surfacing

02510.20 Asphalt Surfaces

Asphalt Surfaces	UNIT	MAN/ HOURS
Asphalt wearing surface, for flexible pavement		
1" thick	S.Y.	0.016
1-1/2" thick	S.Y.	0.019
2" thick	S.Y.	0.024
3" thick	S.Y.	0.032
Binder course		
1-1/2" thick	S.Y.	0.018
2" thick	S.Y.	0.022
3" thick	S.Y.	0.029
4" thick	S.Y.	0.032
5" thick	S.Y.	0.036
6" thick	S.Y.	0.040
Bituminous sidewalk, no base		
2" thick	S.Y.	0.028
3" thick	S.Y.	0.030

02520.10 Concrete Paving

Concrete Paving	UNIT	MAN/ HOURS
Concrete paving, reinforced, 5000 psi concrete		
6" thick	S.Y.	0.150
7" thick	S.Y.	0.160
8" thick	S.Y.	0.171
9" thick	S.Y.	0.185
10" thick	S.Y.	0.200
11" thick	S.Y.	0.218
12" thick	S.Y.	0.240
15" thick	S.Y.	0.300
Concrete paving, for pipe trench, reinforced		
7" thick	S.Y.	0.240
8" thick	S.Y.	0.267
9" thick	S.Y.	0.300
10" thick	S.Y.	0.343
Fibrous concrete		
5" thick	S.Y.	0.185
8" thick	S.Y.	0.200
Roller compacted concrete, (RCC), place and compact		
8" thick	S.Y.	0.240
12" thick	S.Y.	0.300
Steel edge forms up to		
12" deep	L.F.	0.027
15" deep	L.F.	0.032
Paving finishes		
Belt dragged	S.Y.	0.040
Curing	S.Y.	0.008

02545.10 Asphalt Repair

Asphalt Repair	UNIT	MAN/ HOURS
Coal tar emulsion seal coat, rubber additive, fuel resistant	S.Y.	0.011
Bituminous surface treatment, single	S.Y.	0.008
Double	S.Y.	0.001
Bituminous prime coat	S.Y.	0.001
Tack coat	S.Y.	0.001
Crack sealing, concrete paving	L.F.	0.005
Bituminous paving for pipe trench, 4" thick	S.Y.	0.160
Polypropylene, nonwoven paving fabric	S.Y.	0.004
Rubberized asphalt	S.Y.	0.073
Asphalt slurry seal	S.Y.	0.047

02580.10 Pavement Markings

Pavement Markings	UNIT	MAN/ HOURS
Pavement line marking, paint		
4" wide	L.F.	0.002
6" wide	L.F.	0.004
8" wide	L.F.	0.007
Reflective paint, 4" wide	L.F.	0.007
Airfield markings, retro-reflective		

Paving And Surfacing

02580.10 Pavement Markings

Paving And Surfacing	UNIT	MAN/HOURS
White	L.F.	0.007
Yellow	L.F.	0.007
Preformed tape, 4" wide		
Inlaid reflective	L.F.	0.001
Reflective paint	L.F.	0.002
Thermoplastic		
White	L.F.	0.004
Yellow	L.F.	0.004
12" wide, thermoplastic, white	L.F.	0.011
Directional arrows, reflective preformed tape	EA.	0.800
Messages, reflective preformed tape (per letter)	EA.	0.400
Handicap symbol, preformed tape	EA.	0.800
Parking stall painting	EA.	0.160

Utilities

02605.30 Manholes

Utilities	UNIT	MAN/HOURS
Precast sections, 48" dia.		
Base section	EA.	2.000
1'0" riser	EA.	1.600
1'4" riser	EA.	1.714
2'8" riser	EA.	1.846
4'0" riser	EA.	2.000
2'8" cone top	EA.	2.400
Precast manholes, 48" dia.		
4' deep	EA.	4.800
6' deep	EA.	6.000
7' deep	EA.	6.857
8' deep	EA.	8.000
10' deep	EA.	9.600
Cast-in-place, 48" dia., with frame and cover		
5' deep	EA.	12.000
6' deep	EA.	13.714
8' deep	EA.	16.000
10' deep	EA.	19.200
Brick manholes, 48" dia. with cover, 8" thick		
4' deep	EA.	8.000
6' deep	EA.	8.889
8' deep	EA.	10.000
10' deep	EA.	11.429
12' deep	EA.	13.333
14' deep	EA.	16.000
Inverts for manholes		
Single channel	EA.	3.200
Triple channel	EA.	4.000
Frames and covers, 24" diameter		
300 lb	EA.	0.800

Utilities

02605.30 Manholes

Utilities	UNIT	MAN/HOURS
400 lb	EA.	0.889
500 lb	EA.	1.143
Watertight, 350 lb	EA.	2.667
For heavy equipment, 1200 lb	EA.	4.000
Steps for manholes		
7" x 9"	EA.	0.160
8" x 9"	EA.	0.178
Curb inlet, 4' throat, cast in place		
12"-30" pipe	EA.	12.000
36"-48" pipe	EA.	13.714
Raise exist frame and cover, when repaving	EA.	4.800

02610.10 Cast Iron Flanged Pipe

Utilities	UNIT	MAN/HOURS
Cast iron flanged sections		
4" pipe, with one bolt set		
3' section	EA.	0.218
4' section	EA.	0.240
5' section	EA.	0.267
6' section	EA.	0.300
8' section	EA.	0.343
10' section	EA.	0.480
12' section	EA.	0.800
15' section	EA.	1.200
18' section	EA.	1.600
6" pipe, with one bolt set		
3' section	EA.	0.240
4' section	EA.	0.282
5' section	EA.	0.320
6' section	EA.	0.369
8' section	EA.	0.533
10' section	EA.	0.600
12' section	EA.	0.800
15' section	EA.	1.200
18' section	EA.	1.714
8" pipe, with one bolt set		
3' section	EA.	0.300
4' section	EA.	0.343
5' section	EA.	0.400
6' section	EA.	0.480
8' section	EA.	0.686
10' section	EA.	0.800
12' section	EA.	1.200
15' section	EA.	1.600
18' section	EA.	2.000
10" pipe, with one bolt set		
3' section	EA.	0.308
4' section	EA.	0.353
5' section	EA.	0.414
6' section	EA.	0.500
8' section	EA.	0.727
10' section	EA.	0.857
12' section	EA.	1.333
15' section	EA.	1.714
18' section	EA.	2.400

Utilities

02610.10 Cast Iron Flanged Pipe	UNIT	MAN/ HOURS
12" pipe, with one bolt set		
3' section	EA.	0.333
4' section	EA.	0.387
5' section	EA.	0.462
6' section	EA.	0.545
8' section	EA.	0.800
10' section	EA.	0.923
12' section	EA.	1.500
15' section	EA.	2.000
18' section	EA.	2.667

02610.11 Cast Iron Fittings	UNIT	MAN/ HOURS
Mechanical joint, with 2 bolt kits		
90 deg bend		
4"	EA.	0.533
6"	EA.	0.615
8"	EA.	0.800
10"	EA.	1.143
12"	EA.	1.600
14"	EA.	2.000
16"	EA.	2.667
45 deg bend		
4"	EA.	0.533
6"	EA.	0.615
8"	EA.	0.800
10"	EA.	1.143
12"	EA.	1.600
14"	EA.	2.000
16"	EA.	2.667
Tee, with 3 bolt kits		
4" x 4"	EA.	0.800
6" x 6"	EA.	1.000
8" x 8"	EA.	1.333
10" x 10"	EA.	2.000
12" x 12"	EA.	2.667
Wye, with 3 bolt kits		
6" x 6"	EA.	1.000
8" x 8"	EA.	1.333
10" x 10"	EA.	2.000
12" x 12"	EA.	2.667
Reducer, with 2 bolt kits		
6" x 4"	EA.	1.000
8" x 6"	EA.	1.333
10" x 8"	EA.	2.000
12" x 10"	EA.	2.667
Flanged, 90 deg bend, 125 lb.		
4"	EA.	0.667
6"	EA.	0.800
8"	EA.	1.000
10"	EA.	1.333
12"	EA.	2.000
14"	EA.	2.667
16"	EA.	2.667
Tee		

Utilities

02610.11 Cast Iron Fittings	UNIT	MAN/ HOURS
4"	EA.	1.000
6"	EA.	1.143
8"	EA.	1.333
10"	EA.	1.600
12"	EA.	2.000
14"	EA.	2.667
16"	EA.	4.000

02610.13 Gate Valves	UNIT	MAN/ HOURS
Gate valve, (AWWA) mechanical joint, with adjustable box		
4" valve	EA.	0.800
6" valve	EA.	0.960
8" valve	EA.	1.200
10" valve	EA.	1.412
12" valve	EA.	1.714
14" valve	EA.	2.000
16" valve	EA.	2.182
18" valve	EA.	2.400
Flanged, with box, post indicator (AWWA)		
4" valve	EA.	0.960
6" valve	EA.	1.091
8" valve	EA.	1.333
10" valve	EA.	1.600
12" valve	EA.	2.000
14" valve	EA.	2.400
16" valve	EA.	3.000

02610.15 Water Meters	UNIT	MAN/ HOURS
Water meter, displacement type		
1"	EA.	0.800
1-1/2"	EA.	0.889
2"	EA.	1.000

02610.17 Corporation Stops	UNIT	MAN/ HOURS
Stop for flared copper service pipe		
3/4"	EA.	0.400
1"	EA.	0.444
1-1/4"	EA.	0.533
1-1/2"	EA.	0.667
2"	EA.	0.800

02610.40 Ductile Iron Pipe	UNIT	MAN/ HOURS
Ductile iron pipe, cement lined, slip-on joints		
4"	L.F.	0.067
6"	L.F.	0.071
8"	L.F.	0.075
10"	L.F.	0.080
12"	L.F.	0.096
14"	L.F.	0.120
16"	L.F.	0.133
18"	L.F.	0.150
20"	L.F.	0.171
Mechanical joint pipe		
4"	L.F.	0.092

Utilities	UNIT	MAN/ HOURS
02610.40 Ductile Iron Pipe		
6"	L.F.	0.100
8"	L.F.	0.109
10"	L.F.	0.120
12"	L.F.	0.160
14"	L.F.	0.185
16"	L.F.	0.218
18"	L.F.	0.240
20"	L.F.	0.267
Fittings, mechanical joint		
90 degree elbow		
4"	EA.	0.533
6"	EA.	0.615
8"	EA.	0.800
10"	EA.	1.143
12"	EA.	1.600
14"	EA.	2.000
16"	EA.	2.667
18"	EA.	3.200
20"	EA.	4.000
45 degree elbow		
4"	EA.	0.533
6"	EA.	0.615
8"	EA.	0.800
10"	EA.	1.143
12"	EA.	1.600
14"	EA.	2.000
16"	EA.	2.667
18"	EA.	4.000
20"	EA.	4.000
Tee		
4"x3"	EA.	1.000
4"x4"	EA.	1.000
6"x3"	EA.	1.143
6"x4"	EA.	1.143
6"x6"	EA.	1.143
8"x4"	EA.	1.333
8"x6"	EA.	1.333
8"x8"	EA.	1.333
10"x4"	EA.	1.600
10"x6"	EA.	1.600
10"x8"	EA.	1.600
10"x10"	EA.	1.600
12"x4"	EA.	2.000
12"x6"	EA.	2.000
12"x8"	EA.	2.000
12"x10"	EA.	2.000
12"x12"	EA.	2.133
14"x4"	EA.	2.286
14"x6"	EA.	2.286
14"x8"	EA.	2.286
14"x10"	EA.	2.286
14"x12"	EA.	2.462
14"x14"	EA.	2.462
16"x4"	EA.	2.667

Utilities	UNIT	MAN/ HOURS
02610.40 Ductile Iron Pipe		
16"x6"	EA.	2.667
16"x8"	EA.	2.667
16"x10"	EA.	2.667
16"x12"	EA.	2.667
16"x14"	EA.	2.667
16"x16"	EA.	2.667
18"x6"	EA.	2.909
18"x8"	EA.	2.909
18"x10"	EA.	2.909
18"x12"	EA.	2.909
18"x14"	EA.	2.909
18"x16"	EA.	2.909
18"x18"	EA.	2.909
20"x6"	EA.	3.200
20"x8"	EA.	3.200
20"x10"	EA.	3.200
20"x12"	EA.	3.200
20"x14"	EA.	3.200
20"x16"	EA.	3.200
20"x18"	EA.	3.200
20"x20"	EA.	3.200
Cross		
4"x3"	EA.	1.333
4"x4"	EA.	1.333
6"x3"	EA.	1.600
6"x4"	EA.	1.600
6"x6"	EA.	1.600
8"x4"	EA.	1.778
8"x6"	EA.	1.778
8"x8"	EA.	1.778
10"x4"	EA.	2.000
10"x6"	EA.	2.000
10"x8"	EA.	2.000
10"x10"	EA.	2.000
12"x4"	EA.	2.286
12"x6"	EA.	2.286
12"x8"	EA.	2.286
12"x10"	EA.	2.462
12"x12"	EA.	2.462
14"x4"	EA.	2.667
14"x6"	EA.	2.667
14"x8"	EA.	2.667
14"x10"	EA.	2.667
14"x12"	EA.	2.909
14"x14"	EA.	2.909
16"x4"	EA.	3.200
16"x6"	EA.	3.200
16"x8"	EA.	3.200
16"x10"	EA.	3.200
16"x12"	EA.	3.200
16"x14"	EA.	3.200
16"x16"	EA.	3.200
18"x6"	EA.	3.556
18"x8"	EA.	3.556

Utilities

02610.40 Ductile Iron Pipe	UNIT	MAN/ HOURS
18"x10"	EA.	3.556
18"x12"	EA.	3.556
18"x14"	EA.	3.556
18"x16"	EA.	3.556
18"x18"	EA.	3.556
20"x6"	EA.	3.810
20"x8"	EA.	3.810
20"x10"	EA.	3.810
20"x12"	EA.	3.810
20"x14"	EA.	3.810
20"x16"	EA.	3.810
20"x18"	EA.	4.000
20"x20"	EA.	4.000

02610.60 Plastic Pipe	UNIT	MAN/ HOURS
PVC, class 150 pipe		
4" dia.	L.F.	0.060
6" dia.	L.F.	0.065
8" dia.	L.F.	0.069
10" dia.	L.F.	0.075
12" dia.	L.F.	0.080
Schedule 40 pipe		
1-1/2" dia.	L.F.	0.047
2" dia.	L.F.	0.050
2-1/2" dia.	L.F.	0.053
3" dia.	L.F.	0.057
4" dia.	L.F.	0.067
6" dia.	L.F.	0.080
90 degree elbows		
1"	EA.	0.133
1-1/2"	EA.	0.133
2"	EA.	0.145
2-1/2"	EA.	0.160
3"	EA.	0.178
4"	EA.	0.200
6"	EA.	0.267
45 degree elbows		
1"	EA.	0.133
1-1/2"	EA.	0.133
2"	EA.	0.145
2-1/2"	EA.	0.160
3"	EA.	0.178
4"	EA.	0.200
6"	EA.	0.267
Tees		
1"	EA.	0.160
1-1/2"	EA.	0.160
2"	EA.	0.178
2-1/2"	EA.	0.200
3"	EA.	0.229
4"	EA.	0.267
6"	EA.	0.320
Couplings		
1"	EA.	0.133

Utilities

02610.60 Plastic Pipe	UNIT	MAN/ HOURS
1-1/2"	EA.	0.133
2"	EA.	0.145
2-1/2"	EA.	0.160
3"	EA.	0.178
4"	EA.	0.200
6"	EA.	0.267
Drainage pipe		
PVC schedule 80		
1" dia.	L.F.	0.047
1-1/2" dia.	L.F.	0.047
ABS, 2" dia.	L.F.	0.050
2-1/2" dia.	L.F.	0.053
3" dia.	L.F.	0.057
4" dia.	L.F.	0.067
6" dia.	L.F.	0.080
8" dia.	L.F.	0.063
10" dia.	L.F.	0.075
12" dia.	L.F.	0.080
90 degree elbows		
1"	EA.	0.133
1-1/2"	EA.	0.133
2"	EA.	0.145
2-1/2"	EA.	0.160
3"	EA.	0.178
4"	EA.	0.200
6"	EA.	0.267
45 degree elbows		
1"	EA.	0.133
1-1/2"	EA.	0.133
2"	EA.	0.145
2-1/2"	EA.	0.160
3"	EA.	0.178
4"	EA.	0.200
6"	EA.	0.267
Tees		
1"	EA.	0.160
1-1/2"	EA.	0.160
2"	EA.	0.178
2-1/2"	EA.	0.200
3"	EA.	0.229
4"	EA.	0.267
6"	EA.	0.320
Couplings		
1"	EA.	0.133
1-1/2"	EA.	0.133
2"	EA.	0.145
2-1/2"	EA.	0.160
3"	EA.	0.178
4"	EA.	0.200
6"	EA.	0.267
Pressure pipe		
PVC, class 200 pipe		
3/4"	L.F.	0.040
1"	L.F.	0.042

Utilities

02610.60 Plastic Pipe

Utilities	UNIT	MAN/ HOURS
1-1/4"	L.F.	0.044
1-1/2"	L.F.	0.047
2"	L.F.	0.050
2-1/2"	L.F.	0.053
3"	L.F.	0.057
4"	L.F.	0.067
6"	L.F.	0.080
8"	L.F.	0.069
90 degree elbows		
3/4"	EA.	0.133
1"	EA.	0.133
1-1/4"	EA.	0.133
1-1/2"	EA.	0.133
2"	EA.	0.145
2-1/2"	EA.	0.160
3"	EA.	0.178
4"	EA.	0.200
6"	EA.	0.267
8"	EA.	0.400
45 degree elbows		
3/4"	EA.	0.133
1"	EA.	0.133
1-1/4"	EA.	0.133
1-1/2"	EA.	0.133
2"	EA.	0.145
2-1/2"	EA.	0.160
3"	EA.	0.178
4"	EA.	0.200
6"	EA.	0.267
8"	EA.	0.400
Tees		
3/4"	EA.	0.160
1"	EA.	0.160
1-1/4"	EA.	0.160
1-1/2"	EA.	0.160
2"	EA.	0.178
2-1/2"	EA.	0.200
3"	EA.	0.229
4"	EA.	0.267
6"	EA.	0.320
8"	EA.	0.444
Couplings		
3/4"	EA.	0.133
1"	EA.	0.133
1-1/4"	EA.	0.133
1-1/2"	EA.	0.133
2"	EA.	0.145
2-1/2"	EA.	0.160
3"	EA.	0.178
4"	EA.	0.178
6"	EA.	0.200
8"	EA.	0.267

Utilities

02610.90 Vitrified Clay Pipe

Utilities	UNIT	MAN/ HOURS
Vitrified clay pipe, extra strength		
6" dia.	L.F.	0.109
8" dia.	L.F.	0.114
10" dia.	L.F.	0.120
12" dia.	L.F.	0.160
15" dia.	L.F.	0.240
18" dia.	L.F.	0.267
24" dia.	L.F.	0.343
30" dia.	L.F.	0.480
36" dia.	L.F.	0.686

02630.10 Tapping Saddles & Sleeves

Utilities	UNIT	MAN/ HOURS
Tapping saddle, tap size to 2"		
4" saddle	EA.	0.400
6" saddle	EA.	0.500
8" saddle	EA.	0.667
10" saddle	EA.	0.800
12" saddle	EA.	1.143
14" saddle	EA.	1.600
Tapping sleeve		
4x4	EA.	0.533
6x4	EA.	0.615
6x6	EA.	0.615
8x4	EA.	0.800
8x6	EA.	0.800
10x4	EA.	0.960
10x6	EA.	0.960
10x8	EA.	0.960
10x10	EA.	1.000
12x4	EA.	1.000
12x6	EA.	1.091
12x8	EA.	1.200
12x10	EA.	1.333
12x12	EA.	1.500
Tapping valve, mechanical joint		
4" valve	EA.	3.000
6" valve	EA.	4.000
8" valve	EA.	6.000
10" valve	EA.	8.000
12" valve	EA.	12.000
Tap hole in pipe		
4" hole	EA.	1.000
6" hole	EA.	1.600
8" hole	EA.	2.667
10" hole	EA.	3.200
12" hole	EA.	4.000

02640.15 Valve Boxes

Utilities	UNIT	MAN/ HOURS
Valve box, adjustable, for valves up to 20"		
3' deep	EA.	0.267
4' deep	EA.	0.320
5' deep	EA.	0.400

Utilities	UNIT	MAN/ HOURS
02640.19 Thrust Blocks		
Thrust block, 3000# concrete		
1/4 c.y.	EA.	1.333
1/2 c.y.	EA.	1.600
3/4 c.y.	EA.	2.667
1 c.y.	EA.	5.333
02645.10 Fire Hydrants		
Standard, 3 way post, 6" mechanical joint		
2' deep	EA.	8.000
4' deep	EA.	9.600
6' deep	EA.	12.000
8' deep	EA.	13.714
02665.10 Chilled Water Systems		
Chilled water pipe, 2" thick insulation, w/casing		
Align and tack weld on sleepers		
1-1/2" dia.	L.F.	0.022
3" dia.	L.F.	0.034
4" dia.	L.F.	0.048
6" dia.	L.F.	0.060
8" dia.	L.F.	0.069
10" dia.	L.F.	0.080
12" dia.	L.F.	0.096
14" dia.	L.F.	0.104
16" dia.	L.F.	0.120
Align and tack weld on trench bottom		
18" dia.	L.F.	0.133
20" dia.	L.F.	0.150
Preinsulated fittings		
Align and tack weld on sleepers		
Elbows		
1-1/2"	EA.	0.500
3"	EA.	0.800
4"	EA.	1.000
6"	EA.	1.333
8"	EA.	1.600
Tees		
1-1/2"	EA.	0.533
3"	EA.	0.889
4"	EA.	1.143
6"	EA.	1.600
8"	EA.	2.000
Reducers		
3"	EA.	0.667
4"	EA.	0.800
6"	EA.	1.000
8"	EA.	1.333
Anchors, not including concrete		
4"	EA.	1.000
6"	EA.	1.000
Align and tack weld on trench bottom		
Elbows		
10"	EA.	1.500
12"	EA.	1.714

Utilities	UNIT	MAN/ HOURS
02665.10 Chilled Water Systems		
14"	EA.	1.846
16"	EA.	2.000
18"	EA.	2.182
20"	EA.	2.400
Tees		
10"	EA.	1.500
12"	EA.	1.714
14"	EA.	1.846
16"	EA.	2.000
18"	EA.	2.182
20"	EA.	2.400
Reducers		
10"	EA.	1.000
12"	EA.	1.091
14"	EA.	1.200
16"	EA.	1.333
18"	EA.	1.500
20"	EA.	1.714
Anchors, not including concrete		
10"	EA.	1.000
12"	EA.	1.091
14"	EA.	1.200
16"	EA.	1.333
18"	EA.	1.500
20"	EA.	1.714
02670.10 Wells		
Domestic water, drilled and cased		
4" dia.	L.F.	0.480
6" dia.	L.F.	0.533
8" dia.	L.F.	0.600
02685.10 Gas Distribution		
Gas distribution lines		
Polyethylene, 60 psi coils		
1-1/4" dia.	L.F.	0.053
1-1/2" dia.	L.F.	0.057
2" dia.	L.F.	0.067
3" dia.	L.F.	0.080
30' pipe lengths		
3" dia.	L.F.	0.089
4" dia.	L.F.	0.100
6" dia.	L.F.	0.133
8" dia.	L.F.	0.160
Steel, schedule 40, plain end		
1" dia.	L.F.	0.067
2" dia.	L.F.	0.073
3" dia.	L.F.	0.080
4" dia.	L.F.	0.160
5" dia.	L.F.	0.171
6" dia.	L.F.	0.200
8" dia.	L.F.	0.218
Natural gas meters, direct digital reading, threaded		
250 cfh @ 5 lbs	EA.	1.600

Utilities

02685.10 Gas Distribution	UNIT	MAN/ HOURS
425 cfh @ 10 lbs	EA.	1.600
800 cfh @ 20 lbs	EA.	2.000
1,000 cfh @ 25 lbs	EA.	2.000
1,400 cfh @ 100 lbs	EA.	2.667
2,300 cfh @ 100 lbs	EA.	4.000
5,000 cfh @ 100 lbs	EA.	8.000
Gas pressure regulators		
Threaded		
3/4"	EA.	1.000
1"	EA.	1.333
1-1/4"	EA.	1.333
1-1/2"	EA.	1.333
2"	EA.	1.600
Flanged		
3"	EA.	2.000
4"	EA.	2.667

02690.10 Storage Tanks	UNIT	MAN/ HOURS
Oil storage tank, underground		
Steel		
500 gals	EA.	3.000
1,000 gals	EA.	4.000
4,000 gals	EA.	8.000
5,000 gals	EA.	12.000
10,000 gals	EA.	24.000
Fiberglass, double wall		
550 gals	EA.	4.000
1,000 gals	EA.	4.000
2,000 gals	EA.	6.000
4,000 gals	EA.	12.000
6,000 gals	EA.	16.000
8,000 gals	EA.	24.000
10,000 gals	EA.	30.000
12,000 gals	EA.	40.000
15,000 gals	EA.	53.333
20,000 gals	EA.	60.000
Above ground		
Steel		
275 gals	EA.	2.400
500 gals	EA.	4.000
1,000 gals	EA.	4.800
1,500 gals	EA.	6.000
2,000 gals	EA.	8.000
5,000 gals	EA.	12.000
Fill cap	EA.	0.800
Vent cap	EA.	0.800
Level indicator	EA.	0.800

02695.80 Steam Meters	UNIT	MAN/ HOURS
In-line turbine, direct reading, 300 lb, flanged		
2"	EA.	1.000
3"	EA.	1.333
4"	EA.	1.600
Threaded, 2"		
5" line	EA.	8.000
6" line	EA.	8.000
8" line	EA.	8.000
10" line	EA.	8.000
12" line	EA.	8.000
14" line	EA.	8.000
16" line	EA.	8.000

Sewerage And Drainage

02720.10 Catch Basins	UNIT	MAN/ HOURS
Standard concrete catch basin		
Cast in place, 3'8" x 3'8", 6" thick wall		
2' deep	EA.	6.000
3' deep	EA.	6.000
4' deep	EA.	8.000
5' deep	EA.	8.000
6' deep	EA.	9.600
4'x4', 8" thick wall, cast in place		
2' deep	EA.	6.000
3' deep	EA.	6.000
4' deep	EA.	8.000
5' deep	EA.	8.000
6' deep	EA.	9.600
Frames and covers, cast iron		
Round		
24" dia.	EA.	2.000
26" dia.	EA.	2.000
28" dia.	EA.	2.000
Rectangular		
23"x23"	EA.	2.000
27"x20"	EA.	2.000
24"x24"	EA.	2.000
26"x26"	EA.	2.000
Curb inlet frames and covers		
27"x27"	EA.	2.000
24"x36"	EA.	2.000
24"x25"	EA.	2.000
24"x22"	EA.	2.000
20"x22"	EA.	2.000
Airfield catch basin frame and grating, galvanized		
2'x4'	EA.	2.000
2'x2'	EA.	2.000

Sewerage And Drainage

02720.40 Storm Drainage

Sewerage And Drainage	UNIT	MAN/ HOURS
Concrete pipe		
Plain, bell and spigot joint, Class II		
6" pipe	L.F.	0.109
8" pipe	L.F.	0.120
10" pipe	L.F.	0.126
12" pipe	L.F.	0.133
15" pipe	L.F.	0.141
18" pipe	L.F.	0.150
21" pipe	L.F.	0.160
24" pipe	L.F.	0.171
Reinforced, class III, tongue and groove joint		
12" pipe	L.F.	0.133
15" pipe	L.F.	0.141
18" pipe	L.F.	0.150
21" pipe	L.F.	0.160
24" pipe	L.F.	0.171
27" pipe	L.F.	0.185
30" pipe	L.F.	0.200
36" pipe	L.F.	0.218
42" pipe	L.F.	0.240
48" pipe	L.F.	0.267
54" pipe	L.F.	0.300
60" pipe	L.F.	0.343
66" pipe	L.F.	0.400
72" pipe	L.F.	0.480
Flared end-section, concrete		
12" pipe	L.F.	0.133
15" pipe	L.F.	0.141
18" pipe	L.F.	0.150
24" pipe	L.F.	0.171
30" pipe	L.F.	0.200
36" pipe	L.F.	0.218
42" pipe	L.F.	0.240
48" pipe	L.F.	0.267
54" pipe	L.F.	0.300
Corrugated metal pipe, coated, paved invert		
16 ga.		
8" pipe	L.F.	0.080
10" pipe	L.F.	0.083
12" pipe	L.F.	0.086
15" pipe	L.F.	0.092
18" pipe	L.F.	0.100
21" pipe	L.F.	0.109
24" pipe	L.F.	0.120
30" pipe	L.F.	0.133
36" pipe	L.F.	0.150
12 ga., 48" pipe	L.F.	0.171
10 ga.		
60" pipe	L.F.	0.200
72" pipe	L.F.	0.240
Galvanized or aluminum, plain		
16 ga.		
8" pipe	L.F.	0.080
10" pipe	L.F.	0.083
12" pipe	L.F.	0.086
15" pipe	L.F.	0.092
18" pipe	L.F.	0.100
24" pipe	L.F.	0.120
30" pipe	L.F.	0.133
36" pipe	L.F.	0.150
12 ga., 48" pipe	L.F.	0.171
10 ga., 60" pipe	L.F.	0.200
Galvanized or aluminum, coated oval arch		
16 ga.		
17" x 13"	L.F.	0.109
21" x 15"	L.F.	0.120
14 ga.		
28" x 20"	L.F.	0.133
35" x 24"	L.F.	0.171
12 ga.		
42" x 29"	L.F.	0.200
57" x 38"	L.F.	0.240
64" x 43"	L.F.	0.253
Oval arch culverts, plain		
16 ga.		
17" x 13"	L.F.	0.109
21" x 15"	L.F.	0.120
14 ga.		
28" x 20"	L.F.	0.133
35" x 24"	L.F.	0.171
12 ga.		
57" x 38"	L.F.	0.200
64" x 43"	L.F.	0.240
71" x 47"	L.F.	0.253
Nestable corrugated metal pipe		
16 ga.		
10" pipe	L.F.	0.083
12" pipe	L.F.	0.086
15" pipe	L.F.	0.092
18" pipe	L.F.	0.100
24" pipe	L.F.	0.120
30" pipe	L.F.	0.133
14 ga., 36" pipe	L.F.	0.150
Headwalls, cast in place, 30 deg wingwall		
12" pipe	EA.	2.000
15" pipe	EA.	2.000
18" pipe	EA.	2.286
24" pipe	EA.	2.286
30" pipe	EA.	2.667
36" pipe	EA.	4.000
42" pipe	EA.	4.000
48" pipe	EA.	5.333
54" pipe	EA.	6.667
60" pipe	EA.	8.000
4" cleanout for storm drain		
4" pipe	EA.	1.000
6" pipe	EA.	1.000
8" pipe	EA.	1.000

Sewerage And Drainage

02720.40 Storm Drainage	UNIT	MAN/ HOURS
Connect new drain line		
To existing manhole	EA.	2.667
To new manhole	EA.	1.600

02720.70 Underdrain	UNIT	MAN/ HOURS
Drain tile, clay		
6" pipe	L.F.	0.053
8" pipe	L.F.	0.056
12" pipe	L.F.	0.060
Porous concrete, standard strength		
6" pipe	L.F.	0.053
8" pipe	L.F.	0.056
12" pipe	L.F.	0.060
15" pipe	L.F.	0.067
18" pipe	L.F.	0.080
Corrugated metal pipe, perforated type		
6" pipe	L.F.	0.060
8" pipe	L.F.	0.063
10" pipe	L.F.	0.067
12" pipe	L.F.	0.071
18" pipe	L.F.	0.075
Perforated clay pipe		
6" pipe	L.F.	0.069
8" pipe	L.F.	0.071
12" pipe	L.F.	0.073
Drain tile, concrete		
6" pipe	L.F.	0.053
8" pipe	L.F.	0.056
12" pipe	L.F.	0.060
Perforated rigid PVC underdrain pipe		
4" pipe	L.F.	0.040
6" pipe	L.F.	0.048
8" pipe	L.F.	0.053
10" pipe	L.F.	0.060
12" pipe	L.F.	0.069
Underslab drainage, crushed stone		
3" thick	S.F.	0.008
4" thick	S.F.	0.009
6" thick	S.F.	0.010
8" thick	S.F.	0.010
Plastic filter fabric for drain lines	S.F.	0.008
Gravel fill in trench, crushed or bank run, 1/2" to 3/4"	C.Y.	0.600

02730.10 Sanitary Sewers	UNIT	MAN/ HOURS
Clay		
6" pipe	L.F.	0.080
8" pipe	L.F.	0.086
10" pipe	L.F.	0.092
12" pipe	L.F.	0.100
PVC		
4" pipe	L.F.	0.060
6" pipe	L.F.	0.063
8" pipe	L.F.	0.067
10" pipe	L.F.	0.071
12" pipe	L.F.	0.075
Cleanout		
4" pipe	EA.	1.000
6" pipe	EA.	1.000
8" pipe	EA.	1.000
Connect new sewer line		
To existing manhole	EA.	2.667
To new manhole	EA.	1.600

02740.10 Drainage Fields	UNIT	MAN/ HOURS
Perforated PVC pipe, for drain field		
4" pipe	L.F.	0.053
6" pipe	L.F.	0.057

02740.50 Septic Tanks	UNIT	MAN/ HOURS
Septic tank, precast concrete		
1000 gals	EA.	4.000
2000 gals	EA.	6.000
5000 gals	EA.	12.000
25,000 gals	EA.	48.000
40,000 gals	EA.	80.000
Leaching pit, precast concrete, 72" diameter		
3' deep	EA.	3.000
6' deep	EA.	3.429
8' deep	EA.	4.000

02760.10 Pipeline Restoration	UNIT	MAN/ HOURS
Relining existing water main		
6" dia.	L.F.	0.240
8" dia.	L.F.	0.253
10" dia.	L.F.	0.267
12" dia.	L.F.	0.282
14" dia.	L.F.	0.300
16" dia.	L.F.	0.320
18" dia.	L.F.	0.343
20" dia.	L.F.	0.369
24" dia.	L.F.	0.400
36" dia.	L.F.	0.480
48" dia.	L.F.	0.533
72" dia.	L.F.	0.600
Replacing in line gate valves		
6" valve	EA.	3.200
8" valve	EA.	4.000
10" valve	EA.	4.800
12" valve	EA.	6.000
16" valve	EA.	6.857
18" valve	EA.	8.000
20" valve	EA.	9.600
24" valve	EA.	12.000
36" valve	EA.	16.000

02 SITEWORK

Site Improvements	UNIT	MAN/ HOURS
02810.40 Lawn Irrigation		
Pipe		
Schedule 40, PVC		
1/2"	L.F.	0.042
3/4"	L.F.	0.044
1"	L.F.	0.046
1-1/4"	L.F.	0.046
1-1/2"	L.F.	0.047
2"	L.F.	0.050
2-1/2"	L.F.	0.053
3"	L.F.	0.057
4"	L.F.	0.067
6"	L.F.	0.080
Fittings		
Tee		
1/2"	EA.	0.133
3/4"	EA.	0.133
1"	EA.	0.133
1-1/4"	EA.	0.145
1-1/2"	EA.	0.160
2"	EA.	0.178
2-1/2"	EA.	0.200
3"	EA.	0.229
4"	EA.	0.267
6"	EA.	0.320
El		
1/2"	EA.	0.123
3/4"	EA.	0.133
1"	EA.	0.133
1-1/4"	EA.	0.133
1-1/2"	EA.	0.133
2"	EA.	0.145
2-1/2"	EA.	0.160
3"	EA.	0.178
4"	EA.	0.200
6"	EA.	0.267
Coupling		
1/2"	EA.	0.123
3/4"	EA.	0.133
1"	EA.	0.133
1-1/4"	EA.	0.133
1-1/2"	EA.	0.133
2"	EA.	0.145
2-1/2"	EA.	0.160
3"	EA.	0.178
4"	EA.	0.200
6"	EA.	0.267
45 El		
1/2"	EA.	0.123
3/4"	EA.	0.133
1"	EA.	0.133
1-1/4"	EA.	0.133
1-1/2"	EA.	0.133
2"	EA.	0.145
2-1/2"	EA.	0.160

Site Improvements	UNIT	MAN/ HOURS
02810.40 Lawn Irrigation		
3"	EA.	0.178
4"	EA.	0.200
6"	EA.	0.267
Riser, 1/2" diameter		
2" (close)	EA.	0.200
3"	EA.	0.200
4"	EA.	0.229
5"	EA.	0.229
6"	EA.	0.229
3/4" diameter		
2" (close)	EA.	0.200
3"	EA.	0.200
4"	EA.	0.229
5"	EA.	0.229
6"	EA.	0.229
1" diameter		
2" (close)	EA.	0.200
3"	EA.	0.200
4"	EA.	0.229
5"	EA.	0.229
6"	EA.	0.229
Valve Box		
Concrete, Square		
12" x 22"	EA.	1.000
18" x 20"	EA.	1.143
24" x 13"	EA.	1.333
Round		
12"	EA.	0.800
Plastic		
Square		
12"	EA.	1.000
18"	EA.	
Round		
6"	EA.	1.000
10"	EA.	1.000
12"	EA.	1.143
Sprinkler, Pop-Up		
Spray		
2" high	EA.	1.333
3" high	EA.	1.333
4" high	EA.	1.600
6" high	EA.	1.600
12" high	EA.	1.600
Rotor		
4" high	EA.	1.333
6" high	EA.	1.600
Impact		
Brass	EA.	1.333
Plastic	EA.	1.600
Shrub Head		
Spray	EA.	1.333
Rotor	EA.	1.600
Time Clocks		
Minimum	EA.	2.000

Site Improvements

02810.40 Lawn Irrigation

Site Improvements	UNIT	MAN/ HOURS
Average	EA.	2.667
Maximum	EA.	4.000
Valves		
Anti-siphon		
Brass		
3/4"	EA.	1.333
1"	EA.	1.333
Plastic		
3/4"	EA.	1.333
1"	EA.	1.333
Ball Valve		
Plastic		
1/2"	EA.	1.333
3/4"	EA.	1.333
1"	EA.	1.333
1-1/2"	EA.	1.600
2"	EA.	1.600
Brass		
1/2"	EA.	1.333
3/4"	EA.	1.333
1"	EA.	1.333
1-1/2"	EA.	1.600
2"	EA.	1.600
Gate valves, Brass		
1/2"	EA.	1.333
3/4"	EA.	1.333
1"	EA.	1.333
1-1/2"	EA.	1.600
2"	EA.	1.600
Vacuum Breakers		
Brass		
3/4"	EA.	2.667
1"	EA.	2.667
1-1/2"	EA.	2.667
2"	EA.	2.667
Plastic		
3/4"	EA.	2.667
1"	EA.	2.667
1-1/2"	EA.	2.667
2"	EA.	2.667
Backflow Preventors, Brass		
3/4"	EA.	26.667
1"	EA.	26.667
1-1/2"	EA.	26.667
2"	EA.	32.000
Pressure Regulators, Brass		
3/4"	EA.	0.800
1"	EA.	0.800
1-1/2"	EA.	0.889
2"	EA.	1.000
Quick Coupler Valve		
3/4"	EA.	1.333
1"	EA.	1.333

Site Improvements

02830.10 Chain Link Fence

Site Improvements	UNIT	MAN/ HOURS
Chain link fence, 9 ga., galvanized, with posts 10' o.c.		
4' high	L.F.	0.057
5' high	L.F.	0.073
6' high	L.F.	0.100
7' high	L.F.	0.123
8' high	L.F.	0.160
For barbed wire with hangers, add		
3 strand	L.F.	0.040
6 strand	L.F.	0.067
Corner or gate post, 3" post		
4' high	EA.	0.267
5' high	EA.	0.296
6' high	EA.	0.348
7' high	EA.	0.400
8' high	EA.	0.444
4" post		
4' high	EA.	0.296
5' high	EA.	0.348
6' high	EA.	0.400
7' high	EA.	0.444
8' high	EA.	0.500
Gate with gate posts, galvanized, 3' wide		
4' high	EA.	2.000
5' high	EA.	2.667
6' high	EA.	2.667
7' high	EA.	4.000
8' high	EA.	4.000
Fabric, galvanized chain link, 2" mesh, 9 ga.		
4' high	L.F.	0.027
5' high	L.F.	0.032
6' high	L.F.	0.040
8' high	L.F.	0.053
Line post, no rail fitting, galvanized, 2-1/2" dia.		
4' high	EA.	0.229
5' high	EA.	0.250
6' high	EA.	0.267
7' high	EA.	0.320
8' high	EA.	0.400
1-7/8" H beam		
4' high	EA.	0.229
5' high	EA.	0.250
6' high	EA.	0.267
7' high	EA.	0.320
8' high	EA.	0.400
2-1/4" H beam		
4' high	EA.	0.229
5' high	EA.	0.250
6' high	EA.	0.267
7' high	EA.	0.320
8' high	EA.	0.400
Vinyl coated, 9 ga., with posts 10' o.c.		
4' high	L.F.	0.057
5' high	L.F.	0.073
6' high	L.F.	0.100

Site Improvements	UNIT	MAN/HOURS
02830.10 Chain Link Fence		
7' high	L.F.	0.123
8' high	L.F.	0.160
For barbed wire w/hangers, add		
3 strand	L.F.	0.040
6 Strand	L.F.	0.067
Corner, or gate post, 4' high		
3" dia.	EA.	0.267
4" dia.	EA.	0.267
6" dia.	EA.	0.320
Gate, with posts, 3' wide		
4' high	EA.	2.000
5' high	EA.	2.667
6' high	EA.	2.667
7' high	EA.	4.000
8' high	EA.	4.000
Line post, no rail fitting, 2-1/2" dia.		
4' high	EA.	0.229
5' high	EA.	0.250
6' high	EA.	0.267
7' high	EA.	0.320
8' high	EA.	0.400
Corner post, no top rail fitting, 4" dia.		
4' high	EA.	0.267
5' high	EA.	0.296
6' high	EA.	0.348
7' high	EA.	0.400
8' high	EA.	0.444
Fabric, vinyl, chain link, 2" mesh, 9 ga.		
4' high	L.F.	0.027
5' high	L.F.	0.032
6' high	L.F.	0.040
8' high	L.F.	0.053
Swing gates, galvanized, 4' high		
Single gate		
3' wide	EA.	2.000
4' wide	EA.	2.000
Double gate		
10' wide	EA.	3.200
12' wide	EA.	3.200
14' wide	EA.	3.200
16' wide	EA.	3.200
18' wide	EA.	4.571
20' wide	EA.	4.571
22' wide	EA.	4.571
24' wide	EA.	5.333
26' wide	EA.	5.333
28' wide	EA.	6.400
30' wide	EA.	6.400
5' high		
Single gate		
3' wide	EA.	2.667
4' wide	EA.	2.667
Double gate		
10' wide	EA.	4.000

Site Improvements	UNIT	MAN/HOURS
02830.10 Chain Link Fence		
12' wide	EA.	4.000
14' wide	EA.	4.000
16' wide	EA.	4.000
18' wide	EA.	4.571
20' wide	EA.	4.571
22' wide	EA.	4.571
24' wide	EA.	5.333
26' wide	EA.	5.333
28' wide	EA.	6.400
30' wide	EA.	6.400
6' high		
Single gate		
3' wide	EA.	2.667
4' wide	EA.	2.667
Double gate		
10' wide	EA.	4.000
12' wide	EA.	4.000
14' wide	EA.	4.000
16' wide	EA.	4.000
18' wide	EA.	4.571
20' wide	EA.	4.571
22' wide	EA.	4.571
24' wide	EA.	5.333
26' wide	EA.	5.333
28' wide	EA.	6.400
30' wide	EA.	6.400
7' high		
Single gate		
3' wide	EA.	4.000
4' wide	EA.	4.000
Double gate		
10' wide	EA.	5.333
12' wide	EA.	5.333
14' wide	EA.	5.333
16' wide	EA.	5.333
18' wide	EA.	6.400
20' wide	EA.	6.400
22' wide	EA.	6.400
24' wide	EA.	8.000
26' wide	EA.	8.000
28' wide	EA.	10.000
30' wide	EA.	10.000
8' high		
Single gate		
3' wide	EA.	4.000
4' wide	EA.	4.000
Double gate		
10' wide	EA.	5.333
12' wide	EA.	5.333
14' wide	EA.	5.333
16' wide	EA.	5.333
18' wide	EA.	6.400
20' wide	EA.	6.400
22' wide	EA.	6.400

Site Improvements

02830.10 Chain Link Fence

Site Improvements	UNIT	MAN/HOURS
24' wide	EA.	8.000
26' wide	EA.	8.000
28' wide	EA.	10.000
30' wide	EA.	10.000
Vinyl coated swing gates, 4' high		
Single gate		
3' wide	EA.	2.000
4' wide	EA.	2.000
Double gate		
10' wide	EA.	3.200
12' wide	EA.	3.200
14' wide	EA.	3.200
16' wide	EA.	3.200
18' wide	EA.	4.571
20' wide	EA.	4.571
22' wide	EA.	4.571
24' wide	EA.	5.333
26' wide	EA.	5.333
28' wide	EA.	6.400
30' wide	EA.	6.400
5' high		
Single gate		
3' wide	EA.	2.667
4' wide	EA.	2.667
Double gate		
10' wide	EA.	4.000
12' wide	EA.	4.000
14' wide	EA.	4.000
16' wide	EA.	4.000
18' wide	EA.	4.571
20' wide	EA.	4.571
22' wide	EA.	4.571
24' wide	EA.	5.333
26' wide	EA.	5.333
28' wide	EA.	6.400
30' wide	EA.	6.400
6' high		
Single gate		
3' wide	EA.	2.667
4' wide	EA.	2.667
Double gate		
10' wide	EA.	4.000
12' wide	EA.	4.000
14' wide	EA.	4.000
16' wide	EA.	4.000
18' wide	EA.	4.571
20' wide	EA.	4.571
22' wide	EA.	4.571
24' wide	EA.	5.333
26' wide	EA.	5.333
28' wide	EA.	6.400
30' wide	EA.	6.400
7' high		
Single gate		

Site Improvements

02830.10 Chain Link Fence

Site Improvements	UNIT	MAN/HOURS
3' wide	EA.	4.000
4' wide	EA.	4.000
Double gate		
10' wide	EA.	5.333
12' wide	EA.	5.333
14' wide	EA.	5.333
16' wide	EA.	5.333
18' wide	EA.	6.400
20' wide	EA.	6.400
22' wide	EA.	6.400
24' wide	EA.	8.000
26' wide	EA.	8.000
28' wide	EA.	10.000
30' wide	EA.	10.000
8' high		
Single gate		
3' wide	EA.	4.000
4' wide	EA.	4.000
Double gate		
10' wide	EA.	5.333
12' wide	EA.	5.333
14' wide	EA.	5.333
16' wide	EA.	5.333
18' wide	EA.	6.400
20' wide	EA.	6.400
22' wide	EA.	6.400
24' wide	EA.	8.000
28' wide	EA.	8.000
30' wide	EA.	10.000
Motor operator for gates, no wiring	EA.	
Drilling fence post holes		
In soil		
By hand	EA.	0.400
By machine auger	EA.	0.200
In rock		
By jackhammer	EA.	2.667
By rock drill	EA.	0.800
Aluminum privacy slats, installed vertically	S.F.	0.020
Post hole, dig by hand	EA.	0.533
Set fence post in concrete	EA.	0.400

02830.70 Recreational Courts

Site Improvements	UNIT	MAN/HOURS
Walls, galvanized steel		
8' high	L.F.	0.160
10' high	L.F.	0.178
12' high	L.F.	0.211
Vinyl coated		
8' high	L.F.	0.160
10' high	L.F.	0.178
12' high	L.F.	0.211
Gates, galvanized steel		
Single, 3' transom		
3'x7'	EA.	4.000
4'x7'	EA.	4.571

Site Improvements

02830.70 Recreational Courts

Site Improvements	UNIT	MAN/ HOURS
5'x7'	EA.	5.333
6'x7'	EA.	6.400
Double, 3' transom		
10'x7'	EA.	16.000
12'x7'	EA.	17.778
14'x7'	EA.	20.000
Double, no transom		
10'x10'	EA.	13.333
12'x10'	EA.	16.000
14'x10'	EA.	17.778
Vinyl coated		
Single, 3' transom		
3'x7'	EA.	4.000
4'x7'	EA.	4.571
5'x7'	EA.	5.333
6'x7'	EA.	6.400
Double, 3'		
10'x7'	EA.	16.000
12'x7'	EA.	17.778
14'x7'	EA.	20.000
Double, no transom		
10'x10'	EA.	13.333
12'x10'	EA.	16.000
14'x10'	EA.	17.778
Wire and miscellaneous metal fences		
Chicken wire, post 4' o.c.		
2" mesh		
4' high	L.F.	0.040
6' high	L.F.	0.053
Galvanized steel		
12 gauge, 2" by 4" mesh, posts 5' o.c.		
3' high	L.F.	0.040
5' high	L.F.	0.050
14 gauge, 1" by 2" mesh, posts 5' o.c.		
3' high	L.F.	0.040
5' high	L.F.	0.050

02840.30 Guardrails

Site Improvements	UNIT	MAN/ HOURS
Pipe bollard, steel pipe, concrete filled, painted		
6" dia.	EA.	0.667
8" dia.	EA.	1.000
12" dia.	EA.	2.667
Corrugated steel, guardrail, galvanized	L.F.	0.040
End section, wrap around or flared	EA.	0.800
Timber guardrail, 4" x 8"	L.F.	0.030
Guard rail, 3 cables, 3/4" dia.		
Steel posts	L.F.	0.120
Wood posts	L.F.	0.096
Steel box beam		
6" x 6"	L.F.	0.133
6" x 8"	L.F.	0.150
Concrete posts	EA.	0.400
Barrel type impact barrier	EA.	0.800
Light shield, 6' high	L.F.	0.160

02840.40 Parking Barriers

Site Improvements	UNIT	MAN/ HOURS
Timber, treated, 4' long		
4" x 4"	EA.	0.667
6" x 6"	EA.	0.800
Precast concrete, 6' long, with dowels		
12" x 6"	EA.	0.400
12" x 8"	EA.	0.444

02840.60 Signage

Site Improvements	UNIT	MAN/ HOURS
Traffic signs		
Reflectorized signs per OSHA standards, including post		
Stop, 24"x24"	EA.	0.533
Yield, 30" triangle	EA.	0.533
Speed limit, 12"x18"	EA.	0.533
Directional, 12"x18"	EA.	0.533
Exit, 12"x18"	EA.	0.533
Entry, 12"x18"	EA.	0.533
Warning, 24"x24"	EA.	0.533
Informational, 12"x18"	EA.	0.533
Handicap parking, 12"x18"	EA.	0.533

02860.40 Recreational Facilities

Site Improvements	UNIT	MAN/ HOURS
Bleachers, outdoor, portable, per seat		
10 tiers		
Minimum	EA.	0.150
Maximum	EA.	0.200
20 tiers		
Minimum	EA.	0.141
Maximum	EA.	0.185
Grandstands, fixed, wood seat, steel frame, per seat		
15 tiers		
Minimum	EA.	0.240
Maximum	EA.	0.400
30 tiers		
Minimum	EA.	0.210
Maximum	EA.	0.343
Seats		
Seat backs only		
Fiberglass	EA.	0.080
Steel and wood seat	EA.	0.080
Seat restoration, fiberglass on wood		
Seats	EA.	0.160
Plain bench, no backs	EA.	0.067
Benches		
Park, precast concrete with backs		
4' long	EA.	2.667
8' long	EA.	4.000
Fiberglass, with backs		
4' long	EA.	2.000
8' long	EA.	2.667
Wood, with backs and fiberglass supports		
4' long	EA.	2.000
8' long	EA.	2.667
Steel frame, 6' long		
All steel	EA.	2.000

Site Improvements

02860.40 Recreational Facilities

Site Improvements	UNIT	MAN/ HOURS
Hardwood boards	EA.	2.000
Players bench (no back), steel frame, fir seat, 10' long	EA.	2.667
Soccer goal posts	PAIR	
Running track		
Gravel and cinders over stone base	S.Y.	0.060
Rubber-cork base resilient pavement	S.Y.	0.480
For colored surfaces, add	S.Y.	0.048
Colored rubberized asphalt	S.Y.	0.600
Artificial resilient mat over asphalt	S.Y.	1.200
Tennis courts		
Bituminous pavement, 2-1/2" thick	S.Y.	0.150
Colored sealer, acrylic emulsion		
3 coats	S.Y.	0.053
For 2 color seal coating, add	S.Y.	0.008
For preparing old courts, add	S.Y.	0.005
Net, nylon, 42' long	EA.	1.000
Paint markings on asphalt, 2 coats	EA.	8.000
Playground equipment		
Basketball backboard		
Minimum	EA.	2.000
Maximum	EA.	2.286
Bike rack, 10' long	EA.	1.600
Golf shelter, fiberglass	EA.	2.000
Ground socket for movable posts		
Minimum	EA.	0.500
Maximum	EA.	0.500
Horizontal monkey ladder, 14' long	EA.	1.333
Posts, tether ball	EA.	0.400
Multiple purpose, 10' long	EA.	0.800
See-saw, steel		
Minimum	EA.	3.200
Average	EA.	4.000
Maximum	EA.	5.333
Slide		
Minimum	EA.	6.400
Maximum	EA.	7.273
Swings, plain seats		
8' high		
Minimum	EA.	5.333
Maximum	EA.	6.154
12' high		
Minimum	EA.	6.154
Maximum	EA.	8.889

02870.10 Prefabricated Planters

Site Improvements	UNIT	MAN/ HOURS
Concrete precast, circular		
24" dia., 18" high	EA.	0.800
42" dia., 30" high	EA.	1.000
Fiberglass, circular		
36" dia., 27" high	EA.	0.400
60" dia., 39" high	EA.	0.444
Tapered, circular		
24" dia., 36" high	EA.	0.364
40" dia., 36" high	EA.	0.400
Square		
2' by 2', 17" high	EA.	0.364
4' by 4', 39" high	EA.	0.444
Rectangular		
4' by 1', 18" high	EA.	0.400

Landscaping

02910.10 Shrub & Tree Maintenance

Landscaping	UNIT	MAN/ HOURS
Moving shrubs on site		
12" ball	EA.	1.000
24" ball	EA.	1.333
3' high	EA.	0.800
4' high	EA.	0.889
5' high	EA.	1.000
18" spread	EA.	1.143
30" spread	EA.	1.333
Moving trees on site		
24" ball	EA.	1.200
48" ball	EA.	1.600
Trees		
3' high	EA.	0.480
6' high	EA.	0.533
8' high	EA.	0.600
10' high	EA.	0.800
Palm trees		
7' high	EA.	0.600
10' high	EA.	0.800
20' high	EA.	2.400
40' high	EA.	4.800
Guying trees		
4" dia.	EA.	0.400
8" dia.	EA.	0.500

02920.10 Topsoil

Landscaping	UNIT	MAN/ HOURS
Spread topsoil, with equipment		
Minimum	C.Y.	0.080
Maximum	C.Y.	0.100
By hand		
Minimum	C.Y.	0.800
Maximum	C.Y.	1.000
Area preparation for seeding (grade, rake and clean)		
Square yard	S.Y.	0.006
By acre	ACRE	32.000
Remove topsoil and stockpile on site		
4" deep	C.Y.	0.067
6" deep	C.Y.	0.062

Landscaping

Landscaping	UNIT	MAN/ HOURS
02920.10 Topsoil		
Spreading topsoil from stock pile		
By loader	C.Y.	0.073
By hand	C.Y.	0.800
Top dress by hand	S.Y.	0.008
Place imported top soil		
By loader		
4" deep	S.Y.	0.008
6" deep	S.Y.	0.009
By hand		
4" deep	S.Y.	0.089
6" deep	S.Y.	0.100
Plant bed preparation, 18" deep		
With backhoe/loader	S.Y.	0.020
By hand	S.Y.	0.133
02930.30 Seeding		
Mechanical seeding, 175 lb/acre		
By square yard	S.Y.	0.002
By acre	ACRE	8.000
450 lb/acre		
By square yard	S.Y.	0.002
By acre	ACRE	10.000
Seeding by hand, 10 lb per 100 s.y.		
By square yard	S.Y.	0.003
By acre	ACRE	13.333
Reseed disturbed areas	S.F.	0.004
02950.10 Plants		
Euonymus coloratus, 18" (Purple Wintercreeper)	EA.	0.133
Hedera Helix, 2-1/4" pot (English ivy)	EA.	0.133
Liriope muscari, 2" clumps	EA.	0.080
Santolina, 12"	EA.	0.080
Vinca major or minor, 3" pot	EA.	0.080
Cortaderia argentia, 2 gallon (Pampas Grass)	EA.	0.080
Ophiopogan japonicus, 1 quart (4" pot)	EA.	0.080
Ajuga reptans, 2-3/4" pot (carpet bugle)	EA.	0.080
Pachysandra terminalis, 2-3/4" pot (Japanese Spurge)	EA.	0.080
02950.30 Shrubs		
Juniperus conferia litoralis, 18"-24" (Shore Juniper)	EA.	0.320
Horizontalis plumosa, 18"-24" (Andorra Juniper)	EA.	0.320
Sabina tamar-iscfolia-tamarix juniper, 18"-24"	EA.	0.320
Chin San Jose, 18"-24" (San Jose Juniper)	EA.	0.320
Sargenti, 18"-24" (Sargent's Juniper)	EA.	0.320
Nandina domestica, 18"-24" (Heavenly Bamboo)	EA.	0.320
Raphiolepis Indica Springtime, 18"-24" (Indian Hawthorn	EA.	0.320
Osmanthus Heterophyllus Gulftide, 18"-24" (Osmanthus)	EA.	0.320
Ilex Cornuta Burfordi Nana, 18"-24" (Dwarf Burford Holly)	EA.	0.320
Glabra, 18"-24" (Inkberry Holly)	EA.	0.320
Azalea, Indica types, 18"-24"	EA.	0.320
Kurume types, 18"-24"	EA.	0.320
Berberis Julianae, 18"-24" (Wintergreen Barberry)	EA.	0.320
Pieris Japonica Japanese, 18"-24" (Japanese Pieris)	EA.	0.320
Ilex Cornuta Rotunda, 18"-24" (Dwarf Chinese Holly)	EA.	0.320

Landscaping	UNIT	MAN/ HOURS
02950.30 Shrubs		
Juniperus Horizontalis Plumosa, 24"-30" (Andorra Juniper)	EA.	0.400
Rhodopendrow Hybrids, 24"-30"	EA.	0.400
Aucuba Japonica Varigata, 24"-30" (Gold Dust Aucuba)	EA.	0.400
Ilex Crenata Willow Leaf, 24"-30" (Japanese Holly)	EA.	0.400
Cleyera Japonica, 30"-36" (Japanese Cleyera)	EA.	0.500
Pittosporum Tobira, 30"-36"	EA.	0.500
Prumus Laurocerasus, 30"-36"	EA.	0.500
Ilex Cornuta Burfordi, 30"-36" (Burford Holly)	EA.	0.500
Abelia Grandiflora, 24"-36" (Yew Podocarpus)	EA.	0.400
Podocarpos Macrophylla, 24"-36" (Yew Podocarpus)	EA.	0.400
Pyracantha Coccinea Lalandi, 3'-4' (Firethorn)	EA.	0.500
Photinia Frazieri, 3'-4' (Red Photinia)	EA.	0.500
Forsythia Suspensa, 3'-4' (Weeping Forsythia)	EA.	0.500
Camellia Japonica, 3'-4' (Common Camellia)	EA.	0.500
Juniperus Chin Torulosa, 3'-4' (Hollywood Juniper)	EA.	0.500
Cupressocyparis Leylandi, 3'-4'	EA.	0.500
Ilex Opaca Fosteri, 5'-6' (Foster's Holly)	EA.	0.667
Opaca, 5'-6' (American Holly)	EA.	0.667
Nyrica Cerifera, 4'-5' (Southern Wax Myrtles)	EA.	0.571
Ligustrum Japonicum, 4'-5' (Japanese Privet)	EA.	0.571
02950.60 Trees		
Cornus Florida, 5'-6' (White flowering Dogwood)	EA.	0.667
Prunus Serrulata Kwanzan, 6'-8' (Kwanzan Cherry)	EA.	0.800
Caroliniana, 6'-8' (Carolina Cherry Laurel)	EA.	0.800
Cercis Canadensis, 6'-8' (Eastern Redbud)	EA.	0.800
Koelreuteria Paniculata, 8'-10' (Goldenrain Tree)	EA.	1.000
Acer Platanoides, 1-3/4"-2" (11'-13') (Norway Maple)	EA.	1.333
Rubrum, 1-3/4"-2" (11'-13') (Red Maple)	EA.	1.333
Saccharum, 1-3/4"-2" (Sugar Maple)	EA.	1.333
Fraxinus Pennsylvanica, 1-3/4"-2" (Laneolata-Green Ash)	EA.	1.333
Celtis Occidentalis, 1-3/4"-2" (American Hackberry)	EA.	1.333
Glenditsia Triacantos Inermis, 2"	EA.	1.333
Prunus Cerasifera 'Thundercloud', 6'-8'	EA.	0.800
Yeodensis, 6'-8' (Yoshino Cherry)	EA.	0.800
Lagerstroemia Indica, 8'-10' (Crapemyrtle)	EA.	1.000
Crataegus Phaenopyrum, 8'-10' (Washington Hawthorn)	EA.	1.000
Quercus Borealis, 1-3/4"-2" (Northern Red Oak)	EA.	1.333
Quercus Acutissima, 1-3/4"-2" (8'-10') (Sawtooth Oak)	EA.	1.333
Saliz Babylonica, 1-3/4"-2" (Weeping Willow)	EA.	1.333
Tilia Cordata Greenspire, 1-3/4"-2" (10'-12')	EA.	1.333
Malus, 2"-2-1/2" (8'-10') (Flowering Crabapple)	EA.	1.333
Platanus Occidentalis, (12'-14')	EA.	1.600
Pyrus Calleryana Bradford, 2"-2-1/2" (Bradford Pear)	EA.	1.333
Quercus Palustris, 2"-2-1/2" (12'-14') (Pin Oak)	EA.	1.333
Phellos, 2-1/2"-3" (Willow Oak)	EA.	1.600
Nigra, 2"-2-1/2" (Water Oak)	EA.	1.333
Magnolia Soulangeana, 4'-5' (Saucer Magnolia)	EA.	0.667
Grandiflora, 6'-8' (Southern Magnolia)	EA.	0.800
Cedrus Deodara, 10'-12' (Deodare Cedar)	EA.	1.333
Gingko Biloba, 10'-12' (2"-2-1/2") (Maidenhair Tree)	EA.	1.333
Pinus Thunbergi, 5'-6' (Japanese Black Pine)	EA.	0.667
Strobus, 6'-8' (White Pine)	EA.	0.800
Taeda, 6'-8' (Loblolly Pine)	EA.	0.800

Landscaping	UNIT	MAN/ HOURS
02950.60 Trees		
Quercus Virginiana, 2"-2-1/2" (Live Oak)	EA.	1.600
02970.10 Fertilizing		
Fertilizing (23#/1000 sf)		
By square yard	S.Y.	0.002
By acre	ACRE	10.000
Liming (70#/1000 sf)		
By square yard	S.Y.	0.003
By acre	ACRE	13.333
02980.10 Landscape Accessories		
Steel edging, 3/16" x 4"	L.F.	0.010
Landscaping stepping stones, 15"x15", white	EA.	0.040
Wood chip mulch	C.Y.	0.533
2" thick	S.Y.	0.016
4" thick	S.Y.	0.023
6" thick	S.Y.	0.029
Gravel mulch, 3/4" stone	C.Y.	0.800
White marble chips, 1" deep	S.F.	0.008
Peat moss		
2" thick	S.Y.	0.018
4" thick	S.Y.	0.027
6" thick	S.Y.	0.033
Landscaping timbers, treated lumber		
4" x 4"	L.F.	0.027
6" x 6"	L.F.	0.029
8" x 8"	L.F.	0.033

Formwork

03110.05 Beam Formwork

Formwork	UNIT	MAN/ HOURS
Beam forms, job built		
Beam bottoms		
1 use	S.F.	0.133
2 uses	S.F.	0.127
3 uses	S.F.	0.123
4 uses	S.F.	0.118
5 uses	S.F.	0.114
Beam sides		
1 use	S.F.	0.089
2 uses	S.F.	0.084
3 uses	S.F.	0.080
4 uses	S.F.	0.076
5 uses	S.F.	0.073

03110.10 Box Culvert Formwork

Formwork	UNIT	MAN/ HOURS
Box culverts, job built		
6' x 6'		
1 use	S.F.	0.080
2 uses	S.F.	0.076
3 uses	S.F.	0.073
4 uses	S.F.	0.070
5 uses	S.F.	0.067
8' x 12'		
1 use	S.F.	0.067
2 uses	S.F.	0.064
3 uses	S.F.	0.062
4 uses	S.F.	0.059
5 uses	S.F.	0.057

03110.15 Column Formwork

Formwork	UNIT	MAN/ HOURS
Column, square forms, job built		
8" x 8" columns		
1 use	S.F.	0.160
2 uses	S.F.	0.154
3 uses	S.F.	0.148
4 uses	S.F.	0.143
5 uses	S.F.	0.138
12" x 12" columns		
1 use	S.F.	0.145
2 uses	S.F.	0.140
3 uses	S.F.	0.136
4 uses	S.F.	0.131
5 uses	S.F.	0.127
16" x 16" columns		
1 use	S.F.	0.133
2 uses	S.F.	0.129
3 uses	S.F.	0.125
4 uses	S.F.	0.121
5 uses	S.F.	0.118
24" x 24" columns		
1 use	S.F.	0.123
2 uses	S.F.	0.119
3 uses	S.F.	0.116
4 uses	S.F.	0.113

Formwork

03110.15 Column Formwork

Formwork	UNIT	MAN/ HOURS
5 uses	S.F.	0.110
36" x 36" columns		
1 use	S.F.	0.114
2 uses	S.F.	0.111
3 uses	S.F.	0.108
4 uses	S.F.	0.105
5 uses	S.F.	0.103
Round fiber forms, 1 use		
10" dia.	L.F.	0.160
12" dia.	L.F.	0.163
14" dia.	L.F.	0.170
16" dia.	L.F.	0.178
18" dia.	L.F.	0.190
24" dia.	L.F.	0.205
30" dia.	L.F.	0.222
36" dia.	L.F.	0.242
42" dia.	L.F.	0.267

03110.18 Curb Formwork

Formwork	UNIT	MAN/ HOURS
Curb forms		
Straight, 6" high		
1 use	L.F.	0.080
2 uses	L.F.	0.076
3 uses	L.F.	0.073
4 uses	L.F.	0.070
5 uses	L.F.	0.067
Curved, 6" high		
1 use	L.F.	0.100
2 uses	L.F.	0.094
3 uses	L.F.	0.089
4 uses	L.F.	0.085
5 uses	L.F.	0.082

03110.20 Elevated Slab Formwork

Formwork	UNIT	MAN/ HOURS
Elevated slab formwork		
Slab, with drop panels		
1 use	S.F.	0.064
2 uses	S.F.	0.062
3 uses	S.F.	0.059
4 uses	S.F.	0.057
5 uses	S.F.	0.055
Floor slab, hung from steel beams		
1 use	S.F.	0.062
2 uses	S.F.	0.059
3 uses	S.F.	0.057
4 uses	S.F.	0.055
5 uses	S.F.	0.053
Floor slab, with pans or domes		
1 use	S.F.	0.073
2 uses	S.F.	0.070
3 uses	S.F.	0.067
4 uses	S.F.	0.064
5 uses	S.F.	0.062
Equipment curbs, 12" high		

03 CONCRETE

Formwork	UNIT	MAN/HOURS
03110.20 Elevated Slab Formwork		
1 use	L.F.	0.080
2 uses	L.F.	0.076
3 uses	L.F.	0.073
4 uses	L.F.	0.070
5 uses	L.F.	0.067
03110.25 Equipment Pad Formwork		
Equipment pad, job built		
1 use	S.F.	0.100
2 uses	S.F.	0.094
3 uses	S.F.	0.089
4 uses	S.F.	0.084
5 uses	S.F.	0.080
03110.35 Footing Formwork		
Wall footings, job built, continuous		
1 use	S.F.	0.080
2 uses	S.F.	0.076
3 uses	S.F.	0.073
4 uses	S.F.	0.070
5 uses	S.F.	0.067
Column footings, spread		
1 use	S.F.	0.100
2 uses	S.F.	0.094
3 uses	S.F.	0.089
4 uses	S.F.	0.084
5 uses	S.F.	0.080
03110.50 Grade Beam Formwork		
Grade beams, job built		
1 use	S.F.	0.080
2 uses	S.F.	0.076
3 uses	S.F.	0.073
4 uses	S.F.	0.070
5 uses	S.F.	0.067
03110.53 Pile Cap Formwork		
Pile cap forms, job built		
Square		
1 use	S.F.	0.100
2 uses	S.F.	0.094
3 uses	S.F.	0.089
4 uses	S.F.	0.084
5 uses	S.F.	0.080
Triangular		
1 use	S.F.	0.114
2 uses	S.F.	0.107
3 uses	S.F.	0.100
4 uses	S.F.	0.094
5 uses	S.F.	0.089

Formwork	UNIT	MAN/HOURS
03110.55 Slab/mat Formwork		
Mat foundations, job built		
1 use	S.F.	0.100
2 uses	S.F.	0.094
3 uses	S.F.	0.089
4 uses	S.F.	0.084
5 uses	S.F.	0.080
Edge forms		
6" high		
1 use	L.F.	0.073
2 uses	L.F.	0.070
3 uses	L.F.	0.067
4 uses	L.F.	0.064
5 uses	L.F.	0.062
12" high		
1 use	L.F.	0.080
2 uses	L.F.	0.076
3 uses	L.F.	0.073
4 uses	L.F.	0.070
5 uses	L.F.	0.067
Formwork for openings		
1 use	S.F.	0.160
2 uses	S.F.	0.145
3 uses	S.F.	0.133
4 uses	S.F.	0.123
5 uses	S.F.	0.114
03110.60 Stair Formwork		
Stairway forms, job built		
1 use	S.F.	0.160
2 uses	S.F.	0.145
3 uses	S.F.	0.133
4 uses	S.F.	0.123
5 uses	S.F.	0.114
Stairs, elevated		
1 use	S.F.	0.160
2 uses	S.F.	0.133
3 uses	S.F.	0.114
4 uses	S.F.	0.107
5 uses	S.F.	0.100
03110.65 Wall Formwork		
Wall forms, exterior, job built		
Up to 8' high wall		
1 use	S.F.	0.080
2 uses	S.F.	0.076
3 uses	S.F.	0.073
4 uses	S.F.	0.070
5 uses	S.F.	0.067
Over 8' high wall		
1 use	S.F.	0.100
2 uses	S.F.	0.094
3 uses	S.F.	0.089
4 uses	S.F.	0.084
5 uses	S.F.	0.080

Formwork

03110.65 Wall Formwork

Description	UNIT	MAN/HOURS
Over 16' high wall		
1 use	S.F.	0.114
2 uses	S.F.	0.107
3 uses	S.F.	0.100
4 uses	S.F.	0.094
5 uses	S.F.	0.089
Radial wall forms		
1 use	S.F.	0.123
2 uses	S.F.	0.114
3 uses	S.F.	0.107
4 uses	S.F.	0.100
5 uses	S.F.	0.094
Retaining wall forms		
1 use	S.F.	0.089
2 uses	S.F.	0.084
3 uses	S.F.	0.080
4 uses	S.F.	0.076
5 uses	S.F.	0.073
Radial retaining wall forms		
1 use	S.F.	0.133
2 uses	S.F.	0.123
3 uses	S.F.	0.114
4 uses	S.F.	0.107
5 uses	S.F.	0.100
Column pier and pilaster		
1 use	S.F.	0.160
2 uses	S.F.	0.145
3 uses	S.F.	0.133
4 uses	S.F.	0.123
5 uses	S.F.	0.114
Interior wall forms		
Up to 8' high		
1 use	S.F.	0.073
2 uses	S.F.	0.070
3 uses	S.F.	0.067
4 uses	S.F.	0.064
5 uses	S.F.	0.062
Over 8' high		
1 use	S.F.	0.089
2 uses	S.F.	0.084
3 uses	S.F.	0.080
4 uses	S.F.	0.076
5 uses	S.F.	0.073
Over 16' high		
1 use	S.F.	0.100
2 uses	S.F.	0.094
3 uses	S.F.	0.089
4 uses	S.F.	0.084
5 uses	S.F.	0.080
Radial wall forms		
1 use	S.F.	0.107
2 uses	S.F.	0.100
3 uses	S.F.	0.094
4 uses	S.F.	0.089

Formwork

03110.65 Wall Formwork

Description	UNIT	MAN/HOURS
5 uses	S.F.	0.084
Curved wall forms, 24" sections		
1 use	S.F.	0.160
2 uses	S.F.	0.145
3 uses	S.F.	0.133
4 uses	S.F.	0.123
5 uses	S.F.	0.114
PVC form liner, per side, smooth finish		
1 use	S.F.	0.067
2 uses	S.F.	0.064
3 uses	S.F.	0.062
4 uses	S.F.	0.057
5 uses	S.F.	0.053

03110.90 Miscellaneous Formwork

Description	UNIT	MAN/HOURS
Keyway forms (5 uses)		
2 x 4	L.F.	0.040
2 x 6	L.F.	0.044
Bulkheads		
Walls, with keyways		
2 piece	L.F.	0.073
3 piece	L.F.	0.080
Elevated slab, with keyway		
2 piece	L.F.	0.067
3 piece	L.F.	0.073
Ground slab, with keyway		
2 piece	L.F.	0.057
3 piece	L.F.	0.062
Chamfer strips		
Wood		
1/2" wide	L.F.	0.018
3/4" wide	L.F.	0.018
1" wide	L.F.	0.018
PVC		
1/2" wide	L.F.	0.018
3/4" wide	L.F.	0.018
1" wide	L.F.	0.018
Radius		
1"	L.F.	0.019
1-1/2"	L.F.	0.019
Reglets		
Galvanized steel, 24 ga.	L.F.	0.032
Metal formwork		
Straight edge forms		
4" high	L.F.	0.050
6" high	L.F.	0.053
8" high	L.F.	0.057
12" high	L.F.	0.062
16" high	L.F.	0.067
Curb form, S-shape		
12" x		
1'-6"	L.F.	0.114
2'	L.F.	0.107
2'-6"	L.F.	0.100

Formwork	UNIT	MAN/ HOURS
03110.90 Miscellaneous Formwork		
3'	L.F.	0.089

Reinforcement	UNIT	MAN/ HOURS
03210.05 Beam Reinforcing		
Beam-girders		
#3 - #4	TON	20.000
#5 - #6	TON	16.000
#7 - #8	TON	13.333
#9 - #10	TON	11.429
#11 - #12	TON	10.667
#13 - #14	TON	10.000
Galvanized		
#3 - #4	TON	20.000
#5 - #6	TON	16.000
#7 - #8	TON	13.333
#9 - #10	TON	11.429
#11 - #12	TON	10.667
#13 - #14	TON	10.000
Epoxy coated		
#3 - #4	TON	22.857
#5 - #6	TON	17.778
#7 - #8	TON	14.545
#9 - #10	TON	12.308
#11 - #12	TON	11.429
#13 - #14	TON	10.667
Bond Beams		
#3 - #4	TON	26.667
#5 - #6	TON	20.000
#7 - #8	TON	17.778
Galvanized		
#3 - #4	TON	26.667
#5 - #6	TON	20.000
#7 - #8	TON	17.778
Epoxy coated		
#3 - #4	TON	32.000
#5 - #6	TON	22.857
#7 - #8	TON	20.000
03210.10 Box Culvert Reinforcing		
Box culverts		
#3 - #4	TON	10.000
#5 - #6	TON	8.889
#7 - #8	TON	8.000
#9 - #10	TON	7.273
#11 - #12	TON	6.667
Galvanized		

Reinforcement	UNIT	MAN/ HOURS
03210.10 Box Culvert Reinforcing		
#3 - #4	TON	10.000
#5 - #6	TON	8.889
#7 - #8	TON	8.000
#9 - #10	TON	7.273
#11 - #12	TON	6.667
Epoxy coated		
#3 - #4	TON	10.667
#5 - #6	TON	9.412
#7 - #8	TON	8.421
#9 - #10	TON	7.619
#11 - #12	TON	6.957
03210.15 Column Reinforcing		
Columns		
#3 - #4	TON	22.857
#5 - #6	TON	17.778
#7 - #8	TON	16.000
#9 - #10	TON	14.545
#11 - #12	TON	13.333
#13 - #14	TON	12.308
#15 - #16	TON	11.429
Galvanized		
#3 - #4	TON	22.857
#5 - #6	TON	17.778
#7 - #8	TON	16.000
#9 - #10	TON	14.545
#11 - #12	TON	13.333
#13 - #14	TON	12.308
#15 - #16	TON	11.429
Epoxy coated		
#3 - #4	TON	26.667
#5 - #6	TON	20.000
#7 - #8	TON	17.778
#9 - #10	TON	16.000
#11 - #12	TON	14.545
#13 - #14	TON	13.333
#15 - #16	TON	12.308
Spirals		
8" to 24" dia.	TON	20.000
24" to 48" dia.	TON	17.778
48" to 84" dia.	TON	16.000
03210.20 Elevated Slab Reinforcing		
Elevated slab		
#3 - #4	TON	10.000
#5 - #6	TON	8.889
#7 - #8	TON	8.000
#9 - #10	TON	7.273
#11 - #12	TON	6.667
Galvanized		
#3 - #4	TON	10.000
#5 - #6	TON	8.889
#7 - #8	TON	8.000
#9 - #10	TON	7.273

Reinforcement	UNIT	MAN/ HOURS
03210.20 Elevated Slab Reinforcing		
#11 - #12	TON	6.667
Epoxy coated		
#3 - #4	TON	10.667
#5 - #6	TON	9.412
#7 - #8	TON	8.421
#9 - #10	TON	7.619
#11 - #12	TON	6.957
03210.25 Equip. Pad Reinforcing		
Equipment pad		
#3 - #4	TON	16.000
#5 - #6	TON	14.545
#7 - #8	TON	13.333
#9 - #10	TON	12.308
#11 - #12	TON	11.429
03210.35 Footing Reinforcing		
Footings		
Grade 50		
#3 - #4	TON	13.333
#5 - #6	TON	11.429
#7 - #8	TON	10.000
#9 - #10	TON	8.889
Grade 60		
#3 - #4	TON	13.333
#5 - #6	TON	11.429
#7 - #8	TON	10.000
#9 - #10	TON	8.889
Grade 70		
#3 - #4	TON	13.333
#5 - #6	TON	11.429
#7 - #8	TON	10.000
#9 - #10	TON	8.889
#11 #12	TON	8.000
Straight dowels, 24" long		
1" dia. (#8)	EA.	0.080
3/4" dia. (#6)	EA.	0.080
5/8" dia. (#5)	EA.	0.067
1/2" dia. (#4)	EA.	0.057
03210.45 Foundation Reinforcing		
Foundations		
#3 - #4	TON	13.333
#5 - #6	TON	11.429
#7 - #8	TON	10.000
#9 - #10	TON	8.889
#11 - #12	TON	8.000
Galvanized		
#3 - #4	TON	13.333
#5 - #6	TON	11.429
#7 - #8	TON	10.000
#9 - #10	TON	8.889
#11 - #12	TON	8.000
Epoxy Coated		

Reinforcement	UNIT	MAN/ HOURS
03210.45 Foundation Reinforcing		
#3 - #4	TON	14.545
#5 - #6	TON	12.308
#7 - #8	TON	10.667
#9 - #10	TON	9.412
#11 - #12	TON	8.421
03210.50 Grade Beam Reinforcing		
Grade beams		
#3 - #4	TON	12.308
#5 - #6	TON	10.667
#7 - #8	TON	9.412
#9 - #10	TON	8.421
#11 - #12	TON	7.619
Galvanized		
#3 - #4	TON	12.308
#5 - #6	TON	10.667
#7 - #8	TON	9.412
#9 - #10	TON	8.421
#11 - #12	TON	7.619
Epoxy coated		
#3 - #4	TON	13.333
#5 - #6	TON	11.429
#7 - #8	TON	10.000
#9 - #10	TON	8.889
#11 - #12	TON	8.000
03210.53 Pile Cap Reinforcing		
Pile caps		
#3 - #4	TON	20.000
#5 - #6	TON	17.778
#7 - #8	TON	16.000
#9 - #10	TON	14.545
#11 - #12	TON	13.333
Galvanized		
#3 - #4	TON	20.000
#5 - #6	TON	17.778
#7 - #8	TON	16.000
#9 - #10	TON	14.545
#11 - #12	TON	13.333
Epoxy coated		
#3 - #4	TON	22.857
#5 - #6	TON	20.000
#7 - #8	TON	17.778
#9 - #10	TON	16.000
#11 - #12	TON	14.545
03210.55 Slab/mat Reinforcing		
Bars, slabs		
#3 - #4	TON	13.333
#5 - #6	TON	11.429
#7 - #8	TON	10.000
#9 - #10	TON	8.889
#11 - #12	TON	8.000
Galvanized		

Reinforcement

03210.55 Slab/mat Reinforcing

Reinforcement	UNIT	MAN/HOURS
#3 - #4	TON	13.333
#5 - #6	TON	11.429
#7 - #8	TON	10.000
#9 - #10	TON	8.889
#11 - #12	TON	8.000
Epoxy coated		
#3 - #4	TON	14.545
#5 - #6	TON	12.308
#7 - #8	TON	10.667
#9 - #10	TON	9.412
#11 - #12	TON	8.421
Wire mesh, slabs		
Galvanized		
4x4		
W1.4xW1.4	S.F.	0.005
W2.0xW2.0	S.F.	0.006
W2.9xW2.9	S.F.	0.006
W4.0xW4.0	S.F.	0.007
6x6		
W1.4xW1.4	S.F.	0.004
W2.0xW2.0	S.F.	0.004
W2.9xW2.9	S.F.	0.005
W4.0xW4.0	S.F.	0.005
Standard		
2x2		
W.9xW.9	S.F.	0.005
4x4		
W1.4xW1.4	S.F.	0.005
W2.0xW2.0	S.F.	0.006
W2.9xW2.9	S.F.	0.006
W4.0xW4.0	S.F.	0.007
6x6		
W1.4xW1.4	S.F.	0.004
W2.0xW2.0	S.F.	0.004
W2.9xW2.9	S.F.	0.005
W4.0xW4.0	S.F.	0.005

03210.60 Stair Reinforcing

Reinforcement	UNIT	MAN/HOURS
Stairs		
#3 - #4	TON	16.000
#5 - #6	TON	13.333
#7 - #8	TON	11.429
#9 - #10	TON	10.000
Galvanized		
#3 - #4	TON	16.000
#5 - #6	TON	13.333
#7 - #8	TON	11.429
#9 - #10	TON	10.000
Epoxy coated		
#3 - #4	TON	17.778
#5 - #6	TON	14.545
#7 - #8	TON	12.308
#9 - #10	TON	10.667

Reinforcement

03210.65 Wall Reinforcing

Reinforcement	UNIT	MAN/HOURS
Walls		
#3 - #4	TON	11.429
#5 - #6	TON	10.000
#7 - #8	TON	8.889
#9 - #10	TON	8.000
Galvanized		
#3 - #4	TON	11.429
#5 - #6	TON	10.000
#7 - #8	TON	8.889
#9 - #10	TON	8.000
Epoxy coated		
#3 - #4	TON	12.308
#5 - #6	TON	10.667
#7 - #8	TON	9.412
#9 - #10	TON	8.421
Masonry wall (horizontal)		
#3 - #4	TON	32.000
#5 - #6	TON	26.667
Galvanized		
#3 - #4	TON	32.000
#5 - #6	TON	26.667
Masonry wall (vertical)		
#3 - #4	TON	40.000
#5 - #6	TON	32.000
Galvanized		
#3 - #4	TON	40.000
#5 - #6	TON	32.000

Accessories

03250.40 Concrete Accessories

Accessories	UNIT	MAN/HOURS
Expansion joint, poured		
Asphalt		
1/2" x 1"	L.F.	0.016
1" x 2"	L.F.	0.017
Liquid neoprene, cold applied		
1/2" x 1"	L.F.	0.016
1" x 2"	L.F.	0.018
Polyurethane, 2 parts		
1/2" x 1"	L.F.	0.027
1" x 2"	L.F.	0.029
Rubberized asphalt, cold		
1/2" x 1"	L.F.	0.016
1" x 2"	L.F.	0.017
Hot, fuel resistant		
1/2" x 1"	L.F.	0.016
1" x 2"	L.F.	0.017

Accessories

03250.40 Concrete Accessories

Accessories	UNIT	MAN/ HOURS
Expansion joint, premolded, in slabs		
Asphalt		
1/2" x 6"	L.F.	0.020
1" x 12"	L.F.	0.027
Cork		
1/2" x 6"	L.F.	0.020
1" x 12"	L.F.	0.027
Neoprene sponge		
1/2" x 6"	L.F.	0.020
1" x 12"	L.F.	0.027
Polyethylene foam		
1/2" x 6"	L.F.	0.020
1" x 12"	L.F.	0.027
Polyurethane foam		
1/2" x 6"	L.F.	0.020
1" x 12"	L.F.	0.027
Polyvinyl chloride foam		
1/2" x 6"	L.F.	0.020
1" x 12"	L.F.	0.027
Rubber, gray sponge		
1/2" x 6"	L.F.	0.020
1" x 12"	L.F.	0.027
Asphalt felt control joints or bond breaker, screed joints		
4" slab	L.F.	0.016
6" slab	L.F.	0.018
8" slab	L.F.	0.020
10" slab	L.F.	0.023
Keyed cold expansion and control joints, 24 ga.		
4" slab	L.F.	0.050
5" slab	L.F.	0.050
6" slab	L.F.	0.053
8" slab	L.F.	0.057
10" slab	L.F.	0.062
Waterstops		
Polyvinyl chloride		
Ribbed		
3/16" thick x		
4" wide	L.F.	0.040
6" wide	L.F.	0.044
1/2" thick x		
9" wide	L.F.	0.050
Ribbed with center bulb		
3/16" thick x 9" wide	L.F.	0.050
3/8" thick x 9" wide	L.F.	0.050
Dumbbell type, 3/8" thick x 6" wide	L.F.	0.044
Plain, 3/8" thick x 9" wide	L.F.	0.050
Center bulb, 3/8" thick x 9" wide	L.F.	0.050
Rubber		
Flat dumbbell		
3/8" thick x		
6" wide	L.F.	0.044
9" wide	L.F.	0.050
Center bulb		
3/8" thick x		
6" wide	L.F.	0.044
9" wide	L.F.	0.050
Vapor barrier		
4 mil polyethylene	S.F.	0.003
6 mil polyethylene	S.F.	0.003
Gravel porous fill, under floor slabs, 3/4" stone	C.Y.	1.333
Reinforcing accessories		
Beam bolsters		
1-1/2" high, plain	L.F.	0.008
Galvanized	L.F.	0.008
3" high		
Plain	L.F.	0.010
Galvanized	L.F.	0.010
Slab bolsters		
1" high		
Plain	L.F.	0.004
Galvanized	L.F.	0.004
2" high		
Plain	L.F.	0.004
Galvanized	L.F.	0.004
Chairs, high chairs		
3" high		
Plain	EA.	0.020
Galvanized	EA.	0.020
5" high		
Plain	EA.	0.021
Galvanized	EA.	0.021
8" high		
Plain	EA.	0.023
Galvanized	EA.	0.023
12" high		
Plain	EA.	0.027
Galvanized	EA.	0.027
Continuous, high chair		
3" high		
Plain	L.F.	0.005
Galvanized	L.F.	0.005
5" high		
Plain	L.F.	0.006
Galvanized	L.F.	0.006
8" high		
Plain	L.F.	0.006
Galvanized	L.F.	0.006
12" high		
Plain	L.F.	0.007
Galvanized	L.F.	0.007

Cast-in-place Concrete

03350.10 Concrete Finishes	UNIT	MAN/HOURS
Floor finishes		
Broom	S.F.	0.011
Screed	S.F.	0.010
Darby	S.F.	0.010
Steel float	S.F.	0.013
Granolithic topping		
1/2" thick	S.F.	0.036
1" thick	S.F.	0.040
2" thick	S.F.	0.044
Wall finishes		
Burlap rub, with cement paste	S.F.	0.013
Float finish	S.F.	0.020
Etch with acid	S.F.	0.013
Sandblast		
Minimum	S.F.	0.016
Maximum	S.F.	0.016
Bush hammer		
Green concrete	S.F.	0.040
Cured concrete	S.F.	0.062
Break ties and patch holes	S.F.	0.016
Carborundum		
Dry rub	S.F.	0.027
Wet rub	S.F.	0.040
Floor hardeners		
Metallic		
Light service	S.F.	0.010
Heavy service	S.F.	0.013
Non-metallic		
Light service	S.F.	0.010
Heavy service	S.F.	0.013
Rusticated concrete finish		
Beveled edge	L.F.	0.044
Square edge	L.F.	0.057
Solid board concrete finish		
Standard	S.F.	0.067
Rustic	S.F.	0.080

03360.10 Pneumatic Concrete	UNIT	MAN/HOURS
Pneumatic applied concrete (gunite)		
2" thick	S.F.	0.030
3" thick	S.F.	0.040
4" thick	S.F.	0.048
Finish surface		
Minimum	S.F.	0.040
Maximum	S.F.	0.080

03370.10 Curing Concrete	UNIT	MAN/HOURS
Sprayed membrane		
Slabs	S.F.	0.002
Walls	S.F.	0.002
Curing paper		
Slabs	S.F.	0.002
Walls	S.F.	0.002
Burlap		
7.5 oz.	S.F.	0.003
12 oz.	S.F.	0.003

Placing Concrete

03380.05 Beam Concrete	UNIT	MAN/HOURS
Beams and girders		
2500# or 3000# concrete		
By crane	C.Y.	0.960
By pump	C.Y.	0.873
By hand buggy	C.Y.	0.800
3500# or 4000# concrete		
By crane	C.Y.	0.960
By pump	C.Y.	0.873
By hand buggy	C.Y.	0.800
5000# concrete		
By crane	C.Y.	0.960
By pump	C.Y.	0.873
By hand buggy	C.Y.	0.800
Bond beam, 3000# concrete		
By pump		
8" high		
4" wide	L.F.	0.019
6" wide	L.F.	0.022
8" wide	L.F.	0.024
10" wide	L.F.	0.027
12" wide	L.F.	0.030
16" high		
8" wide	L.F.	0.030
10" wide	L.F.	0.034
12" wide	L.F.	0.040
By crane		
8" high		
4" wide	L.F.	0.021
6" wide	L.F.	0.023
8" wide	L.F.	0.024
10" wide	L.F.	0.027
12" wide	L.F.	0.030
16" high		
8" wide	L.F.	0.030
10" wide	L.F.	0.032
12" wide	L.F.	0.037

03380.15 Column Concrete	UNIT	MAN/HOURS
Columns		
2500# or 3000# concrete		
By crane	C.Y.	0.873

Placing Concrete

Placing Concrete	UNIT	MAN/ HOURS
03380.15 Column Concrete		
By pump	C.Y.	0.800
3500# or 4000# concrete		
By crane	C.Y.	0.873
By pump	C.Y.	0.800
5000# concrete		
By crane	C.Y.	0.873
By pump	C.Y.	0.800
03380.20 Elevated Slab Concrete		
Elevated slab		
2500# or 3000# concrete		
By crane	C.Y.	0.480
By pump	C.Y.	0.369
By hand buggy	C.Y.	0.800
3500# or 4000# concrete		
By crane	C.Y.	0.480
By pump	C.Y.	0.369
By hand buggy	C.Y.	0.800
5000# concrete		
By crane	C.Y.	0.480
By pump	C.Y.	0.369
By hand buggy	C.Y.	0.800
Topping		
2500# or 3000# concrete		
By crane	C.Y.	0.480
By pump	C.Y.	0.369
By hand buggy	C.Y.	0.800
3500# or 4000# concrete		
By crane	C.Y.	0.480
By pump	C.Y.	0.369
By hand buggy	C.Y.	0.800
5000# concrete		
By crane	C.Y.	0.480
By pump	C.Y.	0.369
By hand buggy	C.Y.	0.800
03380.25 Equipment Pad Concrete		
Equipment pad		
2500# or 3000# concrete		
By chute	C.Y.	0.267
By pump	C.Y.	0.686
By crane	C.Y.	0.800
3500# or 4000# concrete		
By chute	C.Y.	0.267
By pump	C.Y.	0.686
By crane	C.Y.	0.800
5000# concrete		
By chute	C.Y.	0.267
By pump	C.Y.	0.686
By crane	C.Y.	0.800

Placing Concrete	UNIT	MAN/ HOURS
03380.35 Footing Concrete		
Continuous footing		
2500# or 3000# concrete		
By chute	C.Y.	0.267
By pump	C.Y.	0.600
By crane	C.Y.	0.686
3500# or 4000# concrete		
By chute	C.Y.	0.267
By pump	C.Y.	0.600
By crane	C.Y.	0.686
5000# concrete		
By chute	C.Y.	0.267
By pump	C.Y.	0.600
By crane	C.Y.	0.686
Spread footing		
2500# or 3000# concrete		
Under 5 cy		
By chute	C.Y.	0.267
By pump	C.Y.	0.640
By crane	C.Y.	0.738
Over 5 cy		
By chute	C.Y.	0.200
By pump	C.Y.	0.565
By crane	C.Y.	0.640
3500# or 4000# concrete		
Under 5 c.y.		
By chute	C.Y.	0.267
By pump	C.Y.	0.640
By crane	C.Y.	0.738
Over 5 c.y.		
By chute	C.Y.	0.200
By pump	C.Y.	0.565
By crane	C.Y.	0.640
5000# concrete		
Under 5 c.y.		
By chute	C.Y.	0.267
By pump	C.Y.	0.640
By crane	C.Y.	0.738
Over 5 c.y.		
By chute	C.Y.	0.200
By pump	C.Y.	0.565
By crane	C.Y.	0.640
03380.50 Grade Beam Concrete		
Grade beam		
2500# or 3000# concrete		
By chute	C.Y.	0.267
By crane	C.Y.	0.686
By pump	C.Y.	0.600
By hand buggy	C.Y.	0.800
3500# or 4000# concrete		
By chute	C.Y.	0.267
By crane	C.Y.	0.686
By pump	C.Y.	0.600
By hand buggy	C.Y.	0.800

Placing Concrete

03380.50 Grade Beam Concrete

	UNIT	MAN/HOURS
5000# concrete		
By chute	C.Y.	0.267
By crane	C.Y.	0.686
By pump	C.Y.	0.600
By hand buggy	C.Y.	0.800

03380.53 Pile Cap Concrete

	UNIT	MAN/HOURS
Pile cap		
2500# or 3000 concrete		
By chute	C.Y.	0.267
By crane	C.Y.	0.800
By pump	C.Y.	0.686
By hand buggy	C.Y.	0.800
3500# or 4000# concrete		
By chute	C.Y.	0.267
By crane	C.Y.	0.800
By pump	C.Y.	0.686
By hand buggy	C.Y.	0.800
5000# concrete		
By chute	C.Y.	0.267
By crane	C.Y.	0.800
By pump	C.Y.	0.686
By hand buggy	C.Y.	0.800

03380.55 Slab/mat Concrete

	UNIT	MAN/HOURS
Slab on grade		
2500# or 3000# concrete		
By chute	C.Y.	0.200
By crane	C.Y.	0.400
By pump	C.Y.	0.343
By hand buggy	C.Y.	0.533
3500# or 4000# concrete		
By chute	C.Y.	0.200
By crane	C.Y.	0.400
By pump	C.Y.	0.343
By hand buggy	C.Y.	0.533
5000# concrete		
By chute	C.Y.	0.200
By crane	C.Y.	0.400
By pump	C.Y.	0.343
By hand buggy	C.Y.	0.533
Foundation mat		
2500# or 3000# concrete, over 20 cy		
By chute	C.Y.	0.160
By crane	C.Y.	0.343
By pump	C.Y.	0.300
By hand buggy	C.Y.	0.400

03380.58 Sidewalks

	UNIT	MAN/HOURS
Walks, cast in place with wire mesh, base not incl.		
4" thick	S.F.	0.027
5" thick	S.F.	0.032
6" thick	S.F.	0.040

Placing Concrete

03380.60 Stair Concrete

	UNIT	MAN/HOURS
Stairs		
2500# or 3000# concrete		
By chute	C.Y.	0.267
By crane	C.Y.	0.800
By pump	C.Y.	0.686
By hand buggy	C.Y.	0.800
3500# or 4000# concrete		
By chute	C.Y.	0.267
By crane	C.Y.	0.800
By pump	C.Y.	0.686
By hand buggy	C.Y.	0.800
5000# concrete		
By chute	C.Y.	0.267
By crane	C.Y.	0.800
By pump	C.Y.	0.686
By hand buggy	C.Y.	0.800

03380.65 Wall Concrete

	UNIT	MAN/HOURS
Walls		
2500# or 3000# concrete		
To 4'		
By chute	C.Y.	0.229
By crane	C.Y.	0.800
By pump	C.Y.	0.738
To 8'		
By crane	C.Y.	0.873
By pump	C.Y.	0.800
To 16'		
By crane	C.Y.	0.960
By pump	C.Y.	0.873
Over 16'		
By crane	C.Y.	1.067
By pump	C.Y.	0.960
3500# or 4000# concrete		
To 4'		
By chute	C.Y.	0.229
By crane	C.Y.	0.800
By pump	C.Y.	0.738
To 8'		
By crane	C.Y.	0.873
By pump	C.Y.	0.800
To 16'		
By crane	C.Y.	0.960
By pump	C.Y.	0.873
Over 16'		
By crane	C.Y.	1.067
By pump	C.Y.	0.960
5000# concrete		
To 4'		
By chute	C.Y.	0.229
By crane	C.Y.	0.800
By pump	C.Y.	0.738
To 8'		
By crane	C.Y.	0.873

Placing Concrete

03380.65 Wall Concrete

	UNIT	MAN/ HOURS
By pump	C.Y.	0.800
To 16'		
By crane	C.Y.	0.960
By pump	C.Y.	0.873
Filled block (CMU)		
3000# concrete, by pump		
4" wide	S.F.	0.034
6" wide	S.F.	0.040
8" wide	S.F.	0.048
10" wide	S.F.	0.056
12" wide	S.F.	0.069
Pilasters, 3000# concrete	C.F.	0.960
Wall cavity, 2" thick, 3000# concrete	S.F.	0.032

Precast Concrete

03400.10 Precast Beams

	UNIT	MAN/ HOURS
Prestressed, double tee, 24" deep, 8' wide		
35' span		
115 psf	S.F.	0.008
140 psf	S.F.	0.008
40' span		
80 psf	S.F.	0.009
143 psf	S.F.	0.009
45' span		
50 psf	S.F.	0.007
70 psf	S.F.	0.007
100 psf	S.F.	0.007
130 psf	S.F.	0.007
50' span		
75 psf	S.F.	0.007
100 psf	S.F.	0.007
Precast beams, girders and joists		
1000 lb/lf live load		
10' span	L.F.	0.160
20' span	L.F.	0.096
30' span	L.F.	0.080
3000 lb/lf live load		
10' span	L.F.	0.160
20' span	L.F.	0.096
30' span	L.F.	0.080
5000 lb/lf live load		
10' span	L.F.	0.160
20' span	L.F.	0.096
30' span	L.F.	0.080

Precast Concrete

03400.20 Precast Columns

	UNIT	MAN/ HOURS
Prestressed concrete columns		
10" x 10"		
10' long	EA.	0.960
15' long	EA.	1.000
20' long	EA.	1.067
25' long	EA.	1.143
30' long	EA.	1.200
12" x 12"		
20' long	EA.	1.200
25' long	EA.	1.297
30' long	EA.	1.371
16" x 16"		
20' long	EA.	1.200
25' long	EA.	1.297
30' long	EA.	1.371
20" x 20"		
20' long	EA.	1.263
25' long	EA.	1.333
30' long	EA.	1.412
24" x 24"		
20' long	EA.	1.333
25' long	EA.	1.412
30' long	EA.	1.500
28" x 28"		
20' long	EA.	1.500
25' long	EA.	1.600
30' long	EA.	1.714
32" x 32"		
20' long	EA.	1.600
25' long	EA.	1.714
30' long	EA.	1.846
36" x 36"		
20' long	EA.	1.714
25' long	EA.	1.846
30' long	EA.	2.000

03400.30 Precast Slabs

	UNIT	MAN/ HOURS
Prestressed flat slab		
6" thick, 4' wide		
20' span		
80 psf	S.F.	0.020
110 psf	S.F.	0.020
25' span		
80 psf	S.F.	0.019
Cored slab		
6" thick, 4' wide		
20' span		
80 psf	S.F.	0.020
100 psf	S.F.	0.020
130 psf	S.F.	0.020
8" thick, 4' wide		
25' span		
70 psf	S.F.	0.019
125 psf	S.F.	0.019

Precast Concrete	UNIT	MAN/ HOURS
03400.30 Precast Slabs		
170 psf	S.F.	0.019
30' span		
70 psf	S.F.	0.016
90 psf	S.F.	0.016
35' span		
70 psf	S.F.	0.015
10" thick, 4' wide		
30' span		
75 psf	S.F.	0.016
100 psf	S.F.	0.016
130 psf	S.F.	0.016
35' span		
60 psf	S.F.	0.015
80 psf	S.F.	0.015
120 psf	S.F.	0.015
40' span		
65 psf	S.F.	0.012
Slabs, roof and floor members, 4' wide		
6" thick, 25' span	S.F.	0.019
8" thick, 30' span	S.F.	0.015
10" thick, 40' span	S.F.	0.013
Tee members		
Multiple tee, roof and floor		
Minimum	S.F.	0.012
Maximum	S.F.	0.024
Double tee wall member		
Minimum	S.F.	0.014
Maximum	S.F.	0.027
Single tee		
Short span, roof members		
Minimum	S.F.	0.015
Maximum	S.F.	0.030
Long span, roof members		
Minimum	S.F.	0.012
Maximum	S.F.	0.024
03400.40 Precast Walls		
Wall panel, 8' x 20'		
Gray cement		
Liner finish		
4" wall	S.F.	0.014
5" wall	S.F.	0.014
6" wall	S.F.	0.015
8" wall	S.F.	0.015
Sandblast finish		
4" wall	S.F.	0.014
5" wall	S.F.	0.014
6" wall	S.F.	0.015
8" wall	S.F.	0.015
White cement		
Liner finish		
4" wall	S.F.	0.014
5" wall	S.F.	0.014
6" wall	S.F.	0.015

Precast Concrete	UNIT	MAN/ HOURS
03400.40 Precast Walls		
8" wall	S.F.	0.015
Sandblast finish		
4" wall	S.F.	0.014
5" wall	S.F.	0.014
6" wall	S.F.	0.015
8" wall	S.F.	0.015
Double tee wall panel, 24" deep		
Gray cement		
Liner finish	S.F.	0.016
Sandblast finish	S.F.	0.016
White cement		
Form liner finish	S.F.	0.016
Sandblast finish	S.F.	0.016
Partition panels		
4" wall	S.F.	0.016
5" wall	S.F.	0.016
6" wall	S.F.	0.016
8" wall	S.F.	0.016
Cladding panels		
4" wall	S.F.	0.017
5" wall	S.F.	0.017
6" wall	S.F.	0.017
8" wall	S.F.	0.017
Sandwich panel, 2.5" cladding panel, 2" insulation		
5" wall	S.F.	0.017
6" wall	S.F.	0.017
8" wall	S.F.	0.017
03400.90 Precast Specialties		
Precast concrete, coping, 4' to 8' long		
12" wide	L.F.	0.060
10" wide	L.F.	0.069
Splash block, 30"x12"x4"	EA.	0.400
Stair unit, per riser	EA.	0.400
Sun screen and trellis, 8' long, 12" high		
4" thick blades	EA.	0.300
5" thick blades	EA.	0.300
6" thick blades	EA.	0.320
8" thick blades	EA.	0.320
Bearing pads for precast members, 2" wide strips		
1/8" thick	L.F.	0.003
1/4" thick	L.F.	0.003
1/2" thick	L.F.	0.003
3/4" thick	L.F.	0.004
1" thick	L.F.	0.004
1-1/2" thick	L.F.	0.004

03 CONCRETE

Cementitous Toppings

03550.10 Concrete Toppings

Cementitous Toppings	UNIT	MAN/HOURS
Gypsum fill		
2" thick	S.F.	0.005
2-1/2" thick	S.F.	0.005
3" thick	S.F.	0.005
3-1/2" thick	S.F.	0.005
4" thick	S.F.	0.006
Formboard		
Mineral fiber board		
1" thick	S.F.	0.020
1-1/2" thick	S.F.	0.023
Cement fiber board		
1" thick	S.F.	0.027
1-1/2" thick	S.F.	0.031
Glass fiber board		
1" thick	S.F.	0.020
1-1/2" thick	S.F.	0.023
Poured deck		
Vermiculite or perlite		
1 to 4 mix	C.Y.	0.800
1 to 6 mix	C.Y.	0.738
Vermiculite or perlite		
2" thick		
1 to 4 mix	S.F.	0.005
1 to 6 mix	S.F.	0.005
3" thick		
1 to 4 mix	S.F.	0.007
1 to 6 mix	S.F.	0.007
Concrete plank, lightweight		
2" thick	S.F.	0.024
2-1/2" thick	S.F.	0.024
3-1/2" thick	S.F.	0.027
4" thick	S.F.	0.027
Channel slab, lightweight, straight		
2-3/4" thick	S.F.	0.024
3-1/2" thick	S.F.	0.024
3-3/4" thick	S.F.	0.024
4-3/4" thick	S.F.	0.027
Gypsum plank		
2" thick	S.F.	0.024
3" thick	S.F.	0.024
Cement fiber, T and G planks		
1" thick	S.F.	0.022
1-1/2" thick	S.F.	0.022
2" thick	S.F.	0.024
2-1/2" thick	S.F.	0.024
3" thick	S.F.	0.024
3-1/2" thick	S.F.	0.027
4" thick	S.F.	0.027

Grout

03600.10 Grouting

Grout	UNIT	MAN/HOURS
Grouting for bases		
Nonshrink		
Metallic grout		
1" deep	S.F.	0.160
2" deep	S.F.	0.178
Non-metallic grout		
1" deep	S.F.	0.160
2" deep	S.F.	0.178
Fluid type		
Non-metallic		
1" deep	S.F.	0.160
2" deep	S.F.	0.178
Grouting for joints		
Portland cement grout (1 cement to 3 sand, by volume)		
1/2" joint thickness		
6" wide joints	L.F.	0.027
8" wide joints	L.F.	0.032
1" joint thickness		
4" wide joints	L.F.	0.025
6" wide joints	L.F.	0.028
8" wide joints	L.F.	0.033
Nonshrink, nonmetallic grout		
1/2" joint thickness		
4" wide joint	L.F.	0.023
6" wide joint	L.F.	0.027
8" wide joint	L.F.	0.032
1" joint thickness		
4" wide joint	L.F.	0.025
6" wide joint	L.F.	0.028
8" wide joint	L.F.	0.033

Concrete Restoration

03730.10 Concrete Repair

Concrete Restoration	UNIT	MAN/HOURS
Epoxy grout floor patch, 1/4" thick	S.F.	0.080
Epoxy gel grout	S.F.	0.800
Injection valve, 1 way, threaded plastic	EA.	0.160
Grout crack seal, 2 component	C.F.	0.800
Grout, non shrink	C.F.	0.800
Concrete, epoxy modified		
Sand mix	C.F.	0.320
Gravel mix	C.F.	0.296
Concrete repair		
Soffit repair		
16" wide	L.F.	0.160
18" wide	L.F.	0.167
24" wide	L.F.	0.178

Concrete Restoration	UNIT	MAN/ HOURS
03730.10 Concrete Repair		
30" wide	L.F.	0.190
32" wide	L.F.	0.200
Edge repair		
2" spall	L.F.	0.200
3" spall	L.F.	0.211
4" spall	L.F.	0.216
6" spall	L.F.	0.222
8" spall	L.F.	0.235
9" spall	L.F.	0.267
Crack repair, 1/8" crack	L.F.	0.080
Reinforcing steel repair		
1 bar, 4 ft		
#4 bar	L.F.	0.100
#5 bar	L.F.	0.100
#6 bar	L.F.	0.107
#8 bar	L.F.	0.107
#9 bar	L.F.	0.114
#11 bar	L.F.	0.114
Pile repairs		
Polyethylene wrap		
30 mil thick		
60" wide	S.F.	0.267
72" wide	S.F.	0.320
60 mil thick		
60" wide	S.F.	0.267
80" wide	S.F.	0.364
Pile spall, average repair 3'		
18" x 18"	EA.	0.667
20" x 20"	EA.	0.800

04 MASONRY

Mortar And Grout

04100.10 Masonry Grout	UNIT	MAN/ HOURS
Grout, non shrink, non-metallic, trowelable	C.F.	0.016
Grout door frame, hollow metal		
Single	EA.	0.600
Double	EA.	0.632
Grout-filled concrete block (CMU)		
4" wide	S.F.	0.020
6" wide	S.F.	0.022
8" wide	S.F.	0.024
12" wide	S.F.	0.025
Grout-filled individual CMU cells		
4" wide	L.F.	0.012
6" wide	L.F.	0.012
8" wide	L.F.	0.012
10" wide	L.F.	0.014
12" wide	L.F.	0.014
Bond beams or lintels, 8" deep		
6" thick	L.F.	0.022
8" thick	L.F.	0.024
10" thick	L.F.	0.027
12" thick	L.F.	0.030
Cavity walls		
2" thick	S.F.	0.032
3" thick	S.F.	0.032
4" thick	S.F.	0.034
6" thick	S.F.	0.040

04150.10 Masonry Accessories	UNIT	MAN/ HOURS
Foundation vents	EA.	0.320
Bar reinforcing		
Horizontal		
#3 - #4	Lb.	0.032
#5 - #6	Lb.	0.027
Vertical		
#3 - #4	Lb.	0.040
#5 - #6	Lb.	0.032
Horizontal joint reinforcing		
Truss type		
4" wide, 6" wall	L.F.	0.003
6" wide, 8" wall	L.F.	0.003
8" wide, 10" wall	L.F.	0.003
10" wide, 12" wall	L.F.	0.004
12" wide, 14" wall	L.F.	0.004
Ladder type		
4" wide, 6" wall	L.F.	0.003
6" wide, 8" wall	L.F.	0.003
8" wide, 10" wall	L.F.	0.003
10" wide, 12" wall	L.F.	0.003
Rectangular wall ties		
3/16" dia., galvanized		
2" x 6"	EA.	0.013
2" x 8"	EA.	0.013
2" x 10"	EA.	0.013
2" x 12"	EA.	0.013
4" x 6"	EA.	0.016

Mortar And Grout

04150.10 Masonry Accessories	UNIT	MAN/ HOURS
4" x 8"	EA.	0.016
4" x 10"	EA.	0.016
4" x 12"	EA.	0.016
1/4" dia., galvanized		
2" x 6"	EA.	0.013
2" x 8"	EA.	0.013
2" x 10"	EA.	0.013
2" x 12"	EA.	0.013
4" x 6"	EA.	0.016
4" x 8"	EA.	0.016
4" x 10"	EA.	0.016
4" x 12"	EA.	0.016
"Z" type wall ties, galvanized		
6" long		
1/8" dia.	EA.	0.013
3/16" dia.	EA.	0.013
1/4" dia.	EA.	0.013
8" long		
1/8" dia.	EA.	0.013
3/16" dia.	EA.	0.013
1/4" dia.	EA.	0.013
10" long		
1/8" dia.	EA.	0.013
3/16" dia.	EA.	0.013
1/4" dia.	EA.	0.013
Dovetail anchor slots		
Galvanized steel, filled		
24 ga.	L.F.	0.020
20 ga.	L.F.	0.020
16 oz. copper, foam filled	L.F.	0.020
Dovetail anchors		
16 ga.		
3-1/2" long	EA.	0.013
5-1/2" long	EA.	0.013
12 ga.		
3-1/2" long	EA.	0.013
5-1/2" long	EA.	0.013
Dovetail, triangular galvanized ties, 12 ga.		
3" x 3"	EA.	0.013
5" x 5"	EA.	0.013
7" x 7"	EA.	0.013
7" x 9"	EA.	0.013
Brick anchors		
Corrugated, 3-1/2" long		
16 ga.	EA.	0.013
12 ga.	EA.	0.013
Non-corrugated, 3-1/2" long		
16 ga.	EA.	0.013
12 ga.	EA.	0.013
Cavity wall anchors, corrugated, galvanized		
5" long		
16 ga.	EA.	0.013
12 ga.	EA.	0.013
7" long		

Mortar And Grout

04150.10 Masonry Accessories	UNIT	MAN/ HOURS
28 ga.	EA.	0.013
24 ga.	EA.	0.013
22 ga.	EA.	0.013
16 ga.	EA.	0.013
Mesh ties, 16 ga., 3" wide		
8" long	EA.	0.013
12" long	EA.	0.013
20" long	EA.	0.013
24" long	EA.	0.013

04150.20 Masonry Control Joints	UNIT	MAN/ HOURS
Control joint, cross shaped PVC	L.F.	0.020
Closed cell joint filler		
1/2"	L.F.	0.020
3/4"	L.F.	0.020
Rubber, for		
4" wall	L.F.	0.020
6" wall	L.F.	0.021
8" wall	L.F.	0.022
PVC, for		
4" wall	L.F.	0.020
6" wall	L.F.	0.021
8" wall	L.F.	0.022

04150.50 Masonry Flashing	UNIT	MAN/ HOURS
Through-wall flashing		
5 oz. coated copper	S.F.	0.067
0.030" elastomeric	S.F.	0.053

Unit Masonry

04210.10 Brick Masonry	UNIT	MAN/ HOURS
Standard size brick, running bond		
Face brick, red (6.4/sf)		
Veneer	S.F.	0.133
Cavity wall	S.F.	0.114
9" solid wall	S.F.	0.229
Common brick (6.4/sf)		
Select common for veneers	S.F.	0.133
Back-up		
4" thick	S.F.	0.100
8" thick	S.F.	0.160
Firewall		
12" thick	S.F.	0.267
16" thick	S.F.	0.364
Glazed brick (7.4/sf)		
Veneer	S.F.	0.145
Buff or gray face brick (6.4/sf)		
Veneer	S.F.	0.133
Cavity wall	S.F.	0.114
Jumbo or oversize brick (3/sf)		
4" veneer	S.F.	0.080
4" back-up	S.F.	0.067
8" back-up	S.F.	0.114
12" firewall	S.F.	0.200
16" firewall	S.F.	0.267
Norman brick, red face, (4.5/sf)		
4" veneer	S.F.	0.100
Cavity wall	S.F.	0.089
Chimney, standard brick, including flue		
16" x 16"	L.F.	0.800
16" x 20"	L.F.	0.800
16" x 24"	L.F.	0.800
20" x 20"	L.F.	1.000
20" x 24"	L.F.	1.000
20" x 32"	L.F.	1.143
Window sill, face brick on edge	L.F.	0.200

04210.20 Structural Tile	UNIT	MAN/ HOURS
Structural glazed tile		
6T series, 5-1/2" x 12"		
Glazed on one side		
2" thick	S.F.	0.080
4" thick	S.F.	0.080
6" thick	S.F.	0.089
8" thick	S.F.	0.100
Glazed on two sides		
4" thick	S.F.	0.100
6" thick	S.F.	0.114
Special shapes		
Group 1	S.F.	0.160
Group 2	S.F.	0.160
Group 3	S.F.	0.160
Group 4	S.F.	0.160
Group 5	S.F.	0.160
Fire rated		
4" thick, 1 hr rating	S.F.	0.080
6" thick, 2 hr rating	S.F.	0.089
8W series, 8" x 16"		
Glazed on one side		
2" thick	S.F.	0.053
4" thick	S.F.	0.053
6" thick	S.F.	0.062
8" thick	S.F.	0.062
Glazed on two sides		
4" thick	S.F.	0.067
6" thick	S.F.	0.080
8" thick	S.F.	0.080
Special shapes		
Group 1	S.F.	0.114
Group 2	S.F.	0.114

04 MASONRY

Unit Masonry

04210.20 Structural Tile	UNIT	MAN/ HOURS
Group 3	S.F.	0.114
Group 4	S.F.	0.114
Group 5	S.F.	0.114
Fire rated		
4" thick, 1 hr rating	S.F.	0.114
6" thick, 2 hr rating	S.F.	0.114

04210.60 Pavers, Masonry	UNIT	MAN/ HOURS
Brick walk laid on sand, sand joints		
Laid flat, (4.5 per sf)	S.F.	0.089
Laid on edge, (7.2 per sf)	S.F.	0.133
Precast concrete patio blocks		
2" thick		
Natural	S.F.	0.027
Colors	S.F.	0.027
Exposed aggregates, local aggregate		
Natural	S.F.	0.027
Colors	S.F.	0.027
Granite or limestone aggregate	S.F.	0.027
White tumblestone aggregate	S.F.	0.027
Stone pavers, set in mortar		
Bluestone		
1" thick		
Irregular	S.F.	0.200
Snapped rectangular	S.F.	0.160
1-1/2" thick, random rectangular	S.F.	0.200
2" thick, random rectangular	S.F.	0.229
Slate		
Natural cleft		
Irregular, 3/4" thick	S.F.	0.229
Random rectangular		
1-1/4" thick	S.F.	0.200
1-1/2" thick	S.F.	0.222
Granite blocks		
3" thick, 3" to 6" wide		
4" to 12" long	S.F.	0.267
6" to 15" long	S.F.	0.229
Crushed stone, white marble, 3" thick	S.F.	0.016

04220.10 Concrete Masonry Units	UNIT	MAN/ HOURS
Hollow, load bearing		
4"	S.F.	0.059
6"	S.F.	0.062
8"	S.F.	0.067
10"	S.F.	0.073
12"	S.F.	0.080
Solid, load bearing		
4"	S.F.	0.059
6"	S.F.	0.062
8"	S.F.	0.067
10"	S.F.	0.073
12"	S.F.	0.080
Back-up block, 8" x 16"		
2"	S.F.	0.046

Unit Masonry

04220.10 Concrete Masonry Units	UNIT	MAN/ HOURS
4"	S.F.	0.047
6"	S.F.	0.050
8"	S.F.	0.053
10"	S.F.	0.057
12"	S.F.	0.062
Foundation wall, 8" x 16"		
6"	S.F.	0.057
8"	S.F.	0.062
10"	S.F.	0.067
12"	S.F.	0.073
Solid		
6"	S.F.	0.062
8"	S.F.	0.067
10"	S.F.	0.073
12"	S.F.	0.080
Exterior, styrofoam inserts, standard weight, 8" x 16"		
6"	S.F.	0.062
8"	S.F.	0.067
10"	S.F.	0.073
12"	S.F.	0.080
Lightweight		
6"	S.F.	0.062
8"	S.F.	0.067
10"	S.F.	0.073
12"	S.F.	0.080
Acoustical slotted block		
4"	S.F.	0.073
6"	S.F.	0.073
8"	S.F.	0.080
Filled cavities		
4"	S.F.	0.089
6"	S.F.	0.094
8"	S.F.	0.100
Hollow, split face		
4"	S.F.	0.059
6"	S.F.	0.062
8"	S.F.	0.067
10"	S.F.	0.073
12"	S.F.	0.080
Split rib profile		
4"	S.F.	0.073
6"	S.F.	0.073
8"	S.F.	0.080
10"	S.F.	0.080
12"	S.F.	0.080
High strength block, 3500 psi		
2"	S.F.	0.059
4"	S.F.	0.062
6"	S.F.	0.062
8"	S.F.	0.067
10"	S.F.	0.073
12"	S.F.	0.080
Solar screen concrete block		
4" thick		

04 MASONRY

Unit Masonry

04220.10 Concrete Masonry Units

	UNIT	MAN/ HOURS
6" x 6"	S.F.	0.178
8" x 8"	S.F.	0.160
12" x 12"	S.F.	0.123
8" thick		
8" x 16"	S.F.	0.114
Glazed block		
Cove base, glazed 1 side, 2"	L.F.	0.089
4"	L.F.	0.089
6"	L.F.	0.100
8"	L.F.	0.100
Single face		
2"	S.F.	0.067
4"	S.F.	0.067
6"	S.F.	0.073
8"	S.F.	0.080
10"	S.F.	0.089
12"	S.F.	0.094
Double face		
4"	S.F.	0.084
6"	S.F.	0.089
8"	S.F.	0.100
Corner or bullnose		
2"	EA.	0.100
4"	EA.	0.114
6"	EA.	0.114
8"	EA.	0.133
10"	EA.	0.145
12"	EA.	0.160
Gypsum unit masonry		
Partition blocks (12"x30")		
Solid		
2"	S.F.	0.032
Hollow		
3"	S.F.	0.032
4"	S.F.	0.033
6"	S.F.	0.036
Vertical reinforcing		
4' o.c., add 5% to labor		
2'8" o.c., add 15% to labor		
Interior partitions, add 10% to labor		

04220.90 Bond Beams & Lintels

	UNIT	MAN/ HOURS
Bond beam, no grout or reinforcement		
8" x 16" x		
4" thick	L.F.	0.062
6" thick	L.F.	0.064
8" thick	L.F.	0.067
10" thick	L.F.	0.070
12" thick	L.F.	0.073
Beam lintel, no grout or reinforcement		
8" x 16" x		
10" thick	L.F.	0.080
12" thick	L.F.	0.089
Precast masonry lintel		

Unit Masonry

04220.90 Bond Beams & Lintels

	UNIT	MAN/ HOURS
6 lf, 8" high x		
4" thick	L.F.	0.133
6" thick	L.F.	0.133
8" thick	L.F.	0.145
10" thick	L.F.	0.145
10 lf, 8" high x		
4" thick	L.F.	0.080
6" thick	L.F.	0.080
8" thick	L.F.	0.089
10" thick	L.F.	0.089
Steel angles and plates		
Minimum	Lb.	0.011
Maximum	Lb.	0.020
Various size angle lintels		
1/4" stock		
3" x 3"	L.F.	0.050
3" x 3-1/2"	L.F.	0.050
3/8" stock		
3" x 4"	L.F.	0.050
3-1/2" x 4"	L.F.	0.050
4" x 4"	L.F.	0.050
5" x 3-1/2"	L.F.	0.050
6" x 3-1/2"	L.F.	0.050
1/2" stock		
6" x 4"	L.F.	0.050

04240.10 Clay Tile

	UNIT	MAN/ HOURS
Hollow clay tile, for back-up, 12" x 12"		
Scored face		
Load bearing		
4" thick	S.F.	0.057
6" thick	S.F.	0.059
8" thick	S.F.	0.062
10" thick	S.F.	0.064
12" thick	S.F.	0.067
Non-load bearing		
3" thick	S.F.	0.055
4" thick	S.F.	0.057
6" thick	S.F.	0.059
8" thick	S.F.	0.062
12" thick	S.F.	0.067
Partition, 12" x 12"		
In walls		
3" thick	S.F.	0.067
4" thick	S.F.	0.067
6" thick	S.F.	0.070
8" thick	S.F.	0.073
10" thick	S.F.	0.076
12" thick	S.F.	0.080
Clay tile floors		
4" thick	S.F.	0.044
6" thick	S.F.	0.047
8" thick	S.F.	0.050
10" thick	S.F.	0.053

04 MASONRY

Unit Masonry

04240.10 Clay Tile

Description	UNIT	MAN/ HOURS
12" thick	S.F.	0.057
Terra cotta		
Coping, 10" or 12" wide, 3" thick	L.F.	0.160

04270.10 Glass Block

Description	UNIT	MAN/ HOURS
Glass block, 4" thick		
6" x 6"	S.F.	0.267
8" x 8"	S.F.	0.200
12" x 12"	S.F.	0.160

04295.10 Parging/masonry Plaster

Description	UNIT	MAN/ HOURS
Parging		
1/2" thick	S.F.	0.053
3/4" thick	S.F.	0.067
1" thick	S.F.	0.080

Stone

04400.10 Stone

Description	UNIT	MAN/ HOURS
Rubble stone		
Walls set in mortar		
8" thick	S.F.	0.200
12" thick	S.F.	0.320
18" thick	S.F.	0.400
24" thick	S.F.	0.533
Dry set wall		
0" thick	C.F.	0.100
12" thick	S.F.	0.200
18" thick	S.F.	0.267
24" thick	S.F.	0.320
Cut stone		
Imported marble		
Facing panels		
3/4" thick	S.F.	0.320
1-1/2" thick	S.F.	0.364
2-1/4" thick	S.F.	0.444
Base		
1" thick		
4" high	L.F.	0.400
6" high	L.F.	0.400
Columns, solid		
Plain faced	C.F.	5.333
Fluted	C.F.	5.333
Flooring, travertine, minimum	S.F.	0.123
Average	S.F.	0.160
Maximum	S.F.	0.178
Domestic marble		

Stone

04400.10 Stone

Description	UNIT	MAN/ HOURS
Facing panels		
7/8" thick	S.F.	0.320
1-1/2" thick	S.F.	0.364
2-1/4" thick	S.F.	0.444
Stairs		
12" treads	L.F.	0.400
6" risers	L.F.	0.267
Thresholds, 7/8" thick, 3' long, 4" to 6" wide		
Plain	EA.	0.667
Beveled	EA.	0.667
Window sill		
6" wide, 2" thick	L.F.	0.320
Stools		
5" wide, 7/8" thick	L.F.	0.320
Limestone panels up to 12' x 5', smooth finish		
2" thick	S.F.	0.096
3" thick	S.F.	0.096
4" thick	S.F.	0.096
Miscellaneous limestone items		
Steps, 14" wide, 6" deep	L.F.	0.533
Coping, smooth finish	C.F.	0.267
Sills, lintels, jambs, smooth finish	C.F.	0.320
Granite veneer facing panels, polished		
7/8" thick		
Black	S.F.	0.320
Gray	S.F.	0.320
Base		
4" high	L.F.	0.160
6" high	L.F.	0.178
Curbing, straight, 6" x 16"	L.F.	0.400
Radius curbs, radius over 5'	L.F.	0.533
Ashlar veneer		
4" thick, random	S.F.	0.320
Pavers, 4" x 4" split		
Gray	S.F.	0.160
Pink	S.F.	0.160
Black	S.F.	0.160
Slate, panels		
1" thick	S.F.	0.320
2" thick	S.F.	0.364
Sills or stools		
1" thick		
6" wide	L.F.	0.320
10" wide	L.F.	0.348
2" thick		
6" wide	L.F.	0.364
10" wide	L.F.	0.400

Masonry Restoration

Masonry Restoration	UNIT	MAN/ HOURS
04520.10 Restoration And Cleaning		
Masonry cleaning		
Washing brick		
Smooth surface	S.F.	0.013
Rough surface	S.F.	0.018
Steam clean masonry		
Smooth face		
Minimum	S.F.	0.010
Maximum	S.F.	0.015
Rough face		
Minimum	S.F.	0.013
Maximum	S.F.	0.020
Sandblast masonry		
Minimum	S.F.	0.016
Maximum	S.F.	0.027
Pointing masonry		
Brick	S.F.	0.032
Concrete block	S.F.	0.023
Cut and repoint		
Brick		
Minimum	S.F.	0.040
Maximum	S.F.	0.080
Stone work	L.F.	0.062
Cut and recaulk		
Oil base caulks	L.F.	0.053
Butyl caulks	L.F.	0.053
Polysulfides and acrylics	L.F.	0.053
Silicones	L.F.	0.053
Cement and sand grout on walls, to 1/8" thick		
Minimum	S.F.	0.032
Maximum	S.F.	0.040
Brick removal and replacement		
Minimum	EA.	0.100
Average	EA.	0.133
Maximum	EA.	0.400
04550.10 Refractories		
Flue liners		
Rectangular		
8" x 12"	L.F.	0.133
12" x 12"	L.F.	0.145
12" x 18"	L.F.	0.160
16" x 16"	L.F.	0.178
18" x 18"	L.F.	0.190
20" x 20"	L.F.	0.200
24" x 24"	L.F.	0.229
Round		
18" dia.	L.F.	0.190
24" dia.	L.F.	0.229

Metal Fastening

05050.10 Structural Welding

	UNIT	MAN/ HOURS
Welding		
Single pass		
1/8"	L.F.	0.040
3/16"	L.F.	0.053
1/4"	L.F.	0.067
Miscellaneous steel shapes		
Plain	Lb.	0.002
Galvanized	Lb.	0.003
Plates		
Plain	Lb.	0.002
Galvanized	Lb.	0.003

05050.95 Metal Lintels

	UNIT	MAN/ HOURS
Lintels, steel		
Plain	Lb.	0.020
Galvanized	Lb.	0.020

05120.10 Structural Steel

	UNIT	MAN/ HOURS
Beams and girders, A-36		
Welded	TON	4.800
Bolted	TON	4.364
Columns		
Pipe		
6" dia.	Lb.	0.005
12" dia.	Lb.	0.004
Purlins and girts		
Welded	TON	8.000
Bolted	TON	6.857
Column base plates		
Up to 150 lb each	Lb.	0.005
Over 150 lb each	Lb.	0.007
Structural pipe		
3" to 5" o.d.	TON	9.600
6" to 12" o.d.	TON	6.857
Structural tube		
6" square		
Light sections	TON	9.600
Heavy sections	TON	6.857
6" wide rectangular		
Light sections	TON	8.000
Heavy sections	TON	6.000
Greater than 6" wide rectangular		
Light sections	TON	8.000
Heavy sections	TON	6.000
Miscellaneous structural shapes		
Steel angle	TON	12.000
Steel plate	TON	8.000
Trusses, field welded		
60 lb/lf	TON	6.000
100 lb/lf	TON	4.800
150 lb/lf	TON	4.000
Bolted		
60 lb/lf	TON	5.333
100 lb/lf	TON	4.364

Metal Fastening

05120.10 Structural Steel

	UNIT	MAN/ HOURS
150 lb/lf	TON	3.692

05200.10 Metal Joists

	UNIT	MAN/ HOURS
Joist		
DLH series	TON	3.200
K series	TON	3.200
LH series	TON	3.200

05300.10 Metal Decking

	UNIT	MAN/ HOURS
Roof, 1-1/2" deep, non-composite		
16 ga.		
Primed	S.F.	0.008
Galvanized	S.F.	0.008
18 ga.		
Primed	S.F.	0.008
Galvanized	S.F.	0.008
20 ga.		
Primed	S.F.	0.008
Galvanized	S.F.	0.008
22 ga.		
Primed	S.F.	0.008
Galvanized	S.F.	0.008
Open type decking, galvanized		
1-1/2" deep		
18 ga.	S.F.	0.008
20 ga.	S.F.	0.008
22 ga.	S.F.	0.008
3" deep		
16 ga.	S.F.	0.009
18 ga.	S.F.	0.009
20 ga.	S.F.	0.009
22 ga.	S.F.	0.009
4-1/2" deep		
16 ga.	S.F.	0.010
18 ga.	S.F.	0.010
6" deep		
16 ga.	S.F.	0.011
18 ga.	S.F.	0.011
7-1/2" deep		
16 ga.	S.F.	0.011
18 ga.	S.F.	0.011
Cellular type		
1-1/2" deep, galvanized		
18-18 ga.	S.F.	0.010
22-18 ga.	S.F.	0.010
3" deep, galvanized		
16-16 ga.	S.F.	0.011
18-16 ga.	S.F.	0.011
18-18 ga.	S.F.	0.011
20-18 ga.	S.F.	0.011
4-1/2" deep, galvanized		
16-16 ga.	S.F.	0.011
18-16 ga.	S.F.	0.011
18-18 ga.	S.F.	0.011

Metal Fastening	UNIT	MAN/HOURS
05300.10 Metal Decking		
20-18 ga.	S.F.	0.011
Composite deck, non-cellular, galvanized		
1-1/2" deep		
18 ga.	S.F.	0.009
20 ga.	S.F.	0.009
22 ga.	S.F.	0.009
3" deep		
18 ga.	S.F.	0.009
20 ga.	S.F.	0.009
22 ga.	S.F.	0.009
Slab form floor deck		
9/16" deep		
28 ga.	S.F.	0.009
1-5/16" deep		
24 ga.	S.F.	0.009
22 ga.	S.F.	0.009

Cold Formed Framing	UNIT	MAN/HOURS
05410.10 Metal Framing		
Furring channel, galvanized		
Beams and columns, 3/4"		
12" o.c.	S.F.	0.080
16" o.c.	S.F.	0.073
Walls, 3/4"		
12" o.c.	S.F.	0.040
16" o.c.	S.F.	0.033
24" o.c.	S.F.	0.027
1-1/2"		
12" o.c.	S.F.	0.040
16" o.c.	S.F.	0.033
24" o.c.	S.F.	0.027
Stud, load bearing		
16" o.c.		
16 ga.		
2-1/2"	S.F.	0.036
3-5/8"	S.F.	0.036
4"	S.F.	0.036
6"	S.F.	0.040
18 ga.		
2-1/2"	S.F.	0.036
3-5/8"	S.F.	0.036
4"	S.F.	0.036
6"	S.F.	0.040
8"	S.F.	0.040
20 ga.		
2-1/2"	S.F.	0.036

Cold Formed Framing	UNIT	MAN/HOURS
05410.10 Metal Framing		
3-5/8"	S.F.	0.036
4"	S.F.	0.036
6"	S.F.	0.040
8"	S.F.	0.040
24" o.c.		
16 ga.		
2-1/2"	S.F.	0.031
3-5/8"	S.F.	0.031
4"	S.F.	0.031
6"	S.F.	0.033
8"	S.F.	0.033
18 ga.		
2-1/2"	S.F.	0.031
3-5/8"	S.F.	0.031
4"	S.F.	0.031
6"	S.F.	0.033
8"	S.F.	0.033
20 ga.		
2-1/2"	S.F.	0.031
3-5/8"	S.F.	0.031
4"	S.F.	0.031
6"	S.F.	0.033
8"	S.F.	0.033

Metal Fabrications	UNIT	MAN/HOURS
05510.10 Stairs		
Stock unit, steel, complete, per riser		
Tread		
3'-6" wide	EA.	1.000
4' wide	EA.	1.143
5' wide	EA.	1.333
Metal pan stair, cement filled, per riser		
3'-6" wide	EA.	0.800
4' wide	EA.	0.889
5' wide	EA.	1.000
Landing, steel pan	S.F.	0.200
Cast iron tread, steel stringers, stock units, per riser		
Tread		
3'-6" wide	EA.	1.000
4' wide	EA.	1.143
5' wide	EA.	1.333
Stair treads, abrasive, 12" x 3'-6"		
Cast iron		
3/8"	EA.	0.400
1/2"	EA.	0.400
Cast aluminum		

05 METALS

Metal Fabrications

Metal Fabrications	UNIT	MAN/ HOURS
05510.10 Stairs		
5/16"	EA.	0.400
3/8"	EA.	0.400
1/2"	EA.	0.400
05515.10 Ladders		
Ladder, 18" wide		
With cage	L.F.	0.533
Without cage	L.F.	0.400
05520.10 Railings		
Railing, pipe		
1-1/4" diameter, welded steel		
2-rail		
Primed	L.F.	0.160
Galvanized	L.F.	0.160
3-rail		
Primed	L.F.	0.200
Galvanized	L.F.	0.200
Wall mounted, single rail, welded steel		
Primed	L.F.	0.123
Galvanized	L.F.	0.123
1-1/2" diameter, welded steel		
2-rail		
Primed	L.F.	0.160
Galvanized	L.F.	0.160
3-rail		
Primed	L.F.	0.200
Galvanized	L.F.	0.200
Wall mounted, single rail, welded steel		
Primed	L.F.	0.123
Galvanized	L.F.	0.123
2" diameter, welded steel		
2-rail		
Primed	L.F.	0.178
Galvanized	L.F.	0.178
3-rail		
Primed	L.F.	0.229
Galvanized	L.F.	0.229
Wall mounted, single rail, welded steel		
Primed	L.F.	0.133
Galvanized	L.F.	0.133
05530.10 Metal Grating		
Floor plate, checkered, steel		
1/4"		
Primed	S.F.	0.011
Galvanized	S.F.	0.011
3/8"		
Primed	S.F.	0.012
Galvanized	S.F.	0.012
Aluminum grating, pressure-locked bearing bars		
3/4" x 1/8"	S.F.	0.020
1" x 1/8"	S.F.	0.020
1-1/4" x 1/8"	S.F.	0.020

Metal Fabrications	UNIT	MAN/ HOURS
05530.10 Metal Grating		
1-1/4" x 3/16"	S.F.	0.020
1-1/2" x 1/8"	S.F.	0.020
1-3/4" x 3/16"	S.F.	0.020
Miscellaneous expenses		
Cutting		
Minimum	L.F.	0.053
Maximum	L.F.	0.080
Banding		
Minimum	L.F.	0.133
Maximum	L.F.	0.160
Toe plates		
Minimum	L.F.	0.160
Maximum	L.F.	0.200
Steel grating, primed		
3/4" x 1/8"	S.F.	0.027
1" x 1/8"	S.F.	0.027
1-1/4" x 1/8"	S.F.	0.027
1-1/4" x 3/16"	S.F.	0.027
1-1/2" x 1/8"	S.F.	0.027
1-3/4" x 3/16"	S.F.	0.027
Galvanized		
3/4" x 1/8"	S.F.	0.027
1" x 1/8"	S.F.	0.027
1-1/4" x 1/8"	S.F.	0.027
1-1/4" x 3/16"	S.F.	0.027
1-1/2" x 1/8"	S.F.	0.027
1-3/4" x 3/16"	S.F.	0.027
Miscellaneous expenses		
Cutting		
Minimum	L.F.	0.057
Maximum	L.F.	0.089
Banding		
Minimum	L.F.	0.145
Maximum	L.F.	0.178
Toe plates		
Minimum	L.F.	0.178
Maximum	L.F.	0.229
05540.10 Castings		
Miscellaneous castings		
Light sections	Lb.	0.016
Heavy sections	Lb.	0.011
Manhole covers and frames		
Regular, city type		
18" dia.		
100 lb	EA.	1.600
24" dia.		
200 lb	EA.	1.600
300 lb	EA.	1.778
400 lb	EA.	1.778
26" dia., 475 lb	EA.	2.000
30" dia., 600 lb	EA.	2.286
8" square, 75 lb	EA.	0.320
24" square		

Metal Fabrications

05540.10 Castings	UNIT	MAN/ HOURS
126 lb	EA.	1.600
500 lb	EA.	2.000
Watertight type		
20" dia., 200 lb	EA.	2.000
24" dia., 350 lb	EA.	2.667
Steps, cast iron		
7" x 9"	EA.	0.160
8" x 9"	EA.	0.178
Manhole covers and frames, aluminum		
12" x 12"	EA.	0.320
18" x 18"	EA.	0.320
24" x 24"	EA.	0.400
Corner protection		
Steel angle guard with anchors		
2" x 2" x 3/16"	L.F.	0.114
2" x 3" x 1/4"	L.F.	0.114
3" x 3" x 5/16"	L.F.	0.114
3" x 4" x 5/16"	L.F.	0.123
4" x 4" x 5/16"	L.F.	0.123

Misc. Fabrications

05580.10 Metal Specialties	UNIT	MAN/ HOURS
Kick plate		
4" high x 1/4" thick		
Primed	L.F.	0.160
Galvanized	L.F.	0.160
6" high x 1/4" thick		
Primed	L.F.	0.178
Galvanized	L.F.	0.178

05700.10 Ornamental Metal	UNIT	MAN/ HOURS
Railings, vertical square bars, 6" o.c., with shaped top rails		
Steel	L.F.	0.400
Aluminum	L.F.	0.400
Bronze	L.F.	0.533
Stainless steel	L.F.	0.533
Laminated metal or wood handrails with metal supports		
2-1/2" round or oval shape	L.F.	0.400
Grilles and louvers		
Fixed type louvers		
4 through 10 sf	S.F.	0.133
Over 10 sf	S.F.	0.100
Movable type louvers		
4 through 10 sf	S.F.	0.133
Over 10 sf	S.F.	0.100
Aluminum louvers		

Misc. Fabrications

05700.10 Ornamental Metal	UNIT	MAN/ HOURS
Residential use, fixed type, with screen		
8" x 8"	EA.	0.400
12" x 12"	EA.	0.400
12" x 18"	EA.	0.400
14" x 24"	EA.	0.400
18" x 24"	EA.	0.400
30" x 24"	EA.	0.444

05800.10 Expansion Control	UNIT	MAN/ HOURS
Expansion joints with covers, floor assembly type		
With 1" space		
Aluminum	L.F.	0.133
Bronze	L.F.	0.133
Stainless steel	L.F.	0.133
With 2" space		
Aluminum	L.F.	0.133
Bronze	L.F.	0.133
Stainless steel	L.F.	0.133
Ceiling and wall assembly type		
With 1" space		
Aluminum	L.F.	0.160
Bronze	L.F.	0.160
Stainless steel	L.F.	0.160
With 2" space		
Aluminum	L.F.	0.160
Bronze	L.F.	0.160
Stainless steel	L.F.	0.160
Exterior roof and wall, aluminum		
Roof to roof		
With 1" space	L.F.	0.133
With 2" space	L.F.	0.133
Roof to wall		
With 1" space	L.F.	0.145
With 2" space	L.F.	0.145
Flat wall to wall		
With 1" space	L.F.	0.133
With 2" space	L.F.	0.133
Corner to flat wall		
With 1" space	L.F.	0.160
With 2" in space	L.F.	0.160

Fasteners And Adhesives

06050.10 Accessories

Item	UNIT	MAN/ HOURS
Column/post base, cast aluminum		
4" x 4"	EA.	0.200
6" x 6"	EA.	0.200
Bridging, metal, per pair		
12" o.c.	EA.	0.080
16" o.c.	EA.	0.073
Anchors		
Bolts, threaded two ends, with nuts and washers		
1/2" dia.		
4" long	EA.	0.050
7-1/2" long	EA.	0.050
3/4" dia.		
7-1/2" long	EA.	0.050
15" long	EA.	0.050
Framing anchors		
10 gauge	EA.	0.067
Bolts, carriage		
1/4 x 4	EA.	0.080
5/16 x 6	EA.	0.084
3/8 x 6	EA.	0.084
1/2 x 6	EA.	0.084
Joist and beam hangers		
18 ga.		
2 x 4	EA.	0.080
2 x 6	EA.	0.080
2 x 8	EA.	0.080
2 x 10	EA.	0.089
2 x 12	EA.	0.100
16 ga.		
3 x 6	EA.	0.089
3 x 8	EA.	0.089
3 x 10	EA.	0.094
3 x 12	EA.	0.107
3 x 14	EA.	0.114
4 x 6	EA.	0.089
4 x 8	EA.	0.089
4 x 10	EA.	0.094
4 x 12	EA.	0.107
4 x 14	EA.	0.114
Rafter anchors, 18 ga., 1-1/2" wide		
5-1/4" long	EA.	0.067
10-3/4" long	EA.	0.067
Shear plates		
2-5/8" dia.	EA.	0.062
4" dia.	EA.	0.067
Sill anchors		
Embedded in concrete	EA.	0.080
Split rings		
2-1/2" dia.	EA.	0.089
4" dia.	EA.	0.100
Strap ties, 14 ga., 1-3/8" wide		
12" long	EA.	0.067
18" long	EA.	0.073
24" long	EA.	0.080

Fasteners And Adhesives

06050.10 Accessories

Item	UNIT	MAN/ HOURS
36" long	EA.	0.089
Toothed rings		
2-5/8" dia.	EA.	0.133
4" dia.	EA.	0.160

Rough Carpentry

06110.10 Blocking

Item	UNIT	MAN/ HOURS
Steel construction		
Walls		
2x4	L.F.	0.053
2x6	L.F.	0.062
2x8	L.F.	0.067
2x10	L.F.	0.073
2x12	L.F.	0.080
Ceilings		
2x4	L.F.	0.062
2x6	L.F.	0.073
2x8	L.F.	0.080
2x10	L.F.	0.089
2x12	L.F.	0.100
Wood construction		
Walls		
2x4	L.F.	0.044
2x6	L.F.	0.050
2x8	L.F.	0.053
2x10	L.F.	0.057
2x12	L.F.	0.062
Ceilings		
2x4	L.F.	0.050
2x6	L.F.	0.057
2x8	L.F.	0.062
2x10	L.F.	0.067
2x12	L.F.	0.073

06110.20 Ceiling Framing

Item	UNIT	MAN/ HOURS
Ceiling joists		
12" o.c.		
2x4	S.F.	0.019
2x6	S.F.	0.020
2x8	S.F.	0.021
2x10	S.F.	0.022
2x12	S.F.	0.024
16" o.c.		
2x4	S.F.	0.015
2x6	S.F.	0.016
2x8	S.F.	0.017

06 WOOD AND PLASTICS

Rough Carpentry

06110.20 Ceiling Framing

Rough Carpentry	UNIT	MAN/ HOURS
2x10	S.F.	0.017
2x12	S.F.	0.018
24" o.c.		
2x4	S.F.	0.013
2x6	S.F.	0.013
2x8	S.F.	0.014
2x10	S.F.	0.015
2x12	S.F.	0.016
Headers and nailers		
2x4	L.F.	0.026
2x6	L.F.	0.027
2x8	L.F.	0.029
2x10	L.F.	0.031
2x12	L.F.	0.033
Sister joists for ceilings		
2x4	L.F.	0.057
2x6	L.F.	0.067
2x8	L.F.	0.080
2x10	L.F.	0.100
2x12	L.F.	0.133

06110.30 Floor Framing

Rough Carpentry	UNIT	MAN/ HOURS
Floor joists		
12" o.c.		
2x6	S.F.	0.016
2x8	S.F.	0.016
2x10	S.F.	0.017
2x12	S.F.	0.017
2x14	S.F.	0.017
3x6	S.F.	0.017
3x8	S.F.	0.017
3x10	S.F.	0.018
3x12	S.F.	0.019
3x14	S.F.	0.020
4x6	S.F.	0.017
4x8	S.F.	0.017
4x10	S.F.	0.018
4x12	S.F.	0.019
4x14	S.F.	0.020
16" o.c.		
2x6	S.F.	0.013
2x8	S.F.	0.014
2x10	S.F.	0.014
2x12	S.F.	0.014
2x14	S.F.	0.015
3x6	S.F.	0.014
3x8	S.F.	0.014
3x10	S.F.	0.015
3x12	S.F.	0.015
3x14	S.F.	0.016
4x6	S.F.	0.014
4x8	S.F.	0.014
4x10	S.F.	0.015
4x12	S.F.	0.015
4x14	S.F.	0.016
Sister joists for floors		
2x4	L.F.	0.050
2x6	L.F.	0.057
2x8	L.F.	0.067
2x10	L.F.	0.080
2x12	L.F.	0.100
3x6	L.F.	0.080
3x8	L.F.	0.089
3x10	L.F.	0.100
3x12	L.F.	0.114
4x6	L.F.	0.080
4x8	L.F.	0.089
4x10	L.F.	0.100
4x12	L.F.	0.114

06110.40 Furring

Rough Carpentry	UNIT	MAN/ HOURS
Furring, wood strips		
Walls		
On masonry or concrete walls		
1x2 furring		
12" o.c.	S.F.	0.025
16" o.c.	S.F.	0.023
24" o.c.	S.F.	0.021
1x3 furring		
12" o.c.	S.F.	0.025
16" o.c.	S.F.	0.023
24" o.c.	S.F.	0.021
On wood walls		
1x2 furring		
12" o.c.	S.F.	0.018
16" o.c.	S.F.	0.016
24" o.c.	S.F.	0.015
1x3 furring		
12" o.c.	S.F.	0.018
16" o.c.	S.F.	0.016
24" o.c.	S.F.	0.015
Ceilings		
On masonry or concrete ceilings		
1x2 furring		
12" o.c.	S.F.	0.044
16" o.c.	S.F.	0.040
24" o.c.	S.F.	0.036
1x3 furring		
12" o.c.	S.F.	0.044
16" o.c.	S.F.	0.040
24" o.c.	S.F.	0.036
On wood ceilings		
1x2 furring		
12" o.c.	S.F.	0.030
16" o.c.	S.F.	0.027
24" o.c.	S.F.	0.024
1x3		
12" o.c.	S.F.	0.030

06 WOOD AND PLASTICS

Rough Carpentry

06110.40 Furring

Rough Carpentry	UNIT	MAN/ HOURS
16" o.c.	S.F.	0.027
24" o.c.	S.F.	0.024

06110.50 Roof Framing

Rough Carpentry	UNIT	MAN/ HOURS
Roof framing		
Rafters, gable end		
0-2 pitch (flat to 2-in-12)		
12" o.c.		
2x4	S.F.	0.017
2x6	S.F.	0.017
2x8	S.F.	0.018
2x10	S.F.	0.019
2x12	S.F.	0.020
16" o.c.		
2x6	S.F.	0.014
2x8	S.F.	0.015
2x10	S.F.	0.015
2x12	S.F.	0.016
24" o.c.		
2x6	S.F.	0.012
2x8	S.F.	0.013
2x10	S.F.	0.013
2x12	S.F.	0.013
4-6 pitch (4-in-12 to 6-in-12)		
12" o.c.		
2x4	S.F.	0.017
2x6	S.F.	0.018
2x8	S.F.	0.019
2x10	S.F.	0.020
2x12	S.F.	0.021
16" o.c.		
2x6	S.F.	0.015
2x8	S.F.	0.015
2x10	S.F.	0.016
2x12	S.F.	0.017
24" o.c.		
2x6	S.F.	0.013
2x8	S.F.	0.013
2x10	S.F.	0.014
2x12	S.F.	0.015
8-12 pitch (8-in-12 to 12-in-12)		
12" o.c.		
2x4	S.F.	0.018
2x6	S.F.	0.019
2x8	S.F.	0.020
2x10	S.F.	0.021
2x12	S.F.	0.022
16" o.c.		
2x6	S.F.	0.015
2x8	S.F.	0.016
2x10	S.F.	0.017
2x12	S.F.	0.017
24" o.c.		
2x6	S.F.	0.013

Rough Carpentry

06110.50 Roof Framing

Rough Carpentry	UNIT	MAN/ HOURS
2x8	S.F.	0.013
2x10	S.F.	0.014
2x12	S.F.	0.014
Ridge boards		
2x6	L.F.	0.040
2x8	L.F.	0.044
2x10	L.F.	0.050
2x12	L.F.	0.057
Hip rafters		
2x6	L.F.	0.029
2x8	L.F.	0.030
2x10	L.F.	0.031
2x12	L.F.	0.032
Jack rafters		
4-6 pitch (4-in-12 to 6-in-12)		
16" o.c.		
2x6	S.F.	0.024
2x8	S.F.	0.024
2x10	S.F.	0.026
2x12	S.F.	0.027
24" o.c.		
2x6	S.F.	0.018
2x8	S.F.	0.019
2x10	S.F.	0.020
2x12	S.F.	0.020
8-12 pitch (8-in-12 to 12-in-12)		
16" o.c.		
2x6	S.F.	0.025
2x8	S.F.	0.026
2x10	S.F.	0.027
2x12	S.F.	0.028
24" o.c.		
2x6	S.F.	0.019
2x8	S.F.	0.020
2x10	S.F.	0.020
2x12	S.F.	0.021
Sister rafters		
2x4	L.F.	0.057
2x6	L.F.	0.067
2x8	L.F.	0.080
2x10	L.F.	0.100
2x12	L.F.	0.133
Fascia boards		
2x4	L.F.	0.040
2x6	L.F.	0.040
2x8	L.F.	0.044
2x10	L.F.	0.044
2x12	L.F.	0.050
Cant strips		
Fiber		
3x3	L.F.	0.023
4x4	L.F.	0.024
Wood		
3x3	L.F.	0.024

Rough Carpentry

06110.60 Sleepers	UNIT	MAN/ HOURS
Sleepers, over concrete		
12" o.c.		
1x2	S.F.	0.018
1x3	S.F.	0.019
2x4	S.F.	0.022
2x6	S.F.	0.024
16" o.c.		
1x2	S.F.	0.016
1x3	S.F.	0.016
2x4	S.F.	0.019
2x6	S.F.	0.020

06110.65 Soffits	UNIT	MAN/ HOURS
Soffit framing		
2x3	L.F.	0.057
2x4	L.F.	0.062
2x6	L.F.	0.067
2x8	L.F.	0.073

06110.70 Wall Framing	UNIT	MAN/ HOURS
Framing wall, studs		
12" o.c.		
2x3	S.F.	0.015
2x4	S.F.	0.015
2x6	S.F.	0.016
2x8	S.F.	0.017
16" o.c.		
2x3	S.F.	0.013
2x4	S.F.	0.013
2x6	S.F.	0.013
2x8	S.F.	0.014
24" o.c.		
2x3	S.F.	0.011
2x4	S.F.	0.011
2x6	S.F.	0.011
2x8	S.F.	0.012
Plates, top or bottom		
2x3	L.F.	0.024
2x4	L.F.	0.025
2x6	L.F.	0.027
2x8	L.F.	0.029
Headers, door or window		
2x6		
Single		
3' long	EA.	0.400
6' long	EA.	0.500
Double		
3' long	EA.	0.444
6' long	EA.	0.571
2x8		
Single		
4' long	EA.	0.500
8' long	EA.	0.615
Double		

Rough Carpentry

06110.70 Wall Framing	UNIT	MAN/ HOURS
4' long	EA.	0.571
8' long	EA.	0.727
2x10		
Single		
5' long	EA.	0.615
10' long	EA.	0.800
Double		
5' long	EA.	0.667
10' long	EA.	0.800
2x12		
Single		
6' long	EA.	0.615
12' long	EA.	0.800
Double		
6' long	EA.	0.727
12' long	EA.	0.889

06115.10 Floor Sheathing	UNIT	MAN/ HOURS
Sub-flooring, plywood, CDX		
1/2" thick	S.F.	0.010
5/8" thick	S.F.	0.011
3/4" thick	S.F.	0.013
Structural plywood		
1/2" thick	S.F.	0.010
5/8" thick	S.F.	0.011
3/4" thick	S.F.	0.012
Board type subflooring		
1x6		
Minimum	S.F.	0.018
Maximum	S.F.	0.020
1x8		
Minimum	S.F.	0.017
Maximum	S.F.	0.019
1x10		
Minimum	S.F.	0.016
Maximum	S.F.	0.018
Underlayment		
Hardboard, 1/4" tempered	S.F.	0.010
Plywood, CDX		
3/8" thick	S.F.	0.010
1/2" thick	S.F.	0.011
5/8" thick	S.F.	0.011
3/4" thick	S.F.	0.012

06115.20 Roof Sheathing	UNIT	MAN/ HOURS
Sheathing		
Plywood, CDX		
3/8" thick	S.F.	0.010
1/2" thick	S.F.	0.011
5/8" thick	S.F.	0.011
3/4" thick	S.F.	0.012
Structural plywood		
3/8" thick	S.F.	0.010
1/2" thick	S.F.	0.011

06 WOOD AND PLASTICS

Rough Carpentry

06115.20 Roof Sheathing	UNIT	MAN/HOURS
5/8" thick	S.F.	0.011
3/4" thick	S.F.	0.012

06115.30 Wall Sheathing	UNIT	MAN/HOURS
Sheathing		
Plywood, CDX		
3/8" thick	S.F.	0.012
1/2" thick	S.F.	0.012
5/8" thick	S.F.	0.013
3/4" thick	S.F.	0.015
Waferboard		
3/8" thick	S.F.	0.012
1/2" thick	S.F.	0.012
5/8" thick	S.F.	0.013
3/4" thick	S.F.	0.015
Structural plywood		
3/8" thick	S.F.	0.012
1/2" thick	S.F.	0.012
5/8" thick	S.F.	0.013
3/4" thick	S.F.	0.015
Gypsum, 1/2" thick	S.F.	0.012
Asphalt impregnated fiberboard, 1/2" thick	S.F.	0.012

06125.10 Wood Decking	UNIT	MAN/HOURS
Decking, T&G solid		
Cedar		
3" thick	S.F.	0.020
4" thick	S.F.	0.021
Fir		
3" thick	S.F.	0.020
4" thick	S.F.	0.021
Southern yellow pine		
3" thick	S.F.	0.023
4" thick	S.F.	0.025
White pine		
3" thick	S.F.	0.020
4" thick	S.F.	0.021

06130.10 Heavy Timber	UNIT	MAN/HOURS
Mill framed structures		
Beams to 20' long		
Douglas fir		
6x8	L.F.	0.080
6x10	L.F.	0.083
6x12	L.F.	0.089
6x14	L.F.	0.092
6x16	L.F.	0.096
8x10	L.F.	0.083
8x12	L.F.	0.089
8x14	L.F.	0.092
8x16	L.F.	0.096
Southern yellow pine		
6x8	L.F.	0.080
6x10	L.F.	0.083

Rough Carpentry

06130.10 Heavy Timber	UNIT	MAN/HOURS
6x12	L.F.	0.089
6x14	L.F.	0.092
6x16	L.F.	0.096
8x10	L.F.	0.083
8x12	L.F.	0.089
8x14	L.F.	0.092
8x16	L.F.	0.096
Columns to 12' high		
Douglas fir		
6x6	L.F.	0.120
8x8	L.F.	0.120
10x10	L.F.	0.133
12x12	L.F.	0.133
Southern yellow pine		
6x6	L.F.	0.120
8x8	L.F.	0.120
10x10	L.F.	0.133
12x12	L.F.	0.133
Posts, treated		
4x4	L.F.	0.032
6x6	L.F.	0.040

06190.20 Wood Trusses	UNIT	MAN/HOURS
Truss, fink, 2x4 members		
3-in-12 slope		
24' span	EA.	0.686
26' span	EA.	0.686
28' span	EA.	0.727
30' span	EA.	0.727
34' span	EA.	0.774
38' span	EA.	0.774
5-in-12 slope		
24' span	EA.	0.706
28' span	EA.	0.727
30' span	EA.	0.750
32' span	EA.	0.750
40' span	EA.	0.800
Gable, 2x4 members		
5-in-12 slope		
24' span	EA.	0.706
26' span	EA.	0.706
28' span	EA.	0.727
30' span	EA.	0.750
32' span	EA.	0.750
36' span	EA.	0.774
40' span	EA.	0.800
King post type, 2x4 members		
4-in-12 slope		
16' span	EA.	0.649
18' span	EA.	0.667
24' span	EA.	0.706
26' span	EA.	0.706
30' span	EA.	0.750
34' span	EA.	0.750

Rough Carpentry	UNIT	MAN/ HOURS
06190.20 Wood Trusses		
38' span	EA.	0.774
42' span	EA.	0.828

Finish Carpentry	UNIT	MAN/ HOURS
06200.10 Finish Carpentry		
Mouldings and trim		
Apron, flat		
9/16 x 2	L.F.	0.040
9/16 x 3-1/2	L.F.	0.042
Base		
Colonial		
7/16 x 2-1/4	L.F.	0.040
7/16 x 3	L.F.	0.040
7/16 x 3-1/4	L.F.	0.040
9/16 x 3	L.F.	0.042
9/16 x 3-1/4	L.F.	0.042
11/16 x 2-1/4	L.F.	0.044
Ranch		
7/16 x 2-1/4	L.F.	0.040
7/16 x 3-1/4	L.F.	0.040
9/16 x 2-1/4	L.F.	0.042
9/16 x 3	L.F.	0.042
9/16 x 3-1/4	L.F.	0.042
Casing		
11/16 x 2-1/2	L.F.	0.036
11/16 x 3-1/2	L.F.	0.038
Chair rail		
9/16 x 2-1/2	L.F.	0.040
9/16 x 3-1/2	L.F.	0.040
Closet pole		
1-1/8" dia.	L.F.	0.053
1-5/8" dia.	L.F.	0.053
Cove		
9/16 x 1-3/4	L.F.	0.040
11/16 x 2-3/4	L.F.	0.040
Crown		
9/16 x 1-5/8	L.F.	0.053
9/16 x 2-5/8	L.F.	0.062
11/16 x 3-5/8	L.F.	0.067
11/16 x 4-1/4	L.F.	0.073
11/16 x 5-1/4	L.F.	0.080
Drip cap		
1-1/16 x 1-5/8	L.F.	0.040
Glass bead		
3/8 x 3/8	L.F.	0.050
1/2 x 9/16	L.F.	0.050

Finish Carpentry	UNIT	MAN/ HOURS
06200.10 Finish Carpentry		
5/8 x 5/8	L.F.	0.050
3/4 x 3/4	L.F.	0.050
Half round		
1/2	L.F.	0.032
5/8	L.F.	0.032
3/4	L.F.	0.032
Lattice		
1/4 x 7/8	L.F.	0.032
1/4 x 1-1/8	L.F.	0.032
1/4 x 1-3/8	L.F.	0.032
1/4 x 1-3/4	L.F.	0.032
1/4 x 2	L.F.	0.032
Ogee molding		
5/8 x 3/4	L.F.	0.040
11/16 x 1-1/8	L.F.	0.040
11/16 x 1-3/8	L.F.	0.040
Parting bead		
3/8 x 7/8	L.F.	0.050
Quarter round		
1/4 x 1/4	L.F.	0.032
3/8 x 3/8	L.F.	0.032
1/2 x 1/2	L.F.	0.032
11/16 x 11/16	L.F.	0.035
3/4 x 3/4	L.F.	0.035
1-1/16 x 1-1/16	L.F.	0.036
Railings, balusters		
1-1/8 x 1-1/8	L.F.	0.080
1-1/2 x 1-1/2	L.F.	0.073
Screen moldings		
1/4 x 3/4	L.F.	0.067
5/8 x 5/16	L.F.	0.067
Shoe		
7/16 x 11/16	L.F.	0.032
Sash beads		
1/2 x 3/4	L.F.	0.067
1/2 x 7/8	L.F.	0.067
1/2 x 1-1/8	L.F.	0.073
5/8 x 7/8	L.F.	0.073
Stop		
5/8 x 1-5/8		
Colonial	L.F.	0.050
Ranch	L.F.	0.050
Stools		
11/16 x 2-1/4	L.F.	0.089
11/16 x 2-1/2	L.F.	0.089
11/16 x 5-1/4	L.F.	0.100
Exterior trim, casing, select pine, 1x3	L.F.	0.040
Douglas fir		
1x3	L.F.	0.040
1x4	L.F.	0.040
1x6	L.F.	0.044
1x8	L.F.	0.050
Cornices, white pine, #2 or better		
1x2	L.F.	0.040

06 WOOD AND PLASTICS

Finish Carpentry

06200.10 Finish Carpentry

Finish Carpentry	UNIT	MAN/HOURS
1x4	L.F.	0.040
1x6	L.F.	0.044
1x8	L.F.	0.047
1x10	L.F.	0.050
1x12	L.F.	0.053
Shelving, pine		
1x8	L.F.	0.062
1x10	L.F.	0.064
1x12	L.F.	0.067
Plywood shelf, 3/4", with edge band, 12" wide	L.F.	0.080
Adjustable shelf, and rod, 12" wide		
3' to 4' long	EA.	0.200
5' to 8' long	EA.	0.267
Prefinished wood shelves with brackets and supports		
8" wide		
3' long	EA.	0.200
4' long	EA.	0.200
6' long	EA.	0.200
10" wide		
3' long	EA.	0.200
4' long	EA.	0.200
6' long	EA.	0.200

06220.10 Millwork

Millwork	UNIT	MAN/HOURS
Countertop, laminated plastic		
25" x 7/8" thick		
Minimum	L.F.	0.200
Average	L.F.	0.267
Maximum	L.F.	0.320
25" x 1-1/4" thick		
Minimum	L.F.	0.267
Average	L.F.	0.320
Maximum	L.F.	0.400
Add for cutouts	EA.	0.500
Backsplash, 4" high, 7/8" thick	L.F.	0.160
Plywood, sanded, A-C		
1/4" thick	S.F.	0.027
3/8" thick	S.F.	0.029
1/2" thick	S.F.	0.031
A-D		
1/4" thick	S.F.	0.027
3/8" thick	S.F.	0.029
1/2" thick	S.F.	0.031
Base cabinets, 34-1/2" high, 24" deep, hardwood, no tops		
Minimum	L.F.	0.320
Average	L.F.	0.400
Maximum	L.F.	0.533
Wall cabinets		
Minimum	L.F.	0.267
Average	L.F.	0.320
Maximum	L.F.	0.400
Oil borne		
Water borne		

Architectural Woodwork

06420.10 Panel Work

Panel Work	UNIT	MAN/HOURS
Hardboard, tempered, 1/4" thick		
Natural faced	S.F.	0.020
Plastic faced	S.F.	0.023
Pegboard, natural	S.F.	0.020
Plastic faced	S.F.	0.023
Untempered, 1/4" thick		
Natural faced	S.F.	0.020
Plastic faced	S.F.	0.023
Pegboard, natural	S.F.	0.020
Plastic faced	S.F.	0.023
Plywood unfinished, 1/4" thick		
Birch		
Natural	S.F.	0.027
Select	S.F.	0.027
Knotty pine	S.F.	0.027
Cedar (closet lining)		
Standard boards T&G	S.F.	0.027
Particle board	S.F.	0.027
Plywood, prefinished, 1/4" thick, premium grade		
Birch veneer	S.F.	0.032
Cherry veneer	S.F.	0.032
Chestnut veneer	S.F.	0.032
Lauan veneer	S.F.	0.032
Mahogany veneer	S.F.	0.032
Oak veneer (red)	S.F.	0.032
Pecan veneer	S.F.	0.032
Rosewood veneer	S.F.	0.032
Teak veneer	S.F.	0.032
Walnut veneer	S.F.	0.032

06430.10 Stairwork

Stairwork	UNIT	MAN/HOURS
Risers, 1x8, 42" wide		
White oak	EA.	0.400
Pino	EA.	0.400
Treads, 1-1/16" x 9-1/2" x 42"		
White oak	EA.	0.500

06440.10 Columns

Columns	UNIT	MAN/HOURS
Column, hollow, round wood		
12" diameter		
10' high	EA.	0.800
12' high	EA.	0.857
14' high	EA.	0.960
16' high	EA.	1.200
24" diameter		
16' high	EA.	1.200
18' high	EA.	1.263
20' high	EA.	1.263
22' high	EA.	1.333
24' high	EA.	1.333

Moisture Protection

07100.10 Waterproofing

Moisture Protection	UNIT	MAN/HOURS
Membrane waterproofing, elastomeric		
Butyl		
1/32" thick	S.F.	0.032
1/16" thick	S.F.	0.033
Butyl with nylon		
1/32" thick	S.F.	0.032
1/16" thick	S.F.	0.033
Neoprene		
1/32" thick	S.F.	0.032
1/16" thick	S.F.	0.033
Neoprene with nylon		
1/32" thick	S.F.	0.032
1/16" thick	S.F.	0.033
Bituminous membrane waterproofing, asphalt felt, 15 lb.		
One ply	S.F.	0.020
Two ply	S.F.	0.024
Three ply	S.F.	0.029
Four ply	S.F.	0.033
Five ply	S.F.	0.042
Modified asphalt membrane waterproofing, fibrous asphalt		
One ply	S.F.	0.033
Two ply	S.F.	0.040
Three ply	S.F.	0.044
Four ply	S.F.	0.053
Five ply	S.F.	0.064
Asphalt coated protective board		
1/8" thick	S.F.	0.020
1/4" thick	S.F.	0.020
3/8" thick	S.F.	0.020
1/2" thick	S.F.	0.021
Cement protective board		
3/8" thick	S.F.	0.027
1/2" thick	S.F.	0.027
Fluid applied, neoprene		
50 mil	S.F.	0.027
90 mil	S.F.	0.027
Tab extended polyurethane		
.050" thick	S.F.	0.020
Fluid applied rubber based polyurethane		
6 mil	S.F.	0.025
15 mil	S.F.	0.020
Bentonite waterproofing, panels		
3/16" thick	S.F.	0.020
1/4" thick	S.F.	0.020
5/8" thick	S.F.	0.021
Granular admixtures, trowel on, 3/8" thick	S.F.	0.020
Metallic oxide waterproofing, iron compound, troweled		
5/8" thick	S.F.	0.020
3/4" thick	S.F.	0.023

07150.10 Dampproofing

Moisture Protection	UNIT	MAN/HOURS
Silicone dampproofing, sprayed on		
Concrete surface		
1 coat	S.F.	0.004
2 coats	S.F.	0.006
Concrete block		
1 coat	S.F.	0.005
2 coats	S.F.	0.007
Brick		
1 coat	S.F.	0.006
2 coats	S.F.	0.008

07160.10 Bituminous Dampproofing

Moisture Protection	UNIT	MAN/HOURS
Building paper, asphalt felt		
15 lb	S.F.	0.032
30 lb	S.F.	0.033
Asphalt dampproofing, troweled, cold, primer plus		
1 coat	S.F.	0.027
2 coats	S.F.	0.040
3 coats	S.F.	0.050
Fibrous asphalt dampproofing, hot troweled, primer plus		
1 coat	S.F.	0.032
2 coats	S.F.	0.044
3 coats	S.F.	0.057
Asphaltic paint dampproofing, per coat		
Brush on	S.F.	0.011
Spray on	S.F.	0.009

07190.10 Vapor Barriers

Moisture Protection	UNIT	MAN/HOURS
Vapor barrier, polyethylene		
2 mil	S.F.	0.004
6 mil	S.F.	0.004
8 mil	S.F.	0.004
10 mil	S.F.	0.004

Insulation

07210.10 Batt Insulation

Insulation	UNIT	MAN/HOURS
Ceiling, fiberglass, unfaced		
3-1/2" thick, R11	S.F.	0.009
6" thick, R19	S.F.	0.011
9" thick, R30	S.F.	0.012
Suspended ceiling, unfaced		
3-1/2" thick, R11	S.F.	0.009
6" thick, R19	S.F.	0.010
9" thick, R30	S.F.	0.011
Crawl space, unfaced		
3-1/2" thick, R11	S.F.	0.012
6" thick, R19	S.F.	0.013
9" thick, R30	S.F.	0.015
Wall, fiberglass		

Insulation	UNIT	MAN/ HOURS
07210.10 Batt Insulation		
Paper backed		
2" thick, R7	S.F.	0.008
3" thick, R8	S.F.	0.009
4" thick, R11	S.F.	0.009
6" thick, R19	S.F.	0.010
Foil backed, 1 side		
2" thick, R7	S.F.	0.008
3" thick, R11	S.F.	0.009
4" thick, R14	S.F.	0.009
6" thick, R21	S.F.	0.010
Foil backed, 2 sides		
2" thick, R7	S.F.	0.009
3" thick, R11	S.F.	0.010
4" thick, R14	S.F.	0.011
6" thick, R21	S.F.	0.011
Unfaced		
2" thick, R7	S.F.	0.008
3" thick, R9	S.F.	0.009
4" thick, R11	S.F.	0.009
6" thick, R19	S.F.	0.010
Mineral wool batts		
Paper backed		
2" thick, R6	S.F.	0.008
4" thick, R12	S.F.	0.009
6" thick, R19	S.F.	0.010
Fasteners, self adhering, attached to ceiling deck		
2-1/2" long	EA.	0.013
4-1/2" long	EA.	0.015
Capped, self-locking washers for fastening insulation	EA.	0.008
07210.20 Board Insulation		
Insulation, rigid		
Fiberglass, roof		
0.75" thick, R2.78	S.F.	0.007
1.06" thick, R4.17	S.F.	0.008
1.31" thick, R5.26	S.F.	0.008
1.63" thick, R6.67	S.F.	0.008
2.25" thick, R8.33	S.F.	0.009
Composite board, roof		
1-1/2" thick, R6.67	S.F.	0.008
1-5/8" thick, R7.69	S.F.	0.008
2" thick, R10.0	S.F.	0.009
2-1/4" thick, R12.50	S.F.	0.009
2-1/2" thick, R14.29	S.F.	0.010
2-3/4" thick, R16.67	S.F.	0.011
3-1/4" thick, R20.00	S.F.	0.011
Perlite board, roof		
1.00" thick, R2.78	S.F.	0.007
1.50" thick, R4.17	S.F.	0.007
2.00" thick, R5.92	S.F.	0.007
2.50" thick, R6.67	S.F.	0.008
3.00" thick, R8.33	S.F.	0.008
4.00" thick, R10.00	S.F.	0.008
5.25" thick, R14.29	S.F.	0.009

Insulation	UNIT	MAN/ HOURS
07210.20 Board Insulation		
Rigid urethane		
Roof		
1" thick, R6.67	S.F.	0.007
1.20" thick, R8.33	S.F.	0.007
1.50" thick, R11.11	S.F.	0.007
2" thick, R14.29	S.F.	0.007
2.25" thick, R16.67	S.F.	0.008
Wall		
1" thick, R6.67	S.F.	0.008
1.5" thick, R11.11	S.F.	0.009
2" thick, R14.29	S.F.	0.009
Polystyrene		
Roof		
1.0" thick, R4.17	S.F.	0.007
1.5" thick, R6.26	S.F.	0.007
2.0" thick, R8.33	S.F.	0.007
Wall		
1.0" thick, R4.17	S.F.	0.008
1.5" thick, R6.26	S.F.	0.009
2.0" thick, R8.33	S.F.	0.009
Rigid board insulation, deck		
Mineral fiberboard		
1" thick, R3.0	S.F.	0.007
2" thick, R5.26	S.F.	0.007
Fiberglass		
1" thick, R4.3	S.F.	0.007
2" thick, R8.5	S.F.	0.007
Polystyrene		
1" thick, R5.4	S.F.	0.007
2" thick, R10.8	S.F.	0.007
Urethane		
.75" thick, R5.4	S.F.	0.007
1" thick, R6.4	S.F.	0.007
1.5" thick, R10.7	S.F.	0.007
2" thick, R14.3	S.F.	0.007
Foamglass		
1" thick, R1.8	S.F.	0.007
2" thick, R5.26	S.F.	0.007
Wood fiber		
1" thick, R3.85	S.F.	0.007
2" thick, R7.7	S.F.	0.007
Particle board		
3/4" thick, R2.08	S.F.	0.007
1" thick, R2.77	S.F.	0.007
2" thick, R5.50	S.F.	0.007
07210.60 Loose Fill Insulation		
Blown-in type		
Fiberglass		
5" thick, R11	S.F.	0.007
6" thick, R13	S.F.	0.008
9" thick, R19	S.F.	0.011
Rockwool, attic application		
6" thick, R13	S.F.	0.008

Insulation

07210.60 Loose Fill Insulation	UNIT	MAN/HOURS
8" thick, R19	S.F.	0.010
10" thick, R22	S.F.	0.012
12" thick, R26	S.F.	0.013
15" thick, R30	S.F.	0.016
Poured type		
Fiberglass		
1" thick, R4	S.F.	0.005
2" thick, R8	S.F.	0.006
3" thick, R12	S.F.	0.007
4" thick, R16	S.F.	0.008
Mineral wool		
1" thick, R3	S.F.	0.005
2" thick, R6	S.F.	0.006
3" thick, R9	S.F.	0.007
4" thick, R12	S.F.	0.008
Vermiculite or perlite		
2" thick, R4.8	S.F.	0.006
3" thick, R7.2	S.F.	0.007
4" thick, R9.6	S.F.	0.008
Masonry, poured vermiculite or perlite		
4" block	S.F.	0.004
6" block	S.F.	0.005
8" block	S.F.	0.006
10" block	S.F.	0.006
12" block	S.F.	0.007

07210.70 Sprayed Insulation	UNIT	MAN/HOURS
Foam, sprayed on		
Polystyrene		
1" thick, R4	S.F.	0.008
2" thick, R8	S.F.	0.011
Urethane		
1" thick, R4	S.F.	0.008
2" thick, R8	S.F.	0.011

07250.10 Fireproofing	UNIT	MAN/HOURS
Sprayed on		
1" thick		
On beams	S.F.	0.018
On columns	S.F.	0.016
On decks		
Flat surface	S.F.	0.008
Fluted surface	S.F.	0.010
1-1/2" thick		
On beams	S.F.	0.023
On columns	S.F.	0.020
On decks		
Flat surface	S.F.	0.010
Fluted surface	S.F.	0.013

Shingles And Tiles

07310.10 Asphalt Shingles	UNIT	MAN/HOURS
Standard asphalt shingles, strip shingles		
210 lb/square	SQ.	0.800
235 lb/square	SQ.	0.889
240 lb/square	SQ.	1.000
260 lb/square	SQ.	1.143
300 lb/square	SQ.	1.333
385 lb/square	SQ.	1.600
Roll roofing, mineral surface		
90 lb	SQ.	0.571
110 lb	SQ.	0.667
140 lb	SQ.	0.800

07310.30 Metal Shingles	UNIT	MAN/HOURS
Aluminum, .020" thick		
Plain	SQ.	1.600
Colors	SQ.	1.600
Steel, galvanized		
26 ga.		
Plain	SQ.	1.600
Colors	SQ.	1.600
24 ga.		
Plain	SQ.	1.600
Colors	SQ.	1.600
Porcelain enamel, 22 ga.		
Minimum	SQ.	2.000
Average	SQ.	2.000
Maximum	SQ.	2.000

07310.60 Slate Shingles	UNIT	MAN/HOURS
Slate shingles		
Pennsylvania		
Ribbon	SQ.	4.000
Clear	SQ.	4.000
Vermont		
Black	SQ.	4.000
Gray	SQ.	4.000
Green	SQ.	4.000
Red	SQ.	4.000
Replacement shingles		
Small jobs	EA.	0.267
Large jobs	S.F.	0.133

07310.70 Wood Shingles	UNIT	MAN/HOURS
Wood shingles, on roofs		
White cedar, #1 shingles		
4" exposure	SQ.	2.667
5" exposure	SQ.	2.000
#2 shingles		
4" exposure	SQ.	2.667
5" exposure	SQ.	2.000
Resquared and rebutted		
4" exposure	SQ.	2.667
5" exposure	SQ.	2.000
On walls		

Shingles And Tiles

07310.70 Wood Shingles

	UNIT	MAN/ HOURS
White cedar, #1 shingles		
4" exposure	SQ.	4.000
5" exposure	SQ.	3.200
6" exposure	SQ.	2.667
#2 shingles		
4" exposure	SQ.	4.000
5" exposure	SQ.	3.200
6" exposure	SQ.	2.667

07310.80 Wood Shakes

	UNIT	MAN/ HOURS
Shakes, hand split, 24" red cedar, on roofs		
5" exposure	SQ.	4.000
7" exposure	SQ.	3.200
9" exposure	SQ.	2.667
On walls		
6" exposure	SQ.	4.000
8" exposure	SQ.	3.200
10" exposure	SQ.	2.667

Roofing And Siding

07410.10 Manufactured Roofs

	UNIT	MAN/ HOURS
Aluminum roof panels, for structural steel framing		
Corrugated		
Unpainted finish		
.024"	S.F.	0.020
.030"	S.F.	0.020
Painted finish		
.024"	S.F.	0.020
.030"	S.F.	0.020
V-beam		
Unpainted finish		
.032"	S.F.	0.020
.040"	S.F.	0.020
.050"	S.F.	0.020
Painted finish		
.032"	S.F.	0.020
.040"	S.F.	0.020
.050"	S.F.	0.020
Steel roof panels, for structural steel framing		
Corrugated, painted		
18 ga.	S.F.	0.020
20 ga.	S.F.	0.020
22 ga.	S.F.	0.020
Box rib, painted		
18 ga.	S.F.	0.021
20 ga.	S.F.	0.021

Roofing And Siding

07410.10 Manufactured Roofs

	UNIT	MAN/ HOURS
22 ga.	S.F.	0.021
4" rib, painted		
18 ga.	S.F.	0.022
20 ga.	S.F.	0.022
22 ga.	S.F.	0.022
Standing seam roof		
2" high seam, painted		
22 ga.	S.F.	0.032
24 ga.	S.F.	0.032
26 ga.	S.F.	0.032

07410.30 Manufactured Walls

	UNIT	MAN/ HOURS
Sandwich panels with 1-1/2" fiberglass insulation		
Galvanized 18 ga. steel interior panels		
Exterior panels		
16 ga. aluminum	S.F.	0.107
18 ga. galvanized steel	S.F.	0.107
20 ga. painted steel	S.F.	0.107
20 ga. stainless steel	S.F.	0.107
Metal liner panels, 1-3/8" thick, 24" wide		
Galvanized		
22 ga.	S.F.	0.027
20 ga.	S.F.	0.027
18 ga.	S.F.	0.027
Primed		
22 ga.	S.F.	0.027
20 ga.	S.F.	0.027
18 ga.	S.F.	0.027

07440.10 Aggregate Coated Panels

	UNIT	MAN/ HOURS
Dryvit type system		
1" thick	S.F.	0.027
1-1/2" thick	S.F.	0.029
2" thick	S.F.	0.000

07460.10 Metal Siding Panels

	UNIT	MAN/ HOURS
Aluminum siding panels		
Corrugated		
Plain finish		
.024"	S.F.	0.032
.032"	S.F.	0.032
Painted finish		
.024"	S.F.	0.032
.032"	S.F.	0.032
V. beam		
Plain finish		
.032"	S.F.	0.032
.040"	S.F.	0.032
.050"	S.F.	0.032
Painted finish		
.032"	S.F.	0.032
.040"	S.F.	0.032
.050"	S.F.	0.032
4" rib		

Roofing And Siding	UNIT	MAN/ HOURS
07460.10 Metal Siding Panels		
Plain finish		
.032"	S.F.	0.036
.040"	S.F.	0.036
.050"	S.F.	0.036
Painted finish		
.032"	S.F.	0.036
.040"	S.F.	0.036
.050"	S.F.	0.036
Steel siding panels		
Corrugated		
22 ga.	S.F.	0.053
24 ga.	S.F.	0.053
26 ga.	S.F.	0.053
Box rib		
20 ga.	S.F.	0.053
22 ga.	S.F.	0.053
24 ga.	S.F.	0.053
26 ga.	S.F.	0.053
07460.50 Plastic Siding		
Horizontal vinyl siding, solid		
8" wide		
Standard	S.F.	0.031
Insulated	S.F.	0.031
10" wide		
Standard	S.F.	0.029
Insulated	S.F.	0.029
Vinyl moldings for doors and windows	L.F.	0.032
07460.60 Plywood Siding		
Rough sawn cedar, 3/8" thick	S.F.	0.027
Fir, 3/8" thick	S.F.	0.027
Texture 1-11, 5/8" thick		
Cedar	S.F.	0.029
Fir	S.F.	0.029
Redwood	S.F.	0.029
Southern Yellow Pine	S.F.	0.029
07460.70 Steel Siding		
Ribbed, sheets, galvanized		
22 ga.	S.F.	0.032
24 ga.	S.F.	0.032
26 ga.	S.F.	0.032
28 ga.	S.F.	0.032
Primed		
24 ga.	S.F.	0.032
26 ga.	S.F.	0.032
28 ga.	S.F.	0.032
07460.80 Wood Siding		
Beveled siding, cedar		
A grade		
1/2 x 6	S.F.	0.040
1/2 x 8	S.F.	0.032

Roofing And Siding	UNIT	MAN/ HOURS
07460.80 Wood Siding		
3/4 x 10	S.F.	0.027
Clear		
1/2 x 6	S.F.	0.040
1/2 x 8	S.F.	0.032
3/4 x 10	S.F.	0.027
B grade		
1/2 x 6	S.F.	0.040
1/2 x 8	S.F.	0.320
3/4 x 10	S.F.	0.027
Board and batten		
Cedar		
1x6	S.F.	0.040
1x8	S.F.	0.032
1x10	S.F.	0.029
1x12	S.F.	0.026
Pine		
1x6	S.F.	0.040
1x8	S.F.	0.032
1x10	S.F.	0.029
1x12	S.F.	0.026
Redwood		
1x6	S.F.	0.040
1x8	S.F.	0.032
1x10	S.F.	0.029
1x12	S.F.	0.026
Tongue and groove		
Cedar		
1x4	S.F.	0.044
1x6	S.F.	0.042
1x8	S.F.	0.040
1x10	S.F.	0.038
Pine		
1x4	S.F.	0.044
1x6	S.F.	0.042
1x8	S.F.	0.040
1x10	S.F.	0.038
Redwood		
1x4	S.F.	0.044
1x6	S.F.	0.042
1x8	S.F.	0.040
1x10	S.F.	0.038

07 THERMAL AND MOISTURE

Membrane Roofing

07510.10 Built-up Asphalt Roofing

	UNIT	MAN/ HOURS
Built-up roofing, asphalt felt, including gravel		
2 ply	SQ.	2.000
3 ply	SQ.	2.667
4 ply	SQ.	3.200
Walkway, for built-up roofs		
3' x 3' x		
1/2" thick	S.F.	0.027
3/4" thick	S.F.	0.027
1" thick	S.F.	0.027
Cant strip, 4" x 4"		
Treated wood	L.F.	0.023
Foamglass	L.F.	0.020
Mineral fiber	L.F.	0.020
New gravel for built-up roofing, 400 lb/sq	SQ.	1.600
Roof gravel (ballast)	C.Y.	4.000
Aluminum coating, top surfacing, for built-up roofing	SQ.	1.333
Remove 4-ply built-up roof (includes gravel)	SQ.	4.000
Remove & replace gravel, includes flood coat	SQ.	2.667

07530.10 Single-ply Roofing

	UNIT	MAN/ HOURS
Elastic sheet roofing		
Neoprene, 1/16" thick	S.F.	0.010
EPDM rubber		
45 mil	S.F.	0.010
60 mil	S.F.	0.010
PVC		
45 mil	S.F.	0.010
60 mil	S.F.	0.010
Flashing		
Pipe flashing, 90 mil thick		
1" pipe	EA.	0.200
2" pipe	EA.	0.200
3" pipe	EA.	0.211
4" pipe	EA.	0.211
5" pipe	EA.	0.222
6" pipe	EA.	0.222
8" pipe	EA.	0.235
10" pipe	EA.	0.267
12" pipe	EA.	0.267
Neoprene flashing, 60 mil thick strip		
6" wide	L.F.	0.067
12" wide	L.F.	0.100
18" wide	L.F.	0.133
24" wide	L.F.	0.200
Adhesives		
Mastic sealer, applied at joints only		
1/4" bead	L.F.	0.004
Fluid applied roofing		
Urethane, 2 components, elastomeric top membrane		
1" thick	S.F.	0.013
Vinyl liquid roofing, 2 coats, 2 mils per coat	S.F.	0.011
Silicone roofing, 2 coats sprayed, 16 mil per coat	S.F.	0.013
Inverted roof system		
Insulated membrane with coarse gravel ballast		

Membrane Roofing

07530.10 Single-ply Roofing

	UNIT	MAN/ HOURS
3 ply with 2" polystyrene	S.F.	0.013
Ballast, 3/4" through 1-1/2" dia. river gravel, 100lb/sf	S.F.	0.800
Walkway for membrane roofs, 1/2" thick	S.F.	0.027

Flashing And Sheet Metal

07610.10 Metal Roofing

	UNIT	MAN/ HOURS
Sheet metal roofing, copper, 16 oz, batten seam	SQ.	5.333
Standing seam	SQ.	5.000
Aluminum roofing, natural finish		
Corrugated, on steel frame		
.0175" thick	SQ.	2.286
.0215" thick	SQ.	2.286
.024" thick	SQ.	2.286
.032" thick	SQ.	2.286
V-beam, on steel frame		
.032" thick	SQ.	2.286
.040" thick	SQ.	2.286
.050" thick	SQ.	2.286
Ridge cap		
.019" thick	L.F.	0.027
Corrugated galvanized steel roofing, on steel frame		
28 ga.	SQ.	2.286
26 ga.	SQ.	2.286
24 ga.	SQ.	2.286
22 ga.	SQ.	2.286
26 ga., factory insulated with 1" polystyrene	SQ.	3.200
Ridge roll		
10" wide	L.F.	0.027
20" wide	L.F.	0.032

07620.10 Flashing And Trim

	UNIT	MAN/ HOURS
Counter flashing		
Aluminum, .032"	S.F.	0.080
Stainless steel, .015"	S.F.	0.080
Copper		
16 oz.	S.F.	0.080
20 oz.	S.F.	0.080
24 oz.	S.F.	0.080
32 oz.	S.F.	0.080
Valley flashing		
Aluminum, .032"	S.F.	0.050
Stainless steel, .015	S.F.	0.050
Copper		
16 oz.	S.F.	0.050
20 oz.	S.F.	0.067
24 oz.	S.F.	0.050

Flashing And Sheet Metal	UNIT	MAN/ HOURS
07620.10 Flashing And Trim		
32 oz.	S.F.	0.050
Base flashing		
Aluminum, .040"	S.F.	0.067
Stainless steel, .018"	S.F.	0.067
Copper		
16 oz.	S.F.	0.067
20 oz.	S.F.	0.050
24 oz.	S.F.	0.067
32 oz.	S.F.	0.067
Waterstop, "T" section, 22 ga.		
1-1/2" x 3"	L.F.	0.040
2" x 2"	L.F.	0.040
4" x 3"	L.F.	0.040
6" x 4"	L.F.	0.040
8" x 4"	L.F.	0.040
Scupper outlets		
10" x 10" x 4"	EA.	0.200
22" x 4" x 4"	EA.	0.200
8" x 8" x 5"	EA.	0.200
Flashing and trim, aluminum		
.019" thick	S.F.	0.057
.032" thick	S.F.	0.057
.040" thick	S.F.	0.062
Neoprene sheet flashing, .060" thick	S.F.	0.050
Copper, paper backed		
2 oz.	S.F.	0.080
5 oz.	S.F.	0.080
Drainage boots, roof, cast iron		
2 x 3	L.F.	0.100
3 x 4	L.F.	0.100
4 x 5	L.F.	0.107
4 x 6	L.F.	0.107
5 x 7	L.F.	0.114
Pitch pocket, copper, 16 oz.		
4 x 4	EA.	0.200
6 x 6	EA.	0.200
8 x 8	EA.	0.200
8 x 10	EA.	0.200
8 x 12	EA.	0.200
Reglets, copper 10 oz.	L.F.	0.053
Stainless steel, .020"	L.F.	0.053
Gravel stop		
Aluminum, .032"		
4"	L.F.	0.027
6"	L.F.	0.027
8"	L.F.	0.031
10"	L.F.	0.031
Copper, 16 oz.		
4"	L.F.	0.027
6"	L.F.	0.027
8"	L.F.	0.031
10"	L.F.	0.031

Flashing And Sheet Metal	UNIT	MAN/ HOURS
07620.20 Gutters And Downspouts		
Copper gutter and downspout		
Downspouts, 16 oz. copper		
Round		
3" dia.	L.F.	0.053
4" dia.	L.F.	0.053
Rectangular, corrugated		
2" x 3"	L.F.	0.050
3" x 4"	L.F.	0.050
Rectangular, flat surface		
2" x 3"	L.F.	0.053
3" x 4"	L.F.	0.053
Lead-coated copper downspouts		
Round		
3" dia.	L.F.	0.050
4" dia.	L.F.	0.057
Rectangular, corrugated		
2" x 3"	L.F.	0.053
3" x 4"	L.F.	0.053
Rectangular, plain		
2" x 3"	L.F.	0.053
3" x 4"	L.F.	0.053
Gutters, 16 oz. copper		
Half round		
4" wide	L.F.	0.080
5" wide	L.F.	0.089
Type K		
4" wide	L.F.	0.080
5" wide	L.F.	0.089
Lead-coated copper gutters		
Half round		
4" wide	L.F.	0.080
6" wide	L.F.	0.089
Type K		
4" wide	L.F.	0.080
5" wide	L.F.	0.089
Aluminum gutter and downspout		
Downspouts		
2" x 3"	L.F.	0.053
3" x 4"	L.F.	0.057
4" x 5"	L.F.	0.062
Round		
3" dia.	L.F.	0.053
4" dia.	L.F.	0.057
Gutters, stock units		
4" wide	L.F.	0.084
5" wide	L.F.	0.089
Galvanized steel gutter and downspout		
Downspouts, round corrugated		
3" dia.	L.F.	0.053
4" dia.	L.F.	0.053
5" dia.	L.F.	0.057
6" dia.	L.F.	0.057
Rectangular		
2" x 3"	L.F.	0.053

Flashing And Sheet Metal

07620.20 Gutters And Downspouts

Description	UNIT	MAN/ HOURS
3" x 4"	L.F.	0.050
4" x 4"	L.F.	0.050
Gutters, stock units		
5" wide		
Plain	L.F.	0.089
Painted	L.F.	0.089
6" wide		
Plain	L.F.	0.094
Painted	L.F.	0.094

Roofing Specialties

07700.10 Manufactured Specialties

Description	UNIT	MAN/ HOURS
Moisture relief vent		
Aluminum	EA.	0.114
Copper	EA.	0.114
Expansion joint		
Aluminum		
Opening to 2.5"	L.F.	0.057
Opening to 3.5"	L.F.	0.062
Copper, 16 oz.		
Opening to 2.5"	L.F.	0.057
Opening to 3.5"	L.F.	0.062
Butyl or neoprene		
4" wide		
16 oz. copper bellows	L.F.	0.067
28 ga. stainless steel bellows	L.F.	0.067
6" wide		
Copper bellows	L.F.	0.073
Stainless steel		
Opening to 2.5"	L.F.	0.057
Opening to 3.5"	L.F.	0.062
Smoke vent, 48" x 48"		
Aluminum	EA.	2.000
Galvanized steel	EA.	2.000
Heat/smoke vent, 48" x 96"		
Aluminum	EA.	2.667
Galvanized steel	EA.	2.667
Ridge vent strips		
Mill finish	L.F.	0.053
Connectors	EA.	0.200
End cap	EA.	0.229
Soffit vents		
Mill finish		
2-1/2" wide	L.F.	0.032
3" wide	L.F.	0.032
6" wide	L.F.	0.032

Roofing Specialties

07700.10 Manufactured Specialties

Description	UNIT	MAN/ HOURS
Roof hatches		
Steel, plain, primed		
2'6" x 3'0"	EA.	2.000
2'6" x 4'6"	EA.	2.667
2'6" x 8'0"	EA.	4.000
Galvanized steel		
2'6" x 3'0"	EA.	2.000
2'6" x 4'6"	EA.	2.667
2'6" x 8'0"	EA.	4.000
Aluminum		
2'6" x 3'0"	EA.	2.000
2'6" x 4'6"	EA.	2.667
2'6" x 8'0"	EA.	4.000
Ceiling access doors		
Swing up model, metal frame		
Steel door		
2'6" x 2'6"	EA.	0.800
2'6" x 3'0"	EA.	0.800
Aluminum door		
2'6" x 2'6"	EA.	0.800
2'6" x 3'0"	EA.	0.800
Swing down model, metal frame		
Steel door		
2'6" x 2'6"	EA.	0.800
2'6" x 3'0"	EA.	0.800
Aluminum door		
2'6" x 2'6"	EA.	0.800
2'6" x 3'0"	EA.	0.800
Gravity ventilators, with curb, base, damper and screen		
Stationary siphon		
6" dia.	EA.	0.533
12" dia.	EA.	0.533
24" dia.	EA.	0.800
30" dia.	EA.	0.800
Wind driven spinner		
6" dia.	EA.	0.533
12" dia.	EA.	0.533
24" dia.	EA.	0.800
36" dia.	EA.	0.800
Stationary mushroom		
16" dia.	EA.	0.800
30" dia.	EA.	1.000
36" dia.	EA.	1.333
42" dia.	EA.	1.600

Skylights

07810.10 Plastic Skylights	UNIT	MAN/HOURS
Single thickness, not including mounting curb		
2' x 4'	EA.	1.000
4' x 4'	EA.	1.333
5' x 5'	EA.	2.000
6' x 8'	EA.	2.667
Double thickness, not including mounting curb		
2' x 4'	EA.	1.000
4' x 4'	EA.	1.333
5' x 5'	EA.	2.000
6' x 8'	EA.	2.667
Metal framed skylights		
Translucent panels, 2-1/2" thick	S.F.	0.080
Continuous vaults, 8' wide		
Single glazed	S.F.	0.100
Double glazed	S.F.	0.114

Joint Sealers

07920.10 Caulking	UNIT	MAN/HOURS
Caulk exterior, two component		
1/4 x 1/2	L.F.	0.040
3/8 x 1/2	L.F.	0.044
1/2 x 1/2	L.F.	0.050
Caulk interior, single component		
1/4 x 1/2	L.F.	0.038
3/8 x 1/2	L.F.	0.042
1/2 x 1/2	L.F.	0.047
Butyl rubber fillers		
1/4" x 1/4"	L.F.	0.016
1/2" x 1/2"	L.F.	0.027
1/2" x 3/4"	L.F.	0.032
3/4" x 3/4"	L.F.	0.032
1" x 1"	L.F.	0.036
Seals, "O" ring type cord		
1/4" dia.	L.F.	0.020
1/2" dia.	L.F.	0.021
1" dia.	L.F.	0.022
1-1/4" dia.	L.F.	0.024
1-1/2" dia.	L.F.	0.025
1-3/4" dia.	L.F.	0.026
2" dia.	L.F.	0.027
Polyvinyl chloride, closed cell		
1/4" x 2"	L.F.	0.029
1/4" x 6"	L.F.	0.036
Silicon foam penetration seal		
1/4" x 1/2"	L.F.	0.010
1/2" x 1/2"	L.F.	0.013

Joint Sealers

07920.10 Caulking	UNIT	MAN/HOURS
1/2" x 3/4"	L.F.	0.016
3/4" x 3/4"	L.F.	0.020
1/8" x 1"	L.F.	0.010
1/8" x 3"	L.F.	0.016
1/4" x 3"	L.F.	0.020
1/4" x 6"	L.F.	0.027
1/2" x 6"	L.F.	0.062
1/2" x 9"	L.F.	0.100
1/2" x 12"	L.F.	0.145
Oil base sealants and caulking		
1/4" x 1/4"	L.F.	0.020
1/4" x 3/8"	L.F.	0.021
1/4" x 1/2"	L.F.	0.022
3/8" x 3/8"	L.F.	0.023
3/8" x 1/2"	L.F.	0.024
3/8" x 5/8"	L.F.	0.026
3/8" x 3/4"	L.F.	0.028
1/2" x 1/2"	L.F.	0.031
1/2" x 5/8"	L.F.	0.035
1/2" x 3/4"	L.F.	0.040
1/2" x 7/8"	L.F.	0.041
1/2" x 1"	L.F.	0.042
3/4" x 3/4"	L.F.	0.043
1" x 1"	L.F.	0.044
Polyurethane compounds		
1/4" x 1/4"	L.F.	0.020
1/4" x 3/8"	L.F.	0.021
1/4" x 1/2"	L.F.	0.022
3/8" x 3/8"	L.F.	0.023
3/8" x 1/2"	L.F.	0.024
3/8" x 5/8"	L.F.	0.026
3/8" x 3/4"	L.F.	0.028
1/2" x 1/2"	L.F.	0.031
1/2" x 5/8"	L.F.	0.035
1/2" x 3/4"	L.F.	0.040
1/2" x 7/8"	L.F.	0.041
1/2" x 1"	L.F.	0.044
3/4" x 3/4"	L.F.	0.043
3/4" x 1"	L.F.	0.044
Backer rod, polyethylene		
1/4"	L.F.	0.020
1/2"	L.F.	0.021
3/4"	L.F.	0.022
1"	L.F.	0.024

Metal

08110.10 Metal Doors

Metal	UNIT	MAN/HOURS
Flush hollow metal, standard duty, 20 ga., 1-3/8" thick		
2-6 x 6-8	EA.	0.889
2-8 x 6-8	EA.	0.889
3-0 x 6-8	EA.	0.889
1-3/4" thick		
2-6 x 6-8	EA.	0.889
2-8 x 6-8	EA.	0.889
3-0 x 6-8	EA.	0.889
2-6 x 7-0	EA.	0.889
2-8 x 7-0	EA.	0.889
3-0 x 7-0	EA.	0.889
Heavy duty, 20 ga., unrated, 1-3/4"		
2-8 x 6-8	EA.	0.889
3-0 x 6-8	EA.	0.889
2-8 x 7-0	EA.	0.889
3-0 x 7-0	EA.	0.889
3-4 x 7-0	EA.	0.889
18 ga., 1-3/4", unrated door		
2-0 x 7-0	EA.	0.889
2-4 x 7-0	EA.	0.889
2-6 x 7-0	EA.	0.889
2-8 x 7-0	EA.	0.889
3-0 x 7-0	EA.	0.889
3-4 x 7-0	EA.	0.889
2", unrated door		
2-0 x 7-0	EA.	1.000
2-4 x 7-0	EA.	1.000
2-6 x 7-0	EA.	1.000
2-8 x 7-0	EA.	1.000
3-0 x 7-0	EA.	1.000
3-4 x 7-0	EA.	1.000
Galvanized metal door		
3-0 x 7-0	EA.	1.000

08110.40 Metal Door Frames

Metal	UNIT	MAN/HOURS
Hollow metal, stock, 18 ga., 4-3/4" x 1-3/4"		
2-0 x 7-0	EA.	1.000
2-4 x 7-0	EA.	1.000
2-6 x 7-0	EA.	1.000
2-8 x 7-0	EA.	1.000
3-0 x 7-0	EA.	1.000
4-0 x 7-0	EA.	1.333
5-0 x 7-0	EA.	1.333
6-0 x 7-0	EA.	1.333
16 ga., 6-3/4" x 1-3/4"		
2-0 x 7-0	EA.	1.000
2-4 x 7-0	EA.	1.000
2-6 x 7-0	EA.	1.000
2-8 x 7-0	EA.	1.000
3-0 x 7-0	EA.	1.000
4-0 x 7-0	EA.	1.333
6-0 x 7-0	EA.	1.333
Transom frame		
3-4 x 1-6	EA.	1.000
3-8 x 1-6	EA.	1.000
6-4 x 1-6	EA.	1.000
Transom sash		
3-0 x 1-4	EA.	1.000
3-4 x 1-4	EA.	1.000
6-0 x 1-4	EA.	1.000
Sidelights, complete		
1-0 x 7-2	EA.	1.000
1-4 x 7-2	EA.	1.000
1-0 x 8-8	EA.	1.000
1-6 x 8-8	EA.	1.000
16 ga., 4-3/4" x 1-3/4"		
2-0 x 7-0	EA.	1.000
2-4 x 7-0	EA.	1.000
2-6 x 7-0	EA.	1.000
2-8 x 7-0	EA.	1.000
3-0 x 7-0	EA.	1.000
4-0 x 7-0	EA.	1.333
6-0 x 7-0	EA.	1.333
3-4 x 1-6	EA.	1.000
3-8 x 1-6	EA.	1.000
6-4 x 1-6	EA.	1.000
Transom sash		
3-0 x 1-4	EA.	1.000
3-4 x 1-4	EA.	1.000
6-0 x 1-4	EA.	1.000
Sidelights, complete		
1-0 x 7-2	EA.	1.000
1-4 x 7-2	EA.	1.000
1-0 x 8-8	EA.	1.000
1-4 x 8-8	EA.	1.000
16 ga., 5-3/4" x 1-3/4"		
2-0 x 7-0	EA.	1.000
2-4 x 7-0	EA.	1.000
2-6 x 7-0	EA.	1.000
2-8 x 7-0	EA.	1.000
3-0 x 7-0	EA.	1.000
4-0 x 7-0	EA.	1.333
5-0 x 7-0	EA.	1.333
6-0 x 7-0	EA.	1.333
Mullions, vertical		
5-1/4" x 1-3/4"	L.F.	0.100
5-1/4" x 2"	L.F.	0.100
Horizontal		
5-1/4" x 1-3/4"	L.F.	0.100
5-1/4" x 2"	L.F.	0.100

08120.10 Aluminum Doors

Metal	UNIT	MAN/HOURS
Aluminum doors, commercial		
Narrow stile		
2-6 x 7-0	EA.	4.000
3-0 x 7-0	EA.	4.000
3-6 x 7-0	EA.	4.000
Pair		

Metal	UNIT	MAN/ HOURS
08120.10 Aluminum Doors		
5-0 x 7-0	EA.	8.000
6-0 x 7-0	EA.	8.000
7-0 x 7-0	EA.	8.000
Wide stile		
2-6 x 7-0	EA.	4.000
3-0 x 7-0	EA.	4.000
3-6 x 7-0	EA.	4.000
Pair		
5-0 x 7-0	EA.	8.000
6-0 x 7-0	EA.	8.000
7-0 x 7-0	EA.	8.000

Wood And Plastic	UNIT	MAN/ HOURS
08210.10 Wood Doors		
Solid core, 1-3/8" thick		
Birch faced		
2-4 x 7-0	EA.	1.000
2-8 x 7-0	EA.	1.000
3-0 x 7-0	EA.	1.000
3-4 x 7-0	EA.	1.000
2-4 x 6-8	EA.	1.000
2-6 x 6-8	EA.	1.000
2-8 x 6-8	EA.	1.000
3-0 x 6-8	EA.	1.000
Lauan faced		
2-4 x 6-8	EA.	1.000
2-8 x 6-8	EA.	1.000
3-0 x 6-8	EA.	1.000
3-4 x 6-8	EA.	1.000
Tempered hardboard faced		
2-4 x 7-0	EA.	1.000
2-8 x 7-0	EA.	1.000
3-0 x 7-0	EA.	1.000
3-4 x 7-0	EA.	1.000
Hollow core, 1-3/8" thick		
Birch faced		
2-4 x 7-0	EA.	1.000
2-8 x 7-0	EA.	1.000
3-0 x 7-0	EA.	1.000
3-4 x 7-0	EA.	1.000
Lauan faced		
2-4 x 6-8	EA.	1.000
2-6 x 6-8	EA.	1.000
2-8 x 6-8	EA.	1.000
3-0 x 6-8	EA.	1.000
3-4 x 6-8	EA.	1.000

Wood And Plastic	UNIT	MAN/ HOURS
08210.10 Wood Doors		
Tempered hardboard faced		
2-4 x 7-0	EA.	1.000
2-6 x 7-0	EA.	1.000
2-8 x 7-0	EA.	1.000
3-0 x 7-0	EA.	1.000
3-4 x 7-0	EA.	1.000
Solid core, 1-3/4" thick		
Birch faced		
2-4 x 7-0	EA.	1.000
2-6 x 7-0	EA.	1.000
2-8 x 7-0	EA.	1.000
3-0 x 7-0	EA.	1.000
3-4 x 7-0	EA.	1.000
Lauan faced		
2-4 x 7-0	EA.	1.000
2-6 x 7-0	EA.	1.000
2-8 x 7-0	EA.	1.000
3-4 x 7-0	EA.	1.000
3-0 x 7-0	EA.	1.000
Tempered hardboard faced		
2-4 x 7-0	EA.	1.000
2-6 x 7-0	EA.	1.000
2-8 x 7-0	EA.	1.000
3-0 x 7-0	EA.	1.000
3-4 x 7-0	EA.	1.000
Hollow core, 1-3/4" thick		
Birch faced		
2-4 x 7-0	EA.	1.000
2-6 x 7-0	EA.	1.000
2-8 x 7-0	EA.	1.000
3-0 x 7-0	EA.	1.000
3-4 x 7-0	EA.	1.000
Lauan faced		
2-4 x 6-8	EA.	1.000
2-6 x 6-8	EA.	1.000
2-8 x 6-8	EA.	1.000
3-0 x 6-8	EA.	1.000
3-4 x 6-8	EA.	1.000
Tempered hardboard		
2-4 x 7-0	EA.	1.000
2-6 x 7-0	EA.	1.000
2-8 x 7-0	EA.	1.000
3-0 x 7-0	EA.	1.000
3-4 x 7-0	EA.	1.000
Add-on, louver	EA.	0.800
Glass	EA.	0.800
Exterior doors, 3-0 x 7-0 x 2-1/2", solid core		
Carved		
One face	EA.	2.000
Two faces	EA.	2.000
Closet doors, 1-3/4" thick		
Bi-fold or bi-passing, includes frame and trim		
Paneled		
4-0 x 6-8	EA.	1.333

08 DOORS AND WINDOWS

Wood And Plastic	UNIT	MAN/ HOURS
08210.10 Wood Doors		
6-0 x 6-8	EA.	1.333
Louvered		
4-0 x 6-8	EA.	1.333
6-0 x 6-8	EA.	1.333
Flush		
4-0 x 6-8	EA.	1.333
6-0 x 6-8	EA.	1.333
Primed		
4-0 x 6-8	EA.	1.333
6-0 x 6-8	EA.	1.333
08210.90 Wood Frames		
Frame, interior, pine		
2-6 x 6-8	EA.	1.143
2-8 x 6-8	EA.	1.143
3-0 x 6-8	EA.	1.143
5-0 x 6-8	EA.	1.143
6-0 x 6-8	EA.	1.143
2-6 x 7-0	EA.	1.143
2-8 x 7-0	EA.	1.143
3-0 x 7-0	EA.	1.143
5-0 x 7-0	EA.	1.600
6-0 x 7-0	EA.	1.600
Exterior, custom, with threshold, including trim		
Walnut		
3-0 x 7-0	EA.	2.000
6-0 x 7-0	EA.	2.000
Oak		
3-0 x 7-0	EA.	2.000
6-0 x 7-0	EA.	2.000
Pine		
2-4 x 7-0	EA.	1.600
2-6 x 7-0	EA.	1.600
2-8 x 7-0	EA.	1.600
3-0 x 7-0	EA.	1.600
3-4 x 7-0	EA.	1.600
6-0 x 7-0	EA.	2.667
08300.10 Special Doors		
Vault door and frame, class 5, steel	EA.	8.000
Overhead door, coiling insulated		
Chain gear, no frame, 12' x 12'	EA.	10.000
Aluminum, bronze glass panels, 12-9 x 13-0	EA.	8.000
Garage door, flush insulated metal, primed, 9-0 x 7-0	EA.	2.667
Sliding metal fire doors, motorized, fusible link, 3 hr.		
3-0 x 6-8	EA.	16.000
3-8 x 6-8	EA.	16.000
4-0 x 8-0	EA.	16.000
5-0 x 8-0	EA.	16.000
Metal clad doors, including electric motor		
Light duty		
Minimum	S.F.	0.133
Maximum	S.F.	0.320
Heavy duty		

Wood And Plastic	UNIT	MAN/ HOURS
08300.10 Special Doors		
Minimum	S.F.	0.400
Maximum	S.F.	0.500
Hangar doors, based on 150' openings		
To 20' high	S.F.	0.096
20' to 40' high	S.F.	0.060
40' to 60' high	S.F.	0.040
60' to 80' high	S.F.	0.024
Over 80' high	S.F.	0.019
Counter doors, (roll-up shutters), standard, manual		
Opening, 4' high		
4' wide	EA.	6.667
6' wide	EA.	6.667
8' wide	EA.	7.273
10' wide	EA.	10.000
14' wide	EA.	10.000
6' high		
4' wide	EA.	6.667
6' wide	EA.	7.273
8' wide	EA.	8.000
10' wide	EA.	10.000
14' wide	EA.	11.429
Service doors, (roll up shutters), standard, manual		
Opening		
8' high x 8' wide	EA.	4.444
10' high x 10' wide	EA.	6.667
12' high x 12' wide	EA.	10.000
14' high x 14' wide	EA.	13.333
16' high x 14' wide	EA.	13.333
20' high x 14' wide	EA.	20.000
24' high x 16' wide	EA.	17.778
Roll-up doors		
13-0 high x 14-0 wide	EA.	11.429
12-0 high x 14-0 wide	EA.	11.429
Top coiling grilles, manually operated, steel or aluminum		
Opening, 4' high x		
4' wide	EA.	3.200
6' wide	EA.	3.200
8' wide	EA.	4.444
12' wide	EA.	4.444
16' wide	EA.	6.667
6' high x		
4' wide	EA.	6.667
6' wide	EA.	7.273
8' wide	EA.	8.000
12' wide	EA.	8.889
16' wide	EA.	11.429
Side coiling grilles, manually operated, aluminum		
Opening, 8' high x		
18' wide	EA.	60.000
24' wide	EA.	68.571
12' high x		
12' wide	EA.	60.000
18' wide	EA.	68.571
24' wide	EA.	80.000

08 DOORS AND WINDOWS

Wood And Plastic

08300.10 Special Doors

Description	UNIT	MAN/ HOURS
Accordion folding doors, tracks and fittings included		
Vinyl covered, 2 layers	S.F.	0.320
Woven mahogany and vinyl	S.F.	0.320
Economy vinyl	S.F.	0.320
Rigid polyvinyl chloride	S.F.	0.320
Sectional wood overhead doors, frames not included		
Commercial grade, heavy duty, 1-3/4" thick, manual		
8' x 8'	EA.	6.667
10' x 10'	EA.	7.273
12' x 12'	EA.	8.000
Chain hoist		
12' x 16' high	EA.	13.333
14' x 14' high	EA.	10.000
20' x 8' high	EA.	16.000
16' high	EA.	20.000
Sectional metal overhead doors, complete		
Residential grade, manual		
9' x 7'	EA.	3.200
16' x 7'	EA.	4.000
Commercial grade		
8' x 8'	EA.	6.667
10' x 10'	EA.	7.273
12' x 12'	EA.	8.000
20' x 14', with chain hoist	EA.	16.000
Sliding glass doors		
Tempered plate glass, 1/4" thick		
6' wide		
Economy grade	EA.	2.667
Premium grade	EA.	2.667
12' wide		
Economy grade	EA.	4.000
Premium grade	EA.	4.000
Insulating glass, 5/8" thick		
6' wide		
Economy grade	EA.	2.667
Premium grade	EA.	2.667
12' wide		
Economy grade	EA.	4.000
Premium grade	EA.	4.000
1" thick		
6' wide		
Economy grade	EA.	2.667
Premium grade	EA.	2.667
12' wide		
Economy grade	EA.	4.000
Premium grade	EA.	4.000
Vertical lift doors, channel frame construction		
20' high x		
10' wide	EA.	9.600
15' wide	EA.	9.600
20' wide	EA.	17.143
25' wide	EA.	17.143
25' high x		
20' wide	EA.	17.143
25' wide	EA.	20.000
30' high x		
25' wide	EA.	20.000
30' wide	EA.	20.000
35' wide	EA.	20.000
Residential storm door		
Minimum	EA.	1.333
Average	EA.	1.333
Maximum	EA.	2.000

Storefronts

08410.10 Storefronts

Description	UNIT	MAN/ HOURS
Storefront, aluminum and glass		
Minimum	S.F.	0.100
Average	S.F.	0.114
Maximum	S.F.	0.133
Entrance doors, premium, incl. glass, closers, panic dev.,etc.		
1/2" thick glass		
3' x 7'	EA.	6.667
6' x 7'	EA.	10.000
3/4" thick glass		
3' x 7'	EA.	6.667
6' x 7'	EA.	10.000
1" thick glass		
3' x 7'	EA.	6.667
6' x 7'	EA.	10.000
Revolving doors		
7'diameter, 7' high		
Minimum	EA.	60.000
Average	EA.	96.000
Maximum	EA.	120.000

Metal Windows

08510.10 Steel Windows

Description	UNIT	MAN/ HOURS
Steel windows, primed		
Casements		
Operable		
Minimum	S.F.	0.047
Maximum	S.F.	0.053

08 DOORS AND WINDOWS

Metal Windows

08510.10 Steel Windows

	UNIT	MAN/ HOURS
Fixed sash	S.F.	0.040
Double hung	S.F.	0.044
Industrial windows		
Horizontally pivoted sash	S.F.	0.053
Fixed sash	S.F.	0.044
Security sash		
Operable	S.F.	0.053
Fixed	S.F.	0.044
Picture window	S.F.	0.044
Projecting sash		
Minimum	S.F.	0.050
Maximum	S.F.	0.050
Mullions	L.F.	0.040

08520.10 Aluminum Windows

	UNIT	MAN/ HOURS
Jalousie		
3-0 x 4-0	EA.	1.000
3-0 x 5-0	EA.	1.000
Fixed window		
6 sf to 8 sf	S.F.	0.114
12 sf to 16 sf	S.F.	0.089
Projecting window		
6 sf to 8 sf	S.F.	0.200
12 sf to 16 sf	S.F.	0.133
Horizontal sliding		
6 sf to 8 sf	S.F.	0.100
12 sf to 16 sf	S.F.	0.080
Double hung		
6 sf to 8 sf	S.F.	0.160
10 sf to 12 sf	S.F.	0.133
Storm window, 0.5 cfm, up to		
60 u.i. (united inches)	EA.	0.400
70 u.i.	EA.	0.400
80 u.i.	EA.	0.400
90 u.i.	EA.	0.444
100 u.i.	EA.	0.444
2.0 cfm, up to		
60 u.i.	EA.	0.400
70 u.i.	EA.	0.400
80 u.i.	EA.	0.400
90 u.i.	EA.	0.444
100 u.i.	EA.	0.444

Wood And Plastic

08600.10 Wood Windows

	UNIT	MAN/ HOURS
Double hung		
24" x 36"		
Minimum	EA.	0.800
Average	EA.	1.000
Maximum	EA.	1.333
24" x 48"		
Minimum	EA.	0.800
Average	EA.	1.000
Maximum	EA.	1.333
30" x 48"		
Minimum	EA.	0.889
Average	EA.	1.143
Maximum	EA.	1.600
30" x 60"		
Minimum	EA.	0.889
Average	EA.	1.143
Maximum	EA.	1.600
Casement		
1 leaf, 22" x 38" high		
Minimum	EA.	0.800
Average	EA.	1.000
Maximum	EA.	1.333
2 leaf, 50" x 50" high		
Minimum	EA.	1.000
Average	EA.	1.333
Maximum	EA.	2.000
3 leaf, 71" x 62" high		
Minimum	EA.	1.000
Average	EA.	1.333
Maximum	EA.	2.000
4 leaf, 95" x 75" high		
Minimum	EA.	1.143
Average	EA.	1.600
Maximum	EA.	2.667
5 leaf, 119" x 75" high		
Minimum	EA.	1.143
Average	EA.	1.600
Maximum	EA.	2.667
Picture window, fixed glass, 54" x 54" high		
Minimum	EA.	1.000
Average	EA.	1.143
Maximum	EA.	1.333
68" x 55" high		
Minimum	EA.	1.000
Average	EA.	1.143
Maximum	EA.	1.333
Sliding, 40" x 31" high		
Minimum	EA.	0.800
Average	EA.	1.000
Maximum	EA.	1.333
52" x 39" high		
Minimum	EA.	1.000
Average	EA.	1.143
Maximum	EA.	1.333

08 DOORS AND WINDOWS

Wood And Plastic

08600.10 Wood Windows	UNIT	MAN/ HOURS
64" x 72" high		
Minimum	EA.	1.000
Average	EA.	1.333
Maximum	EA.	1.600
Awning windows		
34" x 21" high		
Minimum	EA.	0.800
Average	EA.	1.000
Maximum	EA.	1.333
40" x 21" high		
Minimum	EA.	0.889
Average	EA.	1.143
Maximum	EA.	1.600
48" x 27" high		
Minimum	EA.	0.889
Average	EA.	1.143
Maximum	EA.	1.600
60" x 36" high		
Minimum	EA.	1.000
Average	EA.	1.333
Maximum	EA.	1.600
Window frame, milled		
Minimum	L.F.	0.160
Average	L.F.	0.200
Maximum	L.F.	0.267

Hardware

08710.20 Locksets	UNIT	MAN/ HOURS
Latchset, heavy duty		
Cylindrical	EA.	0.500
Mortise	EA.	0.800
Lockset, heavy duty		
Cylindrical	EA.	0.500
Mortise	EA.	0.800
Mortise locks and latchsets, chrome		
Latchset passage or closet latch	EA.	0.667
Privacy (bath or bedroom)	EA.	0.667
Entry lockset	EA.	0.667
Classroom lockset (outside key operated)	EA.	0.667
Storeroom lock	EA.	0.667
Front door lock	EA.	0.667
Dormitory or exit lock	EA.	0.667
Preassembled locks and latches, brass		
Latchset, passage or closet latch	EA.	0.667
Lockset		
Privacy (bath or bathroom)	EA.	0.667

Hardware

08710.20 Locksets	UNIT	MAN/ HOURS
Entry lock	EA.	0.667
Classroom lock (outside key, operated)	EA.	0.667
Storeroom lock	EA.	0.667
Bored locks and latches, satin chrome plated		
Latchset passage or closet latch	EA.	0.667
Lockset		
Privacy (bath or bedroom)	EA.	0.667
Entry lock	EA.	0.667
Classroom lock	EA.	0.667
Corridor lock	EA.	0.667
Miscellaneous locks		
Exit lock with alarm, single door	EA.	3.200
Electric strike		
Rim mounted wrought steel	EA.	2.000
Mortised, wrought steel with bronze plating	EA.	3.200
Dead bolt		
Bored, wrought brass, keyed both sides	EA.	1.333
Mortised, cast brass	EA.	1.333
Lockset, cipher, mechanical	EA.	0.800

08710.30 Closers	UNIT	MAN/ HOURS
Door closers		
Surface mounted, traditional type, parallel arm		
Standard	EA.	1.000
Heavy duty	EA.	1.000
Modern type, parallel arm, standard duty	EA.	1.000
Overhead, concealed, pivot hung, single acting		
Interior	EA.	1.000
Exterior	EA.	1.000
Floor concealed, single acting, offset, pivoted		
Interior	EA.	2.667
Exterior	EA.	2.667

08710.40 Door Trim	UNIT	MAN/ HOURS
Door bumper, bronze, wall type	EA.	0.160
Wall type, 4" dia. with convex rubber pad, aluminum	EA.	0.160
Floor type		
Aluminum	EA.	0.160
Brass	EA.	0.160
Door holders		
Wall type, bronze	EA.	0.160
Overhead	EA.	0.400
Floor type	EA.	0.400
Plunger type	EA.	0.400
Wall type, aluminum	EA.	0.400
Surface bolt	EA.	0.160
Panic device		
Rim type with thumb piece	EA.	2.000
Mortise	EA.	2.000
Vertical rod	EA.	2.000
Labelled, rim type	EA.	2.000
Mortise	EA.	2.000
Vertical rod	EA.	2.000
Silencers, rubber type	EA.	0.016

08 DOORS AND WINDOWS

Hardware	UNIT	MAN/ HOURS
08710.40 Door Trim		
Dust proof strike with plate, brass	EA.	0.267
Flush bolt, lever extension, brass, rated	EA.	0.160
Surface bolt with strike, brass, 6" long	EA.	0.160
Door coordinator, labeled, brass, satin chrome	EA.	0.571
Door plates		
Kick plate, aluminum, 3 beveled edges		
10" x 28"	EA.	0.400
10" x 30"	EA.	0.400
10" x 34"	EA.	0.400
10" x 38"	EA.	0.400
Push plate, 4" x 16"		
Aluminum	EA.	0.160
Bronze	EA.	0.160
Stainless steel	EA.	0.160
Armor plate, 40" x 34"	EA.	0.320
Pull handle, 4" x 16"		
Aluminum	EA.	0.160
Bronze	EA.	0.160
Stainless steel	EA.	0.160
Hasp assembly		
3"	EA.	0.133
4-1/2"	EA.	0.178
6"	EA.	0.229
Electro-magnetic door holder		
Wall mounted	EA.	2.667
Floor mounted	EA.	2.667
Smoke detector door holder		
Photo electric type	EA.	2.667
Ionization type	EA.	2.667
Pneumatic operators, activated by rubber mats		
Swing		
Single	EA.	6.667
Double	EA.	10.000
Sliding		
Single	EA.	6.667
Double	EA.	10.000
08710.60 Weatherstripping		
Weatherstrip, head and jamb, metal strip, neoprene bulb		
Standard duty	L.F.	0.044
Heavy duty	L.F.	0.050
Spring type		
Metal doors	EA.	2.000
Wood doors	EA.	2.667
Sponge type with adhesive backing	EA.	0.800
Astragal		
1-3/4" x 13 ga., aluminum	L.F.	0.067
1-3/8" x 5/8", oak	L.F.	0.053
Thresholds		
Bronze	L.F.	0.200
Aluminum		
Plain	L.F.	0.200
Vinyl insert	L.F.	0.200
Aluminum with grit	L.F.	0.200

Hardware	UNIT	MAN/ HOURS
08710.60 Weatherstripping		
Steel		
Plain	L.F.	0.200
Interlocking	L.F.	0.667

Glazing	UNIT	MAN/ HOURS
08810.10 Glazing		
Sheet glass, 1/8" thick	S.F.	0.044
Plate glass, bronze or grey, 1/4" thick	S.F.	0.073
Clear	S.F.	0.073
Polished	S.F.	0.073
Plexiglass		
1/8" thick	S.F.	0.073
1/4" thick	S.F.	0.044
Float glass, clear		
3/16" thick	S.F.	0.067
1/4" thick	S.F.	0.073
5/16" thick	S.F.	0.080
3/8" thick	S.F.	0.100
1/2" thick	S.F.	0.133
5/8" thick	S.F.	0.160
3/4" thick	S.F.	0.200
1" thick	S.F.	0.267
Tinted glass, polished plate, twin ground		
3/16" thick	S.F.	0.067
1/4" thick	S.F.	0.073
3/8" thick	S.F.	0.100
1/2" thick	S.F.	0.133
Total, full vision, all glass window system		
To 10' high		
Minimum	S.F.	0.200
Average	S.F.	0.200
Maximum	S.F.	0.200
10' to 20' high		
Minimum	S.F.	0.200
Average	S.F.	0.200
Maximum	S.F.	0.200
Insulated glass, bronze or gray		
1/2" thick	S.F.	0.133
1" thick	S.F.	0.200
Spandrel glass, polished bronze/grey, 1 side, 1/4" thick	S.F.	0.073
Tempered glass (safety)		
Clear sheet glass		
1/8" thick	S.F.	0.044
3/16" thick	S.F.	0.062
Clear float glass		
1/4" thick	S.F.	0.067

Glazing

08810.10 Glazing	UNIT	MAN/ HOURS
5/16" thick	S.F.	0.080
3/8" thick	S.F.	0.100
1/2" thick	S.F.	0.133
5/8" thick	S.F.	0.160
3/4" thick	S.F.	0.267
Tinted float glass		
3/16" thick	S.F.	0.062
1/4" thick	S.F.	0.067
3/8" thick	S.F.	0.100
1/2" thick	S.F.	0.133
Laminated glass		
Float safety glass with polyvinyl plastic interlayer		
1/4", sheet or float		
Two lites, 1/8" thick, clear glass	S.F.	0.067
1/2" thick, float glass		
Two lites, 1/4" thick, clear glass	S.F.	0.133
Tinted glass	S.F.	0.133
Insulating glass, two lites, clear float glass		
1/2" thick	S.F.	0.133
5/8" thick	S.F.	0.160
3/4" thick	S.F.	0.200
7/8" thick	S.F.	0.229
1" thick	S.F.	0.267
Glass seal edge		
3/8" thick	S.F.	0.133
Tinted glass		
1/2" thick	S.F.	0.133
1" thick	S.F.	0.267
Tempered, clear		
1" thick	S.F.	0.267
Wire reinforced	S.F.	0.267
Plate mirror glass		
1/4" thick		
15 sf	S.F.	0.080
Over 15 sf	S.F.	0.073
Door type, 1/4" thick	S.F.	0.080
Transparent, one way vision, 1/4" thick	S.F.	0.080
Sheet mirror glass		
3/16" thick	S.F.	0.080
1/4" thick	S.F.	0.067
Wall tiles, 12" x 12"		
Clear glass	S.F.	0.044
Veined glass	S.F.	0.044
Wire glass, 1/4" thick		
Clear	S.F.	0.267
Hammered	S.F.	0.267
Obscure	S.F.	0.267
Bullet resistant glass, plate glass with inter-leaved vinyl		
1-3/16" thick		
To 15 sf	S.F.	0.400
Over 15 sf	S.F.	0.400
2" thick		
To 15 sf	S.F.	0.667
Over 15 sf	S.F.	0.667

Glazing

08810.10 Glazing	UNIT	MAN/ HOURS
Glazing accessories		
Neoprene glazing gaskets		
1/4" glass	L.F.	0.032
3/8" glass	L.F.	0.033
1/2" glass	L.F.	0.035
3/4" glass	L.F.	0.036
1" glass	L.F.	0.040
Mullion section		
1/4" glass	L.F.	0.016
3/8" glass	L.F.	0.020
1/2" glass	L.F.	0.023
3/4" glass	L.F.	0.027
1" glass	L.F.	0.032
Molded corners	EA.	0.533

Glazed Curtain Walls

08910.10 Glazed Curtain Walls	UNIT	MAN/ HOURS
Curtain wall, aluminum system, framing sections		
2" x 3"		
Jamb	L.F.	0.067
Horizontal	L.F.	0.067
Mullion	L.F.	0.067
2" x 4"		
Jamb	L.F.	0.100
Horizontal	L.F.	0.100
Mullion	L.F.	0.100
3" x 5-1/2"		
Jamb	L.F.	0.100
Horizontal	L.F.	0.100
Mullion	L.F.	0.100
4" corner mullion	L.F.	0.133
Coping sections		
1/8" x 8"	L.F.	0.133
1/8" x 9"	L.F.	0.133
1/8" x 12-1/2"	L.F.	0.160
Sill section		
1/8" x 6"	L.F.	0.080
1/8" x 7"	L.F.	0.080
1/8" x 8-1/2"	L.F.	0.080
Column covers, aluminum		
1/8" x 26"	L.F.	0.200
1/8" x 34"	L.F.	0.211
1/8" x 38"	L.F.	0.211
Doors		
Aluminum framed, standard hardware		
Narrow stile		

Glazed Curtain Walls	UNIT	MAN/ HOURS
08910.10 Glazed Curtain Walls		
2-6 x 7-0	EA.	4.000
3-0 x 7-0	EA.	4.000
3-6 x 7-0	EA.	4.000
Wide stile		
2-6 x 7-0	EA.	4.000
3-0 x 7-0	EA.	4.000
3-6 x 7-0	EA.	4.000
Flush panel doors, to match adjacent wall panels		
2-6 x 7-0	EA.	5.000
3-0 x 7-0	EA.	5.000
3-6 x 7-0	EA.	5.000
Wall panel, insulated		
"U"=.08	S.F.	0.067
"U"=.10	S.F.	0.067
"U"=.15	S.F.	0.067
Window wall system, complete		
Minimum	S.F.	0.080
Average	S.F.	0.089
Maximum	S.F.	0.114

Support Systems

09110.10 Metal Studs

Support Systems	UNIT	MAN/ HOURS
Studs, non load bearing, galvanized		
2-1/2", 20 ga.		
12" o.c.	S.F.	0.017
16" o.c.	S.F.	0.013
25 ga.		
12" o.c.	S.F.	0.017
16" o.c.	S.F.	0.013
24" o.c.	S.F.	0.011
3-5/8", 20 ga.		
12" o.c.	S.F.	0.020
16" o.c.	S.F.	0.016
24" o.c.	S.F.	0.013
25 ga.		
12" o.c.	S.F.	0.020
16" o.c.	S.F.	0.016
24" o.c.	S.F.	0.013
4", 20 ga.		
12" o.c.	S.F.	0.020
16" o.c.	S.F.	0.016
24" o.c.	S.F.	0.013
25 ga.		
12" o.c.	S.F.	0.020
16" o.c.	S.F.	0.016
24" o.c.	S.F.	0.013
6", 20 ga.		
12" o.c.	S.F.	0.025
16" o.c.	S.F.	0.020
24" o.c.	S.F.	0.017
25 ga.		
12" o.c.	S.F.	0.025
16" o.c.	S.F.	0.020
24" o.c.	S.F.	0.017
Load bearing studs, galvanized		
3-5/8", 16 ga.		
12" o.c.	S.F.	0.020
16" o.c.	S.F.	0.016
18 ga.		
12" o.c.	S.F.	0.013
16" o.c.	S.F.	0.016
4", 16 ga.		
12" o.c.	S.F.	0.020
16" o.c.	S.F.	0.016
6", 16 ga.		
12" o.c.	S.F.	0.025
16" o.c.	S.F.	0.020
Furring		
On beams and columns		
7/8" channel	L.F.	0.053
1-1/2" channel	L.F.	0.062
On ceilings		
3/4" furring channels		
12" o.c.	S.F.	0.033
16" o.c.	S.F.	0.032
24" o.c.	S.F.	0.029
1-1/2" furring channels		
12" o.c.	S.F.	0.036
16" o.c.	S.F.	0.033
24" o.c.	S.F.	0.031
On walls		
3/4" furring channels		
12" o.c.	S.F.	0.027
16" o.c.	S.F.	0.025
24" o.c.	S.F.	0.024
1-1/2" furring channels		
12" o.c.	S.F.	0.029
16" o.c.	S.F.	0.027
24" o.c.	S.F.	0.025

Lath And Plaster

09205.10 Gypsum Lath

Lath And Plaster	UNIT	MAN/ HOURS
Gypsum lath, 1/2" thick		
Clipped	S.Y.	0.044
Nailed	S.Y.	0.050

09205.20 Metal Lath

Lath And Plaster	UNIT	MAN/ HOURS
Diamond expanded, galvanized		
2.5 lb., on walls		
Nailed	S.Y.	0.100
Wired	S.Y.	0.114
On ceilings		
Nailed	S.Y.	0.114
Wired	S.Y.	0.133
3.4 lb., on walls		
Nailed	S.Y.	0.100
Wired	S.Y.	0.114
On ceilings		
Nailed	S.Y.	0.114
Wired	S.Y.	0.133
Flat rib		
2.75 lb., on walls		
Nailed	S.Y.	0.100
Wired	S.Y.	0.114
On ceilings		
Nailed	S.Y.	0.114
Wired	S.Y.	0.133
3.4 lb., on walls		
Nailed	S.Y.	0.100
Wired	S.Y.	0.114
On ceilings		
Nailed	S.Y.	0.114

Lath And Plaster	UNIT	MAN/ HOURS
09205.20 Metal Lath		
Wired	S.Y.	0.133
Stucco lath		
1.8 lb.	S.Y.	0.100
3.6 lb.	S.Y.	0.100
Paper backed		
Minimum	S.Y.	0.080
Maximum	S.Y.	0.114
09205.60 Plaster Accessories		
Expansion joint, 3/4", 26 ga., galvanized, one piece	L.F.	0.020
Plaster corner beads, 3/4", galvanized	L.F.	0.023
Casing bead, expanded flange, galvanized	L.F.	0.020
Expanded wing, 1-1/4" wide, galvanized	L.F.	0.020
Joint clips for lath	EA.	0.004
Metal base, galvanized, 2-1/2" high	L.F.	0.027
Stud clips for gypsum lath	EA.	0.004
Sound deadening board, 1/4", nailed or clipped	S.F.	0.013
09210.10 Plaster		
Gypsum plaster, trowel finish, 2 coats		
Ceilings	S.Y.	0.250
Walls	S.Y.	0.235
3 coats		
Ceilings	S.Y.	0.348
Walls	S.Y.	0.308
Vermiculite plaster		
2 coats		
Ceilings	S.Y.	0.381
Walls	S.Y.	0.348
3 coats		
Ceilings	S.Y.	0.471
Walls	S.Y.	0.421
Keenes cement plaster		
2 coats		
Ceilings	S.Y.	0.308
Walls	S.Y.	0.267
3 coats		
Ceilings	S.Y.	0.348
Walls	S.Y.	0.308
On columns, add to installation, 50%	S.Y.	
Chases, fascia, and soffits, add to installation, 50%	S.Y.	
Beams, add to installation, 50%	S.Y.	
09220.10 Portland Cement Plaster		
Stucco, portland, gray, 3 coat, 1" thick		
Sand finish	S.Y.	0.348
Trowel finish	S.Y.	0.364
White cement		
Sand finish	S.Y.	0.364
Trowel finish	S.Y.	0.400
Scratch coat		
For ceramic tile	S.Y.	0.080
For quarry tile	S.Y.	0.080
Portland cement plaster		

Lath And Plaster	UNIT	MAN/ HOURS
09220.10 Portland Cement Plaster		
2 coats, 1/2"	S.Y.	0.160
3 coats, 7/8"	S.Y.	0.200
09250.10 Gypsum Board		
Drywall, plasterboard, 3/8" clipped to		
Metal furred ceiling	S.F.	0.009
Columns and beams	S.F.	0.020
Walls	S.F.	0.008
Nailed or screwed to		
Wood framed ceiling	S.F.	0.008
Columns and beams	S.F.	0.018
Walls	S.F.	0.007
1/2", clipped to		
Metal furred ceiling	S.F.	0.009
Columns and beams	S.F.	0.020
Walls	S.F.	0.008
Nailed or screwed to		
Wood framed ceiling	S.F.	0.008
Columns and beams	S.F.	0.018
Walls	S.F.	0.007
5/8", clipped to		
Metal furred ceiling	S.F.	0.010
Columns and beams	S.F.	0.022
Walls	S.F.	0.009
Nailed or screwed to		
Wood framed ceiling	S.F.	0.010
Columns and beams	S.F.	0.022
Walls	S.F.	0.009
Vinyl faced, clipped to metal studs		
1/2"	S.F.	0.010
5/8"	S.F.	0.010
Taping and finishing joints		
Minimum	S.F.	0.005
Average	S.F.	0.007
Maximum	S.F.	0.008
Casing bead		
Minimum	L.F.	0.023
Average	L.F.	0.027
Maximum	L.F.	0.040
Corner bead		
Minimum	L.F.	0.023
Average	L.F.	0.027
Maximum	L.F.	0.040

Tile	UNIT	MAN/ HOURS
09310.10 Ceramic Tile		
Glazed wall tile, 4-1/4" x 4-1/4"		
Minimum	S.F.	0.057
Average	S.F.	0.067
Maximum	S.F.	0.080
Base, 4-1/4" high		
Minimum	L.F.	0.100
Average	L.F.	0.100
Maximum	L.F.	0.100
Unglazed floor tile		
Portland cement bed, cushion edge, face mounted		
1" x 1"	S.F.	0.073
1" x 2"	S.F.	0.070
2" x 2"	S.F.	0.067
Adhesive bed, with white grout		
1" x 1"	S.F.	0.073
1" x 2"	S.F.	0.070
2" x 2"	S.F.	0.067
Organic adhesive bed, thin set, back mounted		
1" x 1"	S.F.	0.073
1" x 2"	S.F.	0.070
2" x 2"	S.F.	0.067
Unglazed wall tile		
Organic adhesive, face mounted cushion edge		
1" x 1"		
Minimum	S.F.	0.067
Average	S.F.	0.073
Maximum	S.F.	0.080
1" x 2"		
Minimum	S.F.	0.064
Average	S.F.	0.070
Maximum	S.F.	0.076
2" x 2"		
Minimum	S.F.	0.062
Average	S.F.	0.067
Maximum	S.F.	0.073
Back mounted		
1" x 1"		
Minimum	S.F.	0.067
Average	S.F.	0.073
Maximum	S.F.	0.080
1" x 2"		
Minimum	S.F.	0.064
Average	S.F.	0.070
Maximum	S.F.	0.076
2" x 2"		
Minimum	S.F.	0.062
Average	S.F.	0.067
Maximum	S.F.	0.073
Conductive floor tile, unglazed square edged		
Portland cement bed		
1 x 1	S.F.	0.100
1-9/16 x 1-9/16	S.F.	0.100
Dry set		
1 x 1	S.F.	0.100

Tile	UNIT	MAN/ HOURS
09310.10 Ceramic Tile		
1-9/16 x 1-9/16	S.F.	0.100
Epoxy bed with epoxy joints		
1 x 1	S.F.	0.100
1-9/16 x 1-9/16	S.F.	0.100
Ceramic accessories		
Towel bar, 24" long		
Minimum	EA.	0.320
Average	EA.	0.400
Maximum	EA.	0.533
Soap dish		
Minimum	EA.	0.533
Average	EA.	0.667
Maximum	EA.	0.800
09330.10 Quarry Tile		
Floor		
4 x 4 x 1/2"	S.F.	0.107
6 x 6 x 1/2"	S.F.	0.100
6 x 6 x 3/4"	S.F.	0.100
Wall, applied to 3/4" portland cement bed		
4 x 4 x 1/2"	S.F.	0.160
6 x 6 x 3/4"	S.F.	0.133
Cove base		
5 x 6 x 1/2" straight top	L.F.	0.133
6 x 6 x 3/4" round top	L.F.	0.133
Stair treads 6 x 6 x 3/4"	L.F.	0.200
Window sill 6 x 8 x 3/4"	L.F.	0.160
For abrasive surface, add to material, 25%		
09410.10 Terrazzo		
Floors on concrete, 1-3/4" thick, 5/8" topping		
Gray cement	S.F.	0.114
White cement	S.F.	0.114
Sand cushion, 3" thick, 5/8" top, 1/4"		
Gray cement	S.F.	0.133
White cement	S.F.	0.133
Monolithic terrazzo, 3-1/2" base slab, 5/8" topping	S.F.	0.100
Terrazzo wainscot, cast-in-place, 1/2" thick	S.F.	0.200
Base, cast in place, terrazzo cove type, 6" high	L.F.	0.114
Curb, cast in place, 6" wide x 6" high, polished top	L.F.	0.400
Stairs, cast-in-place, topping on concrete or metal		
1-1/2" thick treads, 12" wide	L.F.	0.400
Combined tread and riser	L.F.	1.000
Precast terrazzo, thin set		
Terrazzo tiles, non-slip surface		
9" x 9" x 1" thick	S.F.	0.114
12" x 12"		
1" thick	S.F.	0.107
1-1/2" thick	S.F.	0.114
18" x 18" x 1-1/2" thick	S.F.	0.114
24" x 24" x 1-1/2" thick	S.F.	0.094
Terrazzo wainscot		
12" x 12" x 1" thick	S.F.	0.200
18" x 18" x 1-1/2" thick	S.F.	0.229

Tile

09410.10 Terrazzo	UNIT	MAN/ HOURS
Base		
6" high		
Straight	L.F.	0.062
Coved	L.F.	0.062
8" high		
Straight	L.F.	0.067
Coved	L.F.	0.067
Terrazzo curbs		
8" wide x 8" high	L.F.	0.320
6" wide x 6" high	L.F.	0.267
Precast terrazzo stair treads, 12" wide		
1-1/2" thick		
Diamond pattern	L.F.	0.145
Non-slip surface	L.F.	0.145
2" thick		
Diamond pattern	L.F.	0.145
Non-slip surface	L.F.	0.160
Stair risers, 1" thick to 6" high		
Straight sections	L.F.	0.080
Cove sections	L.F.	0.080
Combined tread and riser		
Straight sections		
1-1/2" tread, 3/4" riser	L.F.	0.229
3" tread, 1" riser	L.F.	0.229
Curved sections		
2" tread, 1" riser	L.F.	0.267
3" tread, 1" riser	L.F.	0.267
Stair stringers, notched for treads and risers		
1" thick	L.F.	0.200
2" thick	L.F.	0.267
Landings, structural, nonslip		
1-1/2" thick	S.F.	0.133
3" thick	S.F.	0.160
Conductive terrazzo, spark proof industrial floor		
Epoxy terrazzo		
Floor	S.F.	0.050
Base	S.F.	0.067
Polyacrylate		
Floor	S.F.	0.050
Base	S.F.	0.067
Polyester		
Floor	S.F.	0.032
Base	S.F.	0.040
Synthetic latex mastic		
Floor	S.F.	0.050
Base	S.F.	0.067

Acoustical Treatment

09510.10 Ceilings And Walls	UNIT	MAN/ HOURS
Acoustical panels, suspension system not included		
Fiberglass panels		
5/8" thick		
2' x 2'	S.F.	0.011
2' x 4'	S.F.	0.009
3/4" thick		
2' x 2'	S.F.	0.011
2' x 4'	S.F.	0.009
Glass cloth faced fiberglass panels		
3/4" thick	S.F.	0.013
1" thick	S.F.	0.013
Mineral fiber panels		
5/8" thick		
2' x 2'	S.F.	0.011
2' x 4'	S.F.	0.009
3/4" thick		
2' x 2'	S.F.	0.011
2' x 4'	S.F.	0.009
Wood fiber panels		
1/2" thick		
2' x 2'	S.F.	0.011
2' x 4'	S.F.	0.009
5/8" thick		
2' x 2'	S.F.	0.011
2' x 4'	S.F.	0.009
3/4" thick		
2' x 2'	S.F.	0.011
2' x 4'	S.F.	0.009
2" thick		
2' x 2'	S.F.	0.013
2' x 4'	S.F.	0.010
Air distributing panels		
3/4" thick	S.F.	0.020
5/8" thick	S.F.	0.016
Acoustical tiles, suspension system not included		
Fiberglass tile, 12" x 12"		
5/8" thick	S.F.	0.015
3/4" thick	S.F.	0.018
Glass cloth faced fiberglass tile		
3/4" thick	S.F.	0.018
3" thick	S.F.	0.020
Mineral fiber tile, 12" x 12"		
5/8" thick		
Standard	S.F.	0.016
Vinyl faced	S.F.	0.016
3/4" thick		
Standard	S.F.	0.016
Vinyl faced	S.F.	0.016
Fire rated	S.F.	0.016
Aluminum or mylar faced	S.F.	0.016
Wood fiber tile, 12" x 12"		
1/2" thick	S.F.	0.016
3/4" thick	S.F.	0.016
Metal pan units, 24 ga. steel		

Acoustical Treatment

09510.10 Ceilings And Walls	UNIT	MAN/ HOURS
12" x 12"	S.F.	0.032
12" x 24"	S.F.	0.027
Aluminum, .025" thick		
12" x 12"	S.F.	0.032
12" x 24"	S.F.	0.027
Anodized aluminum, 0.25" thick		
12" x 12"	S.F.	0.032
12" x 24"	S.F.	0.027
Stainless steel, 24 ga.		
12" x 12"	S.F.	0.032
12" x 24"	S.F.	0.027
Metal ceiling systems		
.020" thick panels		
10', 12', and 16' lengths	S.F.	0.023
Custom lengths, 3' to 20'	S.F.	0.023
.025" thick panels		
32 sf, 38 sf, and 52 sf pieces	S.F.	0.027
Custom lengths, 10 sf to 65 sf	S.F.	0.027
Sound absorption walls, with fabric cover		
2-6" x 9' x 3/4"	S.F.	0.027
2' x 9' x 1"	S.F.	0.027
Starter spline	L.F.	0.020
Internal spline	L.F.	0.020
Acoustical treatment		
Barriers for plenums		
Leaded vinyl		
0.48 lb per sf	S.F.	0.038
0.87 lb per sf	S.F.	0.040
Aluminum foil, fiberglass reinforcement		
Minimum	S.F.	0.027
Maximum	S.F.	0.040
Aluminum mesh, paper backed	S.F.	0.027
Fibered cement sheet, 3/16" thick	S.F.	0.029
Sheet lead, 1/64" thick	S.F.	0.020
Sound attenuation blanket		
1" thick	S.F.	0.080
1-1/2" thick	S.F.	0.080
2" thick	S.F.	0.080
3" thick	S.F.	0.089
Ceiling suspension systems		
T bar system		
2' x 4'	S.F.	0.008
2' x 2'	S.F.	0.009
Concealed Z bar suspension system, 12" module	S.F.	0.013

Flooring

09550.10 Wood Flooring	UNIT	MAN/ HOURS
Wood strip flooring, unfinished		
Fir floor		
C and better		
Vertical grain	S.F.	0.027
Flat grain	S.F.	0.027
Oak floor		
Minimum	S.F.	0.038
Average	S.F.	0.038
Maximum	S.F.	0.038
Maple floor		
25/32" x 2-1/4"		
Minimum	S.F.	0.038
Maximum	S.F.	0.038
33/32" x 3-1/4"		
Minimum	S.F.	0.038
Maximum	S.F.	0.038
Wood block industrial flooring		
Creosoted		
2" thick	S.F.	0.021
2-1/2" thick	S.F.	0.025
3" thick	S.F.	0.027
Parquet, 5/16", white oak		
Finished	S.F.	0.040
Unfinished	S.F.	0.040
Gym floor, 2 ply felt, 25/32" maple, finished, in mastic	S.F.	0.044
Over wood sleepers	S.F.	0.050
Finishing, sand, fill, finish, and wax	S.F.	0.020
Refinish sand, seal, and 2 coats of polyurethane	S.F.	0.027
Clean and wax floors	S.F.	0.004
09630.10 Unit Masonry Flooring		
Clay brick		
9 x 4-1/2 x 3" thick		
Glazed	S.F.	0.067
Unglazed	S.F.	0.067
8 x 4 x 3/4" thick		
Glazed	S.F.	0.070
Unglazed	S.F.	0.070
09660.10 Resilient Tile Flooring		
Solid vinyl tile, 1/8" thick, 12" x 12"		
Marble patterns	S.F.	0.020
Solid colors	S.F.	0.020
Travertine patterns	S.F.	0.020
Conductive resilient flooring, vinyl tile		
1/8" thick, 12" x 12"	S.F.	0.023
09665.10 Resilient Sheet Flooring		
Vinyl sheet flooring		
Minimum	S.F.	0.008
Average	S.F.	0.010
Maximum	S.F.	0.013
Cove, to 6"	L.F.	0.016
Fluid applied resilient flooring		

Flooring

09665.10 Resilient Sheet Flooring	UNIT	MAN/HOURS
Polyurethane, poured in place, 3/8" thick	S.F.	0.067
Vinyl sheet goods, backed		
0.070" thick	S.F.	0.010
0.093" thick	S.F.	0.010
0.125" thick	S.F.	0.010
0.250" thick	S.F.	0.010
Wall base, vinyl		
Group 1		
4" high	L.F.	0.027
6" high	L.F.	0.027
Group 2		
4" high	L.F.	0.027
6" high	L.F.	0.027
Group 3		
4" high	L.F.	0.027
6" high	L.F.	0.027
Stair accessories		
Treads, 1/4" x 12", rubber diamond surface		
Marbled	L.F.	0.067
Plain	L.F.	0.067
Grit strip safety tread, 12" wide, colors		
3/16" thick	L.F.	0.067
5/16" thick	L.F.	0.067
Risers, 7" high, 1/8" thick, colors		
Flat	L.F.	0.040
Coved	L.F.	0.040
Nosing, rubber		
3/16" thick, 3" wide		
Black	L.F.	0.040
Colors	L.F.	0.040
6" wide		
Black	L.F.	0.067
Colors	L.F.	0.067

Carpet

09680.10 Floor Leveling	UNIT	MAN/HOURS
Repair and level floors to receive new flooring		
Minimum	S.Y.	0.027
Average	S.Y.	0.067
Maximum	S.Y.	0.080

09682.10 Carpet Padding	UNIT	MAN/HOURS
Carpet padding		
Foam rubber, waffle type, 0.3" thick	S.Y.	0.040
Jute padding		
Minimum	S.Y.	0.036
Average	S.Y.	0.040
Maximum	S.Y.	0.044
Sponge rubber cushion		
Minimum	S.Y.	0.036
Average	S.Y.	0.040
Maximum	S.Y.	0.044
Urethane cushion, 3/8" thick		
Minimum	S.Y.	0.036
Average	S.Y.	0.040
Maximum	S.Y.	0.044

09685.10 Carpet	UNIT	MAN/HOURS
Carpet, acrylic		
24 oz., light traffic	S.Y.	0.089
28 oz., medium traffic	S.Y.	0.089
Residential		
Nylon		
15 oz., light traffic	S.Y.	0.089
28 oz., medium traffic	S.Y.	0.089
Commercial		
Nylon		
28 oz., medium traffic	S.Y.	0.089
35 oz., heavy traffic	S.Y.	0.089
Wool		
30 oz., medium traffic	S.Y.	0.089
36 oz., medium traffic	S.Y.	0.089
42 oz., heavy traffic	S.Y.	0.089
Carpet tile		
Foam backed		
Minimum	S.F.	0.016
Average	S.F.	0.018
Maximum	S.F.	0.020
Tufted loop or shag		
Minimum	S.F.	0.016
Average	S.F.	0.018
Maximum	S.F.	0.020
Clean and vacuum carpet		
Minimum	S.Y.	0.004
Average	S.Y.	0.005
Maximum	S.Y.	0.008

09700.10 Special Flooring	UNIT	MAN/HOURS
Epoxy flooring, marble chips		
Epoxy with colored quartz chips in 1/4" base	S.F.	0.044
Heavy duty epoxy topping, 3/16" thick	S.F.	0.044
Epoxy terrazzo		
1/4" thick chemical resistant	S.F.	0.050

Painting

09905.10 Painting Preparation

Painting	UNIT	MAN/HOURS
Dropcloths		
Minimum	S.F.	0.001
Average	S.F.	0.001
Maximum	S.F.	0.001
Masking		
Paper and tape		
Minimum	L.F.	0.008
Average	L.F.	0.010
Maximum	L.F.	0.013
Doors		
Minimum	EA.	0.100
Average	EA.	0.133
Maximum	EA.	0.178
Windows		
Minimum	EA.	0.100
Average	EA.	0.133
Maximum	EA.	0.178
Sanding		
Walls and flat surfaces		
Minimum	S.F.	0.005
Average	S.F.	0.007
Maximum	S.F.	0.008
Doors and windows		
Minimum	EA.	0.133
Average	EA.	0.200
Maximum	EA.	0.267
Trim		
Minimum	L.F.	0.010
Average	L.F.	0.013
Maximum	L.F.	0.018
Puttying		
Minimum	S.F.	0.012
Average	S.F.	0.016
Maximum	S.F.	0.020

09910.05 Ext. Painting, Sitework

Painting	UNIT	MAN/HOURS
Benches		
Brush		
First Coat		
Minimum	S.F.	0.008
Average	S.F.	0.010
Maximum	S.F.	0.013
Second Coat		
Minimum	S.F.	0.005
Average	S.F.	0.006
Maximum	S.F.	0.007
Roller		
First Coat		
Minimum	S.F.	0.004
Average	S.F.	0.004
Maximum	S.F.	0.005
Second Coat		
Minimum	S.F.	0.003
Average	S.F.	0.003
Maximum	S.F.	0.004
Brickwork		
Brush		
First Coat		
Minimum	S.F.	0.005
Average	S.F.	0.007
Maximum	S.F.	0.010
Second Coat		
Minimum	S.F.	0.004
Average	S.F.	0.005
Maximum	S.F.	0.007
Spray		
First Coat		
Minimum	S.F.	0.002
Average	S.F.	0.003
Maximum	S.F.	0.004
Second Coat		
Minimum	S.F.	0.002
Average	S.F.	0.003
Maximum	S.F.	0.003
Concrete Block		
Roller		
First Coat		
Minimum	S.F.	0.004
Average	S.F.	0.005
Maximum	S.F.	0.008
Second Coat		
Minimum	S.F.	0.003
Average	S.F.	0.004
Maximum	S.F.	0.007
Spray		
First Coat		
Minimum	S.F.	0.002
Average	S.F.	0.003
Maximum	S.F.	0.003
Second Coat		
Minimum	S.F.	0.001
Average	S.F.	0.002
Maximum	S.F.	0.003
Fences, Chain Link		
Brush		
First Coat		
Minimum	S.F.	0.008
Average	S.F.	0.009
Maximum	S.F.	0.010
Second Coat		
Minimum	S.F.	0.005
Average	S.F.	0.006
Maximum	S.F.	0.007
Roller		
First Coat		
Minimum	S.F.	0.006
Average	S.F.	0.007
Maximum	S.F.	0.008

Painting	UNIT	MAN/ HOURS
09910.05 Ext. Painting, Sitework		
Second Coat		
Minimum	S.F.	0.003
Average	S.F.	0.004
Maximum	S.F.	0.005
Spray		
First Coat		
Minimum	S.F.	0.003
Average	S.F.	0.003
Maximum	S.F.	0.003
Second Coat		
Minimum	S.F.	0.002
Average	S.F.	0.002
Maximum	S.F.	0.003
Fences, Wood or Masonry		
Brush		
First Coat		
Minimum	S.F.	0.008
Average	S.F.	0.010
Maximum	S.F.	0.013
Second Coat		
Minimum	S.F.	0.005
Average	S.F.	0.006
Maximum	S.F.	0.008
Roller		
First Coat		
Minimum	S.F.	0.004
Average	S.F.	0.005
Maximum	S.F.	0.006
Second Coat		
Minimum	S.F.	0.003
Average	S.F.	0.004
Maximum	S.F.	0.005
Spray		
First Coat		
Minimum	S.F.	0.003
Average	S.F.	0.004
Maximum	S.F.	0.005
Second Coat		
Minimum	S.F.	0.002
Average	S.F.	0.003
Maximum	S.F.	0.003
Storage Tanks		
Roller		
First Coat		
Minimum	S.F.	0.003
Average	S.F.	0.004
Maximum	S.F.	0.005
Second Coat		
Minimum	S.F.	0.003
Average	S.F.	0.003
Maximum	S.F.	0.004
Spray		
First Coat		
Minimum	S.F.	0.002

Painting	UNIT	MAN/ HOURS
09910.05 Ext. Painting, Sitework		
Average	S.F.	0.002
Maximum	S.F.	0.003
Second Coat		
Minimum	S.F.	0.002
Average	S.F.	0.002
Maximum	S.F.	0.002
09910.15 Ext. Painting, Buildings		
Decks, Metal		
Spray		
First Coat		
Minimum	S.F.	0.004
Average	S.F.	0.004
Maximum	S.F.	0.004
Second Coat		
Minimum	S.F.	0.003
Average	S.F.	0.003
Maximum	S.F.	0.003
Decks, Wood, Stained		
Brush		
First Coat		
Minimum	S.F.	0.004
Average	S.F.	0.004
Maximum	S.F.	0.005
Second Coat		
Minimum	S.F.	0.003
Average	S.F.	0.003
Maximum	S.F.	0.003
Roller		
First Coat		
Minimum	S.F.	0.003
Average	S.F.	0.003
Maximum	S.F.	0.003
Second Coat		
Minimum	S.F.	0.003
Average	S.F.	0.003
Maximum	S.F.	0.003
Spray		
First Coat		
Minimum	S.F.	0.003
Average	S.F.	0.003
Maximum	S.F.	0.003
Second Coat		
Minimum	S.F.	0.002
Average	S.F.	0.002
Maximum	S.F.	0.003
Doors, Metal		
Roller		
First Coat		
Minimum	S.F.	0.006
Average	S.F.	0.007
Maximum	S.F.	0.008
Second Coat		
Minimum	S.F.	0.004

Painting	UNIT	MAN/ HOURS
09910.15 Ext. Painting, Buildings		
Average	S.F.	0.004
Maximum	S.F.	0.005
Spray		
First Coat		
Minimum	S.F.	0.005
Average	S.F.	0.006
Maximum	S.F.	0.007
Second Coat		
Minimum	S.F.	0.004
Average	S.F.	0.004
Maximum	S.F.	0.004
Door Frames, Metal		
Brush		
First Coat		
Minimum	L.F.	0.010
Average	L.F.	0.013
Maximum	L.F.	0.015
Second Coat		
Minimum	L.F.	0.006
Average	L.F.	0.007
Maximum	L.F.	0.008
Spray		
First Coat		
Minimum	L.F.	0.004
Average	L.F.	0.006
Maximum	L.F.	0.008
Second Coat		
Minimum	L.F.	0.004
Average	L.F.	0.004
Maximum	L.F.	0.004
Doors, Wood		
Brush		
First Coat		
Minimum	S.F.	0.012
Average	S.F.	0.016
Maximum	S.F.	0.020
Second Coat		
Minimum	S.F.	0.010
Average	S.F.	0.011
Maximum	S.F.	0.013
Roller		
First Coat		
Minimum	S.F.	0.005
Average	S.F.	0.007
Maximum	S.F.	0.010
Second Coat		
Minimum	S.F.	0.004
Average	S.F.	0.004
Maximum	S.F.	0.007
Spray		
First Coat		
Minimum	S.F.	0.003
Average	S.F.	0.003
Maximum	S.F.	0.004

Painting	UNIT	MAN/ HOURS
09910.15 Ext. Painting, Buildings		
Second Coat		
Minimum	S.F.	0.002
Average	S.F.	0.002
Maximum	S.F.	0.003
Gutters and Downspouts		
Brush		
First Coat		
Minimum	L.F.	0.010
Average	L.F.	0.011
Maximum	L.F.	0.013
Second Coat		
Minimum	L.F.	0.007
Average	L.F.	0.008
Maximum	L.F.	0.010
Siding, Metal		
Roller		
First Coat		
Minimum	S.F.	0.003
Average	S.F.	0.004
Maximum	S.F.	0.004
Second Coat		
Minimum	S.F.	0.003
Average	S.F.	0.003
Maximum	S.F.	0.004
Spray		
First Coat		
Minimum	S.F.	0.003
Average	S.F.	0.003
Maximum	S.F.	0.003
Second Coat		
Minimum	S.F.	0.002
Average	S.F.	0.002
Maximum	S.F.	0.003
Siding, Wood		
Roller		
First Coat		
Minimum	S.F.	0.003
Average	S.F.	0.003
Maximum	S.F.	0.004
Second Coat		
Minimum	S.F.	0.003
Average	S.F.	0.004
Maximum	S.F.	0.004
Spray		
First Coat		
Minimum	S.F.	0.003
Average	S.F.	0.003
Maximum	S.F.	0.003
Second Coat		
Minimum	S.F.	0.002
Average	S.F.	0.003
Maximum	S.F.	0.004
Stucco		
Roller		

Painting	UNIT	MAN/ HOURS
09910.15 Ext. Painting, Buildings		
First Coat		
Minimum	S.F.	0.004
Average	S.F.	0.004
Maximum	S.F.	0.005
Second Coat		
Minimum	S.F.	0.003
Average	S.F.	0.003
Maximum	S.F.	0.004
Spray		
First Coat		
Minimum	S.F.	0.003
Average	S.F.	0.003
Maximum	S.F.	0.003
Second Coat		
Minimum	S.F.	0.002
Average	S.F.	0.002
Maximum	S.F.	0.003
Trim		
Brush		
First Coat		
Minimum	L.F.	0.003
Average	L.F.	0.004
Maximum	L.F.	0.005
Second Coat		
Minimum	L.F.	0.003
Average	L.F.	0.003
Maximum	L.F.	0.005
Walls		
Roller		
First Coat		
Minimum	S.F.	0.003
Average	S.F.	0.003
Maximum	S.F.	0.003
Second Coat		
Minimum	S.F.	0.003
Average	S.F.	0.003
Maximum	S.F.	0.003
Spray		
First Coat		
Minimum	S.F.	0.001
Average	S.F.	0.002
Maximum	S.F.	0.002
Second Coat		
Minimum	S.F.	0.001
Average	S.F.	0.001
Maximum	S.F.	0.002
Windows		
Brush		
First Coat		
Minimum	S.F.	0.013
Average	S.F.	0.016
Maximum	S.F.	0.020
Second Coat		
Minimum	S.F.	0.011

Painting	UNIT	MAN/ HOURS
09910.15 Ext. Painting, Buildings		
Average	S.F.	0.013
Maximum	S.F.	0.016
09910.25 Ext. Painting, Misc.		
Gratings, Metal		
Roller		
First Coat		
Minimum	S.F.	0.023
Average	S.F.	0.027
Maximum	S.F.	0.032
Second Coat		
Minimum	S.F.	0.016
Average	S.F.	0.020
Maximum	S.F.	0.027
Spray		
First Coat		
Minimum	S.F.	0.011
Average	S.F.	0.013
Maximum	S.F.	0.016
Second Coat		
Minimum	S.F.	0.009
Average	S.F.	0.010
Maximum	S.F.	0.011
Ladders		
Brush		
First Coat		
Minimum	L.F.	0.020
Average	L.F.	0.023
Maximum	L.F.	0.027
Second Coat		
Minimum	L.F.	0.016
Average	L.F.	0.018
Maximum	L.F.	0.020
Spray		
First Coat		
Minimum	L.F.	0.013
Average	L.F.	0.015
Maximum	L.F.	0.016
Second Coat		
Minimum	L.F.	0.011
Average	L.F.	0.012
Maximum	L.F.	0.013
Shakes		
Spray		
First Coat		
Minimum	S.F.	0.003
Average	S.F.	0.004
Maximum	S.F.	0.004
Second Coat		
Minimum	S.F.	0.003
Average	S.F.	0.003
Maximum	S.F.	0.004
Shingles, Wood		
Roller		

Painting	UNIT	MAN/ HOURS
09910.25 Ext. Painting, Misc.		
First Coat		
Minimum	S.F.	0.004
Average	S.F.	0.005
Maximum	S.F.	0.006
Second Coat		
Minimum	S.F.	0.003
Average	S.F.	0.003
Maximum	S.F.	0.004
Spray		
First Coat		
Minimum	L.F.	0.003
Average	L.F.	0.003
Maximum	L.F.	0.004
Second Coat		
Minimum	L.F.	0.002
Average	L.F.	0.003
Maximum	L.F.	0.003
Shutters and Louvres		
Brush		
First Coat		
Minimum	EA.	0.160
Average	EA.	0.200
Maximum	EA.	0.267
Second Coat		
Minimum	EA.	0.100
Average	EA.	0.123
Maximum	EA.	0.160
Spray		
First Coat		
Minimum	EA.	0.053
Average	EA.	0.064
Maximum	EA.	0.080
Second Coat		
Minimum	EA.	0.040
Average	EA.	0.053
Maximum	EA.	0.064
Stairs, metal		
Brush		
First Coat		
Minimum	S.F.	0.009
Average	S.F.	0.010
Maximum	S.F.	0.011
Second Coat		
Minimum	S.F.	0.005
Average	S.F.	0.006
Maximum	S.F.	0.007
Spray		
First Coat		
Minimum	S.F.	0.004
Average	S.F.	0.006
Maximum	S.F.	0.006
Second Coat		
Minimum	S.F.	0.003
Average	S.F.	0.004

Painting	UNIT	MAN/ HOURS
09910.25 Ext. Painting, Misc.		
Maximum	S.F.	0.005
Steel, Structural, Light		
Brush		
First Coat		
Minimum	S.F.	0.016
Average	S.F.	0.020
Maximum	S.F.	0.027
Second Coat		
Minimum	S.F.	0.011
Average	S.F.	0.013
Maximum	S.F.	0.016
Roller		
First Coat		
Minimum	S.F.	0.011
Average	S.F.	0.013
Maximum	S.F.	0.016
Second Coat		
Minimum	S.F.	0.007
Average	S.F.	0.008
Maximum	S.F.	0.010
Spray		
First Coat		
Minimum	S.F.	0.007
Average	S.F.	0.008
Maximum	S.F.	0.010
Second Coat		
Minimum	S.F.	0.006
Average	S.F.	0.007
Maximum	S.F.	0.008
Steel, Medium to Heavy		
Brush		
First Coat		
Minimum	S.F.	0.008
Average	S.F.	0.009
Maximum	S.F.	0.010
Second Coat		
Minimum	S.F.	0.007
Average	S.F.	0.008
Maximum	S.F.	0.009
Roller		
First Coat		
Minimum	S.F.	0.007
Average	S.F.	0.008
Maximum	S.F.	0.009
Second Coat		
Minimum	S.F.	0.005
Average	S.F.	0.006
Maximum	S.F.	0.007
Spray		
First Coat		
Minimum	S.F.	0.004
Average	S.F.	0.005
Maximum	S.F.	0.006
Second Coat		

Painting	UNIT	MAN/HOURS
09910.25 Ext. Painting, Misc.		
Minimum	S.F.	0.004
Average	S.F.	0.004
Maximum	S.F.	0.004
09910.35 Int. Painting, Buildings		
Acoustical Ceiling		
Roller		
First Coat		
Minimum	S.F.	0.005
Average	S.F.	0.007
Maximum	S.F.	0.010
Second Coat		
Minimum	S.F.	0.004
Average	S.F.	0.005
Maximum	S.F.	0.007
Spray		
First Coat		
Minimum	S.F.	0.002
Average	S.F.	0.003
Maximum	S.F.	0.003
Second Coat		
Minimum	S.F.	0.002
Average	S.F.	0.002
Maximum	S.F.	0.002
Cabinets and Casework		
Brush		
First Coat		
Minimum	S.F.	0.008
Average	S.F.	0.009
Maximum	S.F.	0.010
Second Coat		
Minimum	S.F.	0.007
Average	S.F.	0.007
Maximum	S.F.	0.008
Spray		
First Coat		
Minimum	S.F.	0.004
Average	S.F.	0.005
Maximum	S.F.	0.006
Second Coat		
Minimum	S.F.	0.003
Average	S.F.	0.003
Maximum	S.F.	0.004
Ceilings		
Roller		
First Coat		
Minimum	S.F.	0.003
Average	S.F.	0.004
Maximum	S.F.	0.004
Second Coat		
Minimum	S.F.	0.003
Average	S.F.	0.003
Maximum	S.F.	0.003
Spray		

Painting	UNIT	MAN/HOURS
09910.35 Int. Painting, Buildings		
First Coat		
Minimum	S.F.	0.002
Average	S.F.	0.002
Maximum	S.F.	0.003
Second Coat		
Minimum	S.F.	0.002
Average	S.F.	0.002
Maximum	S.F.	0.002
Doors, Metal		
Roller		
First Coat		
Minimum	L.F.	0.005
Average	L.F.	0.006
Maximum	L.F.	0.007
Second Coat		
Minimum	L.F.	0.004
Average	L.F.	0.004
Maximum	L.F.	0.005
Spray		
First Coat		
Minimum	L.F.	0.004
Average	L.F.	0.005
Maximum	L.F.	0.006
Second Coat		
Minimum	L.F.	0.003
Average	L.F.	0.004
Maximum	L.F.	0.004
Doors, Wood		
Brush		
First Coat		
Minimum	S.F.	0.011
Average	S.F.	0.015
Maximum	S.F.	0.018
Second Coat		
Minimum	S.F.	0.009
Average	S.F.	0.010
Maximum	S.F.	0.011
Spray		
First Coat		
Minimum	S.F.	0.002
Average	S.F.	0.003
Maximum	S.F.	0.004
Second Coat		
Minimum	S.F.	0.002
Average	S.F.	0.002
Maximum	S.F.	0.003
Ductwork		
Brush		
Minimum	L.F.	0.010
Average	L.F.	0.011
Maximum	L.F.	0.013
Roller		
Minimum	S.F.	0.007
Average	S.F.	0.007

Painting

09910.35 Int. Painting, Buildings

	UNIT	MAN/ HOURS
Maximum	S.F.	0.008
Spray		
Minimum	S.F.	0.003
Average	S.F.	0.003
Maximum	S.F.	0.003
Floors		
Roller		
First Coat		
Minimum	S.F.	0.003
Average	S.F.	0.003
Maximum	S.F.	0.003
Second Coat		
Minimum	S.F.	0.002
Average	S.F.	0.002
Maximum	S.F.	0.002
Spray		
First Coat		
Minimum	S.F.	0.002
Average	S.F.	0.002
Maximum	S.F.	0.002
Second Coat		
Minimum	S.F.	0.002
Average	S.F.	0.002
Maximum	S.F.	0.002
Pipes to 6" diameter		
Brush		
Minimum	L.F.	0.010
Average	L.F.	0.011
Maximum	L.F.	0.013
Spray		
Minimum	L.F.	0.003
Average	L.F.	0.004
Maximum	L.F.	0.005
Pipes to 12" diameter		
Brush		
Minimum	L.F.	0.020
Average	L.F.	0.023
Maximum	L.F.	0.027
Spray		
Minimum	L.F.	0.007
Average	L.F.	0.008
Maximum	L.F.	0.010
Trim		
Brush		
First Coat		
Minimum	L.F.	0.003
Average	L.F.	0.004
Maximum	L.F.	0.004
Second Coat		
Minimum	L.F.	0.002
Average	L.F.	0.003
Maximum	L.F.	0.004
Walls		
Roller		

Painting

09910.35 Int. Painting, Buildings

	UNIT	MAN/ HOURS
First Coat		
Minimum	S.F.	0.003
Average	S.F.	0.003
Maximum	S.F.	0.003
Second Coat		
Minimum	S.F.	0.003
Average	S.F.	0.003
Maximum	S.F.	0.003
Spray		
First Coat		
Minimum	S.F.	0.001
Average	S.F.	0.002
Maximum	S.F.	0.002
Second Coat		
Minimum	S.F.	0.001
Average	S.F.	0.001
Maximum	S.F.	0.002

09955.10 Wall Covering

	UNIT	MAN/ HOURS
Vinyl wall covering		
Medium duty	S.F.	0.011
Heavy duty	S.F.	0.013
Over pipes and irregular shapes		
Lightweight, 13 oz.	S.F.	0.016
Medium weight, 25 oz.	S.F.	0.018
Heavy weight, 34 oz.	S.F.	0.020
Cork wall covering		
1' x 1' squares		
1/4" thick	S.F.	0.020
1/2" thick	S.F.	0.020
3/4" thick	S.F.	0.020
Wall fabrics		
Natural fabrics, grass cloths		
Minimum	S.F.	0.012
Average	S.F.	0.013
Maximum	S.F.	0.016
Flexible gypsum coated wall fabric, fire resistant	S.F.	0.008
Vinyl corner guards		
3/4" x 3/4" x 8'	EA.	0.100
2-3/4" x 2-3/4" x 4'	EA.	0.100

Specialties	UNIT	MAN/ HOURS
10110.10 Chalkboards		
Chalkboard, metal frame, 1/4" thick		
48"x60"	EA.	0.800
48"x96"	EA.	0.889
48"x144"	EA.	1.000
48"x192"	EA.	1.143
Liquid chalkboard		
48"x60"	EA.	0.800
48"x96"	EA.	0.889
48"x144"	EA.	1.000
48"x192"	EA.	1.143
Map rail, deluxe	L.F.	0.040
Average	PCT.	
10165.10 Toilet Partitions		
Toilet partition, plastic laminate		
Ceiling mounted	EA.	2.667
Floor mounted	EA.	2.000
Metal		
Ceiling mounted	EA.	2.667
Floor mounted	EA.	2.000
Wheel chair partition, plastic laminate		
Ceiling mounted	EA.	2.667
Floor mounted	EA.	2.000
Painted metal		
Ceiling mounted	EA.	2.667
Floor mounted	EA.	2.000
Urinal screen, plastic laminate		
Wall hung	EA.	1.000
Floor mounted	EA.	1.000
Porcelain enameled steel, floor mounted	EA.	1.000
Painted metal, floor mounted	EA.	1.000
Stainless steel, floor mounted	EA.	1.000
Metal toilet partitions		
Toilet partitions, front door and side divider, floor mounted		
Porcelain enameled steel	EA.	2.000
Painted steel	EA.	2.000
Stainless steel	EA.	2.000
10185.10 Shower Stalls		
Shower receptors		
Precast, terrazzo		
32" x 32"	EA.	0.667
32" x 48"	EA.	0.800
Concrete		
32" x 32"	EA.	0.667
48" x 48"	EA.	0.889
Shower door, trim and hardware		
Economy, 24" wide, chrome frame, tempered glass	EA.	0.800
Porcelain enameled steel, flush	EA.	0.800
Baked enameled steel, flush	EA.	0.800
Aluminum frame, tempered glass, 48" wide, sliding	EA.	1.000
Folding	EA.	1.000
Aluminum frame and tempered glass, molded plastic		
Complete with receptor and door		

Specialties	UNIT	MAN/ HOURS
10185.10 Shower Stalls		
32" x 32"	EA.	2.000
36" x 36"	EA.	2.000
40" x 40"	EA.	2.286
Shower compartment, precast concrete receptor		
Single entry type		
Porcelain enameled steel	EA.	8.000
Baked enameled steel	EA.	8.000
Stainless steel	EA.	8.000
Double entry type		
Porcelain enameled steel	EA.	10.000
Baked enameled steel	EA.	10.000
Stainless steel	EA.	10.000
10190.10 Cubicles		
Hospital track		
Ceiling hung	L.F.	0.089
Suspended	L.F.	0.114
Hospital metal dividers, galvanized steel		
Baked enamel finish		
54" high		
10" glass light	L.F.	0.400
14" glass light	L.F.	0.400
24" glass light	L.F.	0.400
60" high		
10" glass light	L.F.	0.444
14" glass light	L.F.	0.444
24" glass light	L.F.	0.444
Stainless steel		
54" high		
10" glass light	L.F.	0.444
14" glass light	L.F.	0.444
24" glass light	L.F.	0.444
60" high		
14" glass light	L.F.	0.500
14" glass light	L.F.	0.500
24" glass light	L.F.	0.500
10210.10 Vents And Wall Louvres		
Block vent, 8"x16"x4" aluminum, w/screen, mill finish	EA.	0.267
Standard	EA.	0.250
Vents w/screen, 4" deep, 8" wide, 5" high		
Modular	EA.	0.250
Aluminum gable louvers	S.F.	0.133
Vent screen aluminum, 4" wide, continuous	L.F.	0.027
Louvers, aluminum, anodized, fixed blade		
Horizontal line	S.F.	0.200
Vertical line	S.F.	0.200
Wall louvre, aluminum mill finish		
Under, 2 sf	S.F.	0.100
2 to 4 sf	S.F.	0.089
5 to 10 sf	S.F.	0.089
Galvanized steel		
Under 2 sf	S.F.	0.100
2 to 4 sf	S.F.	0.089

10 SPECIALTIES

Specialties	UNIT	MAN/ HOURS
10210.10 Vents And Wall Louvres		
5 to 10 sf	S.F.	0.089
10225.10 Door Louvres		
Fixed, 1" thick, enameled steel		
8"x8"	EA.	0.100
12"x8"	EA.	0.100
12"x12"	EA.	0.114
16"x12"	EA.	0.123
18"x12"	EA.	0.200
20"x8"	EA.	0.114
20"x12"	EA.	0.229
20"x16"	EA.	0.267
20"x20"	EA.	0.320
24"x12"	EA.	0.267
24"x16"	EA.	0.286
24"x18"	EA.	0.308
24"x20"	EA.	0.333
24"x24"	EA.	0.364
26"x26"	EA.	0.500
10270.40 Access & Pedestal Floor		
Panels, no covering, 2'x2'		
Plain	S.F.	0.010
Perforated	S.F.	0.400
Pedestals		
For 6" to 12" clearance	EA.	0.080
Stringers		
2'	L.F.	0.038
6'	L.F.	0.027
Accessories		
Ramp assembly	S.F.	0.032
Elevated floor assembly	S.F.	0.030
Handrail	L.F.	0.400
Fascia plate	L.F.	0.200
RF shielding components, floor liner		
Hot rolled steel sheet		
14 ga.	S.F.	0.020
11 ga.	S.F.	0.062
10290.10 Pest Control		
Termite control		
Under slab spraying		
Minimum	S.F.	0.002
Average	S.F.	0.004
Maximum	S.F.	0.008
10350.10 Flagpoles		
Installed in concrete base		
Fiberglass		
25' high	EA.	5.333
50' high	EA.	13.333
Aluminum		
25' high	EA.	5.333
50' high	EA.	13.333

Specialties	UNIT	MAN/ HOURS
10350.10 Flagpoles		
Bonderized steel		
25' high	EA.	6.154
50' high	EA.	16.000
Freestanding tapered, fiberglass		
30' high	EA.	5.714
40' high	EA.	7.273
50' high	EA.	8.000
60' high	EA.	9.412
Wall mounted, with collar, brushed aluminum finish		
15' long	EA.	4.000
18' long	EA.	4.000
20' long	EA.	4.211
24' long	EA.	4.706
Outrigger, wall, including base		
10' long	EA.	5.333
20' long	EA.	6.667
10400.10 Identifying Devices		
Directory and bulletin boards		
Open face boards		
Chrome plated steel frame	S.F.	0.400
Aluminum framed	S.F.	0.400
Bronze framed	S.F.	0.400
Stainless steel framed	S.F.	0.400
Tack board, aluminum framed	S.F.	0.400
Visual aid board, aluminum framed	S.F.	0.400
Glass encased boards, hinged and keyed		
Aluminum framed	S.F.	1.000
Bronze framed	S.F.	1.000
Stainless steel framed	S.F.	1.000
Chrome plated steel framed	S.F.	1.000
Metal plaque		
Cast bronze	S.F.	0.667
Aluminum	S.F.	0.667
Metal engraved plaque		
Porcelain steel	S.F.	0.667
Stainless steel	S.F.	0.667
Brass	S.F.	0.667
Aluminum	S.F.	0.667
Metal built-up plaque		
Bronze	S.F.	0.800
Copper and bronze	S.F.	0.800
Copper and aluminum	S.F.	0.800
Metal nameplate plaques		
Cast bronze	S.F.	0.500
Cast aluminum	S.F.	0.500
Engraved, 1-1/2" x 6"		
Bronze	EA.	0.500
Aluminum	EA.	0.500
Letters, on masonry or concrete, aluminum, satin finish		
1/2" thick		
2" high	EA.	0.320
4" high	EA.	0.400
6" high	EA.	0.444

Specialties	UNIT	MAN/HOURS
10400.10 Identifying Devices		
3/4" thick		
8" high	EA.	0.500
10" high	EA.	0.571
1" thick		
12" high	EA.	0.667
14" high	EA.	0.800
16" high	EA.	1.000
3/8" thick		
2" high	EA.	0.320
4" high	EA.	0.400
1/2" thick, 6" high	EA.	0.444
5/8" thick, 8" high	EA.	0.500
1" thick		
10" high	EA.	0.571
12" high	EA.	0.667
14" high	EA.	0.800
16" high	EA.	1.000
Interior door signs, adhesive, flexible		
2" x 8"	EA.	0.200
4" x 4"	EA.	0.200
6" x 7"	EA.	0.200
6" x 9"	EA.	0.200
10" x 9"	EA.	0.200
10" x 12"	EA.	0.200
Hard plastic type, no frame		
3" x 8"	EA.	0.200
4" x 4"	EA.	0.200
4" x 12"	EA.	0.200
Hard plastic type, with frame		
3" x 8"	EA.	0.200
4" x 4"	EA.	0.200
4" x 12"	EA.	0.200
10450.10 Control		
Access control, 7' high, indoor or outdoor impenetrability		
Remote or card control, type B	EA.	10.667
Free passage, type B	EA.	10.667
Remote or card control, type AA	EA.	10.667
Free passage, type AA	EA.	10.667
10500.10 Lockers		
Locker bench, floor mounted, laminated maple		
4'	EA.	0.667
6'	EA.	0.667
Wardrobe locker, 12" x 60" x 15", baked on enamel		
1-tier	EA.	0.400
2-tier	EA.	0.400
3-tier	EA.	0.421
4-tier	EA.	0.421
12" x 72" x 15", baked on enamel		
1-tier	EA.	0.400
2-tier	EA.	0.400
4-tier	EA.	0.421
5-tier	EA.	0.421

Specialties	UNIT	MAN/HOURS
10500.10 Lockers		
15" x 60" x 15", baked on enamel		
1-tier	EA.	0.400
4-tier	EA.	0.421
Wardrobe locker, single tier type		
12" x 15" x 72"	EA.	0.800
18" x 15" x 72"	EA.	0.842
12" x 18" x 72"	EA.	0.889
18" x 18" x 72"	EA.	0.941
Double tier type		
12" x 15" x 36"	EA.	0.400
18" x 15" x 36"	EA.	0.400
12" x 18" x 36"	EA.	0.400
18" x 18" x 36"	EA.	0.400
Two person unit		
18" x 15" x 72"	EA.	1.333
18" x 18" x 72"	EA.	1.600
Duplex unit		
15" x 15" x 72"	EA.	0.800
15" x 21" x 72"	EA.	0.800
Basket lockers, basket sets with baskets		
24 basket set	SET	4.000
30 basket set	SET	5.000
36 basket set	SET	6.667
42 basket set	SET	8.000
10520.10 Fire Protection		
Portable fire extinguishers		
Water pump tank type		
2.5 gal.		
Red enameled galvanized	EA.	0.533
Red enameled copper	EA.	0.533
Polished copper	EA.	0.533
Carbon dioxide type, red enamel steel		
Squeeze grip with hose and horn		
2.5 lb	EA.	0.533
5 lb	EA.	0.615
10 lb	EA.	0.800
15 lb	EA.	1.000
20 lb	EA.	1.000
Wheeled type		
125 lb	EA.	1.600
250 lb	EA.	1.600
500 lb	EA.	1.600
Dry chemical, pressurized type		
Red enameled steel		
2.5 lb	EA.	0.533
5 lb	EA.	0.615
10 lb	EA.	0.800
20 lb	EA.	1.000
30 lb	EA.	1.000
Chrome plated steel, 2.5 lb	EA.	0.533
Other type extinguishers		
2.5 gal, stainless steel, pressurized water tanks	EA.	0.533
Soda and acid type	EA.	0.533

Specialties	UNIT	MAN/HOURS
10520.10 Fire Protection		
Cartridge operated, water type	EA.	0.533
Loaded stream, water type	EA.	0.533
Foam type	EA.	0.533
40 gal, wheeled foam type	EA.	1.600
Fire extinguisher cabinets		
Enameled steel		
8" x 12" x 27"	EA.	1.600
8" x 16" x 38"	EA.	1.600
Aluminum		
8" x 12" x 27"	EA.	1.600
8" x 16" x 38"	EA.	1.600
8" x 12" x 27"	EA.	1.600
Stainless steel		
8" x 16" x 38"	EA.	1.600
10550.10 Postal Specialties		
Mail chutes		
Single mail chute		
Finished aluminum	L.F.	2.000
Bronze	L.F.	2.000
Single mail chute receiving box		
Finished aluminum	EA.	4.000
Bronze	EA.	4.000
Twin mail chute, double parallel		
Finished aluminum	FLR	4.000
Bronze	FLR	4.000
Receiving box, 36" x 20" x 12"		
Finished aluminum	EA.	6.667
Bronze	EA.	6.667
Locked receiving mail box		
Finished aluminum	EA.	4.000
Bronze	EA.	4.000
Commercial postal accessories for mail chutes		
Letter slot, brass	EA.	1.333
Bulk mail slot, brass	EA.	1.333
Mail boxes		
Residential postal accessories		
Letter slot	EA.	0.400
Rural letter box	EA.	1.000
Apartment house, keyed, 3.5" x 4.5" x 16"	EA.	0.267
Ranch style	EA.	0.400
Commercial postal accessories		
Letter box, with combination lock	EA.	0.286
Key lock	EA.	0.286
Mail box, aluminum w/glass front, 4x5		
Horizontal rear load	EA.	0.229
Vertical front load	EA.	0.229
10600.10 Movable Partitions		
Partition, movable office type, 2-1/2" thick, vinyl-gypsum	S.F.	0.040
Enameled steel frame, with 1/4" thick clear glass	S.F.	0.040
Door frame and hardware for movable partitions	EA.	2.667
Add for acoustic movable partition	S.F.	
Accordion partition, 12' high		

Specialties	UNIT	MAN/HOURS
10600.10 Movable Partitions		
Vinyl	S.F.	0.133
Acoustical	S.F.	0.133
Standard office cubicles, 8' high, steel framed		
Baked enamel finish		
100% flush	L.F.	0.200
75% flush and 25% glass	L.F.	0.222
50% flush and 50% glass	L.F.	0.222
100% glass	L.F.	0.267
Natural hardwood panels		
100% flush	L.F.	0.267
50% flush and 50% glass	L.F.	0.286
Plastic laminated panels		
100% flush	L.F.	0.267
75% flush and 25% glass	L.F.	0.286
50% flush and 50% glass	L.F.	0.286
Vinyl covered panels		
100% flush	L.F.	0.276
75% flush and 25% glass	L.F.	0.296
50% and 50% glass	L.F.	0.296
Aluminum framed		
Enameled or anodized aluminum panels		
100% flush	L.F.	0.200
75% flush and 25% glass	L.F.	0.222
50% flush and 50% glass	L.F.	0.222
Vinyl covered panels		
100% flush	L.F.	0.276
75% flush and 25% glass	L.F.	0.296
50% flush and 50% glass	L.F.	0.296
60" high partitions, steel framed		
Enameled panels	L.F.	0.178
Natural hardwood panels, two sides	L.F.	0.186
Plastic laminated panels	L.F.	0.186
Vinyl covered panels	L.F.	0.178
Aluminum framed		
Anodized or baked enamel panels	L.F.	0.178
Natural hardwood panels	L.F.	0.186
Plastic laminated panels	L.F.	0.186
Vinyl covered panels	L.F.	0.178
Wire mesh partitions		
Wall panels		
4' x 7'	EA.	0.500
4' x 8'	EA.	0.533
4' x 10'	EA.	0.615
Wall filler panels		
1' x 7'	EA.	0.500
1' x 8'	EA.	0.533
1' x 10'	EA.	0.571
2' x 7'	EA.	0.500
2' x 8'	EA.	0.533
2' x 10'	EA.	0.571
3' x 7'	EA.	0.500
3' x 8'	EA.	0.533
3' x 10'	EA.	0.571
Ceiling panels		

Specialties	UNIT	MAN/ HOURS
10600.10 Movable Partitions		
10' x 2'	EA.	1.143
10' x 4'	EA.	1.600
Wall panel with service window		
5' wide		
7' high	EA.	0.500
8' high	EA.	0.533
10' high	EA.	0.571
Doors		
Sliding		
3' x 7'	EA.	2.000
3' x 8'	EA.	2.286
3' x 10'	EA.	3.200
4' x 7'	EA.	2.286
4' x 8'	EA.	3.200
4' x 10'	EA.	4.000
5' x 7'	EA.	3.200
5' x 8'	EA.	4.000
5' x 10'	EA.	4.000
Swing door		
3' x 7'	EA.	2.000
4' x 7'	EA.	2.286
Swing door, with 1' transom		
3' x 7'	EA.	2.286
4' x 7'	EA.	3.200
Swing door, with 3' transom		
3' x 7'	EA.	3.200
4' x 7'	EA.	4.000
10670.10 Shelving		
Shelving, enamel, closed side and back, 12" x 36"		
5 shelves	EA.	1.333
8 shelves	EA.	1.778
Open		
5 shelves	EA.	1.333
8 shelves	EA.	1.778
Metal storage shelving, baked enamel		
7 shelf unit, 72" or 84" high		
10" shelf	L.F.	0.800
12" shelf	L.F.	0.842
15" shelf	L.F.	0.889
18" shelf	L.F.	0.941
24" shelf	L.F.	1.000
30" shelf	L.F.	1.067
36" shelf	L.F.	1.143
4 shelf unit, 40" high		
10" shelf	L.F.	0.667
12" shelf	L.F.	0.727
15" shelf	L.F.	0.800
18" shelf	L.F.	0.842
24" shelf	L.F.	0.889
3 shelf unit, 32" high		
10" shelf	L.F.	0.400
12" shelf	L.F.	0.421
15" shelf	L.F.	0.444

Specialties	UNIT	MAN/ HOURS
10670.10 Shelving		
18" shelf	L.F.	0.471
24" shelf	L.F.	0.500
Single shelf unit, attached to masonry		
10" shelf	L.F.	0.133
12" shelf	L.F.	0.145
15" shelf	L.F.	0.154
18" shelf	L.F.	0.163
24" shelf	L.F.	0.174
Built-in wood shelves		
Posts and trimmed plywood	L.F.	0.114
Solid clear pine	L.F.	0.123
Closet shelf, pine with rod	L.F.	0.123
10750.10 Telephone Enclosures		
Telephone enclosure, wall mounted, shelf, 28" x 30" x 15"	EA.	2.000
Directory shelf, stainless steel, 3 binders	EA.	1.333
10800.10 Bath Accessories		
Ash receiver, wall mounted, aluminum	EA.	0.400
Grab bar, 1-1/2" dia., stainless steel, wall mounted		
24" long	EA.	0.400
36" long	EA.	0.421
42" long	EA.	0.444
48" long	EA.	0.471
52" long	EA.	0.500
1" dia., stainless steel		
12" long	EA.	0.348
18" long	EA.	0.364
24" long	EA.	0.400
30" long	EA.	0.421
36" long	EA.	0.444
48" long	EA.	0.471
Hand dryer, surface mounted, 110 volt	EA.	1.000
Medicine cabinet, 16 x 22, baked enamel, steel, lighted	EA.	0.320
With mirror, lighted	EA.	0.533
Mirror, 1/4" plate glass, up to 10 sf	S.F.	0.080
Mirror, stainless steel frame		
18"x24"	EA.	0.267
18"x32"	EA.	0.320
18"x36"	EA.	0.400
24"x30"	EA.	0.400
24"x36"	EA.	0.444
24"x48"	EA.	0.667
24"x60"	EA.	0.800
30"x30"	EA.	0.800
30"x72"	EA.	1.000
48"x72"	EA.	1.333
With shelf, 18"x24"	EA.	0.320
Sanitary napkin dispenser, stainless steel, wall mounted	EA.	0.533
Shower rod, 1" diameter		
Chrome finish over brass	EA.	0.400
Stainless steel	EA.	0.400
Soap dish, stainless steel, wall mounted	EA.	0.533
Toilet tissue dispenser, stainless, wall mounted		

Specialties	UNIT	MAN/ HOURS
10800.10 Bath Accessories		
Single roll	EA.	0.200
Double roll	EA.	0.229
Towel dispenser, stainless steel		
Flush mounted	EA.	0.444
Surface mounted	EA.	0.400
Combination towel dispenser and waste receptacle	EA.	0.533
Towel bar, stainless steel		
18" long	EA.	0.320
24" long	EA.	0.364
30" long	EA.	0.400
36" long	EA.	0.444
Toothbrush and tumbler holder	EA.	0.267
Waste receptacle, stainless steel, wall mounted	EA.	0.667
10900.10 Wardrobe Specialties		
Hospital wardrobe units, 24" x 24" x 76", with door		
Baked enameled steel	EA.	4.444
Hardwood	EA.	4.444
Stainless steel	EA.	4.444
Plastic laminated	EA.	4.444
Dormitory wardrobe units, 24" x 76", with door		
Hardwood	EA.	4.444
Plastic laminated	EA.	4.444
Hat and coat rack		
Single tier		
Baked enameled steel	L.F.	0.200
Stainless steel	L.F.	0.200
Aluminum	L.F.	0.200
Double tier		
Baked enameled steel	L.F.	0.229
Stainless steel	L.F.	0.229
Aluminum	L.F.	0.229

Architectural Equipment

11010.10 Maintenance Equipment

Architectural Equipment	UNIT	MAN/HOURS
Vacuum cleaning system		
3 valves		
1.5 hp	EA.	8.889
2.5 hp	EA.	11.429
5 valves	EA.	16.000
7 valves	EA.	20.000

11020.10 Security Equipment

Architectural Equipment	UNIT	MAN/HOURS
Bulletproof teller window		
4' x 4'	EA.	13.333
5' x 4'	EA.	16.000
Bulletproof partitions		
Up to 12' high, 2.5" thick	S.F.	0.053
Counter for banks		
Minimum	L.F.	1.600
Maximum	L.F.	2.667
Drive-up window		
Minimum	EA.	11.429
Maximum	EA.	26.667
Night depository		
Minimum	EA.	11.429
Maximum	EA.	26.667
Office safes, 30" x 20" x 20", 1 hr rating	EA.	2.000
30" x 16" x 15", 2 hr rating	EA.	1.600
30" x 28" x 20", H&G rating	EA.	1.000
Service windows, pass through painted steel		
24" x 36"	EA.	8.000
48" x 40"	EA.	10.000
72" x 40"	EA.	16.000
Special doors and windows		
3' x 7' bulletproof door with frame	EA.	11.429
12" x 12" vision panel	EA.	5.714
Surveillance system		
Minimum	EA.	16.000
Maximum	EA.	80.000
Vault door, 3' wide, 6'6" high		
3-1/2" thick	EA.	100.000
7" thick	EA.	133.333
10" thick	EA.	160.000
Insulated vault door		
2 hr rating		
32" wide	EA.	8.000
40" wide	EA.	8.421
4 hr rating		
32" wide	EA.	8.889
40" wide	EA.	10.000
6 hr rating		
32" wide	EA.	8.889
40" wide	EA.	10.000
Insulated file room door		
1 hr rating		
32" wide	EA.	8.000
40" wide	EA.	8.889

11060.10 Theater Equipment

Architectural Equipment	UNIT	MAN/HOURS
Roll out stage, steel frame, wood floor		
Manual	S.F.	0.050
Electric	S.F.	0.080
Portable stages		
8" high	S.F.	0.040
18" high	S.F.	0.044
36" high	S.F.	0.047
48" high	S.F.	0.050
Band risers		
Minimum	S.F.	0.040
Maximum	S.F.	0.040
Chairs for risers		
Minimum	EA.	0.036
Maximum	EA.	0.036

11080.10 Police Equipment

Architectural Equipment	UNIT	MAN/HOURS
Firing range equipment, rifle		
3 position	EA.	26.667
4 position	EA.	40.000
5 position	EA.	44.444
6 position	EA.	47.059

11090.10 Checkroom Equipment

Architectural Equipment	UNIT	MAN/HOURS
Motorized checkroom equipment		
No shelf system, 6'4" height		
7'6" length	EA.	8.000
14'6" length	EA.	8.000
28' length	EA.	8.000
One shelf, 6'8" height		
7'6" length	EA.	8.000
14'6" length	EA.	8.000
28' length	EA.	8.000
Two shelves, 7'5" height		
7'6" length	EA.	8.000
14'6" length	EA.	8.000
28' length	EA.	8.000
Three shelves, 8' height		
7'6" length	EA.	16.000
14'6" length	EA.	16.000
28' length	EA.	16.000
Four shelves, 8'7" height		
7'6" length	EA.	16.000
14'6" length	EA.	16.000
28' length	EA.	16.000

11110.10 Laundry Equipment

Architectural Equipment	UNIT	MAN/HOURS
High capacity, heavy duty		
Washer extractors		
135 lb		
Standard	EA.	6.667
Pass through	EA.	6.667
200 lb		
Standard	EA.	6.667
Pass through	EA.	6.667

11 EQUIPMENT

Architectural Equipment

11110.10 Laundry Equipment

Architectural Equipment	UNIT	MAN/ HOURS
110 lb dryer	EA.	6.667
Hand operated presser	EA.	8.889
Mushroom press	EA.	8.889
Spreader feeders		
2 station	EA.	8.889
4 station	EA.	16.000
Delivery carts		
12 bushel	EA.	0.100
16 bushel	EA.	0.107
18 bushel	EA.	0.114
30 bushel	EA.	0.133
40 bushel	EA.	0.160
Low capacity		
Pressers		
Air operated	EA.	3.200
Hand operated	EA.	3.200
Extractor, low capacity	EA.	3.200
Ironer, 48"	EA.	1.600
Coin washers		
10 lb capacity	EA.	1.600
20 lb capacity	EA.	1.600
Coin dryer	EA.	1.000
Coin dry cleaner, 20 lb	EA.	3.200

11161.10 Loading Dock Equipment

Architectural Equipment	UNIT	MAN/ HOURS
Dock leveler, 10 ton capacity		
6' x 8'	EA.	8.000
7' x 8'	EA.	8.000
Bumpers, laminated rubber		
4-1/2" thick		
6" x 14"	EA.	0.160
6" x 36"	EA.	0.178
10" x 14"	EA.	0.200
10" x 24"	EA.	0.229
10" x 36"	EA.	0.267
12" x 14"	EA.	0.211
12" x 24"	EA.	0.250
12" x 36"	EA.	0.296
6" thick		
10" x 14"	EA.	0.229
10" x 24"	EA.	0.276
10" x 36"	EA.	0.400
Extruded rubber bumpers		
T-section, 22" x 22" x 3"	EA.	0.160
Molded rubber bumpers		
24" x 12" x 3" thick	EA.	0.400
Door seal, 12" x 12", vinyl covered	L.F.	0.200
Dock boards, heavy duty, 5' x 5'		
5000 lb		
Minimum	EA.	6.667
Maximum	EA.	6.667
9000 lb		
Minimum	EA.	6.667
Maximum	EA.	7.273

Architectural Equipment

11161.10 Loading Dock Equipment

Architectural Equipment	UNIT	MAN/ HOURS
15,000 lb	EA.	7.273
Truck shelters		
Minimum	EA.	6.154
Maximum	EA.	11.429

11170.10 Waste Handling

Architectural Equipment	UNIT	MAN/ HOURS
Incinerator, electric		
100 lb/hr		
Minimum	EA.	8.000
Maximum	EA.	8.000
400 lb/hr		
Minimum	EA.	16.000
Maximum	EA.	16.000
1000 lb/hr		
Minimum	EA.	24.242
Maximum	EA.	24.242
Incinerator, medical-waste		
25 lb/hr, 2-7 x 4-0	EA.	16.000
50 lb/hr, 2-11 x 4-11	EA.	16.000
75 lb/hr, 3-8 x 5-0	EA.	32.000
100 lb/hr, 3-8 x 6-0	EA.	32.000
Industrial compactor		
1 cy	EA.	8.889
3 cy	EA.	11.429
5 cy	EA.	16.000
Trash chutes steel, including sprinklers		
18" dia.	L.F.	4.000
24" dia.	L.F.	4.211
30" dia.	L.F.	4.444
36" dia.	L.F.	4.706
Refuse bottom hopper	EA.	4.444

11400.10 Food Service Equipment

Architectural Equipment	UNIT	MAN/ HOURS
Unit kitchens		
30" compact kitchen		
Refrigerator, with range, sink	EA.	4.000
Sink only	EA.	2.667
Range only	EA.	2.000
Cabinet for upper wall section	EA.	1.143
Stainless shield, for rear wall	EA.	0.320
Side wall	EA.	0.320
42" compact kitchen		
Refrigerator with range, sink	EA.	4.444
Sink only	EA.	4.000
Cabinet for upper wall section	EA.	1.333
Stainless shield, for rear wall	EA.	0.333
Side wall	EA.	0.333
54" compact kitchen		
Refrigerator, oven, range, sink	EA.	5.714
Cabinet for upper wall section	EA.	1.600
Stainless shield, for		
Rear wall	EA.	0.364
Side wall	EA.	0.364
60" compact kitchen		

Architectural Equipment

11400.10 Food Service Equipment

Architectural Equipment	UNIT	MAN/ HOURS
Refrigerator, oven, range, sink	EA.	5.714
Cabinet for upper wall section	EA.	1.600
Stainless shield, for		
Rear wall	EA.	0.364
Side wall	EA.	0.364
72" compact kitchen		
Refrigerator, oven, range, sink	EA.	6.667
Cabinet for upper wall section	EA.	1.600
Stainless shield for		
Rear wall	EA.	0.400
Side wall	EA.	0.400
Bake oven		
Single deck		
Minimum	EA.	1.000
Maximum	EA.	2.000
Double deck		
Minimum	EA.	1.333
Maximum	EA.	2.000
Triple deck		
Minimum	EA.	1.333
Maximum	EA.	2.667
Convection type oven, electric, 40" x 45" x 57"		
Minimum	EA.	1.000
Maximum	EA.	2.000
Broiler, without oven, 69" x 26" x 39"		
Minimum	EA.	1.000
Maximum	EA.	1.333
Coffee urns, 10 gallons		
Minimum	EA.	2.667
Maximum	EA.	4.000
Fryer, with submerger		
Single		
Minimum	EA.	1.600
Maximum	EA.	2.667
Double		
Minimum	EA.	2.000
Maximum	EA.	2.667
Griddle, counter		
3' long		
Minimum	EA.	1.333
Maximum	EA.	1.600
5' long		
Minimum	EA.	2.000
Maximum	EA.	2.667
Kettles, steam, jacketed		
20 gallons		
Minimum	EA.	2.000
Maximum	EA.	4.000
40 gallons		
Minimum	EA.	2.000
Maximum	EA.	4.000
60 gallons		
Minimum	EA.	2.000
Maximum	EA.	4.000

Architectural Equipment

11400.10 Food Service Equipment

Architectural Equipment	UNIT	MAN/ HOURS
Range		
Heavy duty, single oven, open top		
Minimum	EA.	1.000
Maximum	EA.	2.667
Fry top		
Minimum	EA.	1.000
Maximum	EA.	2.667
Steamers, electric		
27 kw		
Minimum	EA.	2.000
Maximum	EA.	2.667
18 kw		
Minimum	EA.	2.000
Maximum	EA.	2.667
Dishwasher, rack type		
Single tank, 190 racks/hr	EA.	4.000
Double tank		
234 racks/hr	EA.	4.444
265 racks/hr	EA.	5.333
Dishwasher, automatic 100 meals/hr	EA.	2.667
Disposals		
100 gal/hr	EA.	2.667
120 gal/hr	EA.	2.759
250 gal/hr	EA.	2.857
Exhaust hood for dishwasher, gutter 4 sides, s-steel		
4'x4'x2'	EA.	2.963
4'x7'x2'	EA.	3.200
Food preparation machines		
Vertical cutter mixers		
25 quart	EA.	2.667
40 quart	EA.	2.667
80 quart	EA.	4.000
130 quart	EA.	6.667
Choppers		
5 lb	EA.	2.000
16 lb	EA.	2.667
40 lb	EA.	4.000
Mixers, floor models		
20 quart	EA.	1.000
60 quart	EA.	1.000
80 quart	EA.	1.143
140 quart	EA.	1.600
Ice cube maker		
50 lb per day		
Minimum	EA.	8.000
Maximum	EA.	8.000
500 lb per day		
Minimum	EA.	13.333
Maximum	EA.	13.333
Ice flakers		
300 lb per day	EA.	8.000
600 lb per day	EA.	13.333
1000 lb per day	EA.	17.778
2000 lb per day	EA.	20.000

Architectural Equipment	UNIT	MAN/ HOURS
11400.10 Food Service Equipment		
Refrigerated cases		
Dairy products		
Multi deck type	L.F.	0.533
Delicatessen case, service deli		
Single deck	L.F.	4.000
Multi deck	L.F.	5.000
Meat case		
Single deck	L.F.	4.706
Multi deck	L.F.	5.000
Produce case		
Single deck	L.F.	4.706
Multi deck	L.F.	5.000
Bottle coolers		
6' long		
Minimum	EA.	16.000
Maximum	EA.	16.000
10' long		
Minimum	EA.	26.667
Maximum	EA.	26.667
Frozen food cases		
Chest type	L.F.	4.706
Reach-in, glass door	L.F.	5.000
Island case, single	L.F.	4.706
Multi deck	L.F.	5.000
Ice storage bins		
500 lb capacity	EA.	11.429
1000 lb capacity	EA.	22.857
11450.10 Residential Equipment		
Compactor, 4 to 1 compaction	EA.	2.000
Dishwasher, built-in		
2 cycles	EA.	4.000
4 or more cycles	EA.	4.000
Disposal		
Garbage disposer	EA.	2.667
Heaters, electric, built-in		
Ceiling type	EA.	2.667
Wall type		
Minimum	EA.	2.000
Maximum	EA.	2.667
Hood for range, 2-speed, vented		
30" wide	EA.	2.667
42" wide	EA.	2.667
Ice maker, automatic		
30 lb per day	EA.	1.143
50 lb per day	EA.	4.000
Folding access stairs, disappearing metal stair		
8' long	EA.	1.143
11' long	EA.	1.143
12' long	EA.	1.143
Wood frame, wood stair		
22" x 54" x 8'9" long	EA.	0.800
25" x 54" x 10' long	EA.	0.800
Ranges electric		

Architectural Equipment	UNIT	MAN/ HOURS
11450.10 Residential Equipment		
Built-in, 30", 1 oven	EA.	2.667
2 oven	EA.	2.667
Counter top, 4 burner, standard	EA.	2.000
With grill	EA.	2.000
Free standing, 21", 1 oven	EA.	2.667
30", 1 oven	EA.	1.600
2 oven	EA.	1.600
Water softener		
30 grains per gallon	EA.	2.667
70 grains per gallon	EA.	4.000
11470.10 Darkroom Equipment		
Dryers		
36" x 25" x 68"	EA.	4.000
48" x 25" x 68"	EA.	4.000
Processors, film		
Black and white	EA.	4.000
Color negatives	EA.	4.000
Prints	EA.	4.000
Transparencies	EA.	4.000
Sinks with cabinet and/or stand		
5" sink with stand		
24" x 48"	EA.	2.000
32" x 64"	EA.	2.667
38" x 52"	EA.	2.667
42" x 132"	EA.	4.000
48" x 52"	EA.	4.000
5" sink with cabinet		
24" x 48"	EA.	2.000
32" x 64"	EA.	2.667
38" x 52"	EA.	2.667
42" x 132"	EA.	4.000
48" x 52"	EA.	4.000
10" sink with stand		
24" x 48"	EA.	2.000
32" x 64"	EA.	2.667
38" x 52"	EA.	2.667
10" sink with cabinet		
24" x 48"	EA.	2.000
38" x 52"	EA.	2.667
11480.10 Athletic Equipment		
Basketball backboard		
Fixed	EA.	10.000
Swing-up	EA.	16.000
Portable, hydraulic	EA.	4.000
Suspended type, standard	EA.	16.000
Bleacher, telescoping, manual		
15 tier, minimum	SEAT	0.160
Maximum	SEAT	0.160
20 tier, minimum	SEAT	0.178
Maximum	SEAT	0.178
30 tier, minimum	SEAT	0.267
Maximum	SEAT	0.267

Architectural Equipment	UNIT	MAN/ HOURS
11480.10 Athletic Equipment		
Boxing ring elevated, complete, 22' x 22'	EA.	114.286
Gym divider curtain		
Minimum	S.F.	0.011
Maximum	S.F.	0.011
Scoreboards, single face		
Minimum	EA.	8.000
Maximum	EA.	40.000
Parallel bars		
Minimum	EA.	8.000
Maximum	EA.	13.333
11500.10 Industrial Equipment		
Vehicular paint spray booth, solid back, 14'4" x 9'6"		
24' deep	EA.	8.000
26'6" deep	EA.	8.000
28'6" deep	EA.	8.000
Drive through, 14'9" x 9'6"		
24' deep	EA.	8.000
26'6" deep	EA.	8.000
28'6" deep	EA.	8.000
Water wash, paint spray booth		
5' x 11'2" x 10'8"	EA.	8.000
6' x 11'2" x 10'8"	EA.	8.000
8' x 11'2" x 10'8"	EA.	8.000
10' x 11'2" x 11'2"	EA.	8.000
12' x 12'2" x 11'2"	EA.	8.000
14' x 12'2" x 11'2"	EA.	8.000
16' x 12'2" x 11'2"	EA.	8.000
20' x 12'2" x 11'2"	EA.	8.000
Dry type spray booth, with paint arrestors		
5'4" x 7'2" x 6'8"	EA.	8.000
6'4" x 7'2" x 6'8"	EA.	8.000
8'4" x 7'2" x 9'2"	EA.	8.000
10'4" x 7'2" x 9'2"	EA.	8.000
12'4" x 7'6" x 9'2"	EA.	8.000
14'4" x 7'6" x 9'8"	EA.	8.000
16'4" x 7'7" x 9'8"	EA.	8.000
20'4" x 7'7" x 10'8"	EA.	8.000
Air compressor, electric		
1 hp		
115 volt	EA.	5.333
5 hp		
115 volt	EA.	8.000
230 volt	EA.	8.000
Hydraulic lifts		
8,000 lb capacity	EA.	20.000
11,000 lb capacity	EA.	32.000
24,000 lb capacity	EA.	53.333
Power tools		
Band saws		
10"	EA.	0.667
14"	EA.	0.800
Motorized shaper	EA.	0.615
Motorized lathe	EA.	0.667

Architectural Equipment	UNIT	MAN/ HOURS
11500.10 Industrial Equipment		
Bench saws		
9" saw	EA.	0.533
10" saw	EA.	0.571
12" saw	EA.	0.667
Electric grinders		
1/3 hp	EA.	0.320
1/2 hp	EA.	0.348
3/4 hp	EA.	0.348
11600.10 Laboratory Equipment		
Cabinets, base		
Minimum	L.F.	0.667
Maximum	L.F.	0.667
Full storage, 7' high		
Minimum	L.F.	0.667
Maximum	L.F.	0.667
Wall		
Minimum	L.F.	0.800
Maximum	L.F.	0.800
Counter tops		
Minimum	S.F.	0.100
Average	S.F.	0.114
Maximum	S.F.	0.133
Tables		
Open underneath	S.F.	0.400
Doors underneath	S.F.	0.500
Medical laboratory equipment		
Analyzer		
Chloride	EA.	0.400
Blood	EA.	0.667
Bath, water, utility, countertop unit	EA.	0.800
Hot plate, lab, countertop	EA.	0.727
Stirrer	EA.	0.727
Incubator, anaerobic, 23x23x36"	EA.	4.000
Dry heat bath	EA.	1.333
Incinerator, for sterilizing	EA.	0.080
Meter, serum protein	EA.	0.100
Ph analog, general purpose	EA.	0.114
Refrigerator, blood bank, undercounter type 153 litres	EA.	1.333
5.4 cf, undercounter type	EA.	1.333
Refrigerator/freezer, 4.4 cf, undercounter type	EA.	1.333
Sealer, impulse, free standing, 20x12x4"	EA.	0.267
Timer, electric, 1-60 minutes, bench or wall mounted	EA.	0.444
Glassware washer - dryer, undercounter	EA.	10.000
Balance, torsion suspension, tabletop, 4.5 lb capacity	EA.	0.444
Binocular microscope, with in base illuminator	EA.	0.308
Centrifuge, table model, 19x16x13"	EA.	0.320
Clinical model, with four place head	EA.	0.178
11700.10 Medical Equipment		
Hospital equipment, lights		
Examination, portable	EA.	0.667
Meters		
Air flow meter	EA.	0.444

Architectural Equipment

11700.10 Medical Equipment

Architectural Equipment	UNIT	MAN/ HOURS
Oxygen flow meters	EA.	0.333
Racks		
40 chart, revolving open frame; mobile caddy	EA.	0.667
Scales.		
Clinical, metric with measure rod, 350 lb	EA.	0.727
Physical therapy		
Chair, hydrotherapy	EA.	0.133
Diathermy, shortwave, portable, on casters	EA.	0.320
Exercise bicycle, floor standing, 35" x 15"	EA.	0.267
Hydrocollator, 4 pack, portable, 129 x 90 x 160"	EA.	0.114
Lamp, infrared, mobile with variable heat control	EA.	0.615
Ultra violet, base mounted	EA.	0.615
Mirror, posture training, 27" wide and 72" high	EA.	0.200
Parallel bars, adjustable	EA.	1.000
Platform mat 10'x6', 1" thick	EA.	0.200
Pulley, duplex, wall mounted	EA.	2.667
Rack, crutch, wall mounted, 66 x 16 x 13"	EA.	0.800
Stimulator, galvanic-faradic, hand held	EA.	0.053
Ultrasound, muscle stimulator, portable, 13x13x8"	EA.	0.067
Sandbag set, velcro straps, saddle bag type	EA.	0.114
Whirlpool, 85 gallon	EA.	4.000
65 gallon capacity	EA.	4.000
Radiology		
Radiographic table, motor driven tilting table	EA.	80.000
Fluoroscope image/tv system	EA.	160.000
Processor for washing and drying radiographs		
Water filter unit, 30" x 48-1/2" x 37-1/2"	EA.	13.333
Cassette transfer cabinet	EA.	0.667
Base storage cabinets, sectional design		
With back splash, 24" deep and 35" high	L.F.	0.667
Wall storage cabinets	L.F.	1.000
Steam sterilizers		
For heat and moisture stable materials	EA.	0.800
For fast drying after sterilization	EA.	1.000
Compact unit	EA.	1.000
Semi-automatic	EA.	4.000
Floor loading		
Single door	EA.	6.667
Double door	EA.	8.000
Utensil washer, sanitizer	EA.	6.154
Automatic washer/sterilizer	EA.	16.000
16 x 16 x 26", including generator & accessories	EA.	26.667
Steam generator, elec., 10 kw to 180 kw	EA.	16.000
Surgical scrub		
Minimum	EA.	2.667
Maximum	EA.	2.667
Gas sterilizers		
Automatic, free standing, 21x19x29"	EA.	8.000
Surgical tables		
Minimum	EA.	11.429
Maximum	EA.	16.000
Surgical lights, ceiling mounted		
Minimum	EA.	13.333
Maximum	EA.	16.000

Architectural Equipment

11700.10 Medical Equipment

Architectural Equipment	UNIT	MAN/ HOURS
Water stills		
4 liters/hr	EA.	2.667
8 liters/hr	EA.	2.667
19 liters/hr	EA.	6.667
X-ray equipment		
Mobile unit		
Minimum	EA.	4.000
Maximum	EA.	8.000
Film viewers		
Minimum	EA.	1.333
Maximum	EA.	2.667
Autopsy table		
Minimum	EA.	8.000
Maximum	EA.	8.000
Incubators		
15 cf	EA.	4.000
29 cf	EA.	6.667
Infant transport, portable	EA.	4.211
Beds		
Stretcher, with pad, 30" x 78"	EA.	2.000
Transfer, for patient transport	EA.	2.000
Headwall		
Aluminum, with back frame and console	EA.	4.000
Hospital ground detection system		
Power ground module	EA.	2.286
Ground slave module	EA.	1.739
Master ground module	EA.	1.509
Remote indicator	EA.	1.600
X-ray indicator	EA.	1.739
Micro ammeter	EA.	2.000
Supervisory module	EA.	1.739
Ground cords	EA.	0.296
Hospital isolation monitors, 5 ma		
120v	EA.	3.478
208v	EA.	3.478
240v	EA.	3.478
Digital clock-timers separate display	EA.	1.600
One display	EA.	1.600
Remote control	EA.	1.250
Battery pack	EA.	1.250
Surgical chronometer clock and 3 timers	EA.	2.500
Auxilary control	EA.	1.159

11700.20 Dental Equipment

Architectural Equipment	UNIT	MAN/ HOURS
Dental care equipment		
Drill console with accessories	EA.	13.333
Amalgamator	EA.	0.400
Lathe	EA.	0.267
Finish polisher	EA.	0.533
Model trimmer	EA.	0.364
Motor, wall mounted	EA.	0.364
Cleaner, ultrasonic	EA.	0.800
Curing unit, bench mounted	EA.	1.333
Oral evacuation system, dual pump	EA.	1.000

Architectural Equipment	UNIT	MAN/ HOURS
11700.20 Dental Equipment		
Sterilizer, table top, self contained	EA.	0.444
Dental lights		
Light, floor or ceiling mounted	EA.	4.000
X-ray unit		
Portable	EA.	2.000
Wall mounted with remote control	EA.	6.667
Illuminator, single panel	EA.	11.429
X-ray film processor	EA.	6.667
Shield, portable x-ray, lead lined	EA.	0.533

Interior	UNIT	MAN/ HOURS
12302.10 Casework		
Kitchen base cabinet, prefinished, 24" deep, 35" high		
12"wide	EA.	0.800
18" wide	EA.	0.800
24" wide	EA.	0.889
27" wide	EA.	0.889
36" wide	EA.	1.000
48" wide	EA.	1.000
Corner cabinet, 36" wide	EA.	1.000
Wall cabinet, 12" deep, 12" high		
30" wide	EA.	0.800
36" wide	EA.	0.800
15" high		
30" wide	EA.	0.889
36" wide	EA.	0.889
24" high		
30" wide	EA.	0.889
36" wide	EA.	0.889
30" high		
12" wide	EA.	1.000
18" wide	EA.	1.000
24" wide	EA.	1.000
27" wide	EA.	1.000
30" wide	EA.	1.143
36" wide	EA.	1.143
Corner cabinet, 30" high		
24" wide	EA.	1.333
30" wide	EA.	1.333
36" wide	EA.	1.333
Wardrobe	EA.	2.000
Vanity with top, laminated plastic		
24" wide	EA.	2.000
30" wide	EA.	2.000
36" wide	EA.	2.667
48" wide	EA.	3.200
12390.10 Counter Tops		
Stainless steel, counter top, with backsplash	S.F.	0.200
Acid-proof, kemrock surface	S.F.	0.133
12500.10 Window Treatment		
Drapery tracks, wall or ceiling mounted		
Basic traverse rod		
50 to 90"	EA.	0.400
84 to 156"	EA.	0.444
136 to 250"	EA.	0.444
165 to 312"	EA.	0.500
Traverse rod with stationary curtain rod		
30 to 50"	EA.	0.400
50 to 90"	EA.	0.400
84 to 156"	EA.	0.444
136 to 250"	EA.	0.500
Double traverse rod		
30 to 50"	EA.	0.400
50 to 84"	EA.	0.400

Interior	UNIT	MAN/ HOURS
12500.10 Window Treatment		
84 to 156"	EA.	0.444
136 to 250"	EA.	0.500
12510.10 Blinds		
Venetian blinds		
2" slats	S.F.	0.020
1" slats	S.F.	0.020
12690.40 Floor Mats		
Recessed entrance mat, 3/8" thick, aluminum link	S.F.	0.400
Steel, flexible	S.F.	0.400

Construction	UNIT	MAN/ HOURS
13056.10 Vaults		
Floor safes		
Class C		
1.0 cf	EA.	0.667
1.3 cf	EA.	1.000
1.9 cf	EA.	1.333
5.2 cf	EA.	1.333
13121.10 Pre-engineered Buildings		
Pre-engineered metal building, 40'x100'		
14' eave height	S.F.	0.032
16' eave height	S.F.	0.037
20' eave height	S.F.	0.048
60'x100'		
14' eave height	S.F.	0.032
16' eave height	S.F.	0.037
20' eave height	S.F.	0.048
80'x100'		
14' eave height	S.F.	0.032
16' eave height	S.F.	0.037
20' eave height	S.F.	0.048
100'x100'		
14' eave height	S.F.	0.032
16' eave height	S.F.	0.037
20' eave height	S.F.	0.048
100'x150'		
14' eave height	S.F.	0.032
16' eave height	S.F.	0.037
20' eave height	S.F.	0.048
120'x150'		
14' eave height	S.F.	0.032
16' eave height	S.F.	0.037
20' eave height	S.F.	0.048
140'x150'		
14' eave height	S.F.	0.032
16' eave height	S.F.	0.037
20' eave height	S.F.	0.048
160'x200'		
14' eave height	S.F.	0.032
16' eave height	S.F.	0.037
20' eave height	S.F.	0.048
200'x200'		
14' eave height	S.F.	0.032
16' eave height	S.F.	0.037
20' eave height	S.F.	0.048
Liner panel, 26 ga, painted steel	S.F.	0.020
Wall panel insulated, 26 ga. steel, foam core	S.F.	0.020
Roof panel, 26 ga. painted steel	S.F.	0.011
Plastic (sky light)	S.F.	0.011
Insulation, 3-1/2" thick blanket, R11	S.F.	0.005
13152.10 Swimming Pool Equipment		
Diving boards		
14' long		
Aluminum	EA.	4.444

Construction	UNIT	MAN/ HOURS
13152.10 Swimming Pool Equipment		
Fiberglass	EA.	4.444
Ladders, heavy duty		
2 steps		
Minimum	EA.	1.600
Maximum	EA.	1.600
4 steps		
Minimum	EA.	2.000
Maximum	EA.	2.000
Lifeguard chair		
Minimum	EA.	8.000
Maximum	EA.	8.000
Lights, underwater		
12 volt, with transformer	EA.	2.000
110 volt		
Minimum	EA.	2.000
Maximum	EA.	2.000
Ground fault interrupter for 110 volt, each light	EA.	0.667
Pool covers		
Reinforced polyethylene	S.F.	0.062
Vinyl water tube		
Minimum	S.F.	0.062
Maximum	S.F.	0.062
Slides with water tube		
Minimum	EA.	6.667
Maximum	EA.	6.667

Elevators

14210.10 Elevators

Elevators	UNIT	MAN/ HOURS
Passenger elevators, electric, geared		
Based on a shaft of 6 stops and 6 openings		
50 fpm, 2000 lb	EA.	24.000
100 fpm, 2000 lb	EA.	26.667
150 fpm		
2000 lb	EA.	30.000
3000 lb	EA.	34.286
4000 lb	EA.	40.000
200 fpm		
2500 lb	EA.	34.286
3000 lb	EA.	36.923
4000 lb	EA.	40.000
250 fpm		
2500 lb	EA.	34.286
3000 lb	EA.	36.923
4000 lb	EA.	40.000
300 fpm		
2500 lb	EA.	34.286
3000 lb	EA.	36.923
4000 lb	EA.	24.000
Based on a shaft of 8 stops and 8 openings		
300 fpm		
3000 lb	EA.	48.000
3500 lb	EA.	48.000
4000 lb	EA.	53.333
5000 lb	EA.	57.143
400 fpm		
3000 lb	EA.	48.000
3500 lb	EA.	48.000
4000 lb	EA.	53.333
5000 lb	EA.	57.143
600 fpm		
3000 lb	EA.	53.333
3500 lb	EA.	57.143
4000 lb	EA.	58.537
5000 lb	EA.	60.000
800 fpm		
3000 lb	EA.	53.333
3500 lb	EA.	57.143
4000 lb	EA.	58.537
5000 lb	EA.	60.000
Hydraulic, based on a shaft of 3 stops, 3 openings		
50 fpm		
2000 lb	EA.	20.000
2500 lb	EA.	20.000
3000 lb	EA.	20.870
100 fpm		
2000 lb	EA.	20.000
2500 lb	EA.	20.870
3000 lb	EA.	21.818
150 fpm		
2000 lb	EA.	20.000
2500 lb	EA.	20.870
3000 lb	EA.	22.857

Elevators

14210.10 Elevators

Elevators	UNIT	MAN/ HOURS
Small elevators, 4 to 6 passenger capacity		
Electric, push		
2 stops	EA.	20.000
3 stops	EA.	21.818
4 stops	EA.	24.000
Freight elevators, electric		
Based on a shaft of 6 stops and 6 openings		
50 fpm		
3500 lb	EA.	26.667
4000 lb	EA.	26.667
5000 lb	EA.	30.000
100 fpm		
3500 lb	EA.	30.000
4000 lb	EA.	30.000
5000 lb	EA.	34.286
200 fpm		
3500 lb	EA.	34.286
4000 lb	EA.	34.286
5000 lb	EA.	40.000
Based on shaft of 8 stops and 8 openings		
100 fpm		
4000 lb	EA.	30.000
6000 lb	EA.	30.769
8000 lb	EA.	32.432
150 fpm		
4000 lb	EA.	34.286
6000 lb	EA.	35.294
8000 lb	EA.	37.500
200 fpm		
4000 lb	EA.	40.000
6000 lb	EA.	41.379
8000 lb	EA.	43.636
Hydraulic, based on 3 stops and 3 openings		
50 fpm		
3000 lb	EA.	17.143
4000 lb	EA.	17.778
6000 lb	EA.	18.462
100 fpm		
3000 lb	EA.	17.143
4000 lb	EA.	17.778
6000 lb	EA.	18.462
150 fpm		
3000 lb	EA.	17.143
4000 lb	EA.	17.778
6000 lb	EA.	18.462

14300.10 Escalators

Escalators	UNIT	MAN/ HOURS
Escalators		
32" wide, floor to floor		
12' high	EA.	40.000
15' high	EA.	48.000
18' high	EA.	60.000
22' high	EA.	80.000
25' high	EA.	96.000

Elevators

14300.10 Escalators	UNIT	MAN/ HOURS
48" wide		
12' high	EA.	41.379
15' high	EA.	50.000
18' high	EA.	63.158
22' high	EA.	85.714
25' high	EA.	96.000

Lifts

14410.10 Personnel Lifts	UNIT	MAN/ HOURS
Residential stair climber, per story	EA.	6.667

14450.10 Vehicle Lifts	UNIT	MAN/ HOURS
Automotive hoist, one post, semi-hydraulic, 8,000 lb	EA.	24.000
Full hydraulic, 8,000 lb	EA.	24.000
2 post, semi-hydraulic, 10,000 lb	EA.	34.286
Full hydraulic		
10,000 lb	EA.	34.286
13,000 lb	EA.	60.000
18,500 lb	EA.	60.000
24,000 lb	EA.	60.000
26,000 lb	EA.	60.000
Pneumatic hoist, fully hydraulic		
11,000 lb	EA.	80.000
24,000 lb	EA.	80.000

Material Handling

14560.10 Chutes	UNIT	MAN/ HOURS
Linen chutes, stainless steel, with supports		
18" dia.	L.F.	0.057
24" dia.	L.F.	0.062
30" dia.	L.F.	0.067
Hopper	EA.	0.533
Skylight	EA.	0.800
Sprinkler unit at top	EA.	0.889

14580.10 Pneumatic Systems	UNIT	MAN/ HOURS
Pneumatic message tube system		
Average, 20 station job		
3" round system	E.A.	72.727
4" round system	E.A.	80.000
6" round system	E.A.	88.889
4" x 7" oval system	E.A.	160.000
Trash and linen tube system		
10 stations	EA.	120.000
15 stations	EA.	160.000
20 stations	EA.	184.615
30 stations	EA.	218.182

Hoists And Cranes

14600.10 Industrial Hoists	UNIT	MAN/ HOURS
Industrial hoists, electric, light to medium duty		
500 lb	EA.	4.000
1000 lb	EA.	4.211
2000 lb	EA.	4.444
3000 lb	EA.	4.706
4000 lb	EA.	5.000
5000 lb	EA.	5.333
6000 lb	EA.	5.517
7500 lb	EA.	5.714
10,000 lb	EA.	5.926
15,000 lb	EA.	6.154
20,000 lb	EA.	6.667
25,000 lb	EA.	7.273
30,000 lb	EA.	8.000
Heavy duty		
500 lb	EA.	4.000
1000 lb	EA.	4.211
2000 lb	EA.	4.444
3000 lb	EA.	4.706
4000 lb	EA.	5.000
5000 lb	EA.	5.333
6000 lb	EA.	5.517
7500 lb	EA.	5.714
10,000 lb	EA.	5.926
15,000 lb	EA.	6.154
20,000 lb	EA.	6.667
25,000 lb	EA.	7.273
30,000 lb	EA.	8.000
Air powered hoists		
500 lb	EA.	4.000
1000 lb	EA.	4.000
2000 lb	EA.	4.211
4000 lb	EA.	4.706
6000 lb	EA.	6.154
Overhead traveling bridge crane		

Hoists And Cranes

Hoists And Cranes	UNIT	MAN/ HOURS
14600.10 Industrial Hoists		
Single girder, 20' span		
3 ton	EA.	12.000
5 ton	EA.	12.000
7.5 ton	EA.	12.000
10 ton	EA.	15.000
15 ton	EA.	15.000
30' span		
3 ton	EA.	12.000
5 ton	EA.	12.000
10 ton	EA.	15.000
15 ton	EA.	15.000
Double girder, 40' span		
3 ton	EA.	26.667
5 ton	EA.	26.667
7.5 ton	EA.	26.667
10 ton	EA.	34.286
15 ton	EA.	34.286
25 ton	EA.	34.286
50' span		
3 ton	EA.	26.667
5 ton	EA.	26.667
7.5 ton	EA.	26.667
10 ton	EA.	34.286
15 ton	EA.	34.286
25 ton	EA.	34.286
14650.10 Jib Cranes		
Self supporting, swinging 8' boom, 200 deg rotation		
1000 lb	EA.	6.667
2000 lb	EA.	6.667
3000 lb	EA.	13.333
4000 lb	EA.	13.333
6000 lb	EA.	13.333
10,000 lb	EA.	13.333
Wall mounted, 180 deg rotation		
2000 lb	EA.	6.667
3000 lb	EA.	6.667
4000 lb	EA.	13.333
6000 lb	EA.	13.333
10,000 lb	EA.	13.333

15 MECHANICAL

Basic Materials	UNIT	MAN/ HOURS
15100.10 Specialties		
Wall penetration		
Concrete wall, 6" thick		
2" dia.	EA.	0.267
4" dia.	EA.	0.400
8" dia.	EA.	0.571
12" thick		
2" dia.	EA.	0.364
4" dia.	EA.	0.571
8" dia.	EA.	0.889
Non-destructive testing, piping systems		
X-ray of welds		
3" dia. pipe	EA.	0.800
4" dia. pipe	EA.	0.800
6" dia. pipe	EA.	0.800
8" dia. pipe	EA.	1.000
10" dia. pipe	EA.	1.000
Liquid penetration of welds		
2" dia. pipe	EA.	0.500
3" dia. pipe	EA.	0.500
4" dia. pipe	EA.	0.500
6" dia. pipe	EA.	0.500
8" dia. pipe	EA.	0.500
10" dia. pipe	EA.	0.500
15120.10 Backflow Preventers		
Backflow preventer, flanged, cast iron, with valves		
3" pipe	EA.	4.000
4" pipe	EA.	4.444
6" pipe	EA.	6.667
8" pipe	EA.	8.000
Threaded		
3/4" pipe	EA.	0.500
2" pipe	EA.	0.800
Reduced pressure assembly, bronze, threaded		
3/4"	EA.	0.500
1"	EA.	0.571
1-1/4"	EA.	0.667
1-1/2"	EA.	0.800
15140.10 Pipe Hangers, Heavy		
Hangers		
1/2" pipe, clevis pipe hanger		
Black steel	EA.	0.267
Galvanized	EA.	0.267
U bolt	EA.	0.080
3/4" pipe, clevis pipe hanger		
Black steel	EA.	0.267
Galvanized	EA.	0.267
1" pipe, clevis pipe hanger		
Black steel	EA.	0.267
Galvanized	EA.	0.267
U bolt	EA.	0.080
2" pipe, clevis pipe hanger		
Black steel	EA.	0.267

Basic Materials	UNIT	MAN/ HOURS
15140.10 Pipe Hangers, Heavy		
Galvanized	EA.	0.267
3" pipe, clevis pipe hanger		
Black steel	EA.	0.267
Galvanized	EA.	0.267
4" pipe, clevis pipe hanger		
Black steel	EA.	0.267
Galvanized	EA.	0.267
6" pipe, clevis pipe hanger		
Black steel	EA.	0.320
Galvanized	EA.	0.320
12" pipe, clevis pipe hanger		
Black steel	EA.	0.320
Galvanized	EA.	0.320
Threaded rod, galvanized		
C-clamp, steel, with lock nut		
3/8"	EA.	0.100
1/2"	EA.	0.100
15140.11 Pipe Hangers, Light		
A band, black iron		
1/2"	EA.	0.057
1"	EA.	0.059
1-1/4"	EA.	0.062
1-1/2"	EA.	0.067
2"	EA.	0.073
2-1/2"	EA.	0.080
3"	EA.	0.089
4"	EA.	0.100
5"	EA.	0.107
6"	EA.	0.114
8"	EA.	0.133
Copper		
1/2"	EA.	0.057
3/4"	EA.	0.059
1"	EA.	0.059
1-1/4"	EA.	0.062
1-1/2"	EA.	0.067
2"	EA.	0.073
2-1/2"	EA.	0.080
3"	EA.	0.089
4"	EA.	0.100
Black riser friction hangers		
3/4"	EA.	0.067
1"	EA.	0.070
1-1/4"	EA.	0.073
1-1/2"	EA.	0.076
2"	EA.	0.080
2-1/2"	EA.	0.089
3"	EA.	0.100
4"	EA.	0.114
Short pattern black riser clamps		
1-1/2"	EA.	0.073
2"	EA.	0.076
3"	EA.	0.080

Basic Materials	UNIT	MAN/ HOURS
15140.11 Pipe Hangers, Light		
4"	EA.	0.089
Copper riser friction hanger		
1/2"	EA.	0.062
3/4"	EA.	0.064
1"	EA.	0.067
1-1/4"	EA.	0.070
1-1/2"	EA.	0.073
2"	EA.	0.076
2-1/2"	EA.	0.080
3"	EA.	0.080
4"	EA.	0.089
Auto grip hangers, galvanized		
1/2"	EA.	0.057
1"	EA.	0.064
2"	EA.	0.073
3"	EA.	0.080
4"	EA.	0.089
Copper		
1/2"	EA.	0.057
1"	EA.	0.064
2"	EA.	0.073
3"	EA.	0.080
4"	EA.	0.089
Split rings (F&M), galvanized		
1/2"	EA.	0.062
1"	EA.	0.067
2"	EA.	0.076
3"	EA.	0.084
4"	EA.	0.089
Copper		
1/2"	EA.	0.062
1"	EA.	0.067
2"	EA.	0.076
3"	EA.	0.084
4"	EA.	0.089
2 hole clips, galvanized		
3/4"	EA.	0.053
1"	EA.	0.055
1-1/4"	EA.	0.057
1-1/2"	EA.	0.059
2"	EA.	0.062
2-1/2"	EA.	0.064
3"	EA.	0.067
4"	EA.	0.073
Perforated strap		
3/4"		
Galvanized, 20 ga.	L.F.	0.040
Copper, 22 ga.	L.F.	0.040
Threaded rod-couplings		
1/4"	EA.	0.050
3/4"	EA.	0.053
1/2"	EA.	0.057
5/8"	EA.	0.062
Reducing rod coupling, 1/2" x 3/8"	EA.	0.057

Basic Materials	UNIT	MAN/ HOURS
15140.11 Pipe Hangers, Light		
C-clamps		
3/4"	EA.	0.080
Top beam clamp		
3/8"	EA.	0.067
1/2"	EA.	0.073
Side beam connector		
3/8"	EA.	0.067
1/2"	EA.	0.073
J-Hooks		
1/2"	EA.	0.036
3/4"	EA.	0.036
1"	EA.	0.038
1-1/4"	EA.	0.039
1-1/2"	EA.	0.040
2"	EA.	0.040
3"	EA.	0.042
4"	EA.	0.042
PVC coated hangers, galvanized, 28 ga.		
1-1/2" x 12"	EA.	0.053
2" x 12"	EA.	0.057
3" x 12"	EA.	0.062
4" x 12"	EA.	0.067
Copper, 30 ga.		
1-1/2" x 12"	EA.	0.053
2" x 12"	EA.	0.057
3" x 12"	EA.	0.062
4" x 12"	EA.	0.067
Wire hook hangers		
Black wire, 1/2" x		
4"	EA.	0.040
6"	EA.	0.042
8"	EA.	0.044
10"	EA.	0.044
12"	EA.	0.047
3/4" x		
4"	EA.	0.042
6"	EA.	0.044
8"	EA.	0.047
10"	EA.	0.050
12"	EA.	0.053
1" x		
4"	EA.	0.044
6"	EA.	0.047
8"	EA.	0.050
10"	EA.	0.053
12"	EA.	0.057
Copper wire hooks		
1/2" x		
4"	EA.	0.040
6"	EA.	0.042
8"	EA.	0.044
10"	EA.	0.047
12"	EA.	0.050
3/4" x		

Basic Materials

15140.11 Pipe Hangers, Light

	UNIT	MAN/HOURS
4"	EA.	0.042
6"	EA.	0.044
8"	EA.	0.047
10"	EA.	0.050
12"	EA.	0.053

15175.60 Expansion Tanks

	UNIT	MAN/HOURS
Expansion tank, 125 psi, steel		
20 gallon	EA.	1.000
80 gallon	EA.	2.286

15240.10 Vibration Control

	UNIT	MAN/HOURS
Vibration isolator, in-line, stainless connector, screwed		
1/2"	EA.	0.444
1"	EA.	0.500
2"	EA.	0.615
3"	EA.	0.727
6"	EA.	0.889

Insulation

15260.10 Fiberglass Pipe Insulation

	UNIT	MAN/HOURS
Fiberglass insulation on 1/2" pipe		
1" thick	L.F.	0.027
1-1/2" thick	L.F.	0.033
3/4" pipe		
1" thick	L.F.	0.027
1-1/2" thick	L.F.	0.033
1" pipe		
1" thick	L.F.	0.027
1-1/2" thick	L.F.	0.033
2" pipe		
1" thick	L.F.	0.033
1-1/2" thick	L.F.	0.036
2" thick	L.F.	0.040
2-1/2" pipe		
1" thick	L.F.	0.033
1-1/2" thick	L.F.	0.036
2" thick	L.F.	0.040
3" pipe		
1" thick	L.F.	0.038
1-1/2" thick	L.F.	0.040
2" thick	L.F.	0.044
4" pipe		
1" thick	L.F.	0.038
1-1/2" thick	L.F.	0.040
2" thick	L.F.	0.044

Insulation

15260.10 Fiberglass Pipe Insulation

	UNIT	MAN/HOURS
5" pipe		
1" thick	L.F.	0.038
2" thick	L.F.	0.040
6" pipe		
1" thick	L.F.	0.042
2" thick	L.F.	0.044
8" pipe		
2" thick	L.F.	0.042
3" thick	L.F.	0.044
10" pipe		
2" thick	L.F.	0.042
3" thick	L.F.	0.044
12" pipe		
2" thick	L.F.	0.042
3" thick	L.F.	0.044

15260.60 Exterior Pipe Insulation

	UNIT	MAN/HOURS
Fiberglass insulation, aluminum jacket		
1/2" pipe		
1" thick	L.F.	0.062
1-1/2" thick	L.F.	0.067
3/4" pipe		
1" thick	L.F.	0.062
1-1/2" thick	L.F.	0.067
1" pipe		
1" thick	L.F.	0.062
1-1/2" thick	L.F.	0.067
1-1/4" pipe		
1" thick	L.F.	0.073
1-1/2" thick	L.F.	0.076
1-1/2" pipe		
1" thick	L.F.	0.073
1-1/2" thick	L.F.	0.076
2" pipe		
1" thick	L.F.	0.073
1-1/2" thick	L.F.	0.076
3" pipe		
1" thick	L.F.	0.080
1-1/2" thick	L.F.	0.084
6" pipe		
1" thick	L.F.	0.089
2" thick	L.F.	0.094
10" pipe		
2" thick	L.F.	0.089
3" thick	L.F.	0.094

15260.90 Pipe Insulation Fittings

	UNIT	MAN/HOURS
Insulation protection saddle		
1" thick covering		
1/2" pipe	EA.	0.320
3/4" pipe	EA.	0.320
1" pipe	EA.	0.320
2" pipe	EA.	0.320
3" pipe	EA.	0.364

Insulation

15260.90 Pipe Insulation Fittings	UNIT	MAN/ HOURS
6" pipe	EA.	0.500
1-1/2" thick covering		
3/4" pipe	EA.	0.320
1" pipe	EA.	0.320
2" pipe	EA.	0.320
3" pipe	EA.	0.320
6" pipe	EA.	0.500
10" pipe	EA.	0.667

15280.10 Equipment Insulation	UNIT	MAN/ HOURS
Equipment insulation, 2" thick, cellular glass	S.F.	0.050
Urethane, rigid, field applied jacket, plastered finish	S.F.	0.100
Fiberglass, rigid, with vapor barrier	S.F.	0.044

15290.10 Ductwork Insulation	UNIT	MAN/ HOURS
Fiberglass duct insulation, plain blanket		
1-1/2" thick	S.F.	0.010
2" thick	S.F.	0.013
With vapor barrier		
1-1/2" thick	S.F.	0.010
2" thick	S.F.	0.013
Rigid with vapor barrier		
2" thick	S.F.	0.027
3" thick	S.F.	0.032
4" thick	S.F.	0.040
6" thick	S.F.	0.053
Weatherproof, polystyrene, 3" thick, w/vapor barrier	S.F.	0.080
Urethane board with vapor barrier	S.F.	0.100

Fire Protection

15330.10 Wet Sprinkler System	UNIT	MAN/ HOURS
Sprinkler head, 212 deg, brass, exposed piping	EA.	0.320
Chrome, concealed piping	EA.	0.444
Water motor alarm	EA.	1.333
Fire department inlet connection	EA.	1.600
Wall plate for fire dept connection	EA.	0.667
Swing check valve flanged iron body, 4"	EA.	2.667
Check valve, 6"	EA.	4.000
Wet pipe valve, flange to groove, 4"	EA.	0.889
Flange to flange		
6"	EA.	1.333
8"	EA.	2.667
Alarm valve, flange to flange, (wet valve)		
4"	EA.	0.889
8"	EA.	6.667
Inspector's test connection	EA.	0.667

Fire Protection

15330.10 Wet Sprinkler System	UNIT	MAN/ HOURS
Wall hydrant, polished brass, 2-1/2" x 2-1/2", single	EA.	0.571
2-way	EA.	0.571
3-way	EA.	0.571
Wet valve trim, includes retard chamber & gauges, 4"-6"	EA.	0.667
Retard pressure switch for wet systems	EA.	1.600
Air maintenance device	EA.	0.007
Wall hydrant non-freeze, 8" thick wall, vacuum breaker	EA.	0.400
12" thick wall	EA.	0.400

15330.50 Dry Sprinkler System	UNIT	MAN/ HOURS
Dry pipe valve, flange to flange		
4"	EA.	1.600
6"	EA.	2.000
Trim, 4" and 6", includes gauges	EA.	0.667
Field testing and flushing	EA.	6.667
Disinfection	EA.	6.667
Pressure switch double circuit, open/close contacts	EA.	2.000
Low air		
Supervisory unit	EA.	1.333
Pressure switch	EA.	0.667

Plumbing

15410.05 C.i. Pipe, Above Ground	UNIT	MAN/ HOURS
No hub pipe		
1-1/2" pipe	L.F.	0.057
2" pipe	L.F.	0.067
3" pipe	L.F.	0.080
4" pipe	L.F.	0.133
6" pipe	L.F.	0.160
8" pipe	L.F.	0.267
10" pipe	L.F.	0.320
No hub fittings, 1-1/2" pipe		
1/4 bend	EA.	0.267
1/8 bend	EA.	0.267
Sanitary tee	EA.	0.400
Wye	EA.	0.400
2" pipe		
1/4 bend	EA.	0.320
1/8 bend	EA.	0.320
Sanitary tee	EA.	0.533
Wye	EA.	0.667
3" pipe		
1/4 bend	EA.	0.400
1/8 bend	EA.	0.400
Sanitary tee	EA.	0.500
Wye	EA.	0.667

Plumbing	UNIT	MAN/ HOURS
15410.05 C.i. Pipe, Above Ground		
4" pipe		
1/4 bend	EA.	0.400
1/8 bend	EA.	0.400
Sanitary tee	EA.	0.667
Wye	EA.	0.667
8" deep	EA.	0.400
6" pipe		
1/4 bend	EA.	0.667
1/8 bend	EA.	0.667
Sanitary tee	EA.	0.800
Wye	EA.	0.800
8" pipe		
1/4 bend	EA.	0.667
1/8 bend	EA.	0.667
Sanitary tee	EA.	1.000
Wye	EA.	0.800
10" pipe		
1/4 bend	EA.	0.667
1/8 bend	EA.	0.667
Wye	EA.	1.333
10x6" wye	EA.	1.333
15410.06 C.i. Pipe, Below Ground		
No hub pipe		
1-1/2" pipe	L.F.	0.040
2" pipe	L.F.	0.044
3" pipe	L.F.	0.050
4" pipe	L.F.	0.067
6" pipe	L.F.	0.073
8" pipe	L.F.	0.089
10" pipe	L.F.	0.100
Fittings, 1-1/2"		
1/4 bend	EA.	0.229
1/8 bend	EA.	0.229
Wye	EA.	0.320
2"		
1/4 bend	EA.	0.267
1/8 bend	EA.	0.267
3"		
1/4 bend	EA.	0.320
1/8 bend	EA.	0.320
Wye	EA.	0.500
4"		
1/4 bend	EA.	0.320
1/8 bend	EA.	0.320
Wye	EA.	0.500
6"		
1/4 bend	EA.	0.500
1/8 bend	EA.	0.500
8"		
1/4 bend	EA.	0.500
1/8 bend	EA.	0.500
Wye	EA.	0.667
10"		

Plumbing	UNIT	MAN/ HOURS
15410.06 C.i. Pipe, Below Ground		
1/4 bend	EA.	0.500
1/8 bend	EA.	0.500
Wye	EA.	1.000
15410.08 Extra Heavy Soil Pipe		
Extra heavy soil pipe, single hub		
2" x 5'	EA.	0.160
3" x 5'	EA.	0.170
4" x 5'	EA.	0.186
6" x 5'	EA.	0.200
Double hub		
2" x 5'	EA.	0.200
4" x 5'	EA.	0.211
5" x 5'	EA.	0.222
6" x 5'	EA.	0.229
Single hub		
3" x 10'	EA.	0.133
4" x 10'	EA.	0.138
6" x 10'	EA.	0.145
15410.09 Service Weight Pipe		
Service weight pipe, single hub		
3" x 5'	EA.	0.170
4" x 5'	EA.	0.178
6" x 5'	EA.	0.200
Double hub		
3" x 5'	EA.	0.216
4" x 5'	EA.	0.229
6" x 5'	EA.	0.267
Single hub		
3" x 10'	EA.	0.216
4" x 10'	EA.	0.229
6" x 10'	EA.	0.267
1/8 bend		
3"	EA.	0.320
4"	EA.	0.364
6"	EA.	0.400
1/4 bend		
3"	EA.	0.320
4"	EA.	0.364
6"	EA.	0.400
Sweep		
3"	EA.	0.320
4"	EA.	0.364
6"	EA.	0.400
Sanitary T		
3"	EA.	0.571
4"	EA.	0.667
6"	EA.	0.727
Wye		
3"	EA.	0.444
4"	EA.	0.471
6"	EA.	0.571

Plumbing

15410.10 Copper Pipe

Plumbing	UNIT	MAN/HOURS
Type "K" copper		
1/2"	L.F.	0.025
3/4"	L.F.	0.027
1"	L.F.	0.029
1-1/4"	L.F.	0.031
1-1/2"	L.F.	0.033
2"	L.F.	0.036
2-1/2"	L.F.	0.040
3"	L.F.	0.042
4"	L.F.	0.044
DWV, copper		
1-1/4"	L.F.	0.033
1-1/2"	L.F.	0.036
2"	L.F.	0.040
3"	L.F.	0.044
4"	L.F.	0.050
6"	L.F.	0.057
Refrigeration tubing, copper, sealed		
1/8"	L.F.	0.032
3/16"	L.F.	0.033
1/4"	L.F.	0.035
5/16"	L.F.	0.036
3/8"	L.F.	0.038
1/2"	L.F.	0.040
7/8"	L.F.	0.046
1-1/8"	L.F.	0.053
1-3/8"	L.F.	0.062
Type "L" copper		
1/4"	L.F.	0.024
3/8"	L.F.	0.024
1/2"	L.F.	0.025
3/4"	L.F.	0.027
1"	L.F.	0.029
1-1/4"	L.F.	0.031
1-1/2"	L.F.	0.033
2"	L.F.	0.036
2-1/2"	L.F.	0.040
3"	L.F.	0.042
3-1/2"	L.F.	0.043
4"	L.F.	0.044
Type "M" copper		
1/2"	L.F.	0.025
3/4"	L.F.	0.027
1"	L.F.	0.029
1-1/4"	L.F.	0.031
2"	L.F.	0.036
2-1/2"	L.F.	0.040
3"	L.F.	0.042
4"	L.F.	0.044

15410.11 Copper Fittings

Plumbing	UNIT	MAN/HOURS
Coupling, with stop		
1/4"	EA.	0.267
3/8"	EA.	0.320
1/2"	EA.	0.348
5/8"	EA.	0.400
3/4"	EA.	0.444
1"	EA.	0.471
3"	EA.	0.800
4"	EA.	1.000
Reducing coupling		
1/4" x 1/8"	EA.	0.320
3/8" x 1/4"	EA.	0.348
1/2" x		
3/8"	EA.	0.400
1/4"	EA.	0.400
1/8"	EA.	0.400
3/4" x		
3/8"	EA.	0.444
1/2"	EA.	0.444
1" x		
3/8"	EA.	0.500
1" x 1/2"	EA.	0.500
1" x 3/4"	EA.	0.500
1-1/4" x		
1/2"	EA.	0.533
3/4"	EA.	0.533
1"	EA.	0.533
1-1/2" x		
1/2"	EA.	0.571
3/4"	EA.	0.571
1"	EA.	0.571
1-1/4"	EA.	0.571
2" x		
1/2"	EA.	0.667
3/4"	EA.	0.667
1"	EA.	0.667
1-1/4"	EA.	0.667
1-1/2"	EA.	0.667
2-1/2" x		
1"	EA.	0.800
1-1/4"	EA.	0.800
1-1/2"	EA.	0.800
2"	EA.	0.800
3" x		
1-1/2"	EA.	1.000
2"	EA.	1.000
2-1/2"	EA.	1.000
4" x		
2"	EA.	1.143
2-1/2"	EA.	1.143
3"	EA.	1.143
Slip coupling		
1/4"	EA.	0.267
1/2"	EA.	0.320
3/4"	EA.	0.400
1"	EA.	0.444
1-1/4"	EA.	0.500

Plumbing	UNIT	MAN/ HOURS
15410.11 Copper Fittings		
1-1/2"	EA.	0.533
2"	EA.	0.667
2-1/2"	EA.	0.667
3"	EA.	0.800
4"	EA.	1.000
Coupling with drain		
1/2"	EA.	0.400
3/4"	EA.	0.444
1"	EA.	0.500
Reducer		
3/8" x 1/4"	EA.	0.320
1/2" x 3/8"	EA.	0.320
3/4" x		
1/4"	EA.	0.364
3/8"	EA.	0.364
1/2"	EA.	0.364
1" x		
1/2"	EA.	0.400
3/4"	EA.	0.400
1-1/4" x		
1/2"	EA.	0.444
3/4"	EA.	0.444
1"	EA.	0.444
1-1/2" x		
1/2"	EA.	0.500
3/4"	EA.	0.500
1"	EA.	0.500
1-1/4"	EA.	0.500
2" x		
1/2"	EA.	0.571
3/4"	EA.	0.571
1"	EA.	0.571
1-1/4"	EA.	0.571
1-1/2"	EA.	0.571
2-1/2" x		
1"	EA.	0.667
1-1/4"	EA.	0.667
1-1/2"	EA.	0.667
2"	EA.	0.667
3" x		
1-1/4"	EA.	0.800
1-1/2"	EA.	0.800
2"	EA.	0.800
2-1/2"	EA.	0.800
4" x		
2"	EA.	1.000
3"	EA.	1.000
Female adapters		
1/4"	EA.	0.320
3/8"	EA.	0.364
1/2"	EA.	0.400
3/4"	EA.	0.444
1"	EA.	0.444
1-1/4"	EA.	0.500

Plumbing	UNIT	MAN/ HOURS
15410.11 Copper Fittings		
1-1/2"	EA.	0.500
2"	EA.	0.533
2-1/2"	EA.	0.571
3"	EA.	0.667
4"	EA.	0.800
Increasing female adapters		
1/8" x		
3/8"	EA.	0.320
1/2"	EA.	0.320
1/4" x 1/2"	EA.	0.348
3/8" x 1/2"	EA.	0.364
1/2" X		
3/4"	EA.	0.400
1"	EA.	0.400
3/4" X		
1"	EA.	0.444
1-1/4"	EA.	0.444
1" x		
1-1/4"	EA.	0.444
1-1/2"	EA.	0.444
1-1/4" x		
1-1/2"	EA.	0.500
2"	EA.	0.500
1-1/2" x 2"	EA.	0.533
Reducing female adapters		
3/8" x 1/4"	EA.	0.364
1/2" x		
1/4"	EA.	0.400
3/8"	EA.	0.400
3/4" x 1/2"	EA.	0.444
1" x		
1/2"	EA.	0.444
3/4"	EA.	0.444
1-1/4" x		
1/2"	EA.	0.500
3/4"	EA.	0.500
1"	EA.	0.500
1-1/2" x		
1"	EA.	0.533
1-1/4"	EA.	0.533
2" x		
1"	EA.	0.571
1-1/4"	EA.	0.571
1-1/2"	EA.	0.571
Female fitting adapters		
1/2"	EA.	0.400
3/4"	EA.	0.400
3/4" x 1/2"	EA.	0.421
1"	EA.	0.444
1-1/4"	EA.	0.471
1-1/2"	EA.	0.500
2"	EA.	0.533
Male adapters		
1/4"	EA.	0.364

Plumbing	UNIT	MAN/ HOURS
15410.11 Copper Fittings		
3/8"	EA.	0.364
3"	EA.	0.667
4"	EA.	0.800
Increasing male adapters		
3/8" x 1/2"	EA.	0.364
1/2" x		
3/4"	EA.	0.400
1"	EA.	0.400
3/4" x		
1"	EA.	0.421
1-1/4"	EA.	0.421
1" x 1-1/4"	EA.	0.444
1-1/2" x		
3/4"	EA.	0.471
1"	EA.	0.471
1-1/4"	EA.	0.471
2" x		
1"	EA.	0.500
1-1/4"	EA.	0.500
1-1/2"	EA.	0.500
2" x 2-1/2"	EA.	0.533
Copper pipe fittings		
1/2"		
90 deg ell	EA.	0.178
45 deg ell	EA.	0.178
Tee	EA.	0.229
Cap	EA.	0.089
Coupling	EA.	0.178
Union	EA.	0.200
3/4"		
90 deg ell	EA.	0.200
45 deg ell	EA.	0.200
Tee	EA.	0.267
Cap	EA.	0.094
Coupling	EA.	0.200
Union	EA.	0.229
1"		
90 deg ell	EA.	0.267
45 deg ell	EA.	0.267
Tee	EA.	0.320
Cap	EA.	0.133
Coupling	EA.	0.267
Union	EA.	0.267
1-1/4"		
90 deg ell	EA.	0.229
45 deg ell	EA.	0.229
Tee	EA.	0.400
Cap	EA.	0.133
Union	EA.	0.286
1-1/2"		
90 deg ell	EA.	0.286
45 deg ell	EA.	0.286
Tee	EA.	0.444
Cap	EA.	0.133

Plumbing	UNIT	MAN/ HOURS
15410.11 Copper Fittings		
Coupling	EA.	0.267
Union	EA.	0.364
2"		
90 deg ell	EA.	0.320
45 deg ell	EA.	0.500
Tee	EA.	0.500
Cap	EA.	0.160
Coupling	EA.	0.320
Union	EA.	0.400
2-1/2"		
90 deg ell	EA.	0.400
45 deg ell	EA.	0.400
Tee	EA.	0.571
Cap	EA.	0.200
Coupling	EA.	0.400
Union	EA.	0.444
15410.15 Brass Fittings		
Compression fittings, union		
3/8"	EA.	0.133
1/2"	EA.	0.133
5/8"	EA.	0.133
Union elbow		
3/8"	EA.	0.133
1/2"	EA.	0.133
5/8"	EA.	0.133
Union tee		
3/8"	EA.	0.133
1/2"	EA.	0.133
5/8"	EA.	0.133
90 deg elbow union		
3/8"	EA.	0.129
1/2"	EA.	0.129
5/8"	EA.	0.129
15410.17 Chrome Plated Fittings		
Fittings		
90 ell		
3/8"	EA.	0.200
1/2"	EA.	0.200
45 ell		
3/8"	EA.	0.200
1/2"	EA.	0.200
Tee		
3/8"	EA.	0.267
1/2"	EA.	0.267
Coupling		
3/8"	EA.	0.200
1/2"	EA.	0.200
Union		
3/8"	EA.	0.200
1/2"	EA.	0.200
Tee		
1/2" x 3/8" x 3/8"	EA.	0.267

Plumbing

15410.17 Chrome Plated Fittings	UNIT	MAN/ HOURS
1/2" x 3/8" x 1/2"	EA.	0.267

15410.18 Glass Pipe	UNIT	MAN/ HOURS
Glass pipe		
1-1/2" dia.	L.F.	0.160
2" dia.	L.F.	0.178
3" dia.	L.F.	0.200
4" dia.	L.F.	0.229
6" dia.	L.F.	0.267

15410.30 Pvc/cpvc Pipe	UNIT	MAN/ HOURS
PVC schedule 40		
1/2" pipe	L.F.	0.033
3/4" pipe	L.F.	0.036
1" pipe	L.F.	0.040
1-1/4" pipe	L.F.	0.044
1-1/2" pipe	L.F.	0.050
2" pipe	L.F.	0.057
2-1/2" pipe	L.F.	0.067
3" pipe	L.F.	0.080
4" pipe	L.F.	0.100
6" pipe	L.F.	0.200
8" pipe	L.F.	0.267
PVC schedule 80 pipe		
1-1/2" pipe	L.F.	0.050
2" pipe	L.F.	0.057
3" pipe	L.F.	0.080
4" pipe	L.F.	0.100
Polypropylene, acid resistant, DWV pipe		
Schedule 40		
1-1/2" pipe	L.F.	0.057
2" pipe	L.F.	0.067
3" pipe	L.F.	0.080
4" pipe	L.F.	0.100
6" pipe	L.F.	0.200
Polyethylene pipe and fittings		
SDR-21		
3" pipe	L.F.	0.100
4" pipe	L.F.	0.133
6" pipe	L.F.	0.200
8" pipe	L.F.	0.229
10" pipe	L.F.	0.267

15410.33 Abs Dwv Pipe	UNIT	MAN/ HOURS
Schedule 40 ABS		
1-1/2" pipe	L.F.	0.040
2" pipe	L.F.	0.044
3" pipe	L.F.	0.057
4" pipe	L.F.	0.080
6" pipe	L.F.	0.100

Plumbing

15410.35 Plastic Pipe	UNIT	MAN/ HOURS
Fiberglass reinforced pipe		
2" pipe	L.F.	0.062
3" pipe	L.F.	0.067
4" pipe	L.F.	0.073
6" pipe	L.F.	0.080
8" pipe	L.F.	0.133
10" pipe	L.F.	0.160
12" pipe	L.F.	0.200

15410.70 Stainless Steel Pipe	UNIT	MAN/ HOURS
Stainless steel, schedule 40, threaded		
1/2" pipe	L.F.	0.114
3/4" pipe	L.F.	0.118
1" pipe	L.F.	0.123
1-1/2" pipe	L.F.	0.133
2" pipe	L.F.	0.145
2-1/2" pipe	L.F.	0.160
3" pipe	L.F.	0.178
4" pipe	L.F.	0.200

15410.80 Steel Pipe	UNIT	MAN/ HOURS
Black steel, extra heavy pipe, threaded		
1/2" pipe	L.F.	0.032
3/4" pipe	L.F.	0.032
1" pipe	L.F.	0.040
1-1/2" pipe	L.F.	0.044
2-1/2" pipe	L.F.	0.100
3" pipe	L.F.	0.133
4" pipe	L.F.	0.160
5" pipe	L.F.	0.200
6" pipe	L.F.	0.200
8" pipe	L.F.	0.267
10" pipe	L.F.	0.320
12" pipe	L.F.	0.400
Fittings, malleable iron, threaded, 1/2" pipe		
90 deg ell	EA.	0.267
45 deg ell	EA.	0.267
Tee	EA.	0.400
3/4" pipe		
90 deg ell	EA.	0.267
45 deg ell	EA.	0.400
Tee	EA.	0.400
1" pipe		
90 deg ell	EA.	0.320
45 deg ell	EA.	0.320
Tee	EA.	0.444
1-1/2" pipe		
90 deg ell	EA.	0.400
45 deg ell	EA.	0.400
Tee	EA.	0.571
2-1/2" pipe		
90 deg ell	EA.	1.000
45 deg ell	EA.	1.000
Tee	EA.	1.333

Plumbing	UNIT	MAN/HOURS
15410.80 Steel Pipe		
3" pipe		
90 deg ell	EA.	1.333
45 deg ell	EA.	1.333
Tee	EA.	2.000
4" pipe		
90 deg ell	EA.	1.600
45 deg ell	EA.	1.600
Tee	EA.	2.667
6" pipe		
90 deg ell	EA.	1.600
45 deg ell	EA.	1.600
Tee	EA.	2.667
8" pipe		
90 deg ell	EA.	3.200
45 deg ell	EA.	3.200
Tee	EA.	5.000
10" pipe		
90 deg ell	EA.	4.000
45 deg ell	EA.	4.000
Tee	EA.	5.000
12" pipe		
90 deg ell	EA.	5.000
45 deg ell	EA.	5.000
Tee	EA.	6.667
15410.82 Galvanized Steel Pipe		
Galvanized pipe		
1/2" pipe	L.F.	0.080
3/4" pipe	L.F.	0.100
1" pipe	L.F.	0.114
1-1/4" pipe	L.F.	0.133
1-1/2" pipe	L.F.	0.160
2" pipe	L.F.	0.200
2-1/2" pipe	L.F.	0.267
3" pipe	L.F.	0.286
4" pipe	L.F.	0.333
6" pipe	L.F.	0.667
15430.23 Cleanouts		
Cleanout, wall		
2"	EA.	0.533
3"	EA.	0.533
4"	EA.	0.667
6"	EA.	0.800
8"	EA.	1.000
Floor		
2"	EA.	0.667
3"	EA.	0.667
4"	EA.	0.800
6"	EA.	1.000
8"	EA.	1.143

Plumbing	UNIT	MAN/HOURS
15430.24 Grease Traps		
Grease traps, cast iron, 3" pipe		
35 gpm, 70 lb capacity	EA.	8.000
50 gpm, 100 lb capacity	EA.	10.000
15430.25 Hose Bibbs		
Hose bibb		
1/2"	EA.	0.267
3/4"	EA.	0.267
15430.60 Valves		
Gate valve, 125 lb, bronze, soldered		
1/2"	EA.	0.200
3/4"	EA.	0.200
1"	EA.	0.267
1-1/2"	EA.	0.320
2"	EA.	0.400
2-1/2"	EA.	0.500
Threaded		
1/4", 125 lb	EA.	0.320
1/2"		
125 lb	EA.	0.320
300 lb	EA.	0.320
3/4"		
125 lb	EA.	0.320
300 lb	EA.	0.320
1"		
125 lb	EA.	0.320
300 lb	EA.	0.400
1-1/2"		
125 lb	EA.	0.400
300 lb	EA.	0.444
2"		
125 lb	EA.	0.571
300 lb	EA.	0.667
Cast iron, flanged		
2", 150 lb	EA.	0.667
2-1/2"		
125 lb	EA.	0.667
250 lb	EA.	0.667
3"		
125 lb	EA.	0.800
250 lb	EA.	0.800
4"		
125 lb	EA.	1.143
250 lb	EA.	1.143
6"		
125 lb	EA.	1.600
250 lb	EA.	1.600
8"		
125 lb	EA.	2.000
250 lb	EA.	2.000
OS&Y, flanged		
2"		
125 lb	EA.	0.667

Plumbing

15430.60 Valves

Description	UNIT	MAN/ HOURS
250 lb	EA.	0.667
2-1/2"		
125 lb	EA.	0.667
250 lb	EA.	0.800
3"		
125 lb	EA.	0.800
250 lb	EA.	0.800
4"		
125 lb	EA.	1.333
250 lb	EA.	1.333
6"		
125 lb	EA.	1.600
250 lb	EA.	1.600
Ball valve, bronze, 250 lb, threaded		
1/2"	EA.	0.320
3/4"	EA.	0.320
1"	EA.	0.400
1-1/4"	EA.	0.444
1-1/2"	EA.	0.500
2"	EA.	0.571
Angle valve, bronze, 150 lb, threaded		
1/2"	EA.	0.286
3/4"	EA.	0.320
1"	EA.	0.320
1-1/4"	EA.	0.400
1-1/2"	EA.	0.444
Balancing valve, with meter connections, circuit setter		
1/2"	EA.	0.320
3/4"	EA.	0.364
1"	EA.	0.400
1-1/4"	EA.	0.444
1-1/2"	EA.	0.533
2"	EA.	0.667
2 1/2"	EA.	0.000
3"	EA.	1.000
4"	EA.	1.333
Pressure reducing valve, bronze, threaded, 250 lb		
1/2"	EA.	0.500
3/4"	EA.	0.500
1"	EA.	0.500
1-1/4"	EA.	0.571
1-1/2"	EA.	0.667
Pressure regulating valve, bronze, class 300		
1"	EA.	0.500
1-1/2"	EA.	0.615
2"	EA.	0.800
3"	EA.	1.143
4"	EA.	1.600
5"	EA.	2.000
6"	EA.	2.667
Solar water temperature regulating valve		
3/4"	EA.	0.667
1"	EA.	0.800
1-1/4"	EA.	0.889

Plumbing

15430.60 Valves

Description	UNIT	MAN/ HOURS
1-1/2"	EA.	1.000
2"	EA.	1.143
2-1/2"	EA.	2.000
Tempering valve, threaded		
3/4"	EA.	0.267
1"	EA.	0.320
1-1/4"	EA.	0.400
1-1/2"	EA.	0.400
2"	EA.	0.500
2-1/2"	EA.	0.667
3"	EA.	0.800
4"	EA.	1.143
Thermostatic mixing valve, threaded		
1/2"	EA.	0.286
3/4"	EA.	0.320
1"	EA.	0.348
1-1/2"	EA.	0.400
2"	EA.	0.500
Sweat connection		
1/2"	EA.	0.286
3/4"	EA.	0.320
Mixing valve, sweat connection		
1/2"	EA.	0.286
3/4"	EA.	0.320
Liquid level gauge, aluminum body		
3/4"	EA.	0.320
4125 psi, pvc body		
3/4"	EA.	0.320
150 psi, crs body		
3/4"	EA.	0.320
1"	EA.	0.320
175 psi, bronze body, 1/2"	EA.	0.286

15430.65 Vacuum Breakers

Description	UNIT	MAN/ HOURS
Vacuum breaker, atmospheric, threaded connection		
3/4"	EA.	0.320
1"	EA.	0.320
Anti-siphon, brass		
3/4"	EA.	0.320
1"	EA.	0.320
1-1/4"	EA.	0.400
1-1/2"	EA.	0.444
2"	EA.	0.500

15430.68 Strainers

Description	UNIT	MAN/ HOURS
Strainer, Y pattern, 125 psi, cast iron body, threaded		
3/4"	EA.	0.286
1"	EA.	0.320
1-1/4"	EA.	0.400
1-1/2"	EA.	0.400
2"	EA.	0.500
250 psi, brass body, threaded		
3/4"	EA.	0.320
1"	EA.	0.320

Plumbing

15430.68 Strainers	UNIT	MAN/ HOURS
1-1/4"	EA.	0.400
1-1/2"	EA.	0.400
2"	EA.	0.500
Cast iron body, threaded		
3/4"	EA.	0.320
1"	EA.	0.320
1-1/4"	EA.	0.400
1-1/2"	EA.	0.400
2"	EA.	0.500

15430.70 Drains, Roof & Floor	UNIT	MAN/ HOURS
Floor drain, cast iron, with cast iron top		
2"	EA.	0.667
3"	EA.	0.667
4"	EA.	0.667
6"	EA.	0.800
Roof drain, cast iron		
2"	EA.	0.667
3"	EA.	0.667
4"	EA.	0.667
5"	EA.	0.800
6"	EA.	0.800

15430.80 Traps	UNIT	MAN/ HOURS
Bucket trap, threaded		
3/4"	EA.	0.500
1"	EA.	0.533
1-1/4"	EA.	0.615
1-1/2"	EA.	0.727
Inverted bucket steam trap, threaded		
3/4"	EA.	0.500
1"	EA.	0.500
1-1/4"	EA.	0.444
1-1/2"	EA.	0.667
Float trap, 15 psi		
3/4"	EA.	0.500
1"	EA.	0.533
1-1/4"	EA.	0.571
1-1/2"	EA.	0.667
2"	EA.	0.800
Float and thermostatic trap, 15 psi		
3/4"	EA.	0.500
1"	EA.	0.533
1-1/4"	EA.	0.571
1-1/2"	EA.	0.667
2"	EA.	0.800
Steam trap, cast iron body, threaded, 125 psi		
3/4"	EA.	0.500
1"	EA.	0.533
1-1/4"	EA.	0.571
1-1/2"	EA.	0.667

Plumbing Fixtures

15440.10 Baths	UNIT	MAN/ HOURS
Bath tub, 5' long		
Minimum	EA.	2.667
Average	EA.	4.000
Maximum	EA.	8.000
6' long		
Minimum	EA.	2.667
Average	EA.	4.000
Maximum	EA.	8.000
Square tub, whirlpool, 4'x4'		
Minimum	EA.	4.000
Average	EA.	8.000
Maximum	EA.	10.000
5'x5'		
Minimum	EA.	4.000
Average	EA.	8.000
Maximum	EA.	10.000
6'x6'		
Minimum	EA.	4.000
Average	EA.	8.000
Maximum	EA.	10.000
For trim and rough-in		
Minimum	EA.	2.667
Average	EA.	4.000
Maximum	EA.	8.000

15440.12 Disposals & Accessories	UNIT	MAN/ HOURS
Continuous feed		
Minimum	EA.	1.600
Maximum	EA.	2.667
Batch feed, 1/2 hp		
Minimum	EA.	1.600
Maximum	EA.	2.667
Hot water dispenser		
Minimum	EA.	1.600
Maximum	EA.	2.667

15440.15 Faucets	UNIT	MAN/ HOURS
Kitchen		
Minimum	EA.	1.333
Average	EA.	1.600
Maximum	EA.	2.000
Bath		
Minimum	EA.	1.333
Average	EA.	1.600
Maximum	EA.	2.000
Lavatory, domestic		
Minimum	EA.	1.333
Average	EA.	1.600
Maximum	EA.	2.000
Hospital, patient rooms		
Minimum	EA.	2.000
Average	EA.	2.667
Maximum	EA.	4.000
Operating room		

Plumbing Fixtures

15440.15 Faucets

Plumbing Fixtures	UNIT	MAN/ HOURS
Minimum	EA.	2.000
Average	EA.	2.667
Maximum	EA.	4.000
Washroom		
Minimum	EA.	1.333
Average	EA.	1.600
Maximum	EA.	2.000
Handicapped		
Minimum	EA.	1.600
Average	EA.	2.000
Maximum	EA.	2.667
Shower		
Minimum	EA.	1.333
Average	EA.	1.600
Maximum	EA.	2.000
For trim and rough-in		
Minimum	EA.	1.600
Average	EA.	2.000
Maximum	EA.	4.000

15440.18 Hydrants

Plumbing Fixtures	UNIT	MAN/ HOURS
Wall hydrant		
8" thick	EA.	1.333
12" thick	EA.	1.600
18" thick	EA.	1.778
24" thick	EA.	2.000
Ground hydrant		
2' deep	EA.	1.000
4' deep	EA.	1.143
6' deep	EA.	1.333
8' deep	EA.	2.000

15440.20 Lavatories

Plumbing Fixtures	UNIT	MAN/ HOURS
Lavatory, counter top, porcelain enamel on cast iron		
Minimum	EA.	1.600
Average	EA.	2.000
Maximum	EA.	2.667
Wall hung, china		
Minimum	EA.	1.600
Average	EA.	2.000
Maximum	EA.	2.667
Handicapped		
Minimum	EA.	2.000
Average	EA.	2.667
Maximum	EA.	4.000
For trim and rough-in		
Minimum	EA.	2.000
Average	EA.	2.667
Maximum	EA.	4.000

15440.30 Showers

Plumbing Fixtures	UNIT	MAN/ HOURS
Shower, fiberglass, 36"x34"x84"		
Minimum	EA.	5.714
Average	EA.	8.000
Maximum	EA.	8.000
Steel, 1 piece, 36"x36"		
Minimum	EA.	5.714
Average	EA.	8.000
Maximum	EA.	8.000
Receptor, molded stone, 36"x36"		
Minimum	EA.	2.667
Average	EA.	4.000
Maximum	EA.	6.667
For trim and rough-in		
Minimum	EA.	3.636
Average	EA.	4.444
Maximum	EA.	8.000

15440.40 Sinks

Plumbing Fixtures	UNIT	MAN/ HOURS
Service sink, 24"x29"		
Minimum	EA.	2.000
Average	EA.	2.667
Maximum	EA.	4.000
Kitchen sink, single, stainless steel, single bowl		
Minimum	EA.	1.600
Average	EA.	2.000
Maximum	EA.	2.667
Double bowl		
Minimum	EA.	2.000
Average	EA.	2.667
Maximum	EA.	4.000
Porcelain enamel, cast iron, single bowl		
Minimum	EA.	1.600
Average	EA.	2.000
Maximum	EA.	2.667
Double bowl		
Minimum	EA.	2.000
Average	EA.	2.007
Maximum	EA.	4.000
Mop sink, 24"x36"x10"		
Minimum	EA.	1.600
Average	EA.	2.000
Maximum	EA.	2.667
Washing machine box		
Minimum	EA.	2.000
Average	EA.	2.667
Maximum	EA.	4.000
For trim and rough-in		
Minimum	EA.	2.667
Average	EA.	4.000
Maximum	EA.	5.333

15440.50 Urinals

Plumbing Fixtures	UNIT	MAN/ HOURS
Urinal, flush valve, floor mounted		
Minimum	EA.	2.000
Average	EA.	2.667
Maximum	EA.	4.000
Wall mounted		

Plumbing Fixtures	UNIT	MAN/ HOURS
15440.50 Urinals		
Minimum	EA.	2.000
Average	EA.	2.667
Maximum	EA.	4.000
For trim and rough-in		
Minimum	EA.	2.000
Average	EA.	4.000
Maximum	EA.	5.333
15440.60 Water Closets		
Water closet flush tank, floor mounted		
Minimum	EA.	2.000
Average	EA.	2.667
Maximum	EA.	4.000
Handicapped		
Minimum	EA.	2.667
Average	EA.	4.000
Maximum	EA.	8.000
Bowl, with flush valve, floor mounted		
Minimum	EA.	2.000
Average	EA.	2.667
Maximum	EA.	4.000
Wall mounted		
Minimum	EA.	2.000
Average	EA.	2.667
Maximum	EA.	4.000
For trim and rough-in		
Minimum	EA.	2.000
Average	EA.	2.667
Maximum	EA.	4.000
15440.70 Water Heaters		
Water heater, electric		
6 gal	EA.	1.333
10 gal	EA.	1.333
15 gal	EA.	1.333
20 gal	EA.	1.600
30 gal	EA.	1.600
40 gal	EA.	1.600
52 gal	EA.	2.000
66 gal	EA.	2.000
80 gal	EA.	2.000
100 gal	EA.	2.667
120 gal	EA.	2.667
Oil fired		
20 gal	EA.	4.000
50 gal	EA.	5.714
15440.90 Miscellaneous Fixtures		
Electric water cooler		
Floor mounted	EA.	2.667
Wall mounted	EA.	2.667
Wash fountain		
Wall mounted	EA.	4.000
Circular, floor supported	EA.	8.000

Plumbing Fixtures	UNIT	MAN/ HOURS
15440.90 Miscellaneous Fixtures		
Deluge shower and eye wash	EA.	4.000
15440.95 Fixture Carriers		
Water fountain, wall carrier		
Minimum	EA.	0.800
Average	EA.	1.000
Maximum	EA.	1.333
Lavatory, wall carrier		
Minimum	EA.	0.800
Average	EA.	1.000
Maximum	EA.	1.333
Sink, industrial, wall carrier		
Minimum	EA.	0.800
Average	EA.	1.000
Maximum	EA.	1.333
Toilets, water closets, wall carrier		
Minimum	EA.	0.800
Average	EA.	1.000
Maximum	EA.	1.333
Floor support		
Minimum	EA.	0.667
Average	EA.	0.800
Maximum	EA.	1.000
Urinals, wall carrier		
Minimum	EA.	0.800
Average	EA.	1.000
Maximum	EA.	1.333
Floor support		
Minimum	EA.	0.667
Average	EA.	0.800
Maximum	EA.	1.000
15450.30 Pumps		
In-line pump, bronze, centrifugal		
5 gpm, 20' head	EA.	0.500
20 gpm, 40' head	EA.	0.500
50 gpm		
50' head	EA.	1.000
Cast iron, centrifugal		
50 gpm, 200' head	EA.	1.000
100 gpm		
100' head	EA.	1.333
Centrifugal, close coupled, c.i., single stage		
50 gpm, 100' head	EA.	1.000
100 gpm, 100' head	EA.	1.333
Base mounted		
50 gpm, 100' head	EA.	1.000
100 gpm, 50' head	EA.	1.333
200 gpm, 100' head	EA.	2.000
300 gpm, 175' head	EA.	2.000
Condensate pump, simplex		
1000 sf EDR, 2 gpm	EA.	6.667
2000 sf EDR, 3 gpm	EA.	6.667
4000 sf EDR, 6 gpm	EA.	7.273

Plumbing Fixtures

Plumbing Fixtures	UNIT	MAN/ HOURS
15450.30 Pumps		
6000 sf EDR, 9 gpm	EA.	7.273
Duplex, bronze		
8000 sf EDR, 12 gpm	EA.	7.273
10,000 sf EDR, 15 gpm	EA.	10.000
15,000 sf EDR, 23 gpm	EA.	11.429
20,000 sf EDR, 30 gpm	EA.	16.000
25,000 sf EDR, 38 gpm	EA.	16.000
15450.40 Storage Tanks		
Hot water storage tank, cement lined		
10 gallon	EA.	2.667
70 gallon	EA.	4.000
200 gallon	EA.	5.714
900 gallon	EA.	10.000
1100 gallon	EA.	10.000
2000 gallon	EA.	10.000
15480.10 Special Systems		
Air compressor, air cooled, two stage		
5.0 cfm, 175 psi	EA.	16.000
10 cfm, 175 psi	EA.	17.778
20 cfm, 175 psi	EA.	19.048
50 cfm, 125 psi	EA.	21.053
80 cfm, 125 psi	EA.	22.857
Single stage, 125 psi		
1.0 cfm	EA.	11.429
1.5 cfm	EA.	11.429
2.0 cfm	EA.	11.429
Automotive compressor, hose reel, air and water, 50' hose	EA.	6.667
Lube equipment, 3 reel, with pumps	EA.	32.000
Tire changer		
Truck	EA.	11.429
Passenger car	EA.	6.154
Air hose reel, includes, 50' hose	EA.	6.154
Hose reel, 5 reel, motor oil, gear oil, lube, air & water	EA.	32.000
Water hose reel, 50' hose	EA.	6.154
Pump, air operated, for motor or gear oil, fits 55 gal drum	EA.	0.800
For chassis lube	EA.	0.800
Fuel dispensing pump, lighted dial, one product		
One hose	EA.	6.667
Two hose	EA.	6.667
Two products, two hose	EA.	6.667

Heating & Ventilating

Heating & Ventilating	UNIT	MAN/ HOURS
15555.10 Boilers		
Cast iron, gas fired, hot water		
115 mbh	EA.	20.000
175 mbh	EA.	21.818
235 mbh	EA.	24.000
940 mbh	EA.	48.000
1600 mbh	EA.	60.000
3000 mbh	EA.	80.000
6000 mbh	EA.	120.000
Steam		
115 mbh	EA.	20.000
175 mbh	EA.	21.818
235 mbh	EA.	24.000
940 mbh	EA.	48.000
1600 mbh	EA.	60.000
3000 mbh	EA.	80.000
6000 mbh	EA.	120.000
Electric, hot water		
115 mbh	EA.	12.000
175 mbh	EA.	12.000
235 mbh	EA.	12.000
940 mbh	EA.	24.000
1600 mbh	EA.	48.000
3000 mbh	EA.	60.000
6000 mbh	EA.	80.000
Steam		
115 mbh	EA.	12.000
175 mbh	EA.	12.000
235 mbh	EA.	12.000
940 mbh	EA.	24.000
1600 mbh	EA.	48.000
3000 mbh	EA.	60.000
6000 mbh	EA.	80.000
Oil fired, hot water		
115 mbh	EA.	16.000
175 mbh	EA.	18.462
235 mbh	EA.	21.818
940 mbh	EA.	40.000
1600 mbh	EA.	48.000
3000 mbh	EA.	60.000
6000 mbh	EA.	120.000
Steam		
115 mbh	EA.	16.000
175 mbh	EA.	18.462
235 mbh	EA.	21.818
940 mbh	EA.	40.000
1600 mbh	EA.	48.000
3000 mbh	EA.	60.000
6000 mbh	EA.	120.000
15610.10 Furnaces		
Electric, hot air		
40 mbh	EA.	4.000
60 mbh	EA.	4.211
80 mbh	EA.	4.444

Heating & Ventilating

15610.10 Furnaces

	UNIT	MAN/ HOURS
100 mbh	EA.	4.706
125 mbh	EA.	4.848
160 mbh	EA.	5.000
200 mbh	EA.	5.161
400 mbh	EA.	5.333
Gas fired hot air		
40 mbh	EA.	4.000
60 mbh	EA.	4.211
80 mbh	EA.	4.444
100 mbh	EA.	4.706
125 mbh	EA.	4.848
160 mbh	EA.	5.000
200 mbh	EA.	5.161
400 mbh	EA.	5.333
Oil fired hot air		
40 mbh	EA.	4.000
60 mbh	EA.	4.211
80 mbh	EA.	4.444
100 mbh	EA.	4.706
125 mbh	EA.	4.848
160 mbh	EA.	5.000
200 mbh	EA.	5.161
400 mbh	EA.	5.333

Refrigeration

15670.10 Condensing Units

	UNIT	MAN/ HOURS
Air cooled condenser, single circuit		
3 ton	EA.	1.333
5 ton	EA.	1.333
7.5 ton	EA.	3.810
20 ton	EA.	4.000
25 ton	EA.	4.000
30 ton	EA.	4.000
40 ton	EA.	5.714
50 ton	EA.	5.714
60 ton	EA.	5.000
With low ambient dampers		
3 ton	EA.	2.000
5 ton	EA.	2.000
7.5 ton	EA.	4.000
20 ton	EA.	5.333
25 ton	EA.	5.333
30 ton	EA.	5.333
40 ton	EA.	6.667
50 ton	EA.	7.273
60 ton	EA.	7.273

Refrigeration

15670.10 Condensing Units

	UNIT	MAN/ HOURS
Dual circuit		
10 ton	EA.	4.000
15 ton	EA.	5.714
20 ton	EA.	5.714
25 ton	EA.	5.714
30 ton	EA.	5.714
40 ton	EA.	6.667
50 ton	EA.	6.667
60 ton	EA.	6.667
80 ton	EA.	8.889
100 ton	EA.	8.889
120 ton	EA.	8.889
With low ambient dampers		
15 ton	EA.	5.714
20 ton	EA.	5.714
25 ton	EA.	5.714
30 ton	EA.	5.714
40 ton	EA.	6.667
50 ton	EA.	6.667
60 ton	EA.	6.667
80 ton	EA.	8.889
100 ton	EA.	8.889
120 ton	EA.	8.889

15680.10 Chillers

	UNIT	MAN/ HOURS
Chiller, reciprocal		
Air cooled, remote condenser, starter		
20 ton	EA.	8.000
25 ton	EA.	8.000
30 ton	EA.	8.000
40 ton	EA.	12.000
50 ton	EA.	13.333
100 ton	EA.	24.000
200 ton	EA.	40.000
Water cooled, with starter		
20 ton	EA.	8.000
25 ton	EA.	8.000
30 ton	EA.	12.000
40 ton	EA.	12.000
50 ton	EA.	13.333
100 ton	EA.	24.000
200 ton	EA.	40.000
Packaged, air cooled, with starter		
20 ton	EA.	6.000
25 ton	EA.	6.000
30 ton	EA.	6.000
40 ton	EA.	6.000
Heat recovery, air cooled, with starter		
50 ton	EA.	12.000
100 ton	EA.	24.000
Water cooled, with starter		
40 ton	EA.	12.000
100 ton	EA.	26.667
Centrifugal, single bundle condenser, with starter		

Refrigeration

15680.10 Chillers	UNIT	MAN/ HOURS
80 ton	EA.	34.286
230 ton	EA.	53.333
460 ton	EA.	80.000
670 ton	EA.	96.000

15710.10 Cooling Towers	UNIT	MAN/ HOURS
Cooling tower, propeller type		
100 ton	EA.	8.000
400 ton	EA.	24.000
1000 ton	EA.	60.000
Centrifugal		
100 ton	EA.	8.000
400 ton	EA.	24.000
1000 ton	EA.	60.000

Heat Transfer

15780.10 Computer Room A/c	UNIT	MAN/ HOURS
Air cooled, alarm, high efficiency filter, elec. heat		
3 ton	EA.	6.154
5 ton	EA.	6.667
7.5 ton	EA.	8.000
10 ton	EA.	10.000
15 ton	EA.	11.429
Steam heat		
3 ton	EA.	6.154
5 ton	EA.	6.667
7.5 ton	EA.	8.000
10 ton	EA.	10.000
15 ton	EA.	11.429
Hot water heat		
3 ton	EA.	6.154
5 ton	EA.	6.667
7.5 ton	EA.	8.000
10 ton	EA.	10.000
15 ton	EA.	11.429
Air cooled condenser, low ambient damper		
3 ton	EA.	1.600
5 ton	EA.	2.000
7.5 ton	EA.	4.000
10 ton	EA.	5.714
15 ton	EA.	4.706
Water cooled, high efficiency filter, alarm, elec. heat		
3 ton	EA.	5.714
5 ton	EA.	6.667
7.5 ton	EA.	10.000
10 ton	EA.	11.429

Heat Transfer

15780.10 Computer Room A/c	UNIT	MAN/ HOURS
15 ton	EA.	13.333
Steam heat		
3 ton	EA.	5.714
5 ton	EA.	6.667
7.5 ton	EA.	10.000
10 ton	EA.	11.429
15 ton	EA.	13.333
Hot water heat		
3 ton	EA.	5.714
5 ton	EA.	6.667
7.5 ton	EA.	10.000
10 ton	EA.	11.429
15 ton	EA.	13.333

15780.20 Rooftop Units	UNIT	MAN/ HOURS
Packaged, single zone rooftop unit, with roof curb		
2 ton	EA.	8.000
3 ton	EA.	8.000
4 ton	EA.	10.000
5 ton	EA.	13.333
7.5 ton	EA.	16.000

15830.20 Fan Coil Units	UNIT	MAN/ HOURS
Fan coil unit, 2 pipe, complete		
200 cfm ceiling hung	EA.	2.667
Floor mounted	EA.	2.000
300 cfm, ceiling hung	EA.	3.200
Floor mounted	EA.	2.667
400 cfm, ceiling hung	EA.	3.810
Floor mounted	EA.	2.667
500 cfm, ceiling hung	EA.	4.000
Floor mounted	EA.	3.077
600 cfm, ceiling hung	EA.	4.420
Floor mounted	EA.	3.636

15830.70 Unit Heaters	UNIT	MAN/ HOURS
Steam unit heater, horizontal		
12,500 btuh, 200 cfm	EA.	1.333
17,000 btuh, 300 cfm	EA.	1.333
40,000 btuh, 500 cfm	EA.	1.333
60,000 btuh, 700 cfm	EA.	1.333
70,000 btuh, 1000 cfm	EA.	2.000
Vertical		
12,500 btuh, 200 cfm	EA.	1.333
17,000 btuh, 300 cfm	EA.	1.333
40,000 btuh, 500 cfm	EA.	1.333
60,000 btuh, 700 cfm	EA.	1.333
70,000 btuh, 1000 cfm	EA.	1.333
Gas unit heater, horizontal		
27,400 btuh	EA.	3.200
38,000 btuh	EA.	3.200
56,000 btuh	EA.	3.200
82,200 btuh	EA.	3.200
103,900 btuh	EA.	5.000

Heat Transfer

15830.70 Unit Heaters	UNIT	MAN/ HOURS
125,700 btuh	EA.	5.000
133,200 btuh	EA.	5.000
149,000 btuh	EA.	5.000
172,000 btuh	EA.	5.000
190,000 btuh	EA.	5.000
225,000 btuh	EA.	5.000
Hot water unit heater, horizontal		
12,500 btuh, 200 cfm	EA.	1.333
17,000 btuh, 300 cfm	EA.	1.333
25,000 btuh, 500 cfm	EA.	1.333
30,000 btuh, 700 cfm	EA.	1.333
50,000 btuh, 1000 cfm	EA.	2.000
60,000 btuh, 1300 cfm	EA.	2.000
Vertical		
12,500 btuh, 200 cfm	EA.	1.333
17,000 btuh, 300 cfm	EA.	1.333
25,000 btuh, 500 cfm	EA.	1.333
30,000 btuh, 700 cfm	EA.	1.333
50,000 btuh, 1000 cfm	EA.	1.333
60,000 btuh, 1300 cfm	EA.	1.333
Cabinet unit heaters, ceiling, exposed, hot water		
200 cfm	EA.	2.667
300 cfm	EA.	3.200
400 cfm	EA.	3.810
600 cfm	EA.	4.211
800 cfm	EA.	5.000
1000 cfm	EA.	5.714
1200 cfm	EA.	6.667
2000 cfm	EA.	8.889

Air Handling

15855.10 Air Handling Units	UNIT	MAN/ HOURS
Air handling unit, medium pressure, single zone		
1500 cfm	EA.	5.000
3000 cfm	EA.	8.889
4000 cfm	EA.	10.000
5000 cfm	EA.	10.667
6000 cfm	EA.	11.429
7000 cfm	EA.	12.308
8500 cfm	EA.	13.333
10,500 cfm	EA.	16.000
12,500 cfm	EA.	17.778
Rooftop air handling units		
4950 cfm	EA.	8.889
7370 cfm	EA.	11.429
9790 cfm	EA.	13.333

Air Handling

15855.10 Air Handling Units	UNIT	MAN/ HOURS
14,300 cfm	EA.	11.429
21,725 cfm	EA.	11.429
33,000 cfm	EA.	13.333

15870.20 Exhaust Fans	UNIT	MAN/ HOURS
Belt drive roof exhaust fans		
640 cfm, 2618 fpm	EA.	1.000
940 cfm, 2604 fpm	EA.	1.000
1050 cfm, 3325 fpm	EA.	1.000
1170 cfm, 2373 fpm	EA.	1.000
2440 cfm, 4501 fpm	EA.	1.000
2760 cfm, 4950 fpm	EA.	1.000
3890 cfm, 6769 fpm	EA.	1.000
2380 cfm, 3382 fpm	EA.	1.000
2880 cfm, 3859 fpm	EA.	1.000
3200 cfm, 4173 fpm	EA.	1.333
3660 cfm, 3437 fpm	EA.	1.333
Direct drive fans		
60 to 390 cfm	EA.	1.000
145 to 590 cfm	EA.	1.000
295 to 860 cfm	EA.	1.000
235 to 1300 cfm	EA.	1.000
415 to 1630 cfm	EA.	1.000
590 to 2045 cfm	EA.	1.000

Air Distribution

15890.10 Metal Ductwork	UNIT	MAN/ HOURS
Rectangular duct		
Galvanized steel		
Minimum	Lb.	0.073
Average	Lb.	0.089
Maximum	Lb.	0.133
Aluminum		
Minimum	Lb.	0.160
Average	Lb.	0.200
Maximum	Lb.	0.267
Fittings		
Minimum	EA.	0.267
Average	EA.	0.400
Maximum	EA.	0.800

15890.30 Flexible Ductwork	UNIT	MAN/ HOURS
Flexible duct, 1.25" fiberglass		
5" dia.	L.F.	0.040
6" dia.	L.F.	0.044
7" dia.	L.F.	0.047

15 MECHANICAL

Air Distribution

Air Distribution	UNIT	MAN/ HOURS
15890.30 Flexible Ductwork		
8" dia.	L.F.	0.050
10" dia.	L.F.	0.057
12" dia.	L.F.	0.062
14" dia.	L.F.	0.067
16" dia.	L.F.	0.073
Flexible duct connector, 3" wide fabric	L.F.	0.133
15895.10 Roof Curbs		
8" high, insulated, with liner and raised can		
15" x 15"	EA.	0.400
21" x 21"	EA.	0.400
36" x 36"	EA.	0.571
48" x 48"	EA.	0.615
60" x 60"	EA.	0.800
72" x 72"	EA.	1.000
15910.10 Dampers		
Horizontal parallel aluminum backdraft damper		
12" x 12"	EA.	0.200
16" x 16"	EA.	0.229
24" x 24"	EA.	0.400
36" x 36"	EA.	0.571
48" x 48"	EA.	0.800
"Up", parallel dampers		
12" x 12"	EA.	0.200
16" x 16"	EA.	0.229
24" x 24"	EA.	0.400
36" x 36"	EA.	0.571
48" x 48"	EA.	0.800
"Down", parallel dampers		
12" x 12"	EA.	0.200
16" x 16"	EA.	0.229
24" x 24"	EA.	0.400
36" x 36"	EA.	0.571
48" x 48"	EA.	0.800
Fire damper, 1.5 hr rating		
12" x 12"	EA.	0.400
16" x 16"	EA.	0.400
24" x 24"	EA.	0.400
36" x 36"	EA.	0.800
48" x 48"	EA.	1.143
15940.10 Diffusers		
Ceiling diffusers, round, baked enamel finish		
6" dia.	EA.	0.267
12" dia.	EA.	0.333
16" dia.	EA.	0.364
20" dia.	EA.	0.400
Rectangular		
6x6"	EA.	0.267
12x12"	EA.	0.400
18x18"	EA.	0.400
24x24"	EA.	0.500
Lay In, flush mounted, perforated face, with grid		

Air Distribution	UNIT	MAN/ HOURS
15940.10 Diffusers		
6x6/24x24	EA.	0.320
12x12/24x24	EA.	0.320
18x18/24x24	EA.	0.320
Two-way slot diffuser with balancing damper, 4'	EA.	0.800
15940.20 Relief Ventilators		
Intake ventilator, aluminum, with screen, no curbs		
12" x 12"	EA.	0.667
16" x 16"	EA.	0.800
36" x 36"	EA.	1.333
48" x 48"	EA.	1.600
15940.40 Registers And Grilles		
Lay in flush mounted, perforated face, return		
6x6/24x24	EA.	0.320
8x8/24x24	EA.	0.320
9x9/24x24	EA.	0.320
10x10/24x24	EA.	0.320
12x12/24x24	EA.	0.320
Rectangular, ceiling return, single deflection		
10x10	EA.	0.400
12x12	EA.	0.400
14x14	EA.	0.400
16x8	EA.	0.400
16x16	EA.	0.400
24x12	EA.	0.400
24x18	EA.	0.400
36x24	EA.	0.444
36x30	EA.	0.444
Wall, return air register		
12x12	EA.	0.200
16x16	EA.	0.200
18x18	EA.	0.200
20x20	EA.	0.200
24x24	EA.	0.200
Ceiling, return air grille		
6x6	EA.	0.267
8x8	EA.	0.320
10x10	EA.	0.320
Ceiling, exhaust grille, aluminum egg crate		
6x6	EA.	0.267
8x8	EA.	0.320
10x10	EA.	0.320
12x12	EA.	0.400
14x14	EA.	0.400
16x16	EA.	0.400
18x18	EA.	0.400
15940.80 Penthouse Louvers		
Penthouse louvers		
12" high, extruded aluminum, 4" louver		
6' perimeter	EA.	2.000
12' perimeter	EA.	2.000
20' perimeter	EA.	5.333

Air Distribution

15940.80 Penthouse Louvers	UNIT	MAN/ HOURS
16" high x 4' perimeter	EA.	2.000
6' perimeter	EA.	2.000
12' perimeter	EA.	2.000
20' perimeter	EA.	5.333
20" high x 4' perimeter	EA.	2.000
6' perimeter	EA.	2.000
12' perimeter	EA.	2.000
20' perimeter	EA.	5.333
24" high x 4' perimeter	EA.	2.000
6' perimeter	EA.	2.000
12' perimeter	EA.	2.000
20' perimeter	EA.	5.333

Controls

15950.10 Hvac Controls	UNIT	MAN/ HOURS
Pressure gauge, direct reading gage cock and siphon	EA.	0.500
Control valve, 1", modulating		
2-way	EA.	0.667
3-way	EA.	1.000
Self contained control valve w/ sensing elmnt, 3/4"	EA.	0.500
Control dampers, round		
6" dia.	EA.	0.320
8" dia	EA.	0.320
10" dia	EA.	0.320
12" dia	EA.	0.320
12" dia	EA.	0.400
18" dia	EA.	0.400
20" dia	EA.	0.400
Rectangular, parallel blade standard leakage		
12" x 12"	EA.	0.400
16" x 16"	EA.	0.400
20" x 20"	EA.	0.400
48" x 48"	EA.	1.143
48" x 60"	EA.	1.333
48" x 72"	EA.	1.333
Low leakage		
12" x 12"	EA.	0.400
16" x 16"	EA.	0.400
36" x 36"	EA.	0.667
48" x 48"	EA.	1.143
48" x 72"	EA.	1.333
Rectangular, opposed horizontal blade		
12" x 12"	EA.	0.400
16" x 16"	EA.	0.400
24" x 24"	EA.	0.400
36" x 36"	EA.	0.667

Controls

15950.10 Hvac Controls	UNIT	MAN/ HOURS
48" x 72"	EA.	1.333

Basic Materials

16110.12 Cable Tray

Basic Materials	UNIT	MAN/HOURS
Cable tray, 6"	L.F.	0.059
Ventilated cover	L.F.	0.030
Solid cover	L.F.	0.030

16110.20 Conduit Specialties

Basic Materials	UNIT	MAN/HOURS
Rod beam clamp, 1/2"	EA.	0.050
Hanger rod		
3/8"	L.F.	0.040
1/2"	L.F.	0.050
All thread rod		
1/4"	L.F.	0.030
3/8"	L.F.	0.040
1/2"	L.F.	0.050
5/8"	L.F.	0.080
Hanger channel, 1-1/2"		
No holes	EA.	0.030
Holes	EA.	0.030
Channel strap		
1/2"	EA.	0.050
3/4"	EA.	0.050
1"	EA.	0.050
1-1/4"	EA.	0.080
1-1/2"	EA.	0.080
2"	EA.	0.080
2-1/2"	EA.	0.123
3"	EA.	0.123
3-1/2"	EA.	0.123
4"	EA.	0.145
5"	EA.	0.145
6"	EA.	0.145
Conduit penetrations, roof and wall, 8" thick		
1/2"	EA.	0.615
3/4"	EA.	0.615
1"	EA.	0.800
1-1/4"	EA.	0.800
1-1/2"	EA.	0.800
2"	EA.	1.600
2-1/2"	EA.	1.600
3"	EA.	1.600
3-1/2"	EA.	2.000
4"	EA.	2.000
Fireproofing, for conduit penetrations		
1/2"	EA.	0.500
3/4"	EA.	0.500
1"	EA.	0.500
1-1/4"	EA.	0.727
1-1/2"	EA.	0.727
2"	EA.	0.727
2-1/2"	EA.	0.899
3"	EA.	0.899
3-1/2"	EA.	1.250
4"	EA.	1.509

Basic Materials

16110.21 Aluminum Conduit

Basic Materials	UNIT	MAN/HOURS
Aluminum conduit		
1/2"	L.F.	0.030
3/4"	L.F.	0.040
1"	L.F.	0.050
1-1/4"	L.F.	0.059
1-1/2"	L.F.	0.080
2"	L.F.	0.089
2-1/2"	L.F.	0.100
3"	L.F.	0.107
3-1/2"	L.F.	0.123
4"	L.F.	0.145
5"	L.F.	0.182
6"	L.F.	0.200

16110.22 Emt Conduit

Basic Materials	UNIT	MAN/HOURS
EMT conduit		
1/2"	L.F.	0.030
3/4"	L.F.	0.040
1"	L.F.	0.050
1-1/4"	L.F.	0.059
1-1/2"	L.F.	0.080
2"	L.F.	0.089
2-1/2"	L.F.	0.100
3"	L.F.	0.123
3-1/2"	L.F.	0.145
4"	L.F.	0.182

16110.23 Flexible Conduit

Basic Materials	UNIT	MAN/HOURS
Flexible conduit, steel		
3/8"	L.F.	0.030
1/2	L.F.	0.030
3/4"	L.F.	0.040
1"	L.F.	0.040
1-1/4"	L.F.	0.050
1-1/2"	L.F.	0.059
2"	L.F.	0.080
2-1/2"	L.F.	0.089
3"	L.F.	0.107

16110.24 Galvanized Conduit

Basic Materials	UNIT	MAN/HOURS
Galvanized rigid steel conduit		
1/2"	L.F.	0.040
3/4"	L.F.	0.050
1"	L.F.	0.059
1-1/4"	L.F.	0.080
1-1/2"	L.F.	0.089
2"	L.F.	0.100
2-1/2"	L.F.	0.145
3"	L.F.	0.182
3-1/2"	L.F.	0.190
4"	L.F.	0.211
5"	L.F.	0.286
6"	L.F.	0.381

Basic Materials

16110.27 Plastic Coated Conduit

	UNIT	MAN/ HOURS
Rigid steel conduit, plastic coated		
1/2"	L.F.	0.050
3/4"	L.F.	0.059
1"	L.F.	0.080
1-1/4"	L.F.	0.100
1-1/2"	L.F.	0.123
2"	L.F.	0.145
2-1/2"	L.F.	0.190
3"	L.F.	0.222
3-1/2"	L.F.	0.250
4"	L.F.	0.308
5"	L.F.	0.381

16110.28 Steel Conduit

	UNIT	MAN/ HOURS
Intermediate metal conduit (IMC)		
1/2"	L.F.	0.030
3/4"	L.F.	0.040
1"	L.F.	0.050
1-1/4"	L.F.	0.059
1-1/2"	L.F.	0.080
2"	L.F.	0.089
2-1/2"	L.F.	0.119
3"	L.F.	0.145
3-1/2"	L.F.	0.182
4"	L.F.	0.190

16110.35 Wiremold

	UNIT	MAN/ HOURS
Wiremold raceway with fittings, surface mounted		
#200	L.F.	0.030
#500	L.F.	0.030
#700	L.F.	0.040
#800	L.F.	0.040
Fittings, #200, 90 degree flat elbow	EA.	0.050
Internal elbow	EA.	0.050
Extension adapter	EA.	0.059
#200, #500, #700		
Single pole switch and box	EA.	0.400
Duplex receptacle with box	EA.	0.308
#500, #700		
90 deg. flat elbow	EA.	0.080
Internal elbow	EA.	0.080
Junction box	EA.	0.133
Fixture box	EA.	0.133
Shallow switch and receptacle box	EA.	0.080
#800		
90 deg. flat elbow	EA.	0.080
Internal elbow	EA.	0.080
Junction box	EA.	0.123

16110.80 Wireways

	UNIT	MAN/ HOURS
Wireway, hinge cover type		
2-1/2" x 2-1/2"		
1' section	EA.	0.154
2'	EA.	0.190
3'	EA.	0.250
5'	EA.	0.381
10'	EA.	0.667
4" x 4"		
1'	EA.	0.250
2'	EA.	0.250
3'	EA.	0.308
4'	EA.	0.308
10'	EA.	0.800

16120.43 Copper Conductors

	UNIT	MAN/ HOURS
Copper conductors, type THW, solid		
#14	L.F.	0.004
#12	L.F.	0.005
#10	L.F.	0.006
Stranded		
#14	L.F.	0.004
#12	L.F.	0.005
#10	L.F.	0.006
#8	L.F.	0.008
#6	L.F.	0.009
#4	L.F.	0.010
#3	L.F.	0.010
#2	L.F.	0.012
#1	L.F.	0.014
1/0	L.F.	0.016
2/0	L.F.	0.020
3/0	L.F.	0.025
4/0	L.F.	0.028
250 MCM	L.F.	0.030
300 MCM	L.F.	0.033
350 MCM	L.F.	0.040
400 MCM	L.F.	0.044
500 MCM	L.F.	0.052
600 MCM	L.F.	0.059
750 MCM	L.F.	0.067
1000 MCM	L.F.	0.076
THHN-THWN, solid		
#14	L.F.	0.004
#12	L.F.	0.005
#10	L.F.	0.006
Stranded		
#14	L.F.	0.004
#12	L.F.	0.005
#10	L.F.	0.006
#8	L.F.	0.008
#6	L.F.	0.009
#4	L.F.	0.010
#2	L.F.	0.012
#1	L.F.	0.014
1/0	L.F.	0.016
2/0	L.F.	0.020
3/0	L.F.	0.025
4/0	L.F.	0.028

Basic Materials

Basic Materials	UNIT	MAN/ HOURS
16120.43 Copper Conductors		
250 MCM	L.F.	0.030
350 MCM	L.F.	0.040
16120.47 Sheathed Cable		
Non-metallic sheathed cable		
Type NM cable with ground		
#14/2	L.F.	0.015
#12/2	L.F.	0.016
#10/2	L.F.	0.018
#8/2	L.F.	0.020
#6/2	L.F.	0.025
#14/3	L.F.	0.026
#12/3	L.F.	0.027
#10/3	L.F.	0.027
#8/3	L.F.	0.028
#6/3	L.F.	0.028
#4/3	L.F.	0.032
#2/3	L.F.	0.035
Type U.F. cable with ground		
#14/2	L.F.	0.016
#12/2	L.F.	0.019
#10/2	L.F.	0.020
#8/2	L.F.	0.023
#6/2	L.F.	0.027
#14/3	L.F.	0.020
#12/3	L.F.	0.022
#10/3	L.F.	0.025
#8/3	L.F.	0.028
#6/3	L.F.	0.032
16130.60 Pull And Junction Boxes		
4"		
Octagon box	EA.	0.114
Box extension	EA.	0.059
Plaster ring	EA.	0.059
Cover blank	EA.	0.059
Square box	EA.	0.114
Box extension	EA.	0.059
Plaster ring	EA.	0.059
Cover blank	EA.	0.059
4-11/16"		
Square box	EA.	0.114
Box extension	EA.	0.059
Plaster ring	EA.	0.059
Cover blank	EA.	0.059
Switch and device boxes		
2 gang	EA.	0.114
3 gang	EA.	0.114
4 gang	EA.	0.160
Device covers		
2 gang	EA.	0.059
3 gang	EA.	0.059
4 gang	EA.	0.059
Handy box	EA.	0.114

Basic Materials

Basic Materials	UNIT	MAN/ HOURS
16130.60 Pull And Junction Boxes		
Extension	EA.	0.059
Switch cover	EA.	0.059
Switch box with knockout	EA.	0.145
Weatherproof cover, spring type	EA.	0.080
Cover plate, dryer receptacle 1 gang plastic	EA.	0.100
For 4" receptacle, 2 gang	EA.	0.100
Duplex receptacle cover plate, plastic	EA.	0.059
4", vertical bracket box, 1-1/2" with		
RMX clamps	EA.	0.145
BX clamps	EA.	0.145
4", octagon device cover		
1 switch	EA.	0.059
1 duplex recept	EA.	0.059
4", square face bracket boxes, 1-1/2"		
RMX	EA.	0.145
BX	EA.	0.145
16130.65 Pull Boxes And Cabinets		
Galvanized pull boxes, screw cover		
4x4x4	EA.	0.190
4x6x4	EA.	0.190
6x6x4	EA.	0.190
6x8x4	EA.	0.190
8x8x4	EA.	0.250
8x10x4	EA.	0.242
8x12x4	EA.	0.250
Screw cover		
10x10x4	EA.	0.308
12x12x6	EA.	0.444
12x15x6	EA.	0.444
12x18x6	EA.	0.500
15x18x6	EA.	0.571
18x24x6	EA.	0.615
18x30x6	EA.	0.727
24x36x6	EA.	0.727
16130.80 Receptacles		
125 volt, 20a, duplex, grounding type, standard grade	EA.	0.200
Ground fault interrupter type	EA.	0.296
250 volt, 20a, 2 pole, single receptacle, ground type	EA.	0.200
120/208v, 4 pole, single receptacle, twist lock		
20a	EA.	0.348
50a	EA.	0.348
125/250v, 3 pole, flush receptacle		
30a	EA.	0.296
50a	EA.	0.296
60a	EA.	0.348
277v, 20a, 2 pole, grounding type, twist lock	EA.	0.200
Dryer receptacle, 250v, 30a/50a, 3 wire	EA.	0.296
Clock receptacle, 2 pole, grounding type	EA.	0.200
125v, 20a single recept. grounding type		
Standard grade	EA.	0.200
Specification	EA.	0.200
Hospital	EA.	0.200

Basic Materials

16130.80 Receptacles

	UNIT	MAN/ HOURS
Isolated ground orange	EA.	0.250
Duplex		
Specification grade	EA.	0.200
Hospital	EA.	0.200
Isolated ground orange	EA.	0.250
GFI hospital grade recepts, 20a, 125v, duplex	EA.	0.296

16198.10 Electric Manholes

	UNIT	MAN/ HOURS
Precast, handhole, 4' deep		
2'x2'	EA.	3.478
3'x3'	EA.	5.556
4'x4'	EA.	10.256
Power manhole, complete, precast, 8' deep		
4'x4'	EA.	14.035
6'x6'	EA.	20.000
8'x8'	EA.	21.053
6' deep, 9' x 12'	EA.	25.000
Cast in place, power manhole, 8' deep		
4'x4'	EA.	14.035
6'x6'	EA.	20.000
8'x8'	EA.	21.053

16199.10 Utility Poles & Fittings

	UNIT	MAN/ HOURS
Wood pole, creosoted		
25'	EA.	2.353
40'	EA.	3.791
55'	EA.	7.547
Treated, wood preservative, 6"x6"		
8'	EA.	0.500
16'	EA.	1.600
20'	EA.	2.000
Aluminum, brushed, no base		
8'	EA.	2.000
20'	EA.	3.200
40'	EA.	6.250
Steel, no base		
10'	EA.	2.500
20'	EA.	3.810
35'	EA.	6.250
Concrete, no base		
13'	EA.	5.517
30'	EA.	12.121
50'	EA.	18.182
60'	EA.	20.000

Power Generation

16210.10 Generators

	UNIT	MAN/ HOURS
Diesel generator, with auto transfer switch		
50kw	EA.	30.769
125kw	EA.	50.000
300kw	EA.	100.000
750kw	EA.	200.000

16320.10 Transformers

	UNIT	MAN/ HOURS
Floor mounted, single phase, int. dry, 480v-120/240v		
3 kva	EA.	1.818
5 kva	EA.	3.077
7.5 kva	EA.	3.478
10 kva	EA.	3.810
15 kva	EA.	4.301
100 kva	EA.	11.594
Three phase, 480v-120/208v		
15 kva	EA.	6.015
30 kva	EA.	9.412
45 kva	EA.	10.811
225 kva	EA.	15.385

16350.10 Circuit Breakers

	UNIT	MAN/ HOURS
Molded case, 240v, 15-60a, bolt-on		
1 pole	EA.	0.250
2 pole	EA.	0.348
70-100a, 2 pole	EA.	0.533
15-60a, 3 pole	EA.	0.400
70-100a, 3 pole	EA.	0.615
480v, 2 pole		
15-60a	EA.	0.296
70-100a	EA.	0.400
3 pole		
15-60a	EA.	0.400
70-100a	EA.	0.444
70-225a	EA.	0.615
Load center circuit breakers, 240v		
1 pole, 10-60a	EA.	0.250
2 pole		
10-60a	EA.	0.400
70-100a	EA.	0.667
110-150a	EA.	0.727
3 pole		
10-60a	EA.	0.500
70-100a	EA.	0.727
Load center, G.F.I. breakers, 240v		
1 pole, 15-30a	EA.	0.296
2 pole, 15-30a	EA.	0.400
Key operated breakers, 240v, 1 pole, 10-30a	EA.	0.296
Tandem breakers, 240v		
1 pole, 15-30a	EA.	0.400
2 pole, 15-30a	EA.	0.533
Bolt-on, G.F.I. breakers, 240v, 1 pole, 15-30a	EA.	0.348

Power Generation

16360.10 Safety Switches

	UNIT	MAN/ HOURS
Fused, 3 phase, 30 amp, 600v, heavy duty		
NEMA 1	EA.	1.143
NEMA 3r	EA.	1.143
NEMA 4	EA.	1.600
NEMA 12	EA.	1.739
60a		
NEMA 1	EA.	1.143
NEMA 3r	EA.	1.143
NEMA 4	EA.	1.600
NEMA 12	EA.	1.739
100a		
NEMA 1	EA.	1.739
NEMA 3r	EA.	1.739
NEMA 4	EA.	2.000
NEMA 12	EA.	2.500
200a		
NEMA 1	EA.	2.500
NEMA 3r	EA.	2.500
NEMA 4	EA.	2.759
NEMA 12	EA.	3.478
Non-fused, 240-600v, heavy duty, 3 phase, 30 amp		
NEMA 1	EA.	1.143
NEMA 3r	EA.	1.143
NEMA 4	EA.	1.739
NEMA 12	EA.	1.739
60a		
NEMA1	EA.	1.143
NEMA 3r	EA.	1.143
NEMA 4	EA.	1.739
NEMA 12	EA.	1.739
100a		
NEMA 1	EA.	1.739
NEMA 3r	EA.	1.739
NEMA 4	EA.	2.500
NEMA 12	EA.	2.500
200a, NEMA 1	EA.	2.500
600a, NEMA 12	EA.	12.308

16365.10 Fuses

	UNIT	MAN/ HOURS
Fuse, one-time, 250v		
30a	EA.	0.050
60a	EA.	0.050
100a	EA.	0.050
200a	EA.	0.050
400a	EA.	0.050
600a	EA.	0.050
600v		
30a	EA.	0.050
60a	EA.	0.050
100a	EA.	0.050
200a	EA.	0.050
400a	EA.	0.050

Power Generation

16395.10 Grounding

	UNIT	MAN/ HOURS
Ground rods, copper clad, 1/2" x		
6'	EA.	0.667
8'	EA.	0.727
10'	EA.	1.000
5/8" x		
5'	EA.	0.615
6'	EA.	0.727
8'	EA.	1.000
10'	EA.	1.250
3/4" x		
8'	EA.	0.727
10'	EA.	0.800
Ground rod clamp		
5/8"	EA.	0.123
3/4"	EA.	0.123

Service And Distribution

16425.10 Switchboards

	UNIT	MAN/ HOURS
Switchboard, 90" high, no main disconnect, 208/120v		
400a	EA.	7.921
600a	EA.	8.000
1000a	EA.	8.000
1200a	EA.	10.000
1600a	EA.	11.940
2000a	EA.	14.035
2500a	EA.	16.000
277/480v		
600a	EA.	8.163
800a	EA.	8.163
1600a	EA.	11.940
2000a	EA.	14.035
2500a	EA.	16.000
3000a	EA.	27.586
4000a	EA.	29.630

16470.10 Panelboards

	UNIT	MAN/ HOURS
3 phase, 480/277v, main lugs only, 120a, 30 circuits	EA.	3.478
277/480v, 4 wire, flush surface		
225a, 30 circuits	EA.	4.000
400a, 30 circuits	EA.	5.000
600a, 42 circuits	EA.	6.015
208/120v, main circuit breaker, 3 phase, 4 wire		
100a		
12 circuits	EA.	5.096
20 circuits	EA.	6.299
30 circuits	EA.	7.018

16 ELECTRICAL

Service And Distribution

16470.10 Panelboards	UNIT	MAN/ HOURS
400a		
30 circuits	EA.	14.815
42 circuits	EA.	16.000
600a, 42 circuits	EA.	18.182
120/208v, flush, 3 ph., 4 wire, main only		
100a		
12 circuits	EA.	5.096
20 circuits	EA.	6.299
30 circuits	EA.	7.018
400a		
30 circuits	EA.	14.815
42 circuits	EA.	16.000
600a, 42 circuits	EA.	18.182

16480.10 Motor Controls	UNIT	MAN/ HOURS
Motor generator set, 3 phase, 480/277v, w/controls		
10kw	EA.	27.586
15kw	EA.	30.769
20kw	EA.	32.000
40kw	EA.	38.095
100kw	EA.	61.538
200kw	EA.	72.727
300kw	EA.	80.000
2 pole, 230 volt starter, w/NEMA-1		
1 hp, 9 amp, size 00	EA.	1.000
2 hp, 18amp, size 0	EA.	1.000
3 hp, 27amp, size 1	EA.	1.000
5 hp, 45amp, size 1p	EA.	1.000
7-1/2 hp, 45a, size 2	EA.	1.000
15 hp, 90a, size 3	EA.	1.000

16490.10 Switches	UNIT	MAN/ HOURS
Fused interrupter load, 35kv		
20A		
1 pole	EA.	16.000
2 pole	EA.	17.021
3 way	EA.	17.021
4 way	EA.	18.182
30a, 1 pole	EA.	16.000
3 way	EA.	17.021
4 way	EA.	18.182
Weatherproof switch, including box & cover, 20a		
1 pole	EA.	16.000
2 pole	EA.	17.021
3 way	EA.	18.182
4 way	EA.	18.182
Specification grade toggle switches, 20a, 120-277v		
Single pole	EA.	0.200
Double pole	EA.	0.296
3 way	EA.	0.250
4 way	EA.	0.296
Switch plates, plastic ivory		
1 gang	EA.	0.080
2 gang	EA.	0.100
3 gang	EA.	0.119
4 gang	EA.	0.145
5 gang	EA.	0.160
6 gang	EA.	0.182
Stainless steel		
1 gang	EA.	0.080
2 gang	EA.	0.100
3 gang	EA.	0.123
4 gang	EA.	0.145
5 gang	EA.	0.160
6 gang	EA.	0.182
Brass		
1 gang	EA.	0.080
2 gang	EA.	0.100
3 gang	EA.	0.123
4 gang	EA.	0.145
5 gang	EA.	0.160
6 gang	EA.	0.182

16490.20 Transfer Switches	UNIT	MAN/ HOURS
Automatic transfer switch 600v, 3 pole		
30a	EA.	3.478
100a	EA.	4.762
400a	EA.	10.000
800a	EA.	18.182
1200a	EA.	22.857
2600a	EA.	42.105

16490.80 Safety Switches	UNIT	MAN/ HOURS
Safety switch, 600v, 3 pole, heavy duty, NEMA-1		
30a	EA.	1.000
60a	EA.	1.143
100a	EA.	1.600
200a	EA.	2.500
400a	EA.	5.517
600a	EA.	8.000
800a	EA.	10.526
1200a	EA.	14.286

Lighting

16510.05 Interior Lighting	UNIT	MAN/ HOURS
Recessed fluorescent fixtures, 2'x2'		
2 lamp	EA.	0.727
4 lamp	EA.	0.727
2 lamp w/flange	EA.	1.000
4 lamp w/flange	EA.	1.000

Lighting

16510.05 Interior Lighting

Lighting	UNIT	MAN/HOURS
1'x4'		
2 lamp	EA.	0.667
3 lamp	EA.	0.667
2 lamp w/flange	EA.	0.727
3 lamp w/flange	EA.	0.727
2'x4'		
2 lamp	EA.	0.727
3 lamp	EA.	0.727
4 lamp	EA.	0.727
2 lamp w/flange	EA.	1.000
3 lamp w/flange	EA.	1.000
4 lamp w/flange	EA.	1.000
4'x4'		
4 lamp	EA.	1.000
6 lamp	EA.	1.000
8 lamp	EA.	1.000
4 lamp w/flange	EA.	1.509
6 lamp w/flange	EA.	1.509
8 lamp, w/flange	EA.	1.509
Surface mounted incandescent fixtures		
40w	EA.	0.667
75w	EA.	0.667
100w	EA.	0.667
150w	EA.	0.667
Pendant		
40w	EA.	0.800
75w	EA.	0.800
100w	EA.	0.800
150w	EA.	0.800
Recessed incandescent fixtures		
40w	EA.	1.509
75w	EA.	1.509
100w	EA.	1.509
150w	EA.	1.509
Exit lights, 120v		
Recessed	EA.	1.250
Back mount	EA.	0.727
Universal mount	EA.	0.727
Emergency battery units, 6v-120v, 50 unit	EA.	1.509
With 1 head	EA.	1.509
With 2 heads	EA.	1.509
Light track single circuit		
2'	EA.	0.500
4'	EA.	0.500
8'	EA.	1.000
12'	EA.	1.509
Fixtures, square		
R-20	EA.	0.145
R-30	EA.	0.145
Mini spot	EA.	0.145

Lighting

16510.10 Lighting Industrial

Lighting	UNIT	MAN/HOURS
Surface mounted fluorescent, wrap around lens		
1 lamp	EA.	0.800
2 lamps	EA.	0.889
4 lamps	EA.	1.000
Wall mounted fluorescent		
2-20w lamps	EA.	0.500
2-30w lamps	EA.	0.500
2-40w lamps	EA.	0.667
Indirect, with wood shielding, 2049w lamps		
4'	EA.	1.000
8'	EA.	1.600
Industrial fluorescent, 2 lamp		
4'	EA.	0.727
8'	EA.	1.333
Strip fluorescent		
4'		
1 lamp	EA.	0.667
2 lamps	EA.	0.667
8'		
1 lamp	EA.	0.727
2 lamps	EA.	0.889
Wire guard for strip fixture, 4' long	EA.	0.348
Strip fluorescent, 8' long, two 4' lamps	EA.	1.333
With four 4' lamps	EA.	1.600
Wet location fluorescent, plastic housing		
4' long		
1 lamp	EA.	1.000
2 lamps	EA.	1.333
8' long		
2 lamps	EA.	1.600
4 lamps	EA.	1.739
Parabolic troffer, 2'x2'		
With 2 "U" lamps	EA.	1.000
With 3 "U" lamps	EA.	1.143
2'x4'		
With 2 40w lamps	EA.	1.143
With 3 40w lamps	EA.	1.333
With 4 40w lamps	EA.	1.333
1'x4'		
With 1 T-12 lamp, 9 cell	EA.	0.727
With 2 T-12 lamps	EA.	0.889
With 1 T-12 lamp, 20 cell	EA.	0.727
With 2 T-12 lamps	EA.	0.889
Steel sided surface fluorescent, 2'x4'		
3 lamps	EA.	1.333
4 lamps	EA.	1.333
Outdoor sign fluor., 1 lamp, remote ballast		
4' long	EA.	6.015
6' long	EA.	8.000
Recess mounted, commercial, 2'x2', 13" high		
100w	EA.	4.000
250w	EA.	4.494
High pressure sodium, hi-bay open		
400w	EA.	1.739

Lighting

16510.10 Lighting Industrial

Description	UNIT	MAN/ HOURS
1000w	EA.	2.424
Enclosed		
400w	EA.	2.424
1000w	EA.	2.963
Metal halide hi-bay, open		
400w	EA.	1.739
1000w	EA.	2.424
Enclosed		
400w	EA.	2.424
1000w	EA.	2.963
High pressure sodium, low bay, surface mounted		
100w	EA.	1.000
150w	EA.	1.143
250w	EA.	1.333
400w	EA.	1.600
Metal halide, low bay, pendant mounted		
175w	EA.	1.333
250w	EA.	1.600
400w	EA.	2.222
Indirect luminare, square, metal halide, freestanding		
175w	EA.	1.000
250w	EA.	1.000
400w	EA.	1.000
High pressure sodium		
150w	EA.	1.000
250w	EA.	1.000
400w	EA.	1.000
Round, metal halide		
175w	EA.	1.000
250w	EA.	1.000
400w	EA.	1.000
High pressure sodium		
150w	EA.	1.000
250w	EA.	1.000
400w	EA.	1.000
Wall mounted, metal halide		
175w	EA.	2.500
250w	EA.	2.500
400w	EA.	3.200
High pressure sodium		
150w	EA.	2.500
250w	EA.	2.500
400w	EA.	3.200
Wall pack lithonia, high pressure sodium		
35w	EA.	0.889
55w	EA.	1.000
150w	EA.	1.600
250w	EA.	1.739
Low pressure sodium		
35w	EA.	1.739
55w	EA.	2.000
Wall pack hubbell, high pressure sodium		
35w	EA.	0.889
150w	EA.	1.600

Lighting

16510.10 Lighting Industrial

Description	UNIT	MAN/ HOURS
250w	EA.	1.739
Compact fluorescent		
2-7w	EA.	1.000
2-13w	EA.	1.333
1-18w	EA.	1.333
Handball & racquet ball court, 2'x2', metal halide		
250w	EA.	2.500
400w	EA.	2.759
High pressure sodium		
250w	EA.	2.500
400w	EA.	2.759
Bollard light, 42" w/found., high pressure sodium		
70w	EA.	2.581
100w	EA.	2.581
150w	EA.	2.581
Light fixture lamps		
Lamp		
20w med. bipin base, cool white, 24"	EA.	0.145
30w cool white, rapid start, 36"	EA.	0.145
40w cool white "U", 3"	EA.	0.145
40w cool white, rapid start, 48"	EA.	0.145
70w high pressure sodium, mogul base	EA.	0.200
75w slimline, 96"	EA.	0.200
100w		
Incandescent, 100a, inside frost	EA.	0.100
Mercury vapor, clear, mogul base	EA.	0.200
High pressure sodium, mogul base	EA.	0.200
150w		
Par 38 flood or spot, incandescent	EA.	0.100
High pressure sodium, 1/2 mogul base	EA.	0.200
175w		
Mercury vapor, clear, mogul base	EA.	0.200
Metal halide, clear, mogul base	EA.	0.200
High pressure sodium, mogul base	EA.	0.200
250w		
Mercury vapor, clear, mogul base	EA.	0.200
Metal halide, clear, mogul base	EA.	0.200
High pressure sodium, mogul base	EA.	0.200
400w		
Mercury vapor, clear, mogul base	EA.	0.200
Metal halide, clear, mogul base	EA.	0.200
High pressure sodium, mogul base	EA.	0.200
1000w		
Mercury vapor, clear, mogul base	EA.	0.250
High pressure sodium, mogul base	EA.	0.250

16510.30 Exterior Lighting

Description	UNIT	MAN/ HOURS
Exterior light fixtures		
Rectangle, high pressure sodium		
70w	EA.	2.500
100w	EA.	2.581
150w	EA.	2.581
250w	EA.	2.759
400w	EA.	3.478

Lighting

16510.30 Exterior Lighting

Lighting	UNIT	MAN/ HOURS
Flood, rectangular, high pressure sodium		
70w	EA.	2.500
100w	EA.	2.581
150w	EA.	2.581
400w	EA.	3.478
1000w	EA.	4.494
Round		
400w	EA.	3.478
1000w	EA.	4.494
Round, metal halide		
400w	EA.	3.478
1000w	EA.	4.494
Light fixture arms, cobra head, 6', high press. sodium		
100w	EA.	2.000
150w	EA.	2.500
250w	EA.	2.500
400w	EA.	2.963
Flood, metal halide		
400w	EA.	3.478
1000w	EA.	4.494
1500w	EA.	6.015
Mercury vapor		
250w	EA.	2.759
400w	EA.	3.478
Incandescent		
300w	EA.	1.739
500w	EA.	2.000
1000w	EA.	3.200

16610.30 Uninterruptible Power

Lighting	UNIT	MAN/ HOURS
Uninterruptible power systems, (U.P.S.), 3kva	EA.	8.000
5 kva	EA.	11.004
7.5 kva	EA.	16.000
10 kva	EA.	21.978
15 kva	EA.	22.857
20 kva	EA.	24.024
25 kva	EA.	25.000
30 kva	EA.	25.974
35 kva	EA.	27.027
40 kva	EA.	27.972
45 kva	EA.	28.986
50 kva	EA.	29.963
62.5 kva	EA.	32.000
75 kva	EA.	34.934
100 kva	EA.	36.036
150 kva	EA.	50.000
200 kva	EA.	55.172
300 kva	EA.	74.766
400 kva	EA.	89.888
500 kva	EA.	109.589

Lighting

16670.10 Lightning Protection

Lighting	UNIT	MAN/ HOURS
Lightning protection		
Copper point, nickel plated, 12'		
1/2" dia.	EA.	1.000
5/8" dia.	EA.	1.000

Communications

16720.10 Fire Alarm Systems

Communications	UNIT	MAN/ HOURS
Master fire alarm box, pedestal mounted	EA.	16.000
Master fire alarm box	EA.	6.015
Box light	EA.	0.500
Ground assembly for box	EA.	0.667
Bracket for pole type box	EA.	0.727
Pull station		
Waterproof	EA.	0.500
Manual	EA.	0.400
Horn, waterproof	EA.	1.000
Interior alarm	EA.	0.727
Coded transmitter, automatic	EA.	2.000
Control panel, 8 zone	EA.	8.000
Battery charger and cabinet	EA.	2.000
Batteries, nickel cadmium or lead calcium	EA.	5.000
CO2 pressure switch connection	EA.	0.727
Annunciator panels		
Fire detection annunciator, remote type, 8 zone	EA.	1.818
12 zone	EA.	2.000
16 zone	EA.	2.500
Fire alarm systems		
Bell	EA.	0.615
Weatherproof bell	EA.	0.667
Horn	EA.	0.727
Siren	EA.	2.000
Chime	EA.	0.615
Audio/visual	EA.	0.727
Strobe light	EA.	0.727
Smoke detector	EA.	0.667
Heat detection	EA.	0.500
Thermal detector	EA.	0.500
Ionization detector	EA.	0.533
Duct detector	EA.	2.759
Test switch	EA.	0.500
Remote indicator	EA.	0.571
Door holder	EA.	0.727
Telephone jack	EA.	0.296
Fireman phone	EA.	1.000
Speaker	EA.	0.800
Remote fire alarm annunciator panel		

Communications

16720.10 Fire Alarm Systems

Item	UNIT	MAN/ HOURS
24 zone	EA.	6.667
48 zone	EA.	13.008
Control panel		
12 zone	EA.	2.963
16 zone	EA.	4.444
24 zone	EA.	6.667
48 zone	EA.	16.000
Power supply	EA.	1.509
Status command	EA.	5.000
Printer	EA.	1.509
Transponder	EA.	0.899
Transformer	EA.	0.667
Transceiver	EA.	0.727
Relays	EA.	0.500
Flow switch	EA.	2.000
Tamper switch	EA.	2.963
End of line resistor	EA.	0.348
Printed ckt. card	EA.	0.500
Central processing unit	EA.	6.154
UPS backup to c.p.u.	EA.	8.999
Smoke detector, fixed temp. & rate of rise comb.	EA.	1.600

16720.50 Security Systems

Item	UNIT	MAN/ HOURS
Sensors		
Balanced magnetic door switch, surface mounted	EA.	0.500
With remote test	EA.	1.000
Flush mounted	EA.	1.860
Mounted bracket	EA.	0.348
Mounted bracket spacer	EA.	0.348
Photoelectric sensor, for fence		
6 beam	EA.	2.759
9 beam	EA.	4.255
Photoelectric sensor, 12 volt dc		
500' range	EA.	1.600
800' range	EA.	2.000
Capacitance wire grid kit		
Surface	EA.	1.000
Duct	EA.	1.600
Tube grid kit	EA.	0.500
Vibration sensor, 30 max per zone	EA.	0.500
Audio sensor, 30 max per zone	EA.	0.500
Inertia sensor		
Outdoor	EA.	0.727
Indoor	EA.	0.500
Ultrasonic transmitter, 20 max per zone		
Omni-directional	EA.	1.600
Directional	EA.	1.333
Transceiver		
Omni-directional	EA.	1.000
Directional	EA.	1.000
Passive infra-red sensor, 20 max per zone	EA.	1.600
Access/secure control unit, balanced magnetic switch	EA.	1.600
Photoelectric sensor	EA.	1.600
Photoelectric fence sensor	EA.	1.600

Communications

16720.50 Security Systems

Item	UNIT	MAN/ HOURS
Capacitance sensor	EA.	1.739
Audio and vibration sensor	EA.	1.600
Inertia sensor	EA.	1.600
Ultrasonic sensor	EA.	1.739
Infra-red sensor	EA.	2.000
Monitor panel, with access/secure tone, standard	EA.	1.739
High security	EA.	2.000
Emergency power indicator	EA.	0.500
Monitor rack with 115v power supply		
1 zone	EA.	1.000
10 zone	EA.	2.500
Monitor cabinet, wall mounted		
1 zone	EA.	1.000
5 zone	EA.	1.600
10 zone	EA.	1.739
20 zone	EA.	2.000
Floor mounted, 50 zone	EA.	4.000
Security system accessories		
Tamper assembly for monitor cabinet	EA.	0.444
Monitor panel blank	EA.	0.348
Audible alarm	EA.	0.500
Audible alarm control	EA.	0.348
Termination screw, terminal cabinet		
25 pair	EA.	1.600
50 pair	EA.	2.500
150 pair	EA.	5.000
Universal termination, cabinets & panel		
Remote test	EA.	1.739
No remote test	EA.	0.727
High security line supervision termination	EA.	1.000
Door cord for capacitance sensor, 12"	EA.	0.500
Insulation block kit for capacitance sensor	EA.	0.348
Termination block for capacitance sensor	EA.	0.348
Guard alert display	EA.	0.615
Uninterrupted power supply	EA.	8.000
Plug-in 40kva transformer		
12 volt	EA.	0.348
18 volt	EA.	0.348
24 volt	EA.	0.348
Test relay	EA.	0.348
Coaxial cable, 50 ohm	L.F.	0.006
Door openers	EA.	0.500
Push buttons		
Standard	EA.	0.348
Weatherproof	EA.	0.444
Bells	EA.	0.727
Horns		
Standard	EA.	1.000
Weatherproof	EA.	1.250
Chimes	EA.	0.667
Flasher	EA.	0.615
Motion detectors	EA.	1.509
Intercom units	EA.	0.727
Remote annunciator	EA.	5.000

Communications	UNIT	MAN/ HOURS
16740.10 Telephone Systems		
Communication cable		
25 pair	L.F.	0.026
100 pair	L.F.	0.029
400 pair	L.F.	0.044
Cable tap in manhole or junction box		
25 pair cable	EA.	3.810
100 pair cable	EA.	15.094
400 pair cable	EA.	61.538
Cable terminations, manhole or junction box		
25 pair cable	EA.	3.756
100 pair cable	EA.	15.094
400 pair cable	EA.	61.538
16780.50 Television Systems		
TV outlet, self terminating, w/cover plate	EA.	0.308
Thru splitter	EA.	1.600
End of line	EA.	1.333
In line splitter multitap		
4 way	EA.	1.818
2 way	EA.	1.702
Equipment cabinet	EA.	1.600
Antenna		
Broad band uhf	EA.	3.478
Lightning arrester	EA.	0.727
TV cable	L.F.	0.005
16850.10 Electric Heating		
Baseboard heater		
2', 375w	EA.	1.000
3', 500w	EA.	1.000
4', 750w	EA.	1.143
5', 935w	EA.	1.333
6', 1125w	EA.	1.600
7', 1310w	EA.	1.010
8', 1500w	EA.	2.000
9', 1680w	EA.	2.222
10', 1875w	EA.	2.286
Unit heater, wall mounted		
1500w	EA.	1.667
2500w	EA.	1.818
4000w	EA.	2.286
Thermostat		
Integral	EA.	0.500
Line voltage	EA.	0.500
Electric heater connection	EA.	0.250

BNi Building News

Supporting Construction Reference Data

This section contains information, text, charts and tables on various aspects of construction. The intent is to provide the user with a better understanding of unfamiliar areas in order to be able to estimate better. This information includes actual takeoff data for some areas and also selected explanations of common construction materials, methods and common practices.

TYPICAL BUILDING COST BROKEN DOWN BY CSI FORMAT

(Commercial Construction)

Division	New Construction	Remodeling Construction
1. General Requirements	6 to 8%	Up to 30%
2. Sitework	4 to 6%	
3. Concrete	15 to 20%	
4. Masonry	8 to 12%	
5. Metals	5 to 7%	
6. Wood and Plastics	1 to 5 %	
7. Thermal and Moisture Protection	4 to 6%	
8. Doors and Windows	5 to 7%	Up to 30% (Divisions 8–9)
9. Finishes	8 to 12 %	
10. Specialties	6 to 10% (Divisions 10–14)	
11. Architectural Equipment		
12. Furnishings		
13. Special Construction		
14. Conveying Systems		
15. Mechanical	15 to 25%	Up to 30% (Divisions 15–16)
16. Electrical	8 to 12%	
TOTAL COST	100%	

CONVERSION FACTORS

Change	To	Multiply By
ATMOSPHERES	POUNDS PER SQUARE INCH	14.696
ATMOSPHERES	INCHES OF MERCURY	29.92
ATMOSPHERES	FEET OF WATER	34
BARRELS, OIL	GALLONS, OF OIL	42
BARRELS, CEMENT	POUNDS OF CEMENT	376
BOGS OR SACKS, CEMENT	POUNDS OF CEMENT	94
BTU/MIN	FOOT-POUNDS/SEC	12.96
BTU/MIN	HORSE-POWER	0.02356
BTU/MIN	KILOWATTS	0.01757
BTU/MM	WATTS	17.57
CENTIMETERS	INCHES	0.3937
CENTIMETERS OF MERCURY	ATMOSPHERES	0.01316
CENTIMETERS OF MERCURY	FEET OF WATER	0.4461
CUBIC INCHES	CUBIC FEET	0.00058
CUBIC FEET	CUBIC INCHES	1728
CUBIC FEET	CUBIC YARDS	0.03703
CUBIC YARDS	CUBIC FEET	27
CUBIC INCHES	GALLONS	0.00433
CUBIC FEET	GALLONS	7.48
FEET	INCHES	12
FEET	YARDS	0.3333
YARDS	FEET	3
FEET OF WATER	ATMOSPHERES	0.02950
FEET OF WATER	INCHES OF MERCURY	0.8826
GALLONS	CUBIC INCHES	231
GALLONS	CUBIC FEET	0.1337
GALLONS	POUNDS OF WATER	8.33
GALLONS	QUARTS	4
GALLONS PER MIN	CUBIC FEET SEC	0.002228
GALLONS PER MIN	CUBIC FEET HOUR	8.0208
GALLONS WATER PER MIN	TONS WATER/24 HOURS	6.0086
HORSE-POWER	FOOT-LBS./SEC	550
INCHES	CENTIMETERS	2.540
INCHES	FEET	0.0833
INCHES	MILLIMETERS	25.4
INCHES OF WATER	POUNDS PER SQ. INCH	0.0361
INCHES OF WATER	INCHES OF MERCURY	0.0735
INCHES OF WATER	OUNCES PER SQUARE INCH	0.578
INCHES OF WATER	OUNCES PER SQUARE FOOT	5.2
INCHES OF MERCURY	INCHES OF WATER	13.6
INCHES OF MERCURY	FEET OF WATER	1.1333
INCHES OF MERCURY	POUNDS PER SQUARE INCH	0.4914
KILOMETERS	MILES	0.6214
METERS	INCHES	39.37
MILES	FEET	5280
MILLIMETERS	CENTIMETERS	0.1
MILLIMETERS	INCHES	0.03937
OUNCES (FLUID)	CUBIC INCHES	1.805
OUNCES	POUNDS	0.0625
POUNDS	OUNCES	16
POUNDS PER SQUARE INCH	INCHES OF WATER	27.72
POUNDS PER SQUARE INCH	FEET OF WATER	2.310
POUNDS PER SQUARE INCH	INCHES OF MERCURY	2.04
POUNDS PER SQUARE INCH	ATMOSPHERES	0.0681
QUARTS	CUBIC INCHES	67.20
SQUARE INCHES	SQUARE FEET	0.00694
SQUARE FEET	SQUARE INCHES	144
SQUARE FEET	SQUARE YARDS	0.11111
SQUARE YARDS	SQUARE FEET	9
SQUARE MILES	ACRES	640
SHORT TONS	POUNDS	2000
SHORT TONS	LONG TONS	0.89285
TONS OF WATER/24 HOURS	GALLONS PER MINUTE	0.16643
YARDS	FEET	3
YARDS	CENTIMETERS	91.44
YARDS	INCHES	36

CONVERSION CALCULATIONS

Commercial Measure

16 grams = 1 ounce
16 ounces = 1 pound
2,000 pounds = 1 ton

Long Measure

12 inches = 1 foot
3 feet = 1 yard
16 ½ feet = 1 rod
40 rods = 1 furlong
8 furlongs (5,280 ft.) = 1 mile
3 miles = 1 league

Square Measure

144 square inches = 1 square foot
9 square feet = 1 square yard
30 ¼ square yards = 1 square rod
160 square rods = 1 acre
4,840 square yards = 1 acre
640 acres = 1 square mile
36 square miles = 1 township

Surveyor's Measure

7.92 inches = 1 link
25 links = 1 rod
4 rods (66 ft.) = 1 chain
10 chains = 1 furlong
8 furlongs = 1 mile
1 square mile = 1 section

Cubic Measure

1728 cubic inches = 1 cubic foot
27 cubic feet = 1 cubic yard
128 cubic feet = 1 cord (wood/stone)
231 cubic inches = 1 U.S. gallon
7.48 U.S. Gallons = 1 cubic foot
2150.4 cubic inches = 1 U.S. bushel

Liquid Measure

4 fluid ounces = 1 gill
4 gills = 1 pint
2 pints = 1 quart
4 quarts = 1 gallon
9 gallons = 1 firkin
31 ½ gallons = 1 barrel
2 barrels = 1 hogshead

Dry Measure

2 pints = 1 quart
8 quarts = 1 peck
4 pecks = 1 bushel
2150.42 cubic inches = 1 bushel

GENERAL / MEASUREMENT | 01025

SQUARE

$A=a^2$

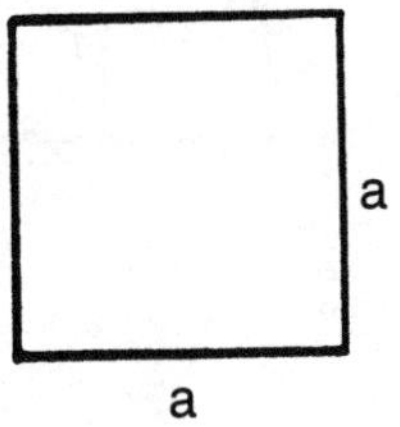

RECTANGLE

$A=bh$

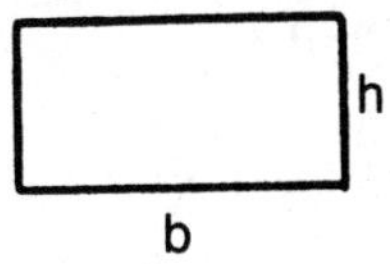

TRIANGLE

$A=\frac{1}{2}bh$

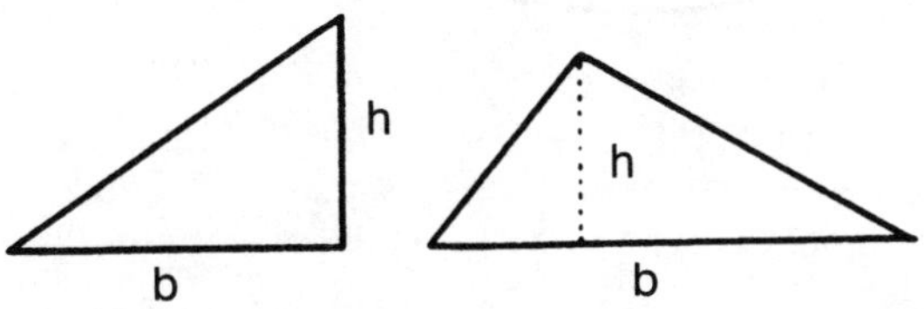

PARALLELOGRAM

$A=bh=ah\ \mathrm{Sin}\varnothing$

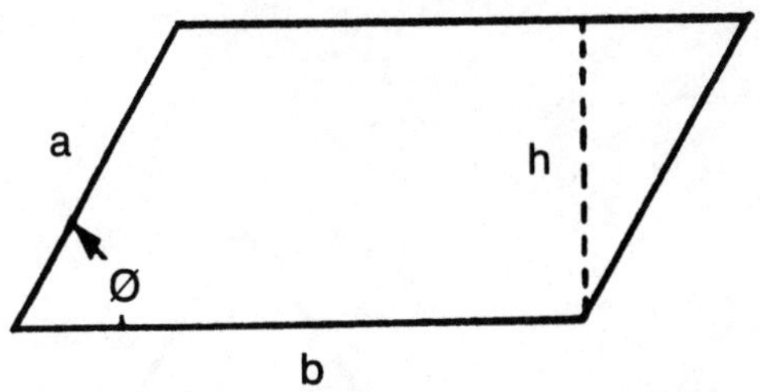

TRAPEZOID

$A=\left(\frac{a+b}{2}\right)h$

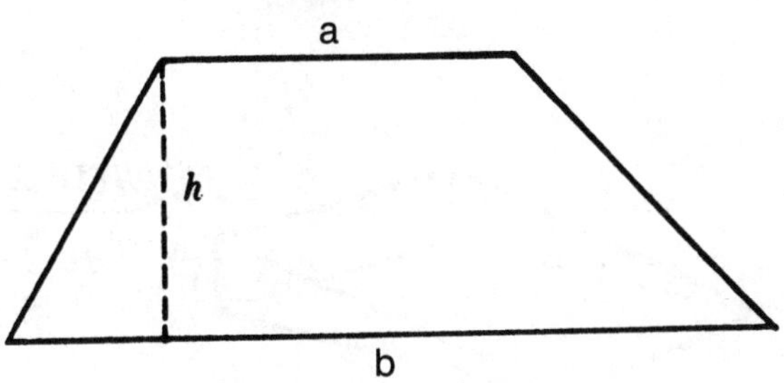

CIRCLE

$A=\pi^2=\frac{\pi d^2}{4}$

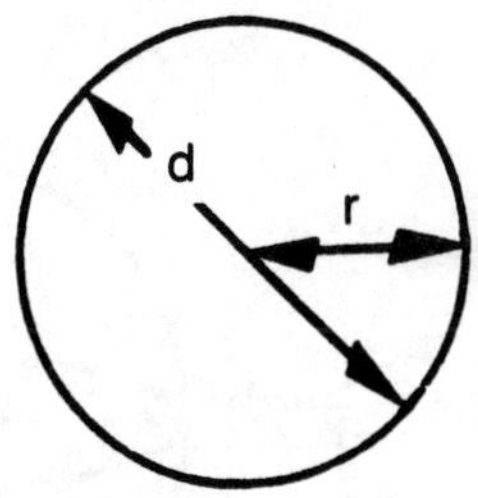

Circumference$=C=2\pi r=\pi d$

ELLIPES

$A=0.7854\ ab$

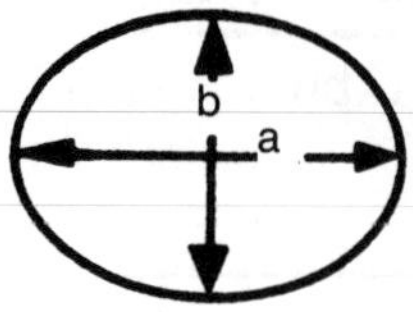

PARABOLA

$A=\frac{2}{3}bh$

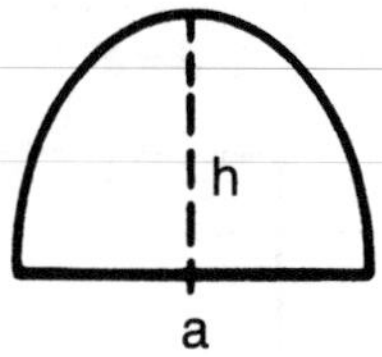

VOLUMES

CUBE

$V=a^3$

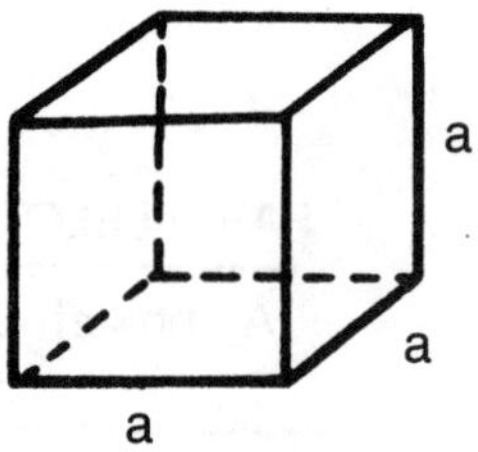

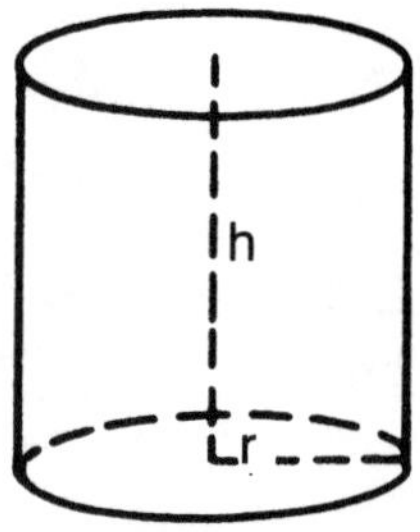

CYLINDER

$V=\pi r^3 h$

PYRAMID

$V=\frac{1}{3}(\text{Base})\ h$

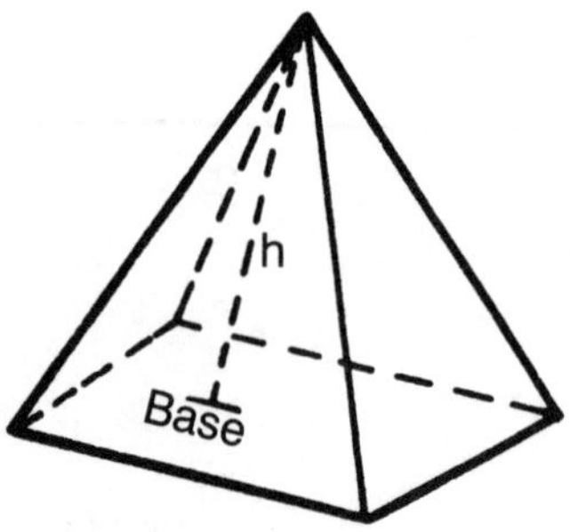

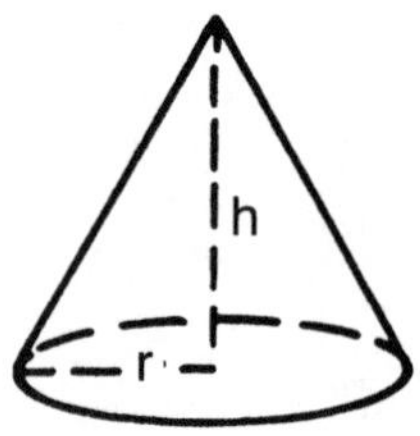

CONE

$V=\frac{1}{3}\pi r^2 h$

SPHERE

$V=\frac{4}{3}\pi r^3=\frac{1}{6}\pi d^3$

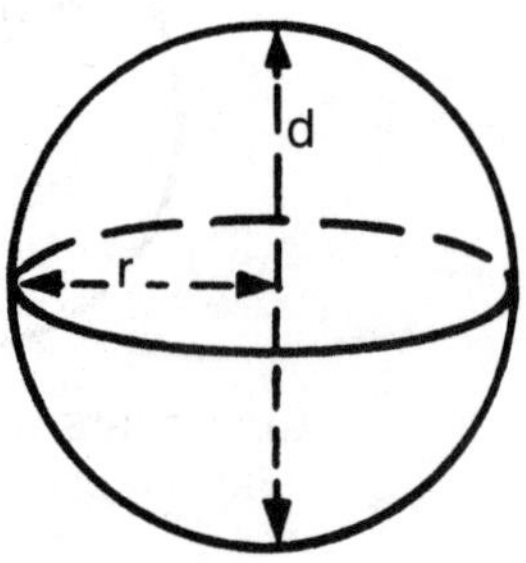

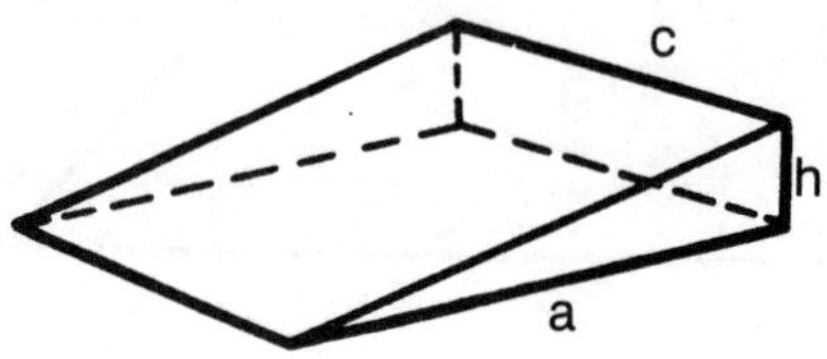

WEDGE

$V=\frac{1}{2}abc$

CONVERSION FACTORS
ENGLISH TO SI (SYSTEM INTERNATIONAL)

To Convert from	To	Multiply by
LENGTH		
Inches	Millimetres	25.4[a]
Feet	Metres	0.3048[a]
Yards	Metres	0.9144[a]
Miles (statute)	Kilometres	1.609
AREA		
Square inches	Square millimetres	645.2
Square feet	Square metres	0.0929
Square yards	Square metres	0.8361
VOLUME		
Cubic inches	Cubic millimetres	16.387
Cubic feet	Cubic metres	0.02832
Cubic yards	Cubic metres	0.7646
Gallons (U.S. liquid) [b]	Cubic metres[c]	0.003785
Gallons (Canadian liquid) [b]	Cubic metres[c]	0.004546
Ounces (U.S. liquid) [b]	Millilitres [c,d]	29.57
Quarts (U.S. liquid) [b]	Litres [c,d]	0.9464
Gallons (U.S. liquid) [b]	Litres [c]	3.785
FORCE		
Kilograms force	Newtons	9.807
Pounds force	Newtons	4.448
Pounds force	Kilograms force [d]	0.4536
Kips	Newtons	4448
Kips	Kilograms force [d]	453.6
PRESSURE, STRESS, STRENGTH (FORCE PER UNIT AREA)		
Kilograms force per sq. centimetre	Megapascals	0.09807
Pounds force per square inch (psi)	Megapascals	6895
Kips per square inch	Megapascals	6.895
Pounds force per square inch (psi)	Kilograms force per square centimetre [d]	0.07031
Pounds force per square foot	Pascals	47.88
Pounds force per square foot	Kilograms force per square metre [d]	4.882
SENDING MOMENT OR TORQUE		
Inch-pounds force	Metre-kilog. force [d]	0.01152
Inch-pounds force	Newton-metres	0.1130
Foot-pounds force	Metre-kilog. force [d]	0.1383
Foot-pounds force	Newton-metres	1.356
Metre-kilograms force	Newton-metres	9.807
MASS		
Ounce (avoirdupois)	Grams	28.35
Pounds (avoirdupois)	Kilograms	0.4536
Tons (metric)	Kilograms	1000[a]
Tons, short (2000 pounds)	Kilograms	907.2
Tons, short (2000 pounds)	Megagrams [e]	0.9072
MASS PER UNIT VOLUME		
Pounds mass per cubic foot	Kilog. per cubic metre	16.02
Pounds mass per cubic yard	Kilog. per cubic metre	0.5933
Pds. mass per gallon (U.S. liquid) [b]	Kilog. per cubic metre	119.8
Pds. mass p/gal. (Canadian liquid) [b]	Kilog. per cubic metre	99.78
TEMPERATURE		
Degrees Fahrenheit	Degrees Celsius	tK = (1F - 32)/1.8
Degrees Fahrenheit	Degrees Kelvin	tK = (1F + 459.67)/1.8
Degree Celsius	Degree Kelvin	tK = 1C + 273.15

[a] The factor given is exact.
[b] One U.S. gallon equals 0.8327 Canadian gallon.
[c] 1 litre = 1000 millilitres =10,000 cubic centimetres = 1 cubic decimetre = 0.001 cubic metre.
[d] Metric but not SI unit.
[e] Called "tonne" in England. Called "metric ton" in other metric systems.

TRENCH BRACING

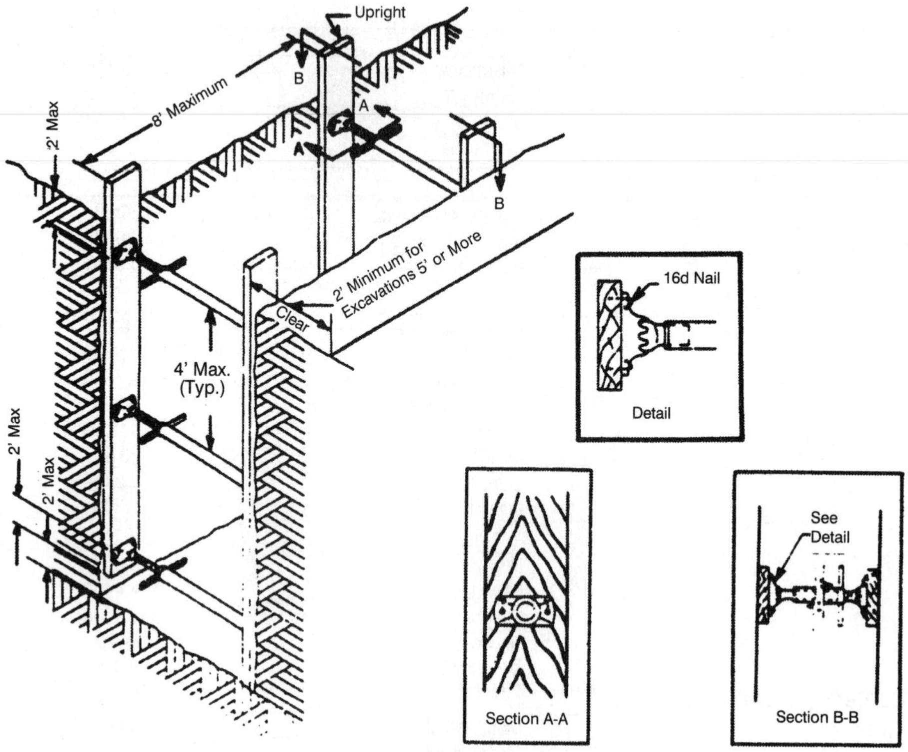

CLOSED VERTICAL SHEETING

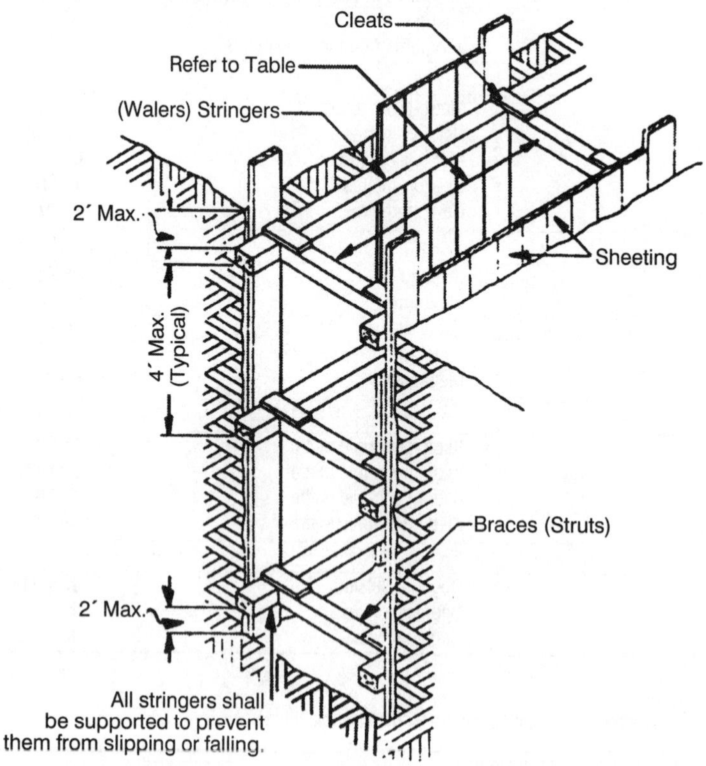

Soil Identification. For most purposes, soils can usually be identified visually and by texture, as described in the chart that follows. For design purposes, however, soils must be formally identified and their performance characteristics determined in a laboratory by skilled soil mechanics.

Classification	Identifying Characteristics
Gravel	Rounded or water-worn pebbles or bulk rock grains. No cohesion or plasticity. Gritty, granular and crunchy underfoot.
Sand	Granular, gritty, loose grains, passing a No. 4 sieve and between .002 and .079 inches in diameter. Individual grains readily seen and felt. No plasticity or cohesion. When dry, it cannot be molded but will crumble when touched. The coarse grains are rounded; the fine grains are visible and angular.
Silt	Fine, barely visible grains passing a No. 200 sieve and between .0002 and .002 inches in diameter. Little or no plasticity and no cohesion. A dried cast is easily crushed. Is permeable and movement of water through the voids occurs easily and is visible. Feels gritty when bitten and will not form a thread.
Clay	Invisible particles under .0002 inches in diameter. Cohesive and highly plastic when moist. Will form a long, thin, flexible thread when rolled between the hands. Does not feel gritty when bitten. Will form hard lumps or clods when dry which resist crushing. Impermeable, with no apparent movement of water through voids.
Muck and Organic Silt	Thoroughly decomposed organic material often mixed with other soils of mineral origin. Usually black with fibrous remains. Odorous when dried and burnt. Found as deposits in swamps, peat bogs and muskeg flats.
Peat	Partly decayed plant material. Mostly organic. Highly fibrous with visible plant remains. Spongy and easily identified.

Classification by Particle Size. Soils can be classified in general terms by the nature of their predominant particle size or the grading of the particle sizes. These particle sizes are usually grouped into gravel, coarse sand, medium sand, fine sand, silt, clay and colloids.

The major divisions of soils are:

Coarse-Grained(Granular)		Fine-Grained		Organic	
Gravel	Sand	Silt	Clay	Muck	Peat

Soils comprised primarily of sand particles are referred to as "granular soils," while fine-grained soils are commonly called "heavy soils." It is accepted practice in the field to refer to a particular soil as a coarse sand, or silt, or by any of the particle size groupings which describe the soil generally from a visual examination.

Classification of Soil Mixtures.

Class	% Sand	% Silt	% Clay
Sandy	80-100	0-20	0-20
Sandy clay loam	50- 80	0- 30	20-30
Sandy loam	50-80	0-50	0-20
Loam	30-50	30-50	0-20
Silty loam	0-50	50-80	0-20
Silt	0-20	80-100	0-20
Silty clay	0-30	50-80	20-30
Silty clay	0-20	50-70	30-50
Clay	20- 50	20- 50	20-30
Sandy clay	50-70	0-20	30-50
Clay	0-50	0-50	30-100

Material	Approx. In-Bank Weight (lbs. per cu. yd.)	Percent Swell
Clay, dry	2300	40
Clay, wet	3000	40
Granite, decomposed	4500	65
Gravel, dry	3250	10-15
Gravel, wet	3600	10-15
Loam, dry	2800	15-35
Loam, wet	3370	25
Rock, well blasted	4200	65
Sand, dry	3250	10-15
Sand, wet	3600	10-15
Shale and soft rock	3000	65
Slate	4700	65

PILES AND PILE DRIVING

General. A pile is a column driven or jetted into the ground which derives its supporting capabilities from end-bearing on the underlying strata, skin friction between the pile surface and the soil, or from a combination of end-bearing and skin friction.

Piles can be divided into two major classes: **Sheet piles** and **load-bearing piles**. Sheet piling is used primarily to restrain lateral forces as in trench sheeting and bulkheads, or to resist the flow of water as in cofferdams. It is prefabricated and is available in steel, wood or concrete. Load-bearing piles are used primarily to transmit loads through soil formations of low bearing values to formations that are capable of supporting the designed loads. If the load is supported predominantly by the action of soil friction on the surface of the pile, it is called a **friction pile**. If the load is transmitted to the soil primarily through the lower tip, it is called an **end-bearing pile**.

There are several load-bearing pile types, which can be classified according to the material from which they are fabricated:

- Timber (Treated and untreated)
- Concrete (Precast and cast in place)
- Steel (H-Section and steel pipe)
- Composite (A combination of two or more materials)

Some of the additional uses of piling are to: eliminate or control settlement of structures, support bridge piers and abutments and protect them from scour, anchor structures against uplift or overturning, and for numerous marine structures such as docks, wharves, fenders, anchorages, piers, trestles and jetties.

Timber Piles. Timber piles, treated or untreated, are the piles most commonly used throughout the world, primarily because they are readily available, economical, easily handled, can be easily cut off to any desired length after driving and can be easily removed if necessary. On the other hand, they have some serious disadvantages which include: difficulty in securing straight piles of long length, problems in driving them into hard formations and difficulty in splicing to increase their length. They are generally not suitable for use as end-bearing piles under heavy load and they are subject to decay and insect attack. Timber piles are resilient and particularly adaptable for use in waterfront structures such as wharves, docks and piers for anchorages since they will bend or give under load or impact where other materials may break. The ease with which they can be worked and their economy makes them popular for trestle construction and for temporary structures such as falsework or centering. Where timber piles can be driven and cut off below the permanent groundwater level, they will last indefinitely; but above this level in the soil, a timber pile will rot or will be attacked by insects and eventually destroyed. In sea water, marine borers and fungus will act to deteriorate timber piles. Treatment of timber piles increases their life but does not protect them indefinitely.

Concrete Piles. Concrete piles are of two general types, precast and cast-in-place. The advantages in the use of concrete piles are that they can be fabricated to meet the most exacting conditions of design, can be cast in any desired shape or length, possess high strength and have excellent resistance to chemical and biological attack. Certain disadvantages are encountered in the use of precast piles, such as:

(a) Their heavy weight and bulk (which introduces problems in handling and driving).
(b) Problems with hair cracks which often develop in the concrete as a result of shrinkage after curing (which may expose the steel reinforcement to deterioration).
(c) Difficulty encountered in cut-off or splicing.
(d) Susceptibility to damage or breakage in handling and driving.
(e) They are more expensive to fabricate, transport and drive.

Precast piles are fabricated in casting yards. Centrifugally spun piles (or piles with square or octagonal cross-sections) are cast in horizontal forms, while round piles are usually cast in vertical forms. With the exception of relatively short lengths, precast piles must be reinforced to provide the designed column strengths and to resist damage or breakage while being transported or driven.

Precast piles can be tapered or have parallel sides. The reinforcement can be of deformed bars or be prestressed or poststressed with high strength steel tendons. Prestressing or prestressing eliminates the problem of open shrinkage cracks in the concrete. Otherwise, the pile must be protected by coating it with a bituminous or plastic material to prevent ultimate deterioration of the reinforcement. Proper curing of the precast concrete in piles is essential.

Cast-in-place pile types are numerous and vary according to the manufacturer of the shell or inventor of the method. In general, they can be classified into two groups: shell-less types and the shell types. The shell-less type is constructed by driving a steel shell into the ground and filling it with concrete as the shell is pulled from the ground. The shell type is constructed by driving a steel shell into the ground and filling it in place with concrete. Some of the advantages of cast-in-place concrete piles are: lightweight shells are handled and driven easily, lengths of the shell may be increased or decreased easily, shells may be transported in short lengths and quickly assembled, the problem of breakage is eliminated and a driven shell may be inspected for shell damage or an uncased hole for "pinching off." Among the disadvantages are problems encountered in the proper centering of the reinforcement cages, in placing and consolidating the concrete without displacement of the reinforcement steel or segregation of the concrete, and shell damage or "pinching-off" of uncased holes.

Shell type piles are fabricated of heavy gage metal or are fluted, corrugated or spirally reinforced with heavy wire to make them strong enough to be driven without a mandrel.

Other thin-shell types are driven with a collapsible steel mandrel or core inside the casing. In addition to making the driving of a long thin shell possible, the mandrel prevents or minimizes damage to the shell from tearing, buckling, collapsing or from hard objects encountered in driving.

Some shell type piles are fabricated of heavy gauge metal with enlargement at the lower end to increase the end bearing.

These enlargements are formed by withdrawing the casing two to three feet after placing concrete in the lower end of the shell. This wet concrete is then struck by a blow of the pile hammer on a core in the casing and the enlargement is formed. As the shell is withdrawn, the core is used to consolidate the concrete after each batch is placed in the shell. The procedure results in completely filling the hole left by the withdrawal of the shell.

Steel Piles. A steel pile is any pile fabricated entirely of steel. They are usually formed of rolled steel H sections, but heavy steel pipe or box piles (fabricated from sections of steel sheet piles welded together) are also used. The advantages of steel piles are that they are readily available, have a thin uniform section and high strength, will take hard driving, will develop high load-bearing values, are easily cut off or extended, are easily adapted to the structure they are to support, and breakage is eliminated. Some disadvantages are: they will rust and deteriorate unless protected from the elements; acid, soils or water will result in corrosion of the pile; and greater lengths may be required than for other types of piles to achieve the same bearing value unless bearing on rock strata. Pipe pile can either be driven open-end or closed-end and can be unfilled, sand filled or concrete filled. After open-end pipe piles are driven, the material from inside can be removed by an earth auger, air or water jets, or other means, inspected, and then filled with concrete. Concrete filled pipe piles are subject to corrosion on the outside surface only.

Composite Piles. Any pile that is fabricated of two or more materials is called a composite pile. There are three general classes of composite piles: wood with concrete, steel with concrete, and wood with steel. Composite piles are usually used for a special purpose or for reasons of economy.

Where a permanent ground-water table exists and a composite pile is to be used, it will generally be of concrete and wood. The wood portion is driven to below the water table level and the concrete upper portion eliminates problems of decay and insect infestation above the water table. Composite piles of steel and concrete are used where high bearing loads are desired or where driving in hard or rocky soils is expected. Composite wood and steel piles are relatively uncommon.

It is important that the pile design provides for a permanent joint between the two materials used, so constructed that the parts do not separate or shift out of axial alignment during driving operations.

Sheet Piles. Sheet piles are made from the same basic materials as other piling: wood, steel and concrete. They are ordinarily designed so as to interlock along the edges of adjacent piles.

Sheet piles are used where support of a vertical wall of earth is required, such as trench walls, bulkheads, waterfront structures or cofferdams. Wood sheet piling is generally used in temporary installations, but is seldom used where water-tightness is required or hard driving expected. Concrete sheet piling has the capability of resisting much larger lateral loads than wood sheet piling, but considerable difficulty is experienced in securing water-tight joints. The type referred to as "fishmouth" type is designed to permit jetting out the joint and filling with grout, but a seal is not always effected unless the adjacent piles are wedged tightly together. Concrete sheet piling has the advantage that it is the most permanent of all types of sheet piling.

Steel sheet piling is manufactured with a tension-type interlock along its edges. Several different shapes are available to permit versatility in its use. It has the advantages that it can take hard driving, has reasonably water-tight joints and can be easily cut, patched, lengthened or reinforced. It can also be easily extracted and reused. Its principal disadvantage is its vulnerability to corrosion.

Types of Pile Driving Hammers. A pile-driving hammer is used to drive load-bearing or sheet piles. The commonly used types are: drop, single-acting, double-acting, differential acting and diesel hammers. The most recent development is a type of hammer that utilizes high-frequency sound and a dead load as the principal sources of driving energy.

Drop Hammers. These hammers employ the principle of lifting a heavy weight by a cable and releasing it to fall on top of the pile. This type of hammer is rapidly disappearing from use, primarily because other types of pile driving hammers are more efficient. Its disadvantages are that it has a slow rate of driving (four to eight blows per minute), that there is some risk of damaging the pile from excessive impact, that damage may occur in adjacent structures from heavy vibration and that it cannot be used directly for driving piles under water. Drop hammers have the advantages of simplicity of operation, ability to vary the energy by changing the height of fall and they represent a small investment in equipment.

Single-Acting Hammers. These hammers can be operated either on steam or compressed air. The driving energy is provided by a free-falling weight (called a ram) which is raised after each stroke by the action of steam or air on a piston. They are manufactured as either open or closed types. Single-acting hammers are best suited for jobs where dense or elastic soil materials must be penetrated or where long heavy timber or precast concrete piles must be driven. The closed type can be used for underwater pile driving. Its advantages include: faster driving (50 blows or more per minute), reduction in skin friction as a result of more frequent blows, lower velocity of the ram which transmits a greater proportion of its energy to the pile and minimizes piles damage during driving, and it has underwater driving capability. Some of its disadvantages are: requires higher investment in equipment (i.e. steam boiler, air compressor, etc.), higher maintenance costs, greater set-up and moving time required, and a larger operating crew.

Double-Acting Hammers. These hammers are similar to the single-acting hammers except that steam or compressed air is used both to lift the ram and to impart energy to the falling ram. While the action is approximately twice as fast as the single-acting hammer (100 blows per minute or more), the ram is much lighter and operates at a greater velocity, thereby making it particularly useful in high production driving of light or medium-weight piles of moderate lengths in granular soils. The hammer is nearly always fully encased by a steel housing which also permits direct driving of piles under water.

Some of its advantages are: faster driving rate, less static skin friction develops between blows, has underwater driving capability and piles can be driven more easily without leads.

Among its disadvantages are: it is less suitable for driving heavy piles in high-friction soils and the more complicated mechanism results in higher maintenance costs.

Differential-Acting Hammers. This type of hammer is, in effect, a modified double-acting hammer with the actuating mechanism having two different diameters. A large-diameter piston operates in an upper cylinder to accelerate the ram on the downstroke and a small-diameter piston operates in a lower cylinder to raise the ram. The additional energy added to the falling ram is the difference in areas of the two pistons multiplied by the unit pressure of the steam or air used. This hammer is a short-stroke, fast-acting hammer with a cycle rate approximately that of the double-acting hammer.

Its advantages are that it has the speed and characteristics of the double-acting hammer with a ram weight comparable to the single-acting type, and it uses 25 to 35 percent less steam or air. It is also more suitable for driving heavy piles under more difficult driving conditions than the double-acting hammer. It is available in the open or closed-type cases, the latter permitting direct underwater pile driving. Its principal disadvantage is higher maintenance costs.

Diesel Hammers. This hammer is a self-contained driving unit which does not require an auxiliary steam boiler or air compressor. It consists essentially of a ram operating as a piston in a cylinder. When the ram is lifted and allowed to fall in the cylinder, diesel fuel is injected in the compression space between the ram and an anvil placed on top of the pile. The continued downstroke of the ram compresses the air and fuel to ignition heat and the resultant explosion drives the pile downward and the ram upward to start another cycle. This hammer is capable of driving at a rate of from 80 to 100 blows per minute. Its advantages are that it has a low equipment investment cost, is easily moved, requires a small crew, has a high driving rate, does not require a steam boiler or air compressor and can be used with or without leads for most work. Its disadvantages are that it is not self-starting (the ram must be mechanically lifted to start the action) and it does not deliver a uniform blow. The latter disadvantage arises from the fact that as the reaction of the pile to driving increases, the reaction to the ram increases correspondingly. That is, when the pile encounters considerable resistance, the rebound of the ram is higher and the energy is increased automatically. The operator is required to observe the driving operations closely to identify changing driving conditions and compensate for such changes with his controls to avoid damaging the pile.

Diesel hammers can be used on all types of piles and they are best suited to jobs where mobility or frequent relocation of the pile driving equipment is necessary.

PILE CHART

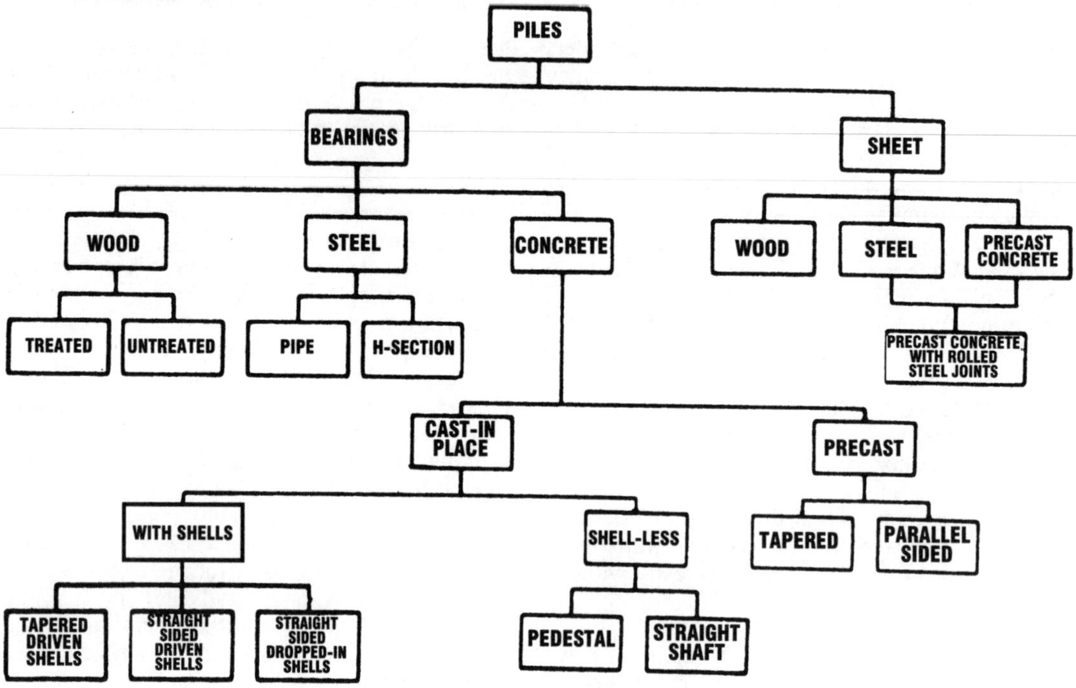

CONCRETE SHEET PILING

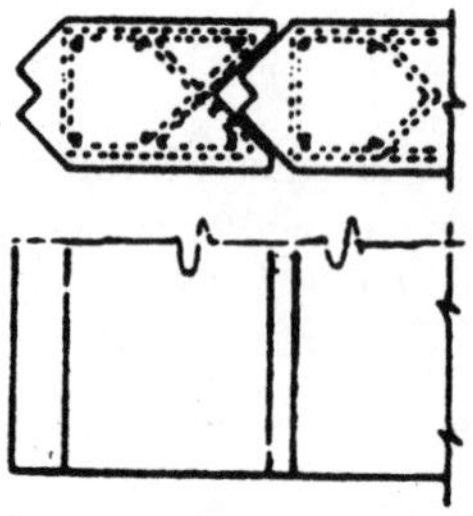

FISH MOUTH JOINTED

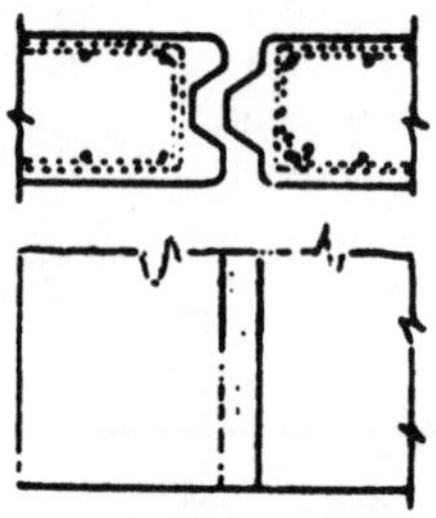

STANDARD JOINT

JOINT GROUTED AFTER PILE HAS BEEN DRIVEN

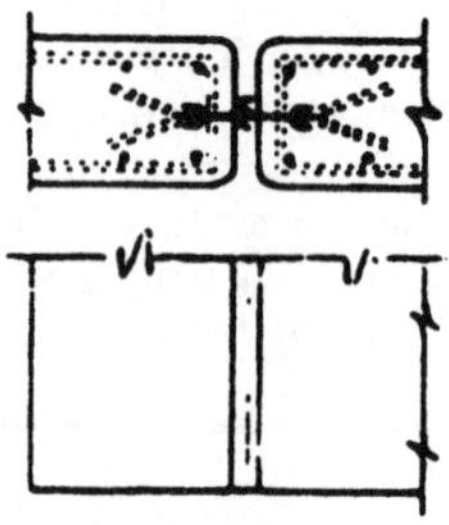

STEEL INTERLOCK

STEEL SHEET PILING

FLAT PILE

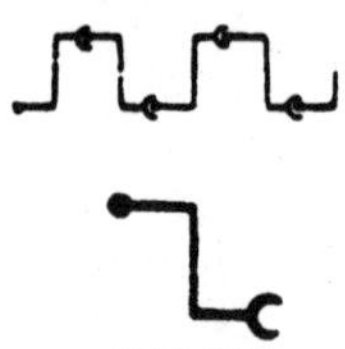

ZEE PILE

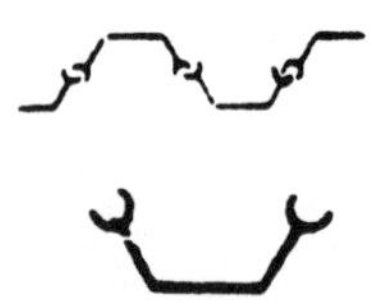

DEEP WEB PILE

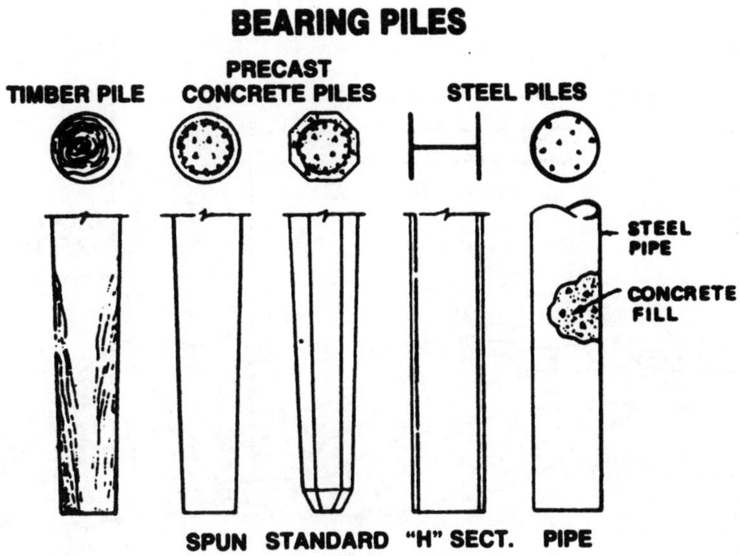

DRIVEN PILES

COMPOSITE WOOD PILE

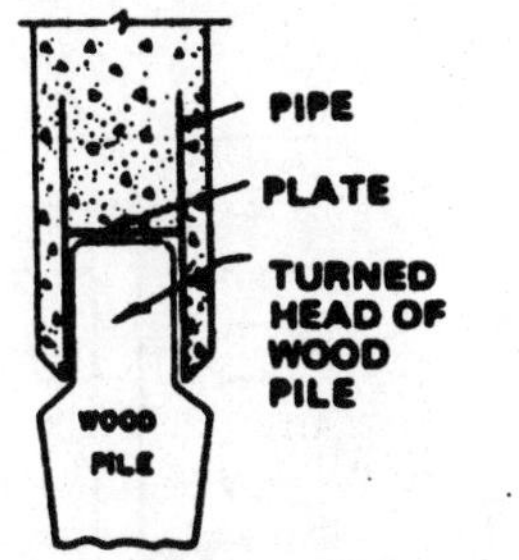

COMPOSITE STEEL PILE

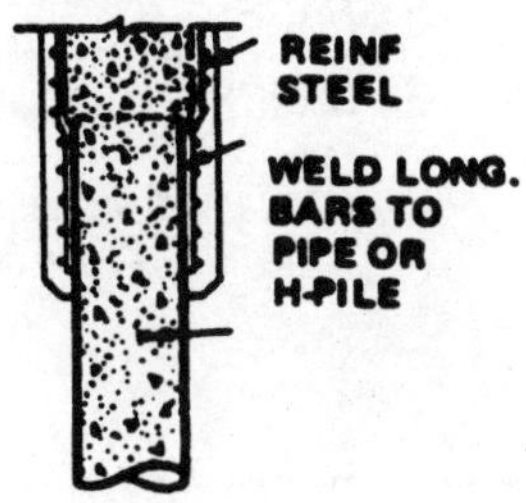

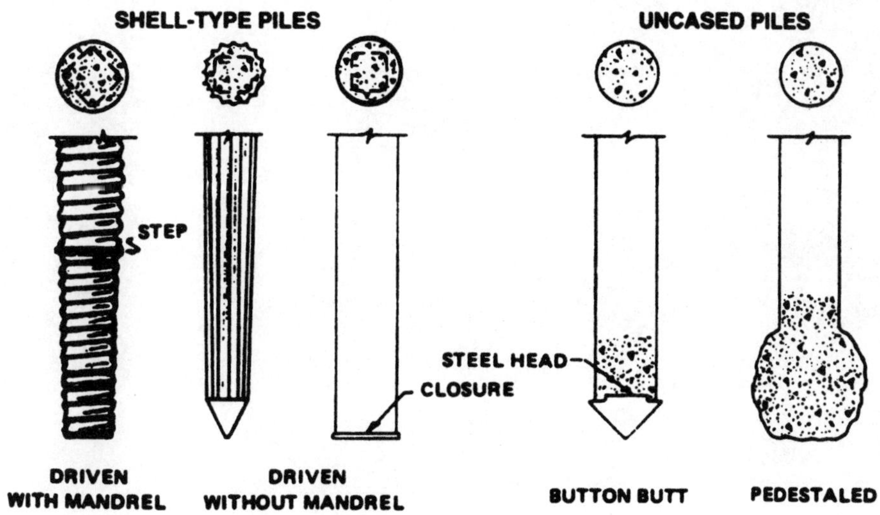

CAST-IN-PLACE PILES

STANDARD NOMENCLATURE FOR STREET CONSTRUCTION

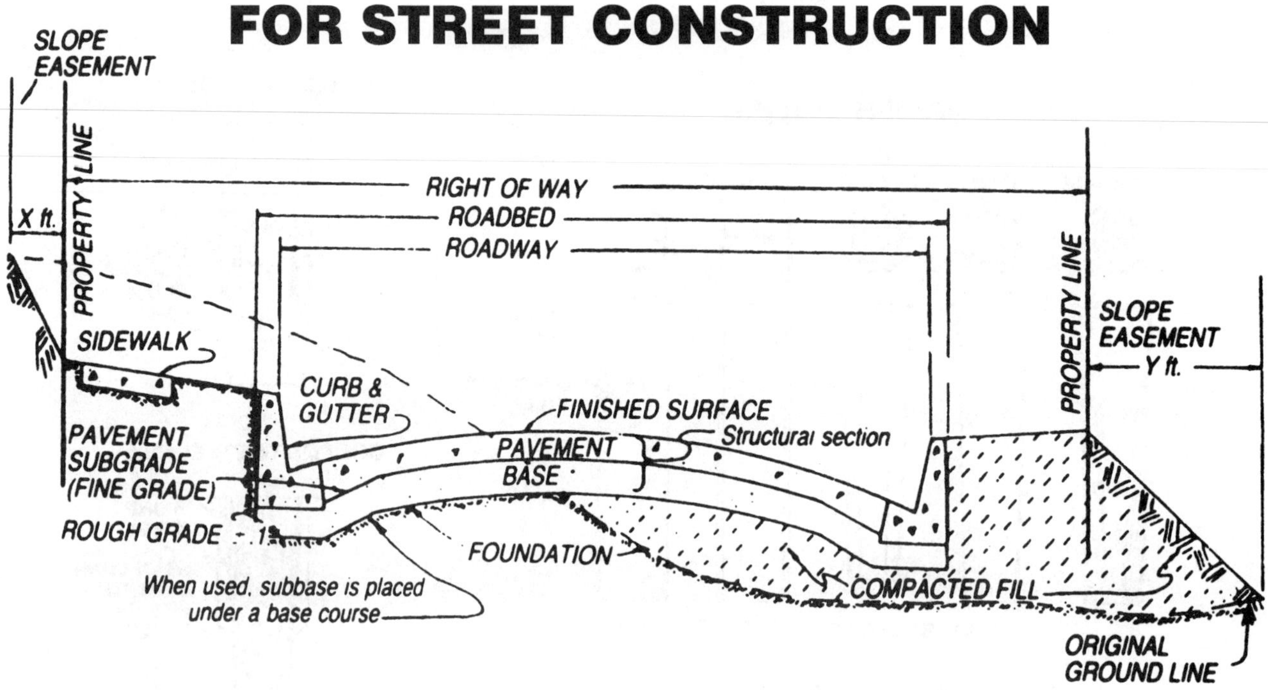

CONCRETE MASONRY PAVING UNITS

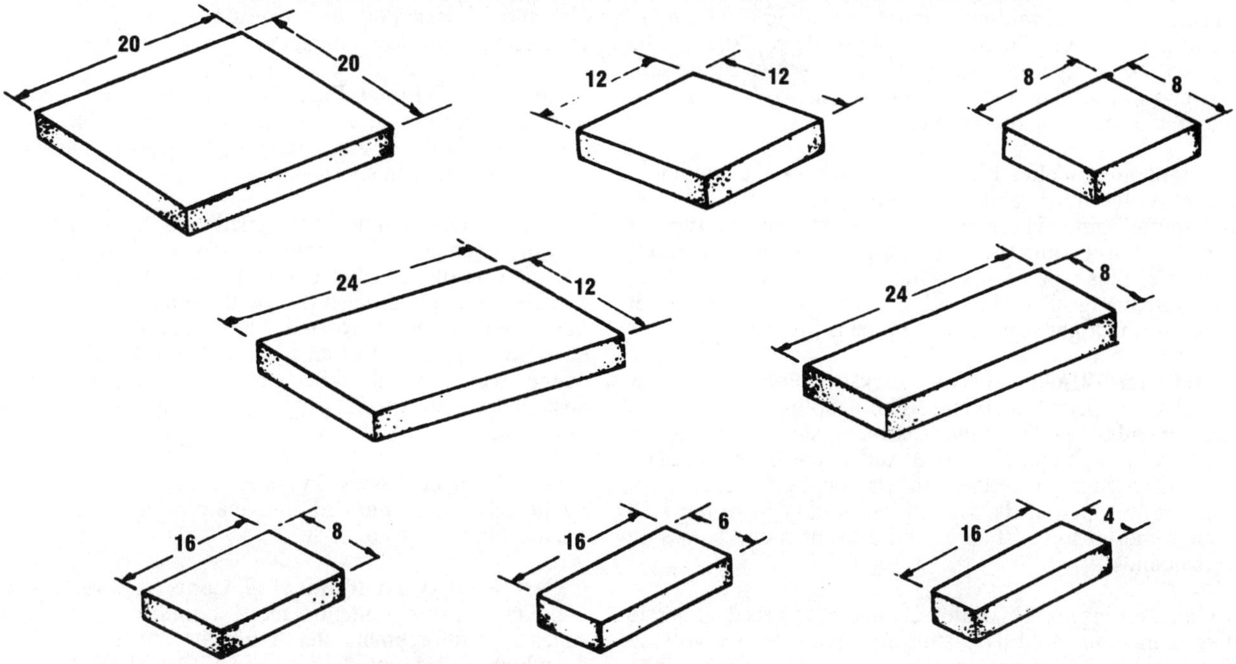

NOTE: Sizes are nominal and will vary by manufacture.

HEXAGON PAVER UNITS
Various Sizes Available

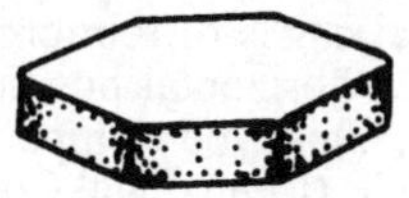

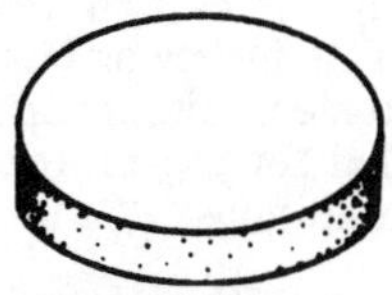

ROUND PAVING UNITS
Various Sizes Available

VEHICULAR PAVING UNITS

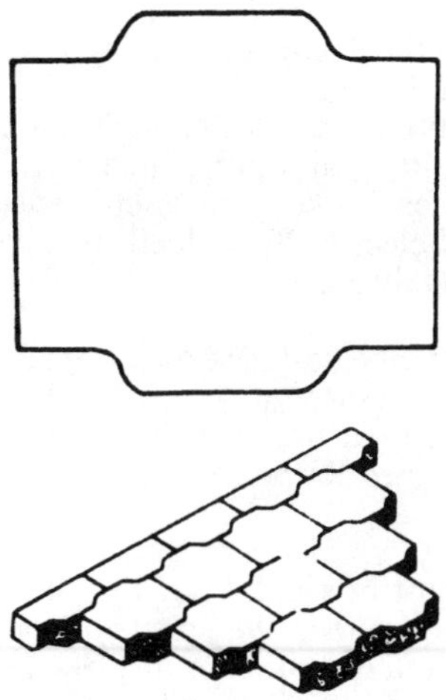

INTERLOCKING PAVER
7¼" X 3" X 8½"

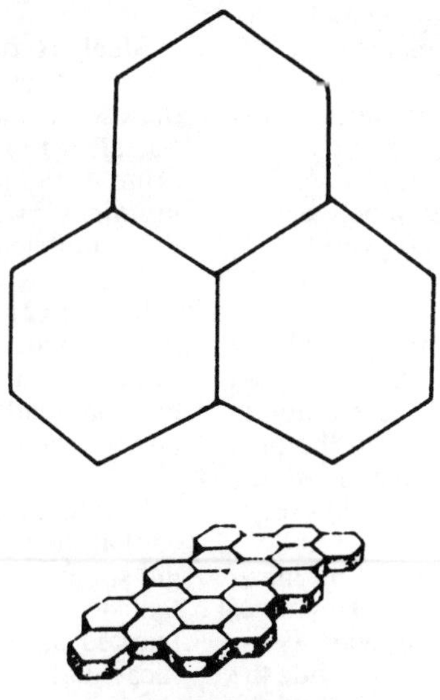

INTERLOCKING PAVER
12" X 3⅝" X 12"

TURF PAVER
24" X 3⅝" X 24"

PIPE

Clay Pipe. Clay pipe is manufactured by blending various clays together, milling, mixing, extruding and firing in a kiln to obtain vitrification. The physical properties of the pipe can be changed by varying the proportions of the several clays used. The pipe is supplied in two basic styles: spigot and socket; and plain end.

Spigot and Socket Pipe has a spigot on one end and a socket on the other, and is commonly referred to as "bell and spigot" pipe. The plans generally specify the type of joint to be used from the several types of jointing methods available. This type of pipe is manufactured with matching polyurethane gaskets molded on the spigot and socket which form a tight seal when the pipe is jointed.

Plain End Pipe is without a socket on either end and is joined with special couplings. The coupling consists of a circular rubber sleeve, two stainless steel compression bands with tightening devices and a corrosion resistant shear ring. Sometimes this joint is supplied with a cardboard form, open at the top, which is filled with portland cement mortar to resist shear and prevent future corrosion of the bands.

Concrete Pipe. Unreinforced and reinforced concrete pipe is manufactured by casting in stationary or revolving metal molds. At the present time, the design practice is to specify reinforced concrete pipe far all purposes.

Unreinforced Concrete Pipe is cast in vertical steel molds, usually in pipe sizes of 21 inches or less, and is of the spigot and socket type. No steel reinforcement is used and the pipe is usually intended for use in irrigation systems and under light loading conditions.

Reinforced Concrete Pipe (RCP) is made in a number of different manufacturing processes and for a wide variety of pressure and non-pressure classes. It is available in standard sizes or it can be made to order to any diameter desired. Some of the larger diameters include diameters of 12 and 14 feet. A large variety of joint details are used with RCP. Tongue and groove joints are used for storm drain pipelines.

Reinforced concrete pipe for wastewater pipeline projects is supplied with gasketed joints and a polyvinyl chloride (PVC) plastic liner cast into the pipe.

(a) **Cast Pipe** is cast vertically in steel forms with the reinforcing cage securely held in place. The reinforcement is generally elliptical in shape to provide the maximum structural strength to resist the loads imposed on the pipe by the backfill and other stresses. Consolidation of the concrete is obtained by the use of external form vibrators.

(b) **Centrifugally Spun Pipe** is manufactured by introducing concrete into a spinning horizontal steel cylinder into which the reinforcement cage has been previously installed and which is equipped with end dams to provide the proper pipe wall thickness. The speed of rotation of the mold is increased and the centrifugal force produces a smooth, dense concrete pipe.

(c) **Pressure Pipe** may be cast or centrifugally spun pipe but it usually has a circular steel reinforcement cage (or cages) designed not only to resist the trench loading, but also the internal pressures exerted on the pipe from the fluid under pressure in the line.

Concrete Cylinder Pipe. This class of pipe is generally used for high pressure water lines and sewer force mains and is available in sizes ranging from 10 inches to 60 inches and larger in special cases.

A sheet steel cylinder is wrapped with the designed steel reinforcement and a concrete lining is centrifugally spun in the interior of the steel cylinder. An exterior coating of concrete is applied generally by the gunite process, while the cylinder is slowly rotated. These coatings vary in thickness from 1/2 to 3/4 of an inch. The joints are commonly of the steel ring and rubber gasket type, but are generally designed for the special purpose for which the pipe line is intended.

Definition of Terms. In general, the terms used to designate types of reinforced concrete pipe refer to the process used in manufacture.

Cast RCP (Cast Reinforced Concrete Pipe). A concrete pipe having one or more cylindrical or elliptical cages of reinforcement steel embedded in it, the concrete for which is cast with the forms in a vertical position.

CSRCP (Centrifugally Spun Reinforced Concrete Pipe). A concrete pipe having one or more cylindrical or elliptical cages of reinforcement steel embedded in it, and cast in a horizontal position while the forms are spinning rapidly. This type of pipe may be designated as Spun RCP or as CCP (Centrifugal Concrete Pipe).

RCP (Reinforced Concrete Pipe). A reinforced concrete pipe manufactured by either the casting or spinning method.

Steel Reinforcement. Steel for reinforcing concrete pipe is generally furnished in large coils which will permit the use of machines to fabricate the "cages." The continuous steel rod is wound spirally at a prescribed pitch on a drum of the proper diameter. Where the rod crosses a longitudinal spacer rod, it is electrically welded to it so that the complete cage is relatively rigid.

Reinforcement cages for pipe designed for external loading are generally elliptical in shape to take full advantage of the steel in tension. Pipe to be used with relatively small external loads or pipe designed for pressure lines will have circular cages.

Reinforcement cages must be rigidly fixed in the forms so that the placement of concrete or the effects of centrifugal spinning will not result in distortion or displacement of the steel The orientation of an elliptical cage must be marked on the forms to assure that the minor axis can be located after the concrete is placed.

CAPACITIES FOR SEPTIC TANKS SERVING AN INDIVIDUAL DWELLING

No. of bedrooms	Capacity of tank (gals.)
2 or less	750
3	900
4	1,000

FORM NOMENCLAURE

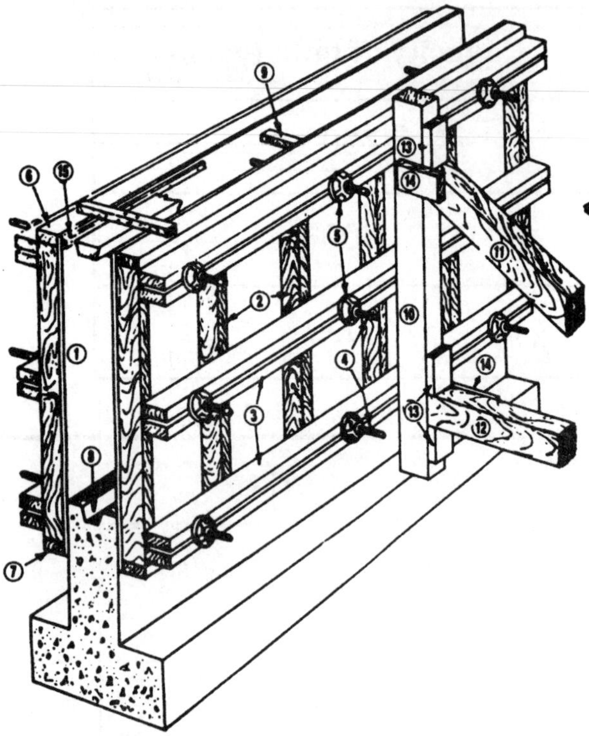

1. SHEATHING
2. STUDS
3. WALES
4. FORM BOLTS
5. NUT WASHER
6. TOP PLATE
7. BOTTOM PLATE
8. KEY-WAY
9. SPREADER
10. STRONGBACK
11. BRACE
12. STRUT
13. CLEATS
14. SCAB
15. POUR STRIP

FALSEWORK NOMENCLATURE

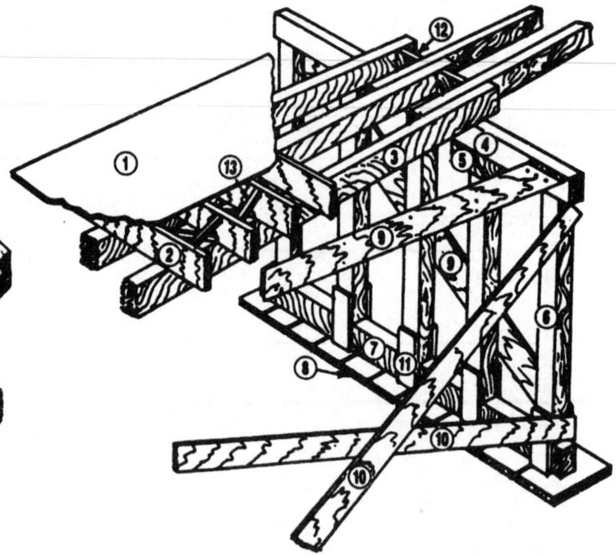

1. SHEATHING
2. JOIST
3. STRINGER
4. CAP
5. CORBEL
6. POST
7. SILL
8. FOOTING
9. SWAY BRACE
10. LONGITUDINAL BRACE
11. SCAB
12. BLOCKING
13. BRIDGING

TYPICAL PAN-JOIST FORM CONSTRUCTION

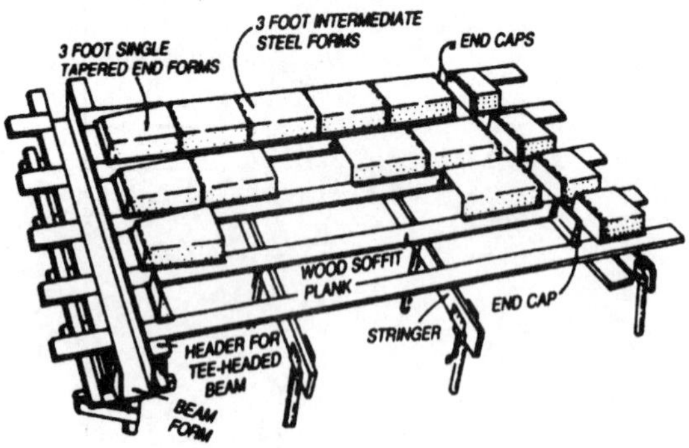

TYPICAL WAFFLE SLAB FORM CONSTRUCTION

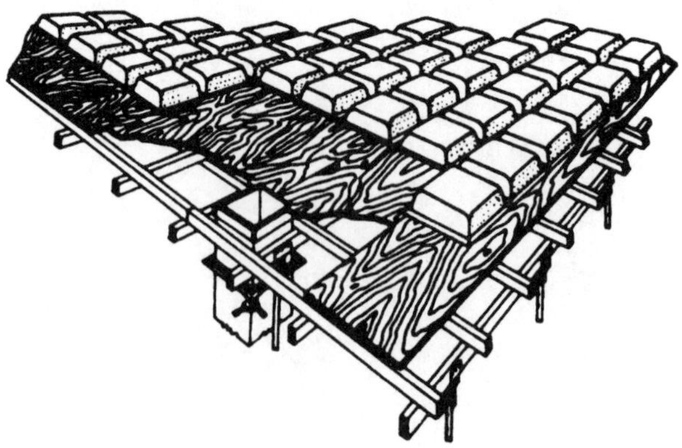

CONCRETE / REINFORCING STEEL 03210

STANDARD SIZES OF STEEL REINFORCEMENT BARS

STANDARD REINFORCEMENT BARS				
		Nominal Dimensions		
Bar Designation Number*	Nominal Weight lb. per ft.	Diameter, in.	Cross Sectional Area, sq. in.	Perimeter, in.
3	0.376	0.375	0.11	1.178
4	0.668	0.500	0.20	1.571
5	1.043	0.625	0.31	1.963
6	1.502	0.750	0.44	2.356
7	2.044	0.875	0.60	2.749
8	2.670	1.000	0.79	3.142
9	3.400	1.128	1.00	3.544
10	4.303	1.270	1.27	3.990
11	5.313	1.410	1.56	4.430
14	7.65	1.693	2.25	5.32

*The bar numbers are based on the number of 1/8 inches included in the nominal diameter of the bar.

Type of Steel & ASTM Specification No.	Size Nos. Inclusive	Grade	Tensile Strength Min., psi.	Yield (a) Min., psi
Billet Steel A 615	3-11	60	90,000	60,000
	11, 14, 18	75	100,000	75,000

CONCRETE / WELDED WIRE FABRIC 03220

Style Designation	Steel Area sq. in per ft.		Weight Approx. lbs. per 100 sq. ft.
	Longit.	Transv.	
Rolls			
6×6 — W1.4×W1.4	.03	.03	21
6×8 — W2×W2	.04	.04	29
6×6 — W2.9×W2.9	.08	.06	42
6×6 — W4×W4	.08	.08	58
4×4 — W1.4×W1.4	.04	.04	31
4×4 — W2×W2	.06	.06	43
4×4 — W2.9×W2.9	.09	.09	62
	.12	.12	
Sheets			
6×6 — W2.9×W2.9	.06	.06	42
6×6 — W4×W4	.08	.08	58
6×6 — W5.5×W5.5	.11	.11	80
	.12	.12	86

COMMON TYPES OF STEEL REINFORCEMENT BARS

ASTM specifications for billet steel reinforcing bars (A 615) require identification marks to be rolled into the surface of one side of the bar to denote the producer's mill designation, bar size and type of steel. For Grade 60 and Grade 75 bars, grade marks indicating yield strength must be show. Grade 40 bars show only three marks (no grade mark) in the following order:
1st — Producing Mill (usually an initial)
2nd — Bar Size Number (#3 through # 18)
3rd — Type (N for New Billet)

NUMBER SYSTEM — GRADE MARKS

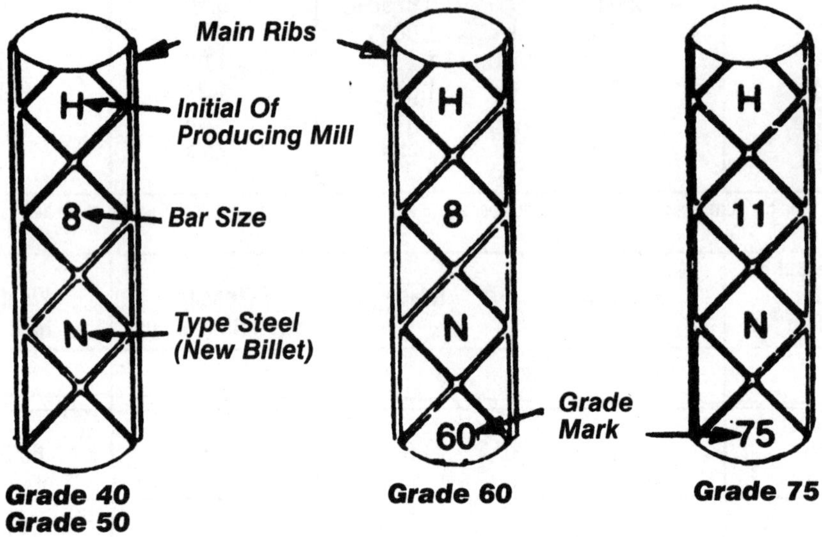

LINE SYSTEM — GRADE MARKS

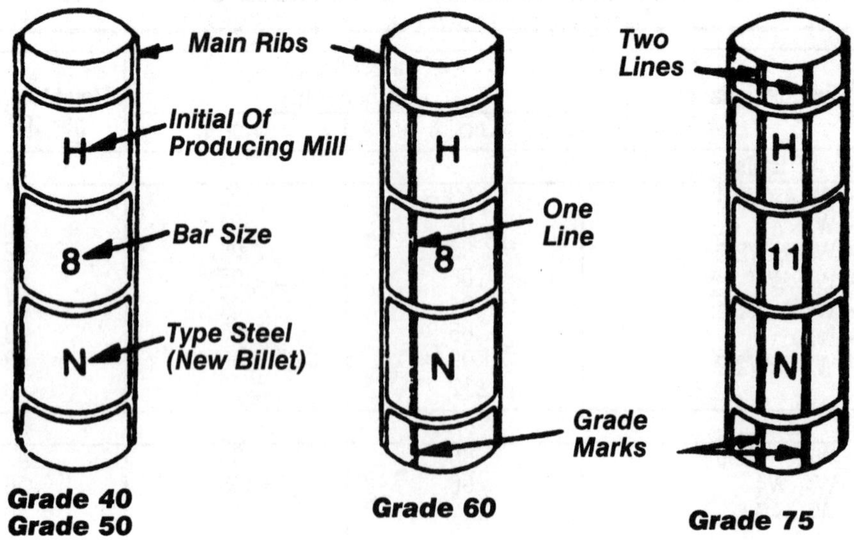

Insofar as is possible, the moisture content should be kept uniform to avoid problems in determining the proper amount of water to be added for mixing. Mixing water must be reduced to compensate for moisture in the aggregate in order to control the slump of the concrete and avoid exceeding the specified water-cement ratio.

Handling Concrete by Pumping Methods. Transportation and placement of concrete by pumping is another method gaining increased popularity. Pumps have several advantages, the primary one being that a pump will high-lift concrete without the need for an expensive crane and bucket. Since the concrete is delivered through pipe and hoses, concrete can be conveyed to remote locations in buildings, in tunnels, to locations otherwise inaccessible on steep hillside slopes for anchor walls, pipe bedding or encasement, or for placing concrete for chain link fence post bases. Concrete pumps have been found to be economical and expedient in the placement of concrete, and this has promoted the use and acceptance of this development. The essence of proper concrete pumping is the placement of the concrete in its final location without segregation.

Modern concrete pumps, depending on the mix design and size of line, can pump to a height of 200 feet or a horizontal distance of 1,000 feet. They can handle, economically, structural mixes, standard mixes, low slump mixes, mixes with two-inch maximum size aggregate and light weight concrete. When a special pump mix is required for structural concrete in a major structure, the mix design must be approved by the Engineer and checked and confirmed by the Supervisor of the Materials Control Group. The Inspector should obtain the pump manufacturer's printed information and evaluate its characteristics and ability to handle the concrete mixture specified for the project.

If concrete is being placed for a major reinforced structure, it is important that the placement continue without interruption. The Inspector should be sure that the contractor has ready access to a back-up pump to be used in the event of a breakdown. In order to further insure the success of the concrete placement by the pumping method, the user should be aware of the following points:

(a) A protective grating over the receiving hopper of the pump is necessary to exclude large pieces of aggregate or foreign material.

(b) The pump and lines require lubrication with a grout of cement and water. All of the excess grout is to be wasted prior to pumping the concrete.

(c) All changes in direction must be made by a large radius bend with a maximum bend of 90 degrees. Wye connections induce segregation and shall not be used.

(d) Pump lines should be made of a material capable of resisting abrasion and with a smooth interior surface having a low coefficient of friction. Steel is commonly used for pump lines, because a chemical reaction occurs between the concrete and the aluminum. Aluminum pipe should not be used for pumping concrete and some of the new plastic or rubber tubing is gaining acceptance. Hydrogen is generated which results in a swelling of the concrete, causing a significant reduction in compressive strength. This reaction is aggravated by any of the following: abrasive coarse aggregate, non-uniformly graded sand, low-slump concrete, low sand-aggregate ratio, high-alkali cement or when no air-entraining agent is used.

(e) During temporary interruptions in pumping, the hopper must remain nearly full, with an occasional turning and pumping to avoid developing a hard slug of concrete in the lines.

(f) Excessive line pressures must be avoided. When this occurs, check these points as the probable cause: segregation caused by too low a slump or too high a slump; large particle contamination caused by large pieces of aggregate or frozen lumps not eliminated by the grating; poor gradation of aggregates or particle shape; rich or lean spots caused by improper mixing.

(g) Corrections must be made to correct excessive slump loss as measured at the transit-mixed concrete truck and as measured at the hose outlet. This may be attributable to porous aggregate, high temperature or rapid setting mixes.

(h) Two transit-mix concrete trucks must be used simultaneously to deliver concrete into the pump hopper. These trucks must be discharged alternately to assure a continuous flow of concrete as trucks are replaced.

(i) Samples of concrete for test specimens prepared to determine the acceptance of the concrete quality are to be taken as required for conventional concrete.

Sampling is done before the concrete is deposited in the pump hopper. However, it is suggested that, where possible, the effect of pumping on the compressive strength be checked by taking companion samples, so identified, from the end of the pump line at the same time. The Record of Test must be properly noted as being a special mix used for pumping purposes. This will enable the Materials Control Group to compile a complete history of mix designs and their respective compressive strengths.

The prudent use of pumped concrete can result in economy and improved quality. However, only the control exercised by the operator will assure continued high standards of quality concrete.

Pump lines must be properly fastened to supports to eliminate excessive vibration. Couplings must be easily and securely fastened in a manner that will prevent mortar leakage. It is preferable to use the flexible hose only at the discharge point. This hose must be moved in such a manner as to avoid kinks or sharp bends. The pump line should be protected from excessive heat during hot weather by water sprinkling or shade.

COMPRESSIVE STRENGTH FOR VARIOUS WATER-CEMENT RATIOS

(The strengths listed are based on the use of normal portland cement)

WATER/CEMENT RATIO		PROBABLE 28-DAY STRENGTH	
WEIGHT	GALS./100#	PSI	MEGAPASCALS*
.40	4.8	5000	34
.45	5.4	4500	31
.50	6.0	4000	28
.55	6.6	3500	24
.60	7.2	3000	21
.65	7.8	2500	17
.70	8.4	2000	14

* International system equivalent.

APPROXIMATE CONTENT OF SAND, CEMENT AND WATER PER CUBIC YARD OF CONCRETE

Based on aggregates of average grading and physical characteristics in concrete mixes having a water-cement ratio (W/C) of about .65 by weight (or 7.8 gallons) per sack of cement; 3-in, slump; and a medium natural sand having a fineness modulus of about 2.75.

COURSE AGGREGATE MAX. SIZE	WATER		CEMENT	% SAND
	POUNDS	GALLONS		
3/8	385	46	590	57
½	365	44	560	50
¾	340	41	525	43
1	325	39	500	39
1 1/2	300	36	460	37

It can be noted from the above chart that, for a given slump, the amount of mixing water increases as the size of the course aggregate decreases. The size of the course aggregate controls the sand content in the same way; that is, the amount of sand required in the mix increases as the size of the course aggregate decreases.

Other typical examples are contained in the pamphlet published by the Portland Cement Association entitled "Design and Control of Concrete Mixtures."

Effects of Temperature on Concrete. Concrete mixtures gain strength rapidly in the first few days after placement. While the rate of gain in strength diminishes, concrete continues to become stronger with time over a period of many years, so long as drying of the concrete is prevented. Its strength at 28 days is considered to be the compressive strength upon which the Engineer bases his calculations. The temperature of the atmosphere has a significant effect upon the development of strength in concrete. Lower temperatures retard and higher temperatures accelerate the gain in strength.

Most destructive of the natural forces is freezing and thawing action. While the concrete is still wet or moist, expansion of the water as it is converted into ice results in severe damage to the fresh concrete. In situations where freezing may be encountered, high early strength cement may be used. Also, the mixing water or the aggregate (or both) may be preheated before mixing. Covering the concrete, and using steam or salamanders to heat the concrete under the covering, will help prevent freezing. Air-entraining agents help to diminish the effects of freezing of fresh concrete as well as in subsequent freezing and thawing cycles throughout the life of the concrete.

Hot weather will present problems of a different nature in placing concrete. Concrete will set up faster and tend to shrink and crack at the surface. To minimize this problem, the concrete should be placed without delay after mixing. Avoid the use of accelerators (perhaps even use a retarding agent), dampen all subgrade and forms, protect the freshly placed concrete from hot dry winds, and provide for adequate curing. Crushed ice or chilled water can be used as part of the mixing water to reduce the temperature of the mix in extremely hot areas.

Admixture	Purpose	Effects on Concrete	Advantages	Disadvantages
Accelerator	Hasten setting.	Improves cement dispersion and increases early strength.	Permits earlier finishing, form removal, and use of the structure.	Increases shrink age, decreases sulfate resistance, tends to clog mixing and handling equipment.
Air-Entraining Agent	Increase workability and reduce mixing water.	Reduces segregation, bleeding and increases freeze-thaw resistance. Increases strength	Increases workability and reduces finishing time.	Excess will reduce strength and increase slump. Bulks concrete volume.
Bonding Agent	Increase bond to old concrete.	Produces a non-dusting, slip resistant finish,	Permits a thin topping without roughening old concrete, self-curing, ready in one day.	Quick setting and susceptible to damage from fats, oils and solvents.
Densifier	To obtain dense concrete.	Increased workability and strength.	Increases workability and increases waterproofing characteristics, more impermeable.	Care must be used to reduce mixing water in proportion to amount used.
Foaming Agent	Reduce weight.	Increases insulating properties.	Produces a more plastic mix, reduces dead weight loads.	Its use must be very carefully regulated — following instructions explicitly.
Retarder	Retard setting.	Increases control of setting.	Provides more time to work and finish concrete.	Performance varies with cement used — adds to slump. Requires stronger forms.
Water Reducer and Retarder	Increase compressive and flexural strength.	Reduces segregation, bleeding, absorption, shrinkage, and increases cement dispersion.	Easier to place work, provides better control.	Performance varies with cement. Of no use in cold weather.
Water Reducer, Retarder and Air-Entraining Agent	Increases workability.	Improves cohesiveness. Reduces bleeding and segregation.	Easier to place and work.	Care must be taken to avoid excessive air entrainment.

ARCHITECTURAL WALL PATTERNS (BONDS)

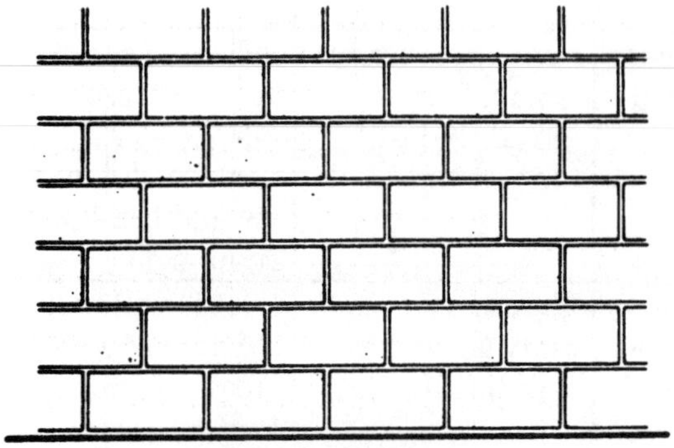

COMMON BOND

8" x 16" UNITS

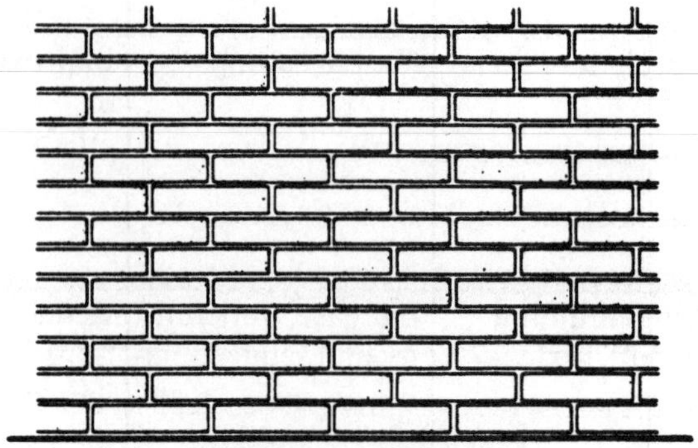

COMMON BOND

4" x 16" UNITS

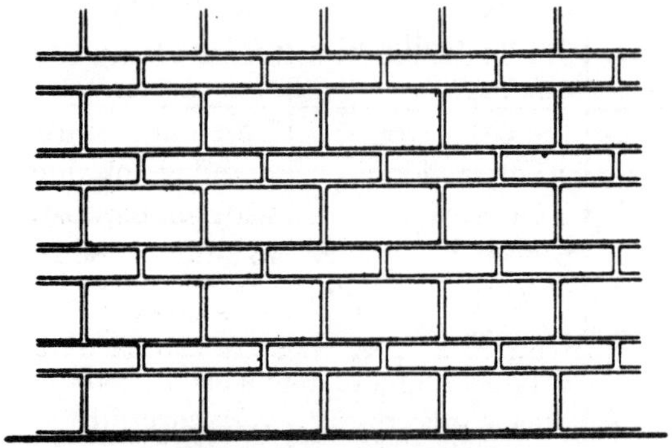

COURSED ASHLER

8" x 16" 8 4" x 16" UNITS

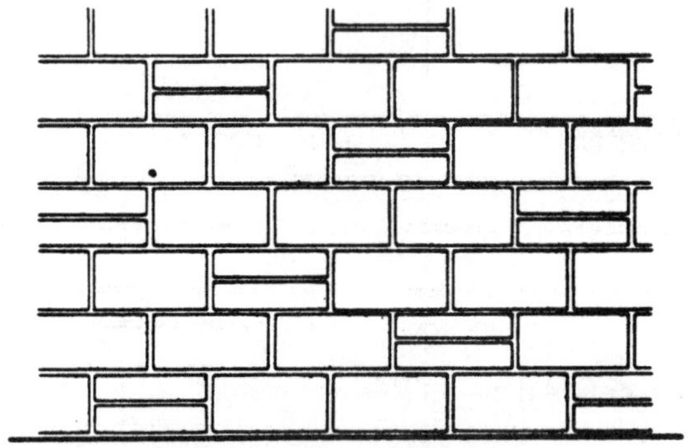

COURSED ASHLER

8" x 16" 8 4" x 16" UNITS

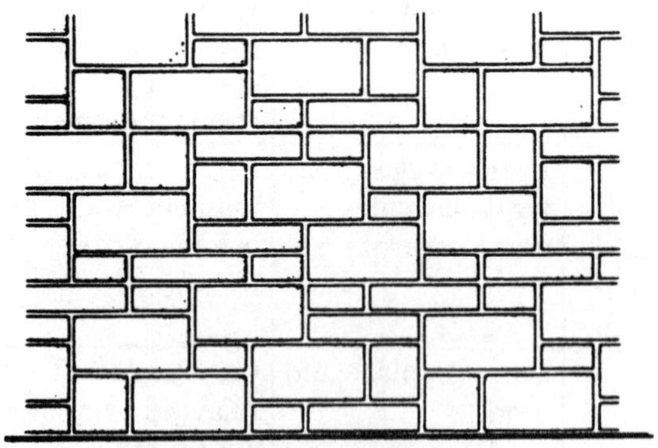

RANDOM ASHLER

8" x 16" 8 4" x 16" UNITS
AND 4" X 8" UNITS

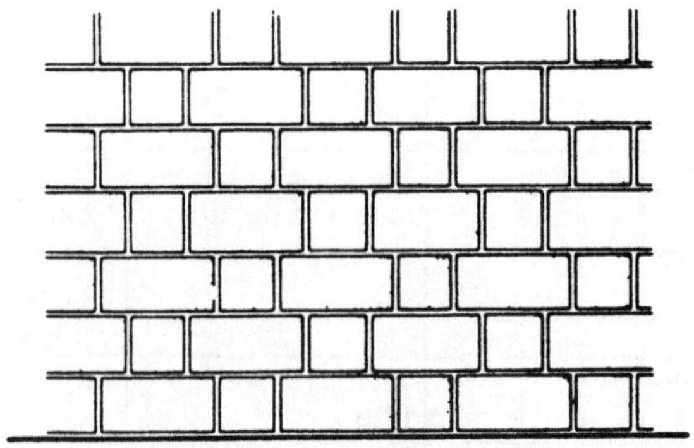

COURSED ASHLER

8" x 16" 8 8" x 8" UNITS

ARCHITECTURAL WALL PATTERNS (BONDS) — (Continued)

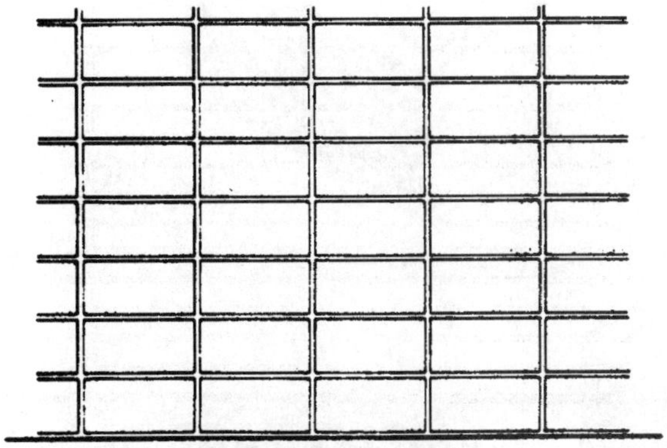

STACKED BOND

8" x 16" UNITS

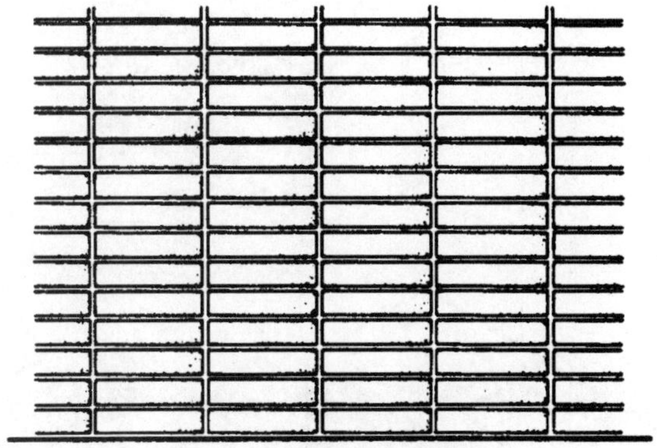

STACKED BOND

8" x 16" 8 4" x 16" UNITS

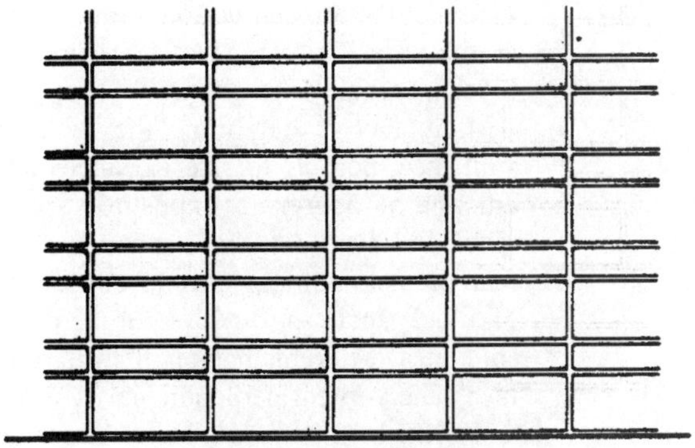

STACKED BOND

8" x 16" 8 4" x 16" UNITS

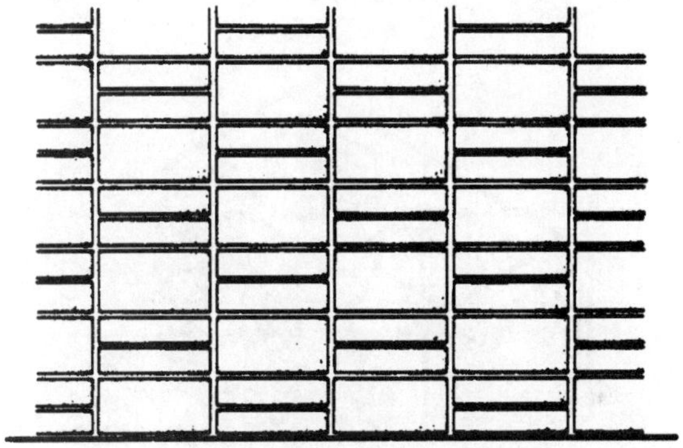

STACKED BOND

8" x 16" 8 4" x 16" UNITS

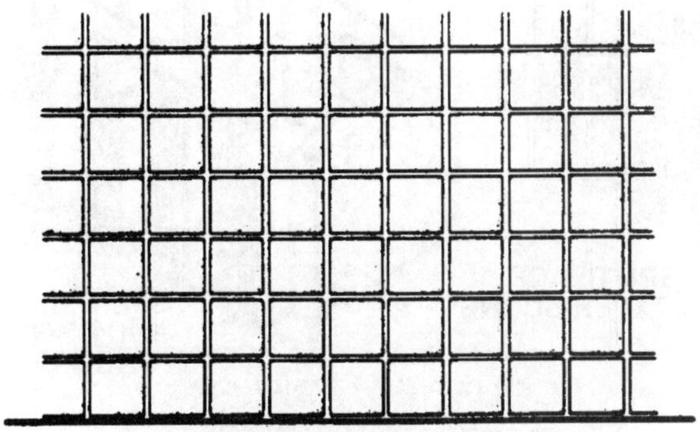

STACKED BOND VERTICAL SCORED UNITS

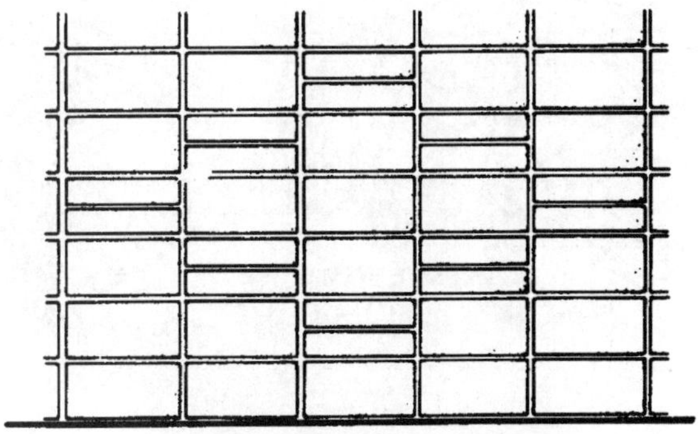

USE OF BLOCK DESIGN IN STACKED BOND

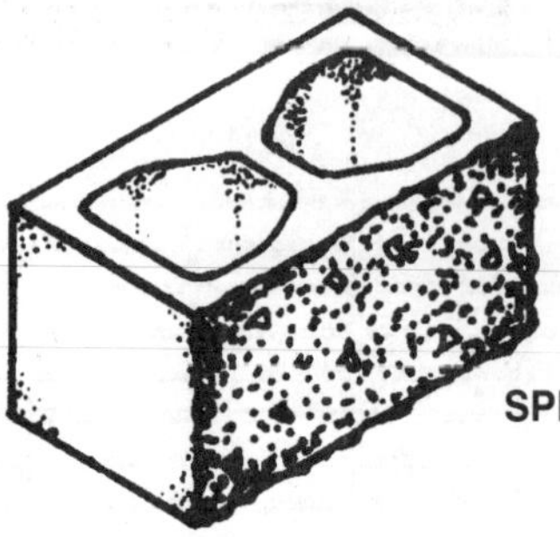

SPLIT FACE

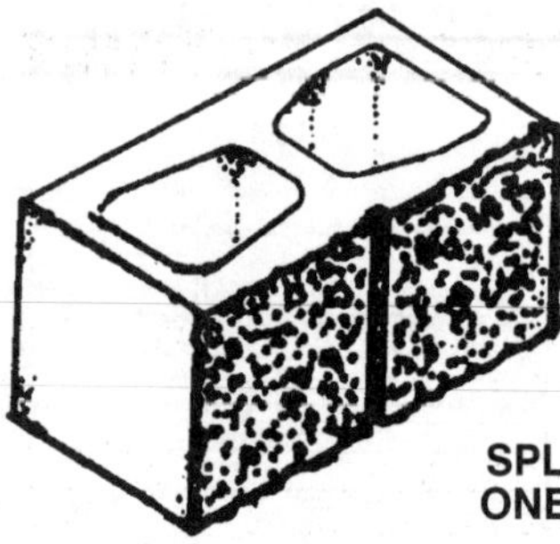

SPLIT FACE ONE-SCORE

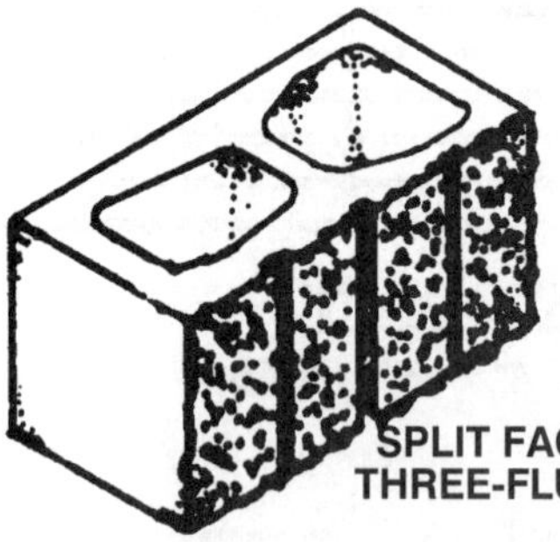

SPLIT FACE THREE-FLUTE

SPLIT FACE FIVE-SCORE

SPLIT FACE BLOCK

Split face block is manufactured as a unit that is normally made double and is literally split apart on a splitter: a machine which resembles a guillotine. The splitter has blades at the top and bottom (and sometimes at the sides) which exert pressure on the blocks, breaking them apart.

Many factors determine the look of the split face, both as to size variances and the amount of aggregate exposure. Split face block is intended to have a rougher texture than precision block. Various configurations of block such as fluted, scored, etc., will split in a different manner than a full split face. The vertical perpendicularity of scored and fluted split face block is subject to variation.

SPLIT FACE THREE-WIDE SCORE

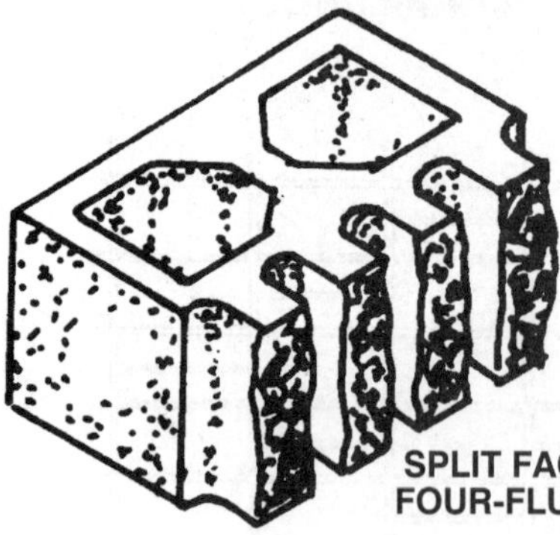

SPLIT FACE FOUR-FLUTE

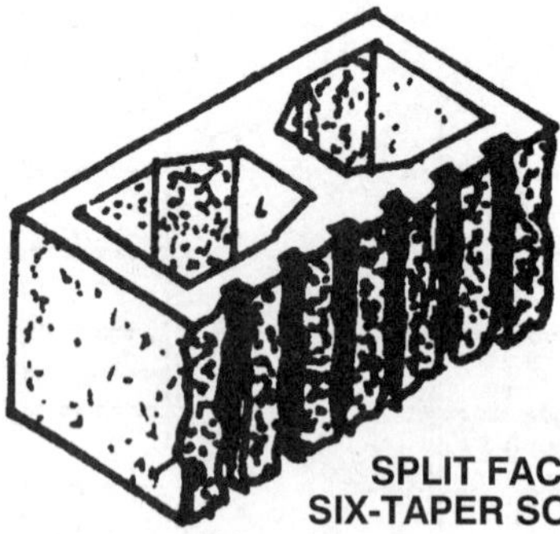

SPLIT FACE SIX-TAPER SCORE

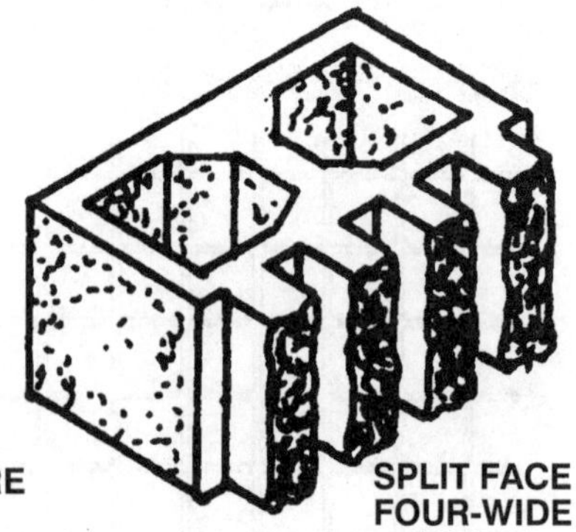

SPLIT FACE FOUR-WIDE SCORE

NOTE: Split face units shown in this manual are a small sampling of the broad range of concrete masonry architectural units available from the industry on special order. Depths and widths of scores vary. Consult a local manufacturer for specific information.

TYPICAL DETAILS — LINTELS AND BOND BEAMS

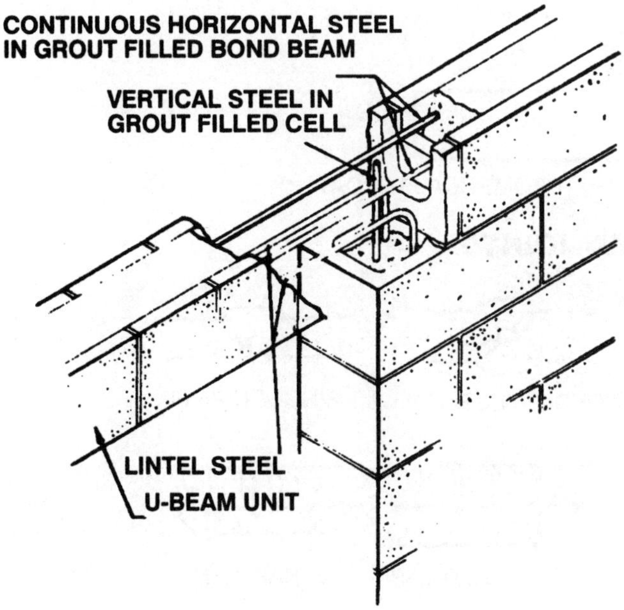

8" x 8" LINTEL & BOND BEAM ON 8" WALL

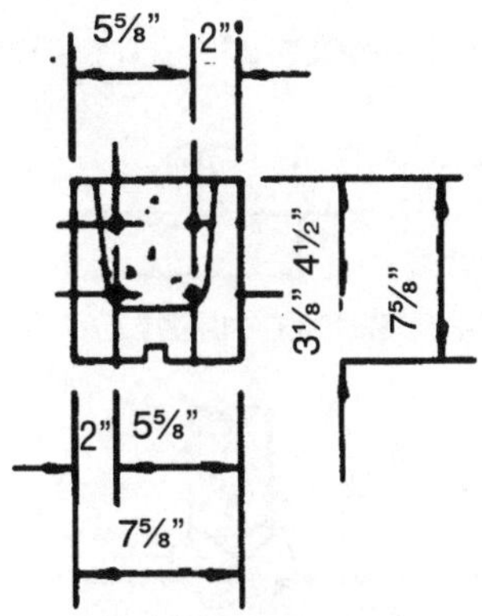

8" x 16" BOND BEAM ON 8" WALL

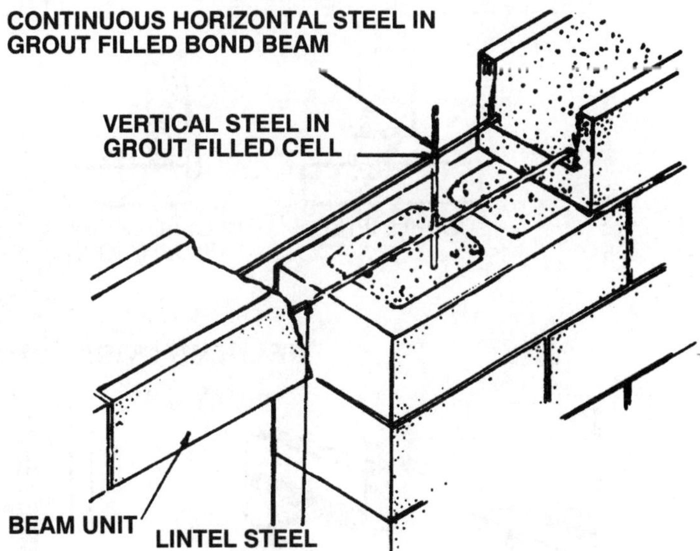

METALS / WELDING 05050

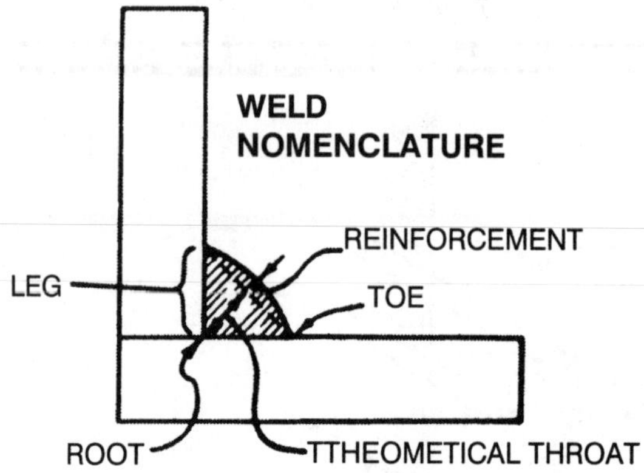

WELDED JOINTS

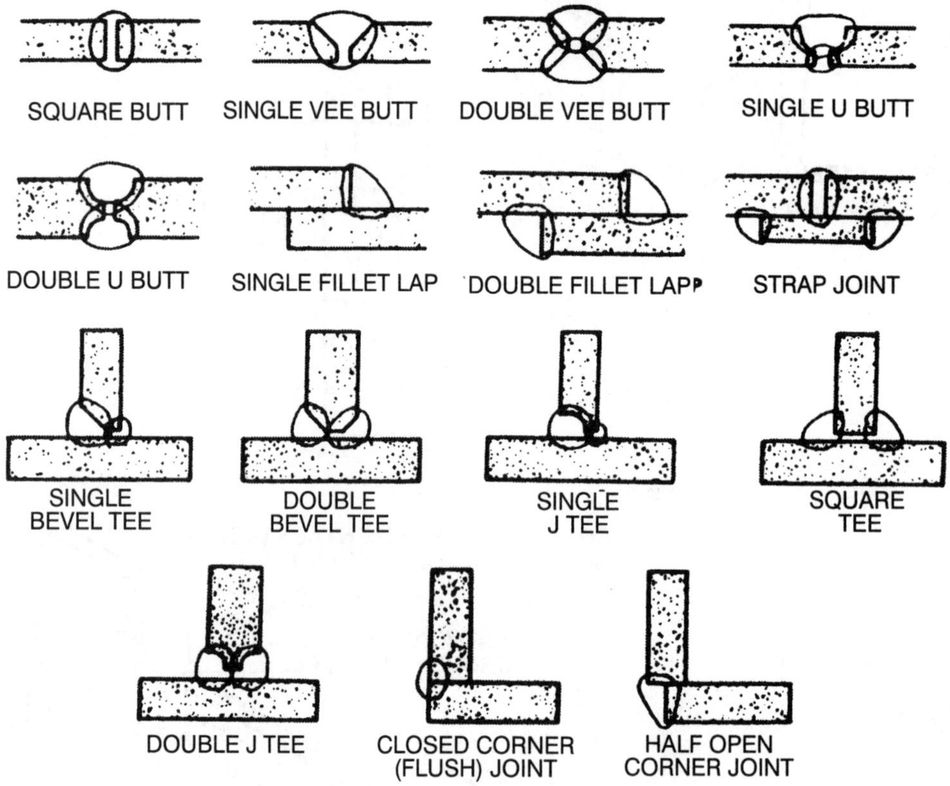

WELDING POSITIONS

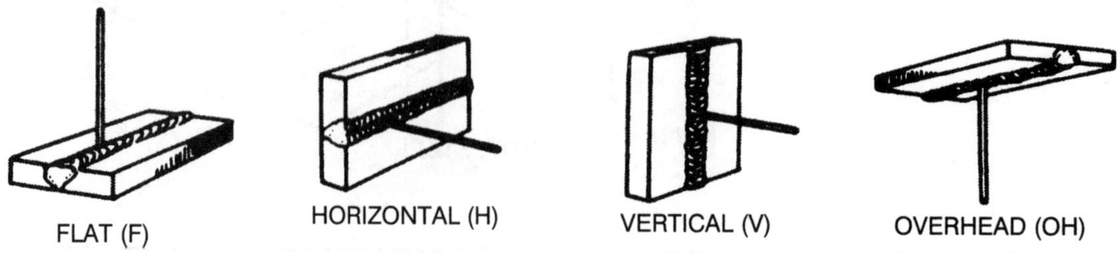

BOLTS IN COMMON USAGE

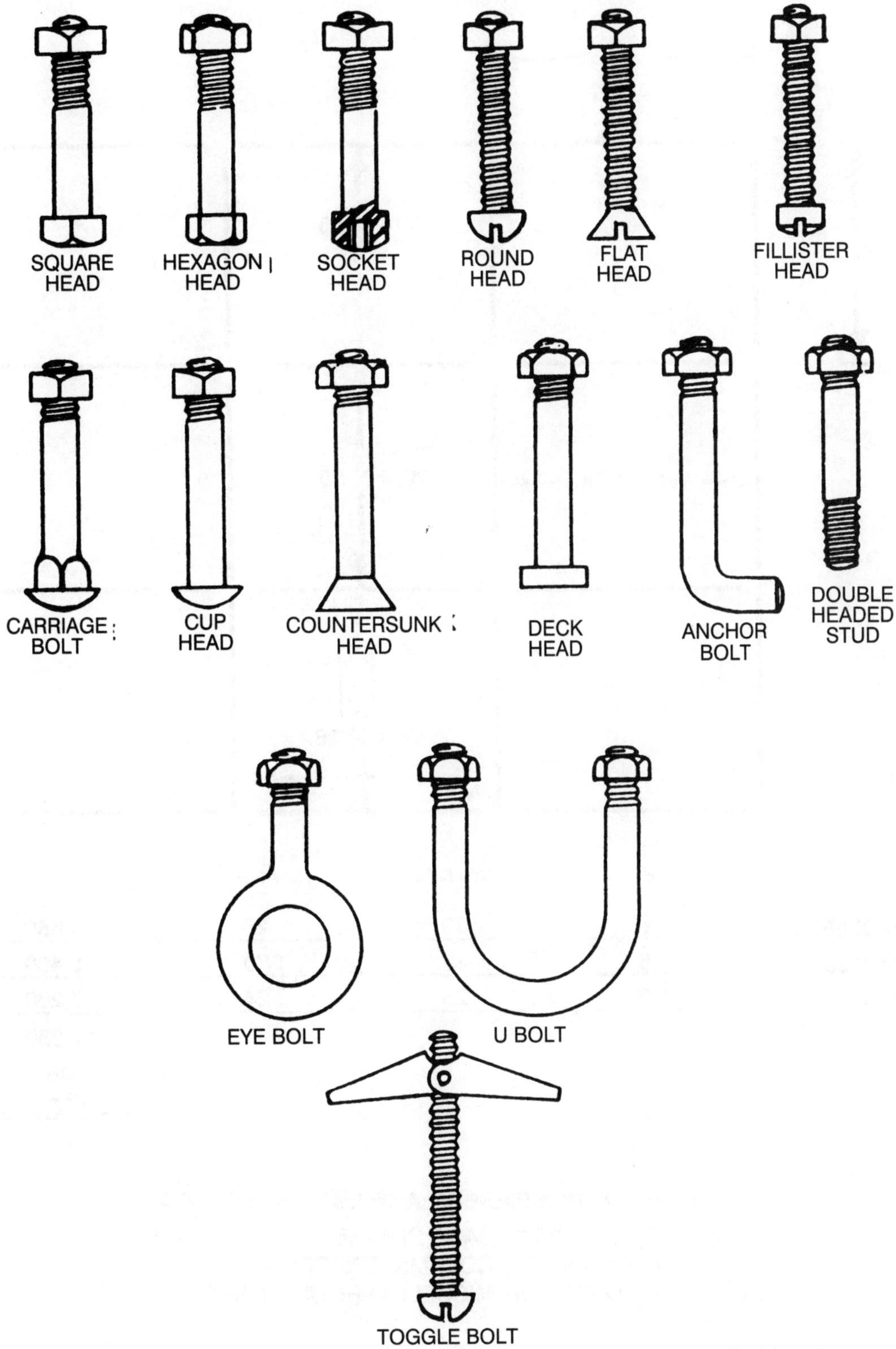

EXAMPLE OF SIMPLIFIED STRUCTURAL STEEL TAKEOFF METHOD

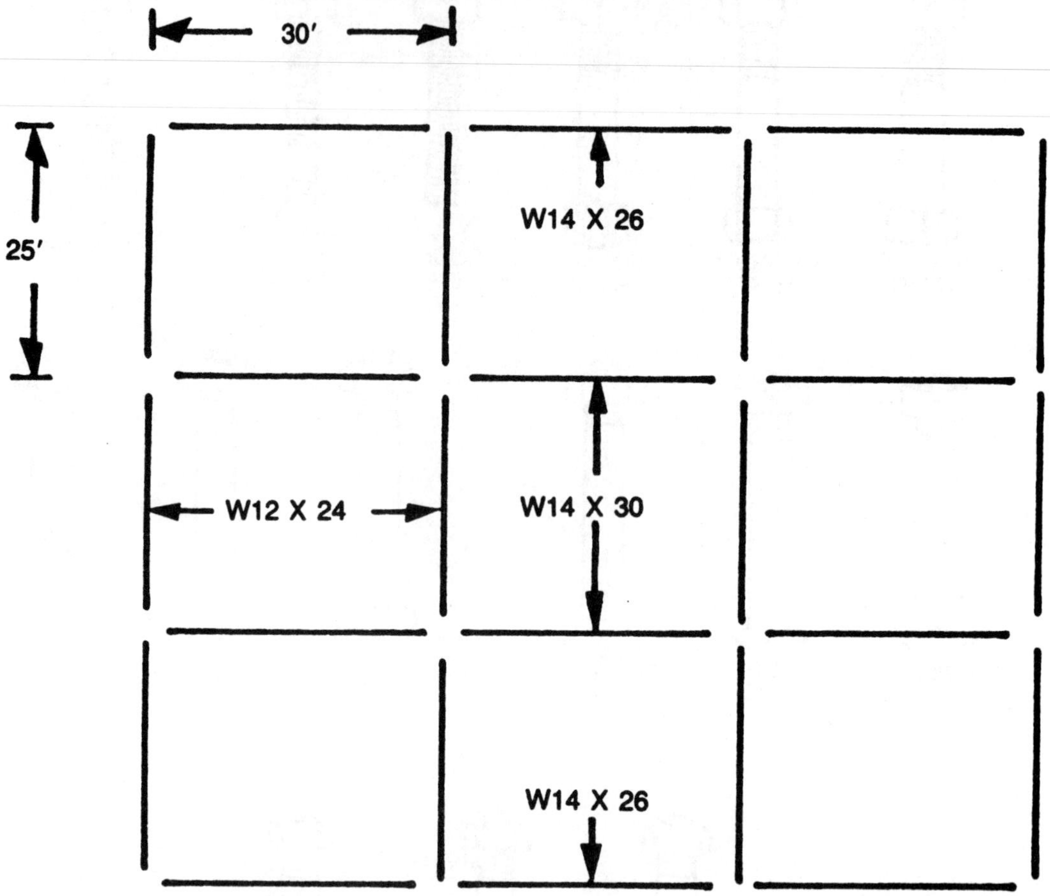

	#		LF. EA.		LBS./LF.			
W14 X 26	6	×	30	×	26	=	4,680	LBS.
W14 X 30	6	×	30	×	30	=	5,400	
W12 X 24	12	×	25	×	24	=	7,200	
							17,280	LBS.
							OR	
							9 ±	TONS

AFTER MAIN MEMBERS ARE ESTIMATED, ADD:

2 TO 3% FOR BASE PLATES
4 TO 5% FOR COLUMN SPLICES
4 TO 5% FOR MISCELLANEOUS COSTS

COMMON WIRE NAILS (ACTUAL SIZE)

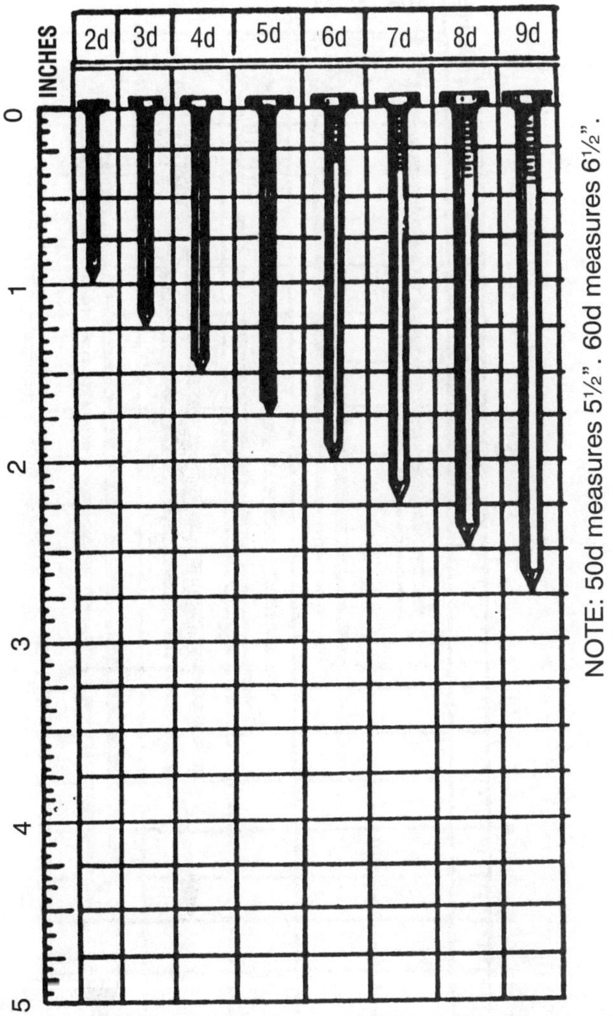

Cut Nails. Cut nails are angular-sided, wedge-shaped with ablunt point.

Wire Nails. Wire nails are round shafted, straight, pointed nails, and are used more generally than cut nails. They are stronger than cut nails and do not buckle as easily when driven into hard wood, but usually split wood more easily than cut nails. Wire nails are available in a variety of sizes varying from two penny to sixty penny.

Nail Finishes. Nails are available with special finishes, Some are galvanized or cadmium plated to resist rust. To increase the resistance to withdrawal, nails are coated with resins or asphalt cement (called cement coated). Nails which are small, sharp-pointed, and often placed in the craftsman's mouth (such as lath or plaster board nails) generally blued and sterilized.

COMMON WIRE NAILS (ACTUAL SIZE) (Cont.)

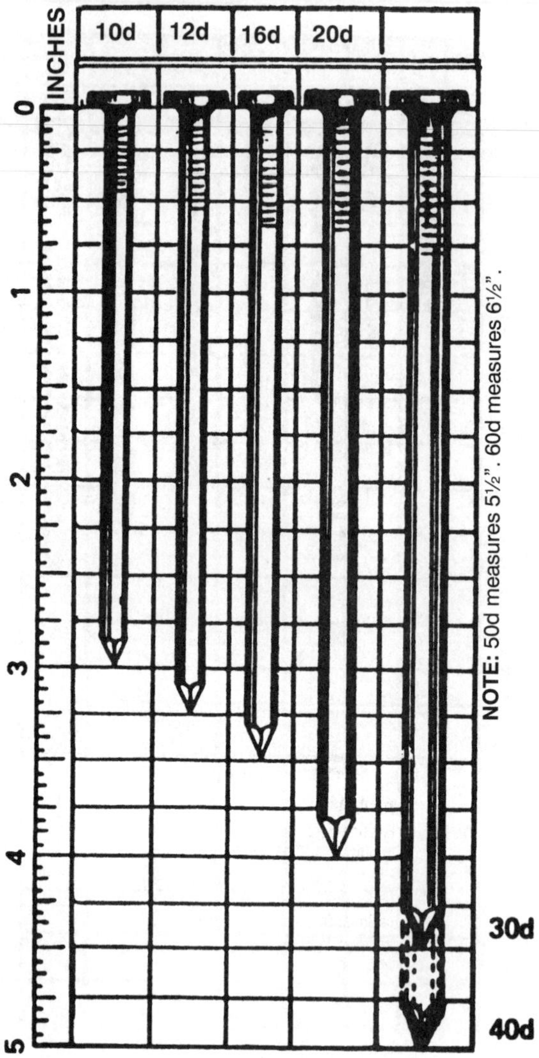

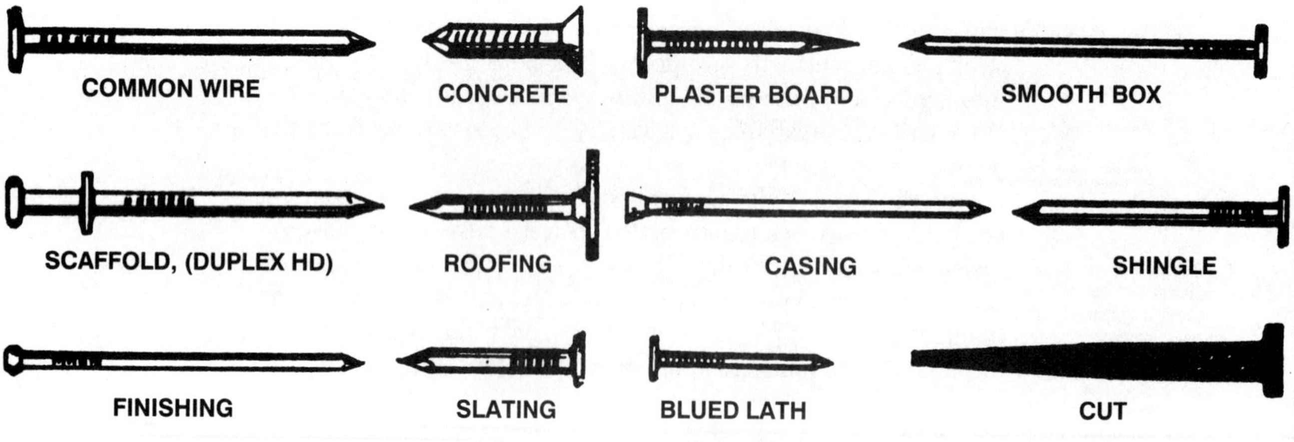

PLYWOOD — BASIC GRADE MARKS
AMERICAN PLYWOOD ASSOCIATION (APA)

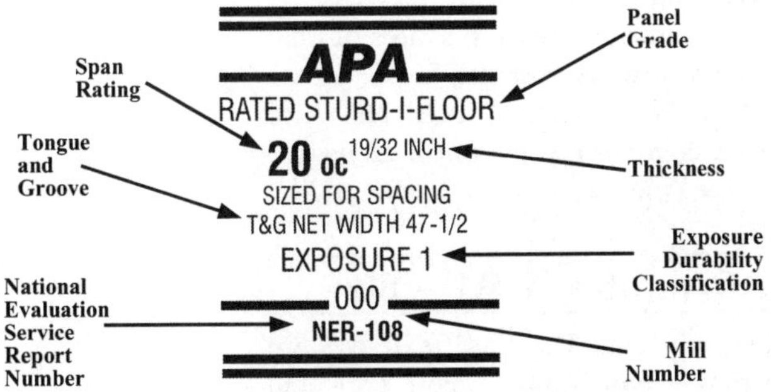

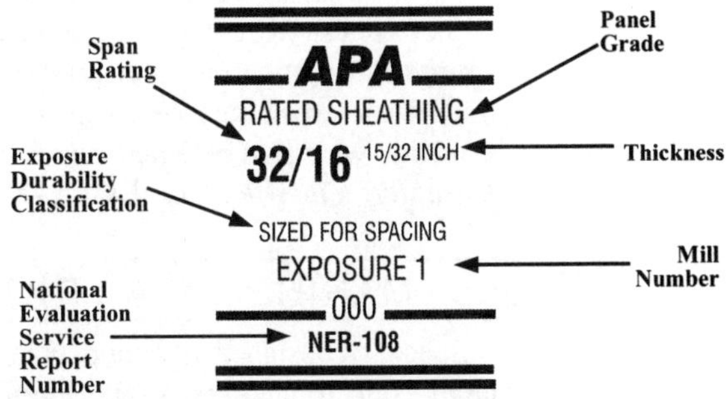

The American Plywood Association's trademarks appear only on products manufactured by APA member mills. The marks signify that the product is manufactured in conformance with APA performance standards and/or U.S. Product Standard PS 1-83 for Construction and Industrial Plywood.

APA
A-C GROUP 1
EXTERIOR
000
PS 1-83

APA A-C
For use where appearance of one side is important in exterior applications such as soffits, fences, structural uses, boxcar and truck linings, farm buildings, tanks, trays, commercial refrigerators, etc. **Exposure Durability Classification: Exterior. Common Thicknesses:** ¼, 11/32, ¾, 15/32, ½, 19/32, 5/8, 23/32, ¾.

APA
A-D GROUP 1
EXPOSURE 1
000
PS 1-83

APA A-D
For use where appearance of only one side is important in interior applications, such as paneling, built-ins, shelving, partitions, flow racks, etc. **Exposure Durability Classifications: Interior, Exposure 1. Common Thicknesses:** ¼, 11/32, 3/8, 15/32, ½, 19/32, 5/8, 23/32, ¾.

APA
B-C GROUP 1
EXTERIOR
000
PS 1-83

APA B-C
Utility panel for farm service and work buildings, boxcar and truck linings, containers, tanks, agricultural equipment, as a base for exterior coatings and other exterior uses. **Exposure Durability Classification: Exterior. Common Thicknesses:** ¼, 11/32, ¾, 15/32, ½, 19/32, 5/8, 23/32, ¾.

APA
B-D GROUP 2
INTERIOR
000
PS 1-83

APA B-D
Utility panel for backing, sides or built-ins, industry shelving, slip sheets, separator boards, bins and other interior or protected applications. **Exposure Durability Classifications: Interior, Exposure 1. Common Thicknesses:** ¼, 11/32, 3/8, 15/32, ½, 19/32, 5/8, 23/32, ¾.

APA
PLYFORM
B-B CLASS 1
EXTERIOR
000
PS 1-83

APA proprietary concrete form panels designed for high reuse. Sanded both sides and mill-oiled unless otherwise specified. Class I, the strongest, stiffest and more commonly available, is limited to Group 1 faces, Group 1 or 2 cross-bands, and Group 1, 2, 3 or 4 inner plies. Class II is limited to Group 1 or 2 faces (Group 3 under certain conditions) and Group 1, 2, 3 or 4 inner plies. Also available in HDO for very smooth concrete finish, in Structural I, and with special overlays. **Exposure Durability Classification: Exterior. Common Thicknesses:** 19/32, ¾, 23/32, ¼.

APA
M. D. OVERLAY
GROUP 1
EXTERIOR
000
PS 1-83

Plywood panel manufactured with smooth, opaque, resin-treated fiber overlay providing ideal base for paint on one or both sides. Excellent material choice for shelving, factory work surfaces, paneling, built-ins, signs and numerous other construction and industrial applications. Also available as a 303 Siding with texture-embossed or smooth surface on one side only and Structural I. **Exposure Durability Classification: Exterior. Common Thicknesses:** 11/32, 3/8, 19/32, ½, 19/32, 5/8, 23/32, ¾.

SPECIALTY PANELS

HDO · A · A · G-1 · EXT – APA · 000 · PS1 – 83

Plywood panel manufactured with a hard, semi-opaque resin-fiber overlay on both sides. Extremely abrasion resistant and ideally suited to scores of punishing construction and industrial applications, such as concrete forms, industrial tanks, work surfaces, signs, agricultural bins, exhaust ducts, etc. Also available with skid-resistant screen-grid surface and in Structural I. ***Exposure Durability Classification:*** **Exterior.** ***Common Thicknesses:*** **3/8, ½, 5/8, 3/4**

MARINE· A · A · EXT – APA · 000 · PS1 – 83

Specialty designed plywood panel made only with Douglas fir or western larch, solid jointed cores, and highly restrictive limitations on core gaps and faces repairs. Ideal for both hulls and other marine applications. Also available with HDO or MDO faces. ***Exposure Durability Classification:*** **Exterior.** ***Common Thicknesses:*** **1/4, 3/8, ½, 5/8, 3/4.**

Unsanded and touch-sanded panels, and panels with "B" or better veneer on one side only, usually carry the APA trademark on the panel back. Panels with both sides of "B" or better veneer, or with special overlaid surfaces (such as Medium Density Overlay), carry the APA trademark on the panel edge, like this:

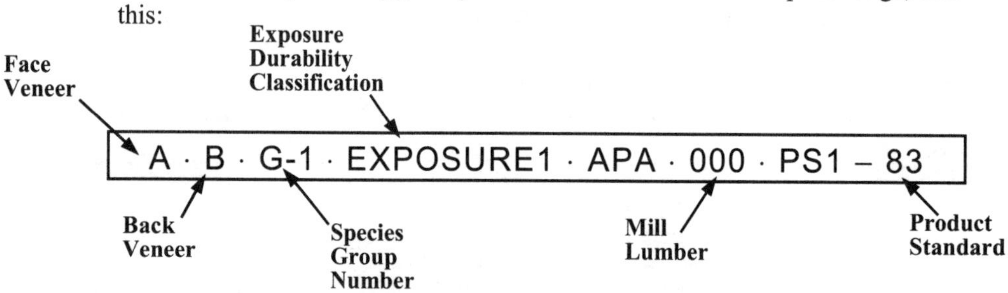

GLOSSARY OF TERMS

Some of the words and terms used in the grading of lumber follow:

Bow. A deviation flatwise from a straight line drawn from end to end of the piece. It is measured at the point of greatest distance from the straight line.

Checks. A separation of the wood which normally occurs across the annual rings and usually as a result of seasoning.

Crook. A deviation edgewise from a straight line drawn from end to end of the piece. It is measured at the point of greatest distance from the straight line.

Cup. A deviation from a straight line drawn across the piece from edge to edge. It is measured at the point of greatest distance from the straight line.

Flat Grain. The annual growth rings pass through the piece at an angle of less than 45 degrees with the flat surface of the piece.

Warp. Any deviation from a true or plane surface, including crook, cup, bow or any combination thereof.

Mixed Grain. The piece may have vertical grain, flat grain or a combination of both vertical and flat grain.

Pitch. An accumulation of resin which occurs in separations in the wood or in the wood cells themselves.

Shake. A separation of the wood which usually occurs between the rings of annual growth.

Splits. A separation of the wood due to tearing apart of the wood cells.

Vertical Grain. The annual growth rings pass through the piece at an angle of 45 degrees or more with the flat surface of the piece.

Wane. Bark or lack of wood from any cause, except eased edges (rounded) on the edge or corner of a piece of lumber.

LUMBER GRADING

GRADING-MARK ABBREVIATIONS

GRADES

(Listed alphabetically — not by quality)

COM	Common
CONST	Construction
ECON	Economy
No. 1	Number One
SEL-MER	Select Merchantable
SEL-STR	Select Structural
STAN	Standard
UTIL	Utility

ALSC TRADEMARKS

CLIS	California Lumber Inspection Service
NELMA	Northeastern Lumber Mfrs. Assoc., Inc.
NH&PMA	Northern Hardwood & Pine Mfrs. Assoc., Inc.
PLIB	Pacific Lumber Inspection Bureau
RIS	Redwood Inspection Service
SPIB	Southern Pine Inspection Bureau
TP	Timber Products Inspection
WCLB	West Coast Lumber Inspection Bureau
WWP	Western Wood Products Association

SPECIES GROUPINGS

AF	Alpine Fir
DF	Douglas Fir
HF	Hem Fir
SP	Sugar Pine
PP	Pondoroca Pipo
LP	Lodgepole Pine
IWP	Idaho White Pine
ES	Engelmann Spruce
WRC	Western Red Cedar
INC CDR	Incense Cedar
L	Larch
LP	Lodgepole Pine
MH	Mountain Hemlock
WW	White Wood

MOISTURE CONTENT

S-GRN	Surfaced at a moisture content of more than 19%.
S-DRY	Surfaced at a moisture content of 19% or less.
MC-15	Surfaced at a moisture content of 15% or less.

FRAMING ESTIMATING RULES OF THUMB

For 16” O.C. stud partitions figure 1 stud for every L.F. of wall; add for top and bottom plates.

For any type of framing, the quantity of basic framing members (in L.F.) can be determined based on spacing and surface area (S.F.):

12” O.C.	1.2 L.F./S.F.
16” O.C.	1.0 L.F./S.F.
24” O.C.	0.8 L.F./S.F.

(Doubled-up members, bands, plates, framed openings, etc., must be added.)

Framing accessories, nails, joist hangers, connectors, etc., should be estimated as separate material costs. Installation should be included with framing. Rule of thumb allowance is 0.5 to 1.5% of lumber cost for rough hardware. Another is 30 to 40 pounds of nails per M.B.F.

BOARD FEET/LINEAR FEET FOR LUMBER

Nominal Size	Actual Size	Board Feet Per Linear Foot	Linear Feet Per 1000 Board Feet
1 × 2	¾ × 1 ½	.167	6000
1 × 3	¾ × 2 ½	.250	4000
1 × 4	¾ × 3 ½	.333	3000
1 × 6	¾ × 5 ½	.500	2000
1 × 8	¾ × 7 ¼	.666	1500
1 × 10	¾ × 9 ¼	.833	1200
1 × 12	¾ × 11 ¼	1.0	1000
2 × 2	1 ½ × 1 ½	.333	3000
2 × 3	1 ½ × 2 ½	.500	2000
2 × 4	1 ½ × 3 ½	.666	1500
2 × 6	1 ½ × 5 ½	1.0	1000
2 × 8	1 ½ × 7 ¼	1.333	750
2 × 10	1 ½ × 9 ¼	1.666	600
2 × 12	1 ½ × 11 ¼	2.0	500

Redwood. Redwood is a fairly strong and moderately lightweight material. The heartwood is red but the sapwood is white. One of the principal advantages of redwood is that the heartwood is highly resistant (but not entirely immune) to decay, fungus and insects. Standard Specifications require that all redwood used in permanent installations shall be "select heart." Grade marking shall be in accordance with the standards established in the California Redwood Association. Grade marking shall be done by, or under the supervision of the Redwood Inspection Service.

Redwood is graded for specific uses as indicated in the following table:

REDWOOD GRADING

Type of Lumber	Grade	Typical Use
Grades for Dimension Only Listed Here	Clear All Heart	Exceptionally fine, knot free, straight-grained timbers. This grade is used primarily for stain finish work of high quality.
	Clear	Same as Clear All Heart except that this grade may contain sound sapwood and medium stain.
	Select Heart	**This grade only is to be used in Agency work, unless otherwise specified in the plans or specifications.** It is sound, live heartwood free from splits or streaks with sound knots. It is generally used where the timber is in contact with the ground, as in posts, mudsills, etc.
	Select Construction Heart	Slightly less quality than Select Heart. It may have some sapwood in the piece. Used for general construction purposes when redwood is needed.
	Construction Common	Same requirement as Construction Heart except that it will contain sapwood and medium stain. Its resistance to decay and insect attack is reduced.
	Merchantable	Used for fence posts, garden stakes, etc.
	Economy	Suitable for crating, bracing and temporary construction.

DOUGLAS FIR GRADING

Type of Lumber	Grade	Typical Use
Select Structural Joists and Planks	Select Structural	Used where strength is the primary consideration, with appearance desirable.
	No. 1	Used where strength is less critical and appearance not a major consideration.
	No. 2	Used for framing elements that will be covered by subsequent construction.
	No. 3	Used for structural framing where strength is required but appearance is not a factor.
Finish Lumber	Superior Prime E	For all types of uses as casings, cabinet, exposed members, etc., where a fine appearance is desired.
Boards (WCLIB)* * Grading is by West Coast Lumber Inspection Bureau rules, but sizes conform to Western Wood Products Assn. rules. These boards are still manufactured by some mills.	Select Merchantable	Intended for use in housing and light construction where a knotty type of lumber with finest appearance is required.
	Construction	Used for sub-flooring, roof and wall sheathing, concrete forms, etc. Has a high degree of serviceability.
	Standard	Used widely for general construction purposes, including subfloors, roof and wall sheathing, concrete forms, etc. Seldom used in exposed construction because appearance.
	Utility	Used in general construction where low cost is a factor and appearance is not important. (Storage shelving, crates, bracing, temporary scaffolding etc.)

BOARD FEET CONVERSION TABLE

Nominal Size (In.)	ACTUAL LENGTH IN FEET								
	8	10	12	14	16	18	20	22	24
1 x 2		1 2/3	2	2 1/3	2 2/3	3	3 ½	3 2/3	4
1 x 3		2 ½	3	3 ½	4	4 ½	5	5 ½	6
1 x 4	2 ¾	3 1/3	4	4 2/3	5 1/3	6	6 2/3	7 1/3	8
1 x 5		4 1/6	5	5 5/6	6 2/3	7 ½	8 1/3	9 1/6	10
1 x 6	4	5	6	7	8	9	10	11	12
1 x 7		5 5/8	7	8 1/6	9 1/3	10 ½	11 2/3	12 5/6	14
1 x 8	5 1/3	6 2/3	8	9 1/3	10 2/3	12	13 1/3	14 2/3	16
1 x 10	6 2/3	8 1/3	10	11 2/3	13 1/3	15	16 2/3	18 1/3	20
1 x 12	8	10	12	14	16	18	20	22	24
1¼ x 4		4 1/6	5	5 5/6	6 2/3	7 ½	8 1/3	9 1/6	10
1¼ x 6		6 ¼	7 ½	8 ¾	10	11 ¼	12 ½	13 ¾	15
1¼ x 8		8 1/3	10	11 2/3	13 1/3	15	16 2/3	18 1/3	20
1¼ x 10		10 5/12	12 ½	14 7/12	16 2/3	18 ¾	20 5/6	22 11/12	25
1¼ x 12		12 ½	15	17 ½	20	22 ½	25	27 ½	30
1½ x 4	4	5	6	7	8	9	10	11	12
1½ x 6	6	7 ½	9	10 ½	12	13 ½	15	16 ½	18
1½ x 8	8	10	12	14	16	18	20	22	24
1½ x 10	10	12 ½	15	17 ½	20	22 ½	25	27 ½	30
1½ x 12	12	15	18	21	24	27	30	33	36
2 x 4	5 1/3	6 2/3	8	9 1/3	10 1/3	12	13 1/3	14 2/3	16
2 x 6	8	10	12	14	16	18	20	22	24
2 x 8	10 2/3	13 1/3	16	18 2/3	21 1/3	24	26 2/3	29 1/3	32
2 x 10	13 1/3	16 2/3	20	23 1/3	26 2/3	30	33 1/3	36 2/3	40
2 x 12	16	20	24	28	32	36	40	44	48
3 x 6	12	15	18	21	24	27	30	33	36
3 x 8	16	20	24	28	32	36	40	44	48
3 x 10	20	25	30	35	40	45	50	55	60
3 x 12	24	30	36	42	48	54	60	66	72
4 x 4	10 2/3	13 1/3	16	18 2/3	21 1/3	24	26 2/3	29 1/3	32
4 x 6	16	20	24	28	32	36	40	44	48
4 x 8	21 1/3	26 2/3	32	37 1/3	42 2/3	48	53 1/3	58 2/3	64
4 x 10	26 2/3	33 1/3	40	46 2/3	53 1/3	60	66 2/3	73 1/3	80
4 x 12	32	40	48	56	64	72	80	88	96

MOISTURE PROTECTION / ROOFING 07000

SLOPE AREA CALCULATIONS

Rise and Run	Multiply Flat Area by	LF of Hips or Valleys per LF of Common Run
2 in 12	1.014	1.424
3 in 12	1.031	1.436
4 in 12	1.054	1.453
5 in 12	1.083	1.474
6 in 12	1.118	1.500
7 in 12	1.158	1.530
8 in 12	1.202	1.564
9 in 12	1.250	1.600
10 in 12	1.302	1.641
11 in 12	1.357	1.685
12 in 12	1.413	1.732

MOISTURE PROTECTION / DOWNSPOUTS 07600

DOWNSPOUT/VERTICAL LEADER CALCULATIONS

Roof Type	Slope	S.F. Roof/ Sq. In. Leader
Gravel	Less than ¼" per foot	300
Gravel	Greater than ¼" per foot	250
Metal or Shingle	Any	200

Alternate calculations:

$$\text{Diameter of downspout/leader} = 1.128\sqrt{\frac{\text{Area of drainage}}{\text{SF Roof/Sq. Inch}}}$$

TYPICAL MINIMUM SIZE OF VERTICAL CONDUCTORS AND LEADERS

Size of leader or conductor (Inches)	Maximum projected roof area (Square feet)
2	544
2 ½	987
3	1,610
4	3,460
5	6,280
6	10,200
8	22,000

TYPICAL MINIMUM SIZE OF ROOF GUTTERS

	MAXIMUM PROJECTED ROOF AREA FOR GUTTERS OF VARIOUS SLOPES			
Diameter gutter (Inches)	1/16 in. Ft. slope (Sq. ft.)	1/8 in. per Ft. slope (Sq. ft.)	¼ in. per Ft. slope (Sq. ft.)	1/2 in. per Ft. /slope (Sq. ft.)
3	170	240	340	480
4	360	510	720	1,020
5	625	880	1,250	1,770
6	960	1,360	1,920	2,770
7	1,380	1,950	2,760	3,900
8	1,990	2,800	3,980	5,600
10	3,600	5,100	7,200	10,000

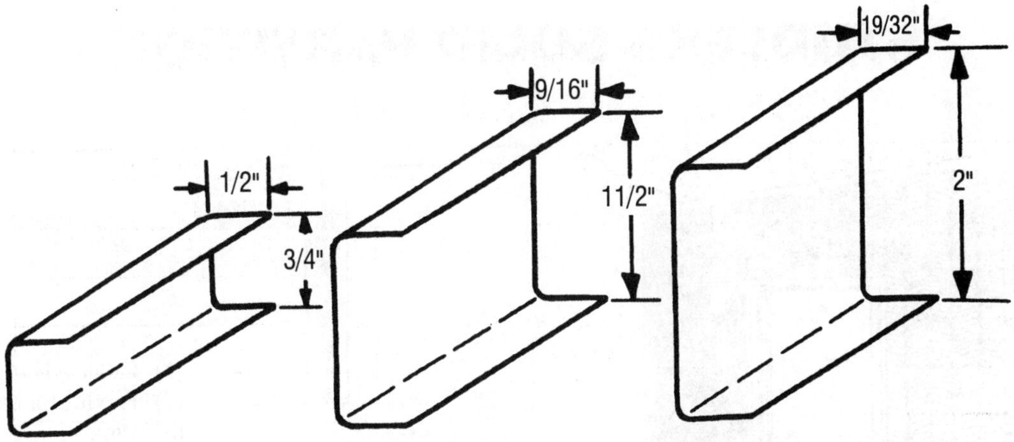

COLD ROLLED CHANNELS

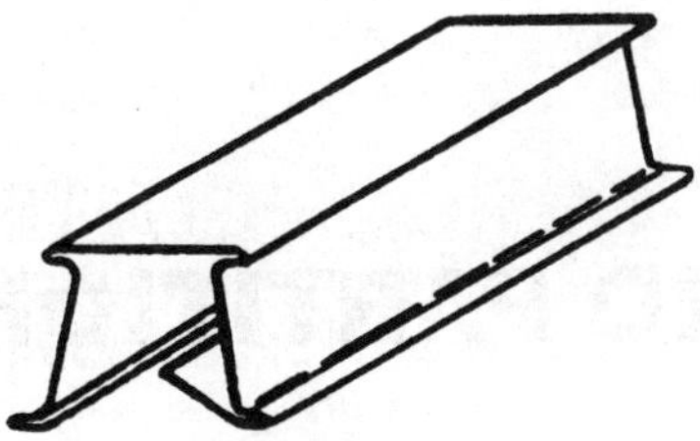

NAILING CHANNEL

SCREW CHANNEL

CHANNEL STUD

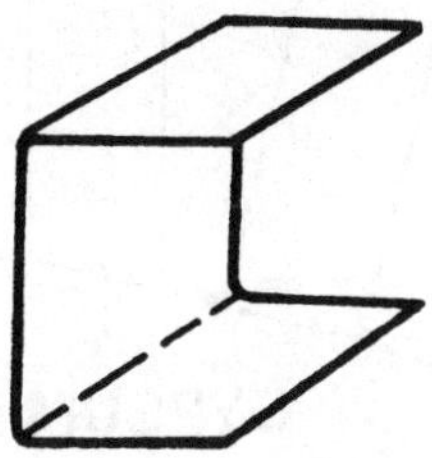

WIDE FLANGE CHANNEL

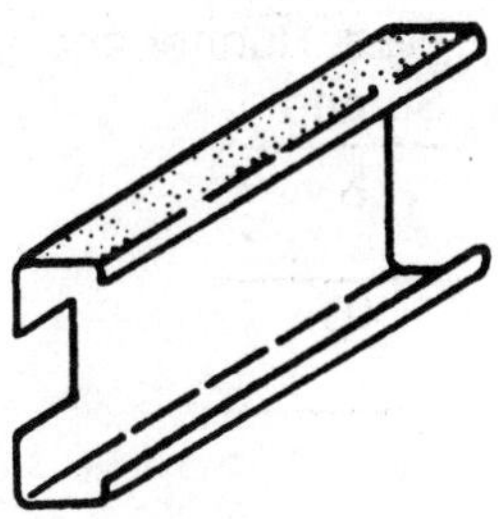

CEE STUD

STUDLESS SOLID PARTITION

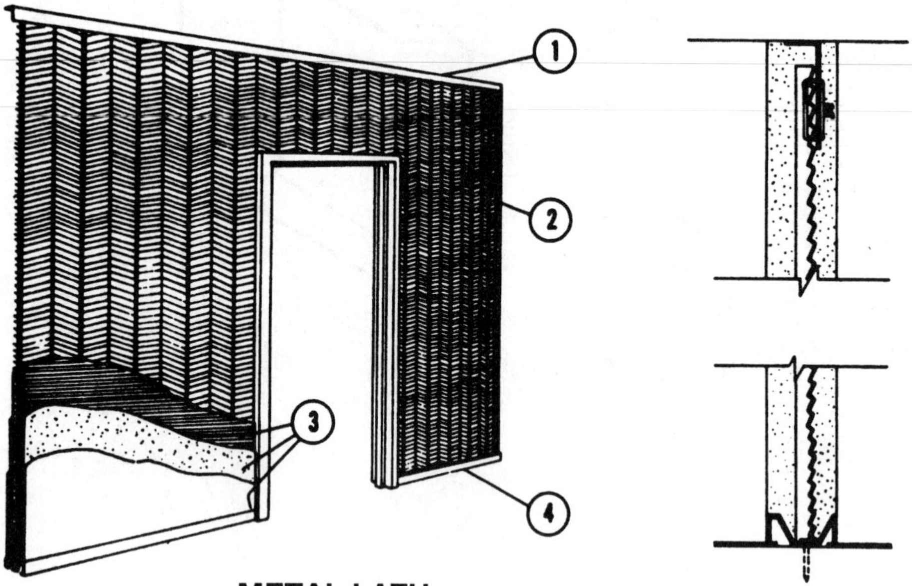

(1) Ceiling Runner
(2) Rib Metal Lath
(3) Plaster
(4) Combination Floor Runner and Screed

METAL LATH

STUDLESS SOLID PARTITION

(1) Ceiling Runner
(2) Long length Gypsum Lath
(3) Plaster
(4) Combination Floor Runner and Screed

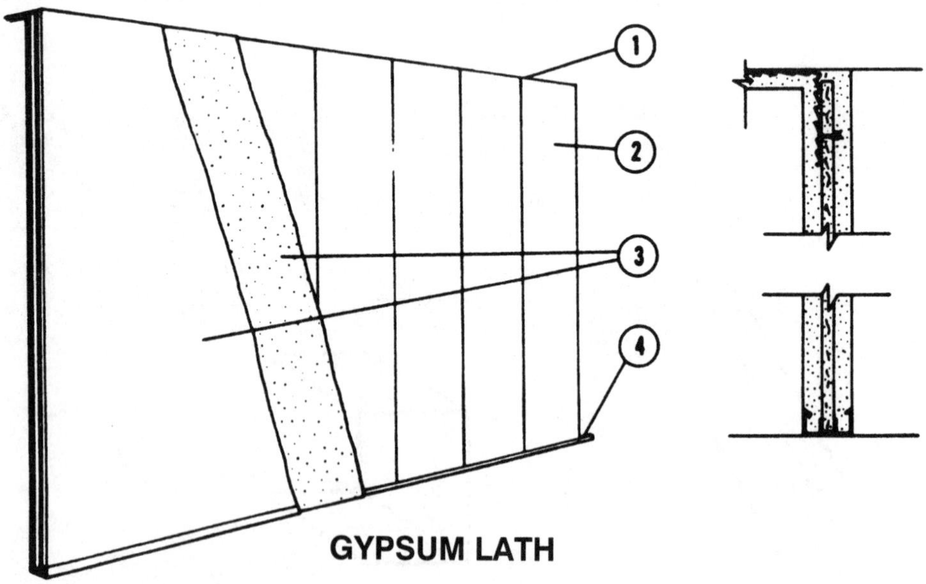

GYPSUM LATH

SUSPENDED CEILINGS

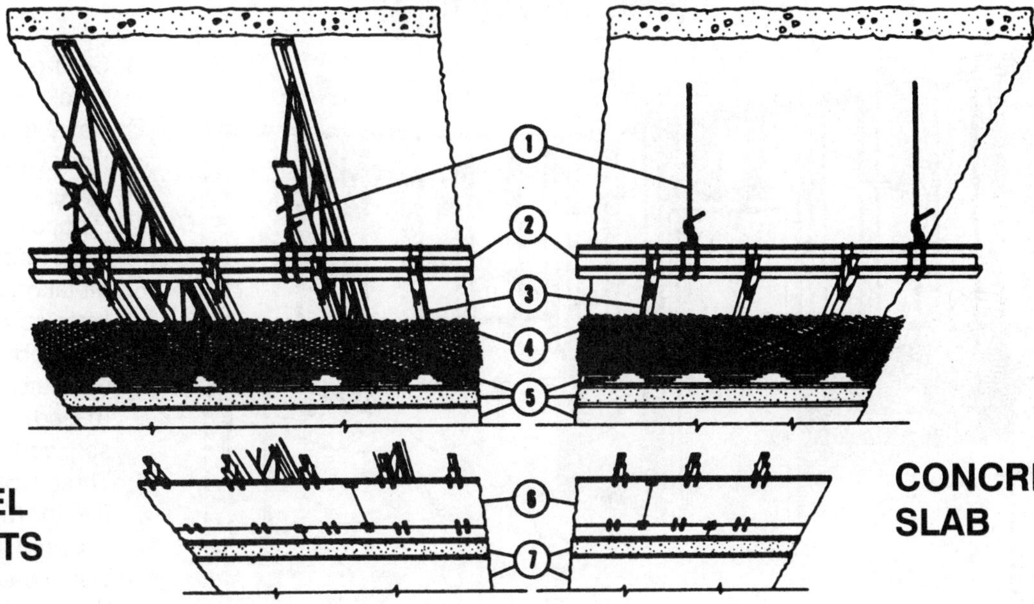

STEEL JOISTS

CONCRETE SLAB

(1) Hanger
(2) Main Runner Channel
(3) Furring Channel
(4) Metal or Wire Fabric Lath
(5) Plaster
(6) Gypsum Lath
(7) Plaster

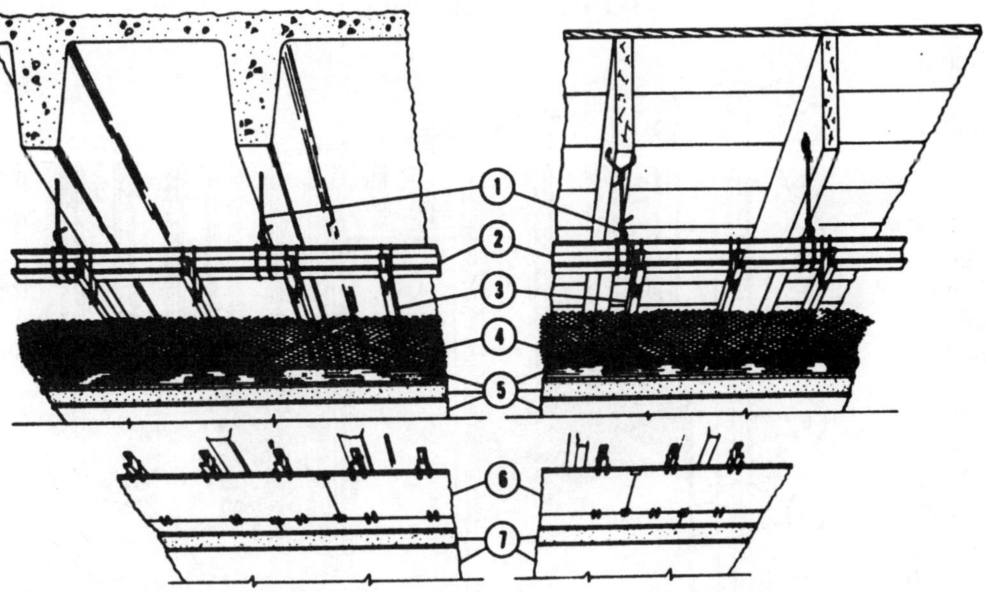

CONCRETE JOISTS

WOOD JOISTS

STEEL STUD

Hollow Partition

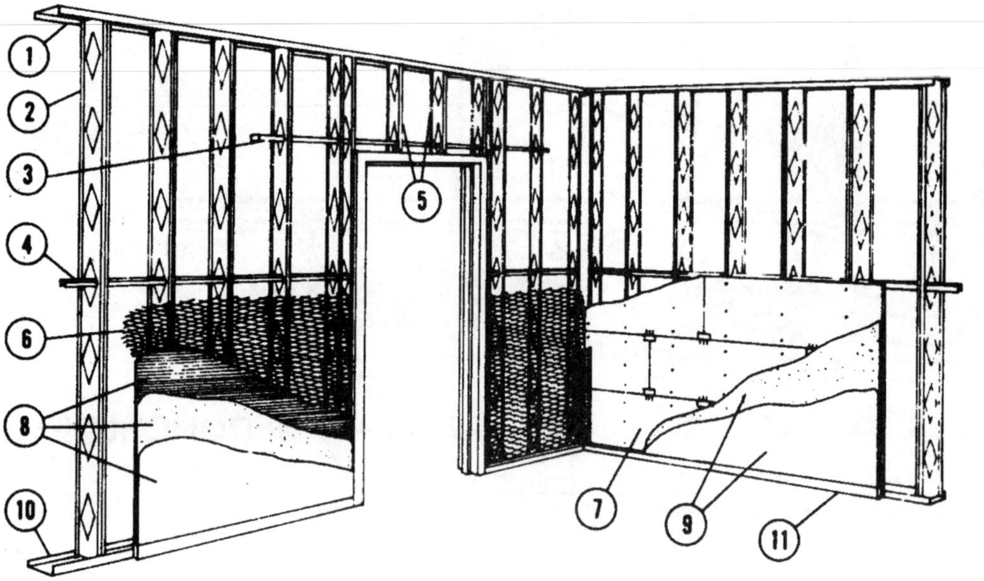

(1) Ceiling Runner Track
(2) Nailable Stud
(3) Door Opening Stiffener
(4) Partition Stiffener
(5) Jack Studs
(6) Metal or Wire Fabric Lath (screwed or wire tied)
(7) Gypsum Lath (nailed, clipped or screwed)
(8) Three Coats of Plaster (Scratch, Brown, Finish)
(9) Two Coats of Plaster (Brown, Finish)
(10) Floor Runner Track
(11) Flush Metal Base

SCREW STUD

Hollow Partition

(1) Ceiling Runner Track
(2) Screwed Stud
(3) Door Opening Stiffener
(4) Partition Stiffener
(5) Jack Studs
(6) Metal or Wire Fabric Lath (screwed on)
(7) Gypsum Lath (screwed on)
(8) Three Coats of Plaster (Scratch, Brown, Finish)
(9) Two Coats of Plaster (Brown, Finish)
(10) Floor Runner Track
(11) Flush Metal Base

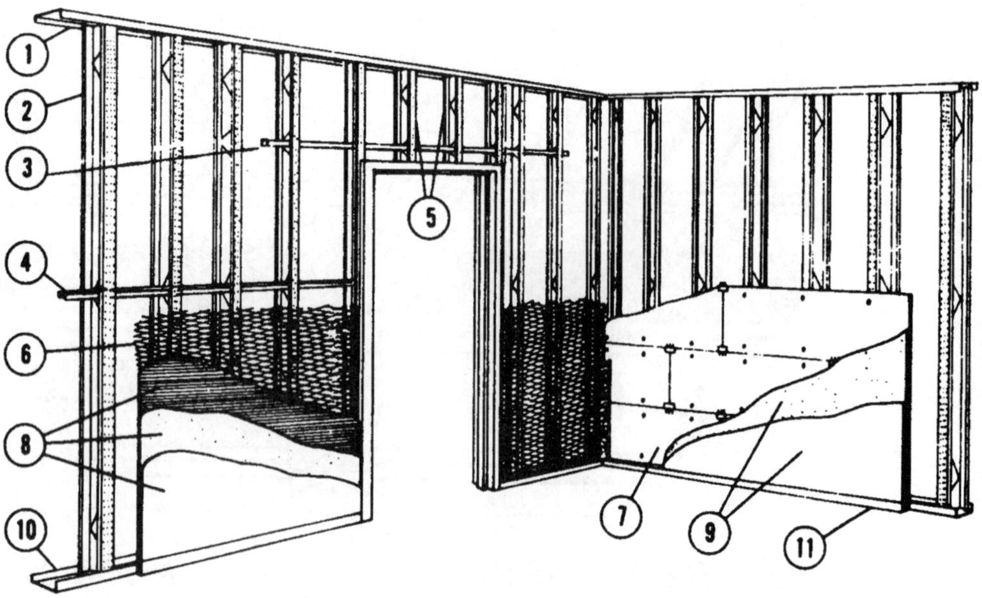

LOAD-BEARING HOLLOW PARTITION

Structural Stud

(1) Ceiling Runner Track
(2) Structural Studs (prefabricated)
(3) Structural Stud (nailable)
(4) Jack Studs
(5) Partition Stiffener (bridging)
(6) Metal or Wire Fabric Lath (wired-tied, nailed or stapled)
(7) Gypsum Lath (nailed or stapled)
(8) Three Coats of Plaster (Scratch, Brown, Finish)
(9) Two Coats of Plaster (Brown, Finish)

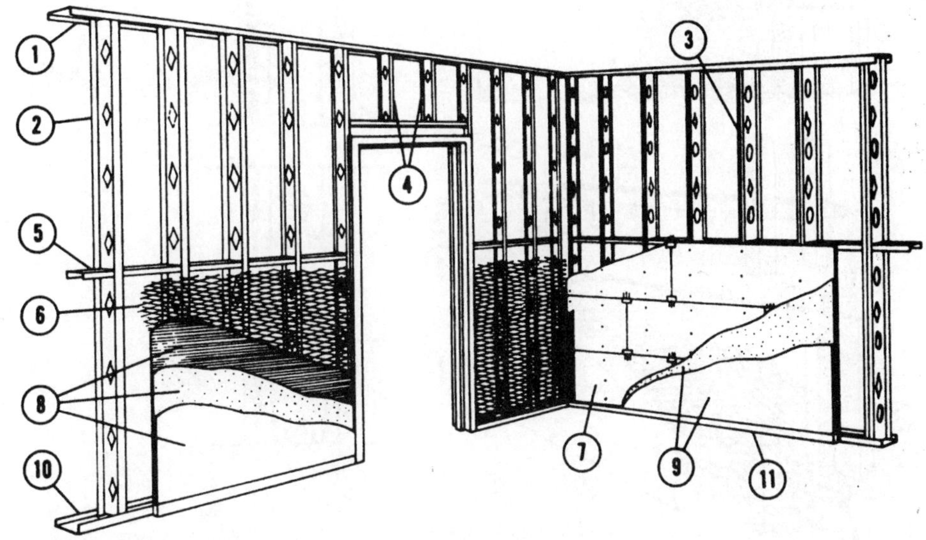

(10) Floor Runner Track (11) Flush Metal Base

VERTICAL FURRING

With Studs

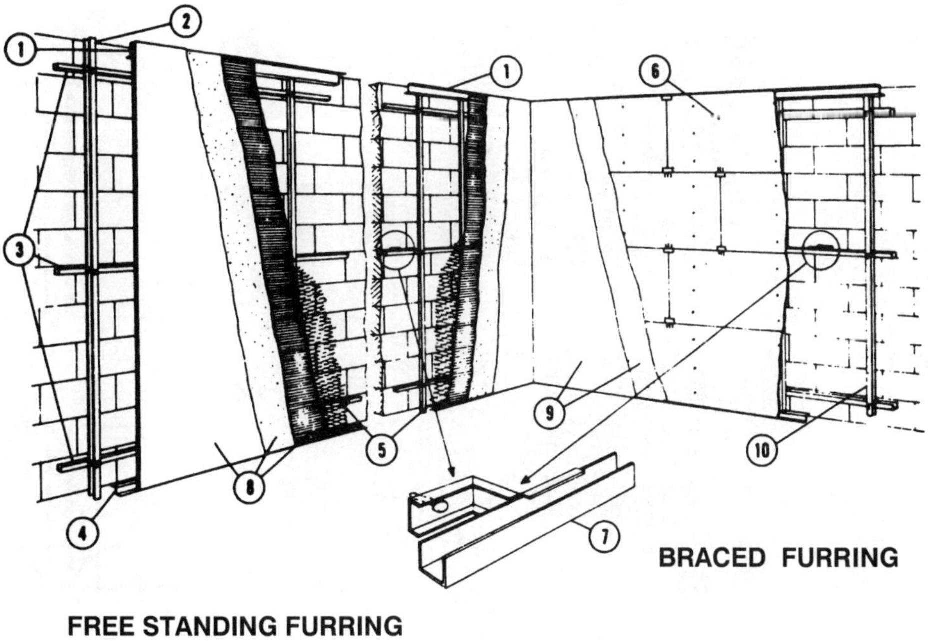

(1) Ceiling Runner Track
(2) Channel Studs
(3) Horizontal Stiffener
(4) Floor Runner
(5) Metal or wire Fabric Lath
(6) Gypsum Lath
(7) Bracing
(8) Three Coats of Plaster (Scratch, Brown, Finish)
(9) Two Coats of Plaster (Brown, Finish)
(10) Screw Channel Studs

DOUBLE CHANNEL STUD

Hollow Partition

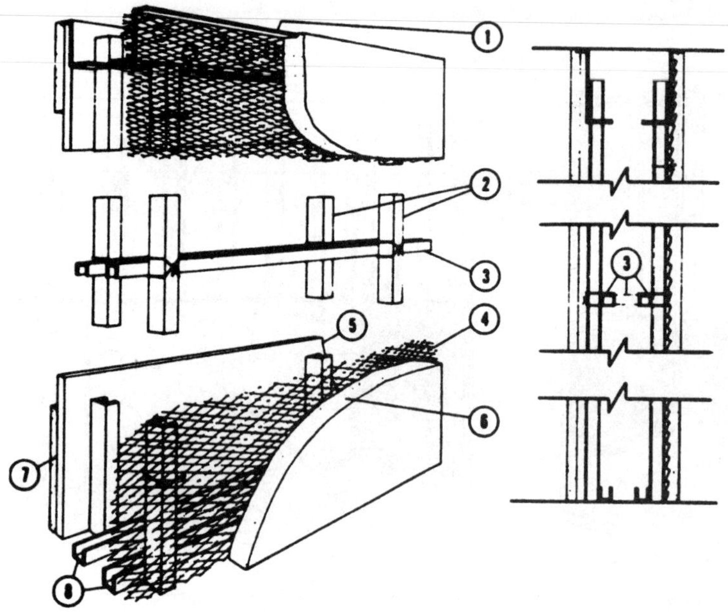

(1) Ceiling Runner
(2) Channel Studs
(3) Partition Stiffener and Channel Spacer
(4) Metal or Wire Fabric Lath (wired-tied)
(5) Gypsum Lath (clipped on)
(6) Three Coats of Plaster (Scratch, Brown, Finish)
(7) Two Coats of Plaster (Brown, Finish)
(8) Floor Runners (channel)

SINGLE CHANNEL STUD

Solid Partition

(1) Ceiling Runner
(2) Channel Stud
(3) Metal or Wire Fabric Lath (wired-tied)
(4) Plaster
(5) Combination Floor Runner and Screed

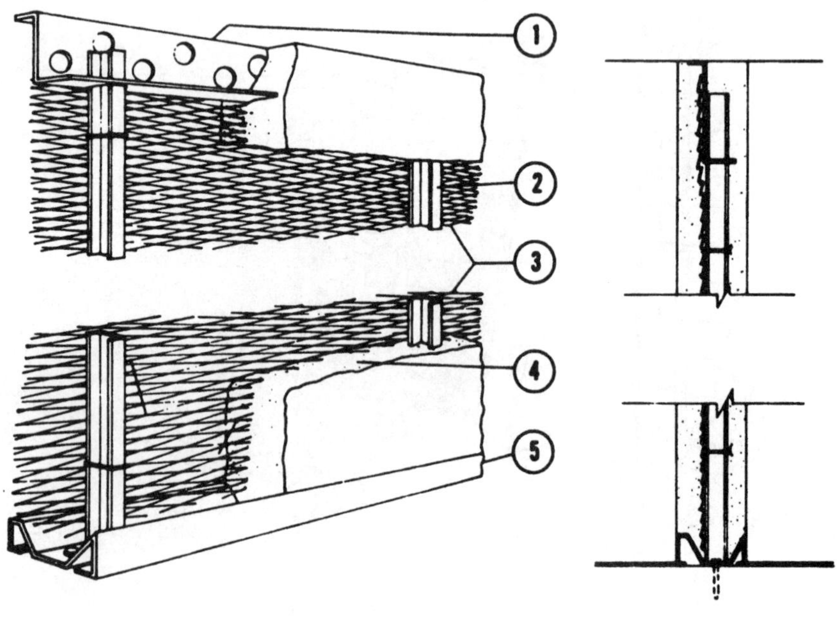

FURRED CEILINGS

STEEL JOISTS

CONCRETE JOISTS

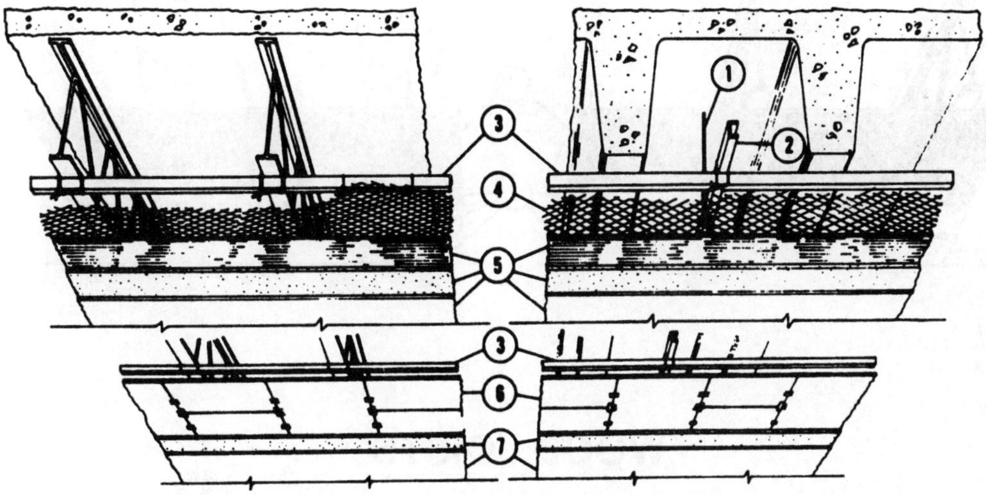

WOOD JOISTS

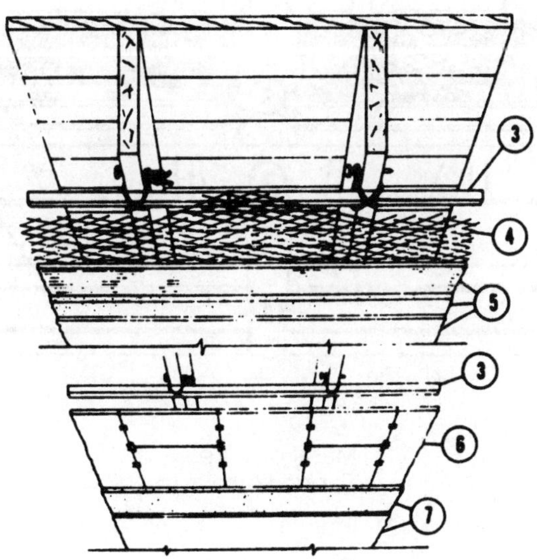

(1) Hanger
(2) ¾-inch Channel
(3) Cross Furring
(4) Metal or Wire Fabric Lath
(5) Plaster
(6) Gypsum Lath
(7) Plaster

CEILINGS

STEEL JOISTS   CONCRETE JOISTS

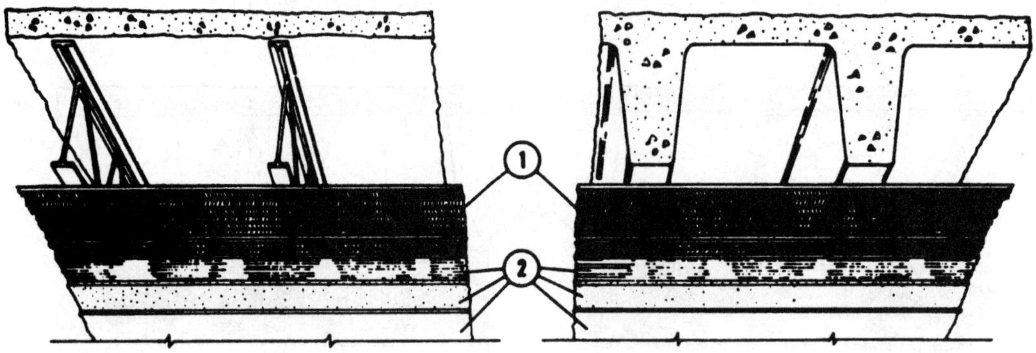

WOOD JOISTS

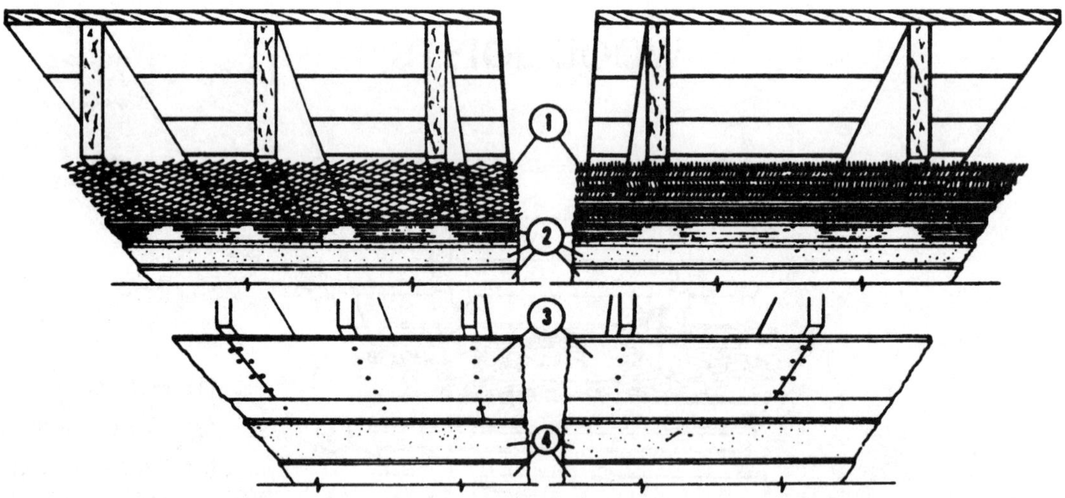

(1) Metal or Wire Fabric Lath

(2) Plaster

(3) Gypsum Lath

(4) Plaster

FINISHES / PLASTER ACCESSORIES | 09205

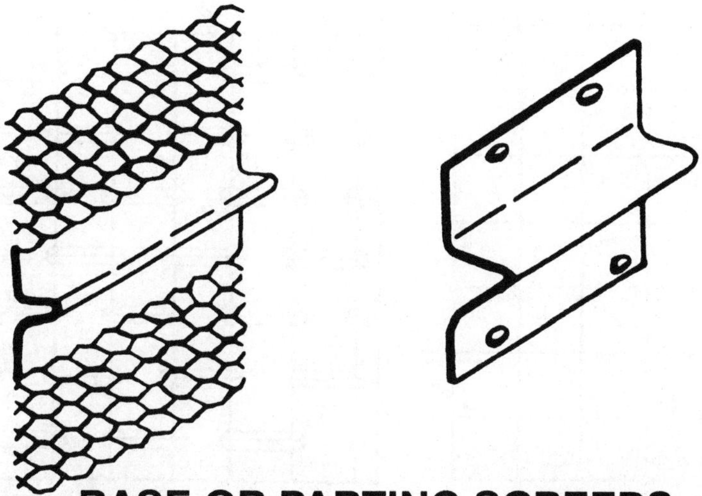

BASE OR PARTING SCREEDS

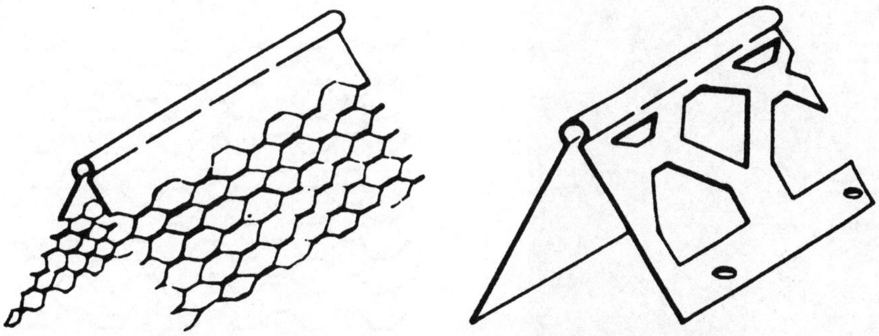

SMALL NOSE CORNER BEADS

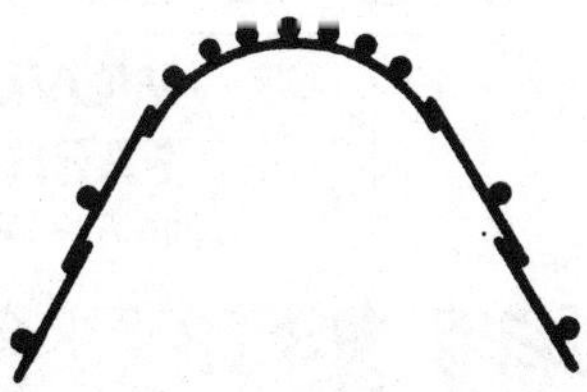

WIRE BULL NOSE CORNER BEADS

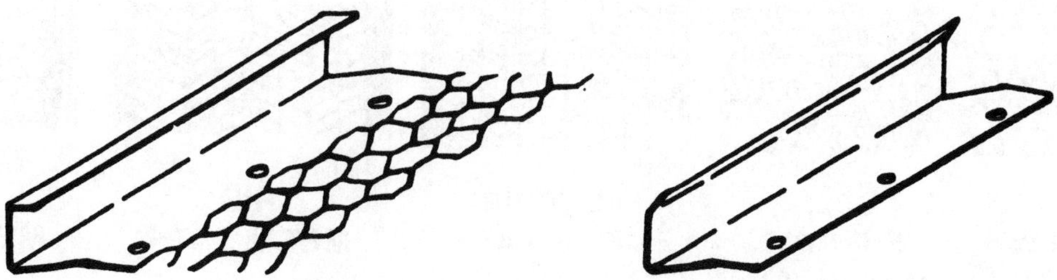

SQUARE CASING BEADS

FINISHES / METAL LATH

09205

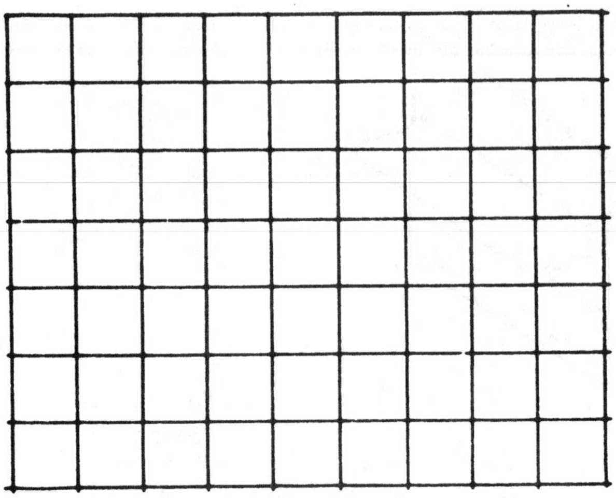

PLAIN WIRE
FABRIC LATH

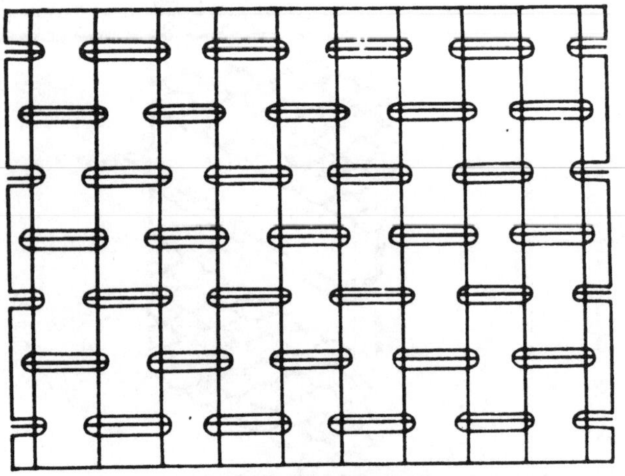

SELF-FURRING
WIRE FABRIC LATH

PAPER BACKED
WOVEN WIRE
FABRIC LATH

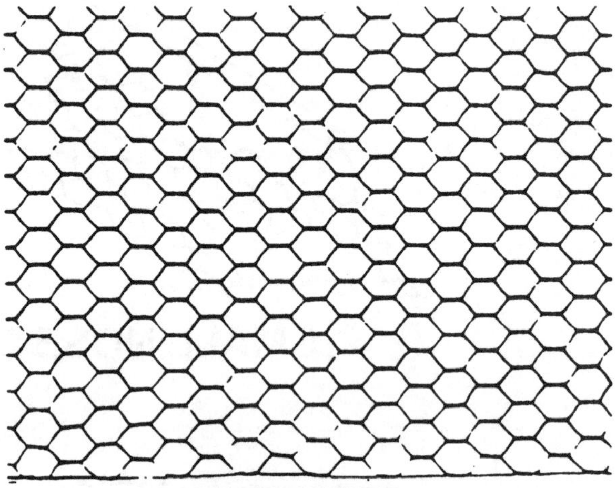

WOVEN WIRE
FABRIC LATH
(Also Available Self-Furred)

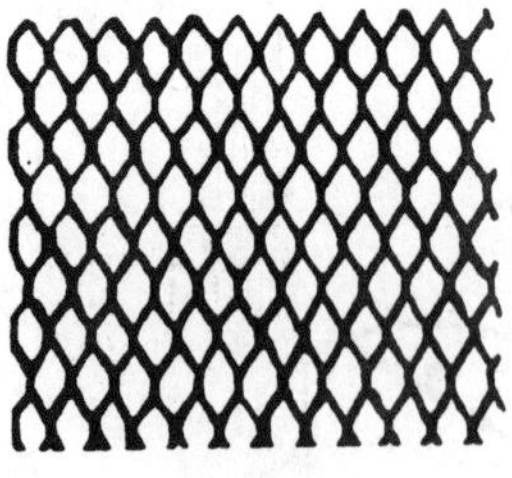

FLAT
DIAMOND MESH
METAL LATH

SELF-
FURRING
METAL LATH

FLAT RIB
METAL LATH

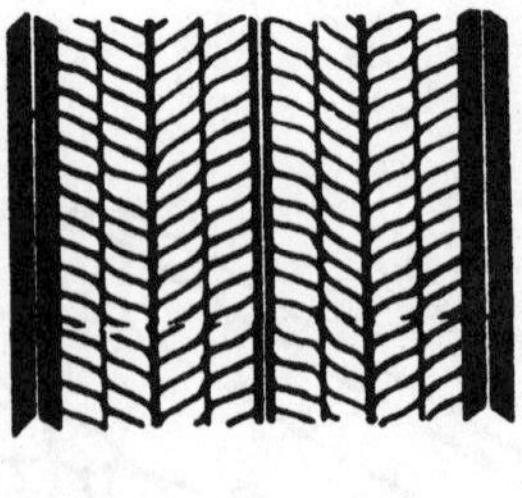

RIB
METAL LATH

RIB
METAL LATH

VERTICAL FURRING

Studless

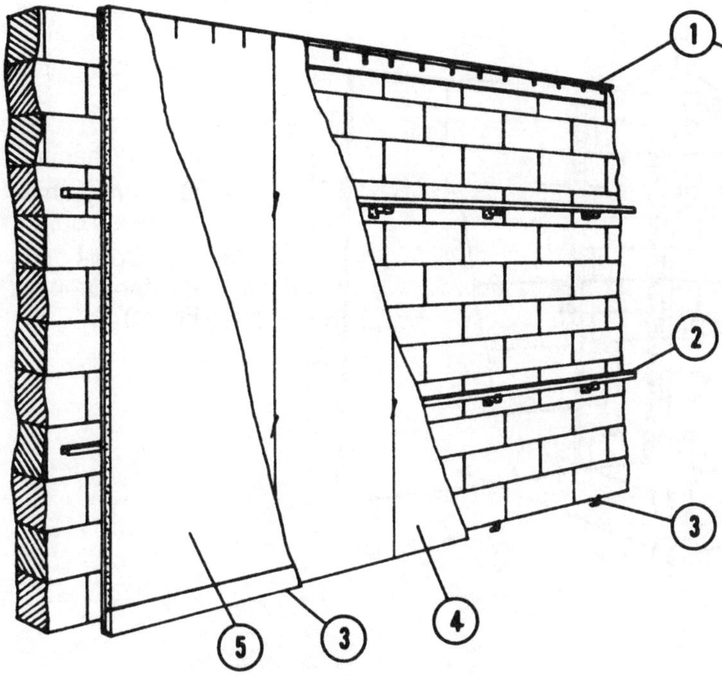

BRACED GYPSUM LATH

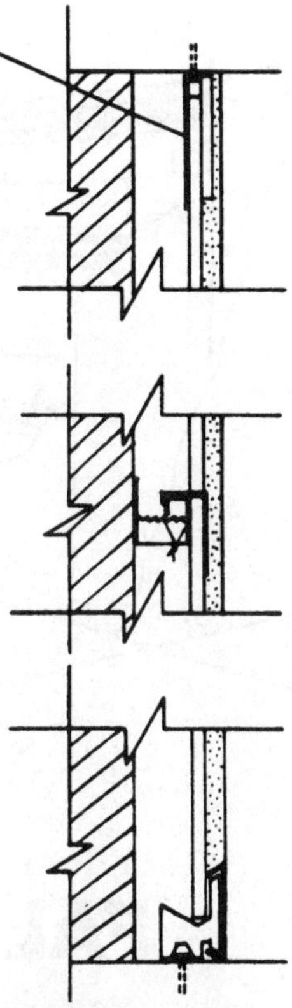

(1) Ceiling Runner

(2) Horizontal Stiffener (secured to bracing attachment)

(3) Metal Base and Clips

(4) Gypsum Lath

(5) Three Coats of Plaster (Scratch, Brown, Finish) (Minimum plaster thickness is ¾ inch.)

COLUMN FURRING

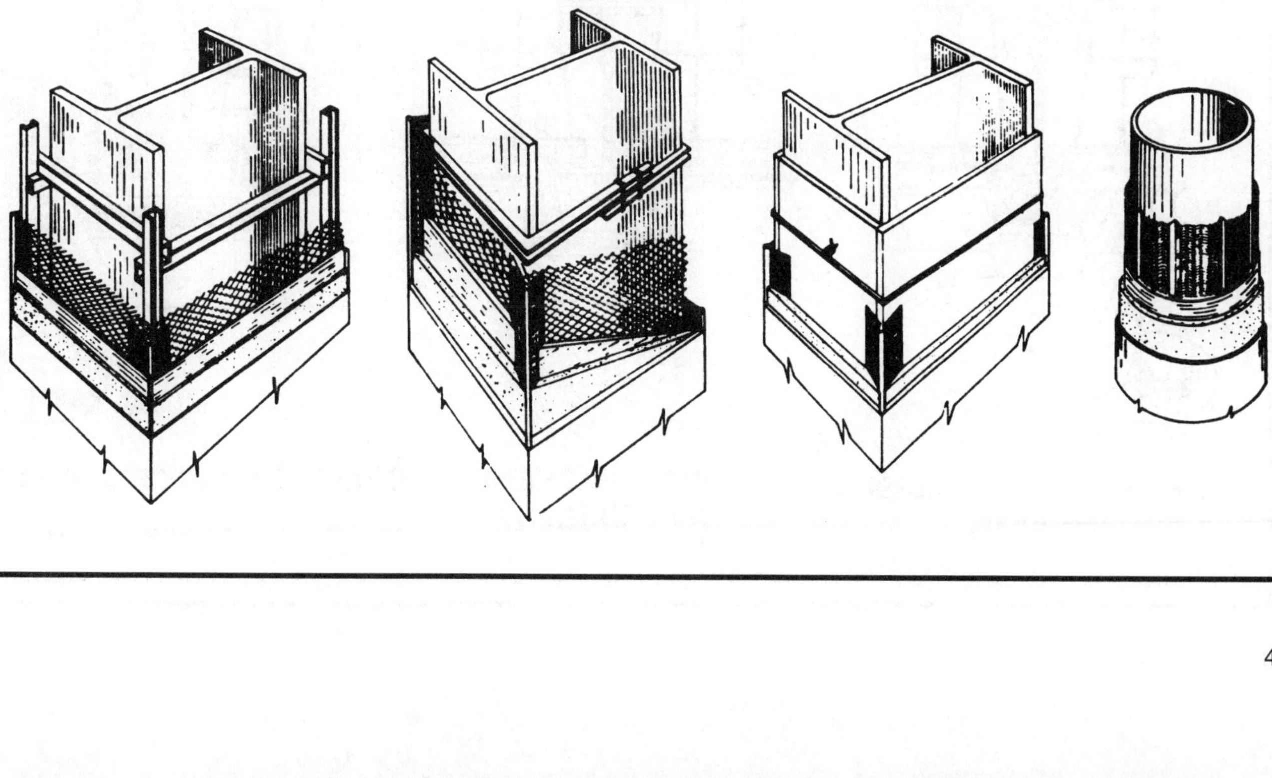

CONTACT FURRING

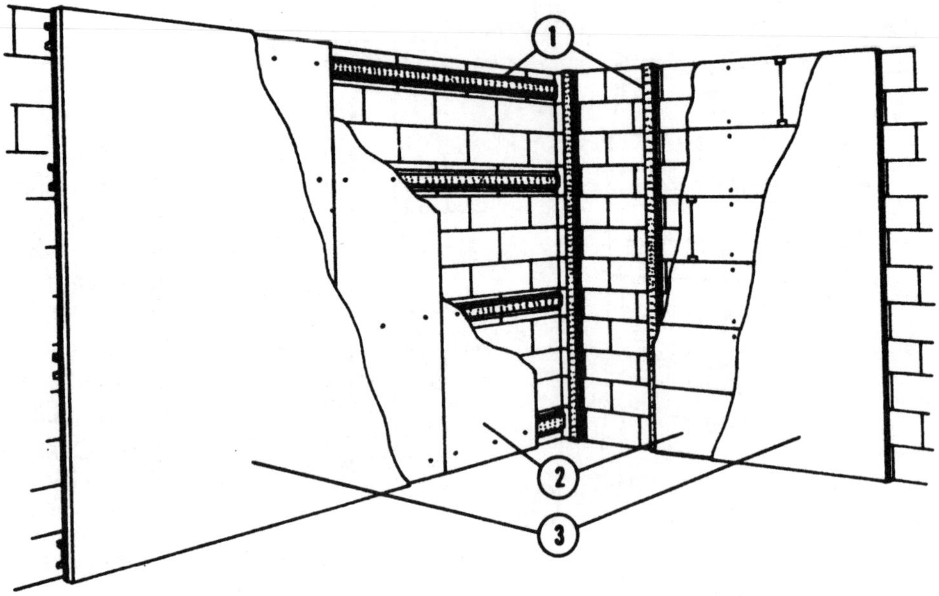

(1) Screw Channel
(2) Gypsum Lath (screwed on)
(3) Two Coats of Plaster (Brown, Finish)

FALSE BEAMS

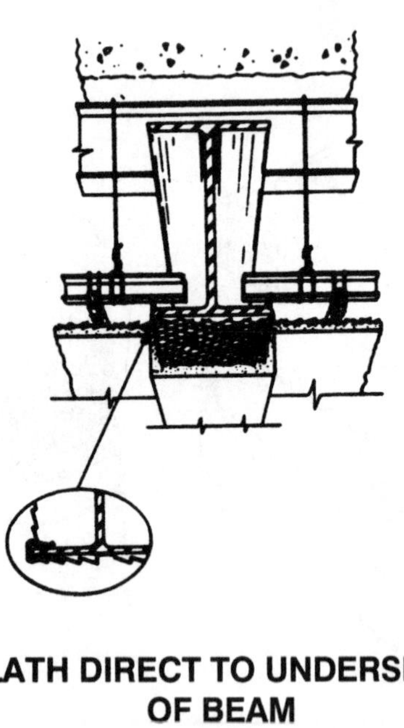

LATH DIRECT TO UNDERSIDE OF BEAM

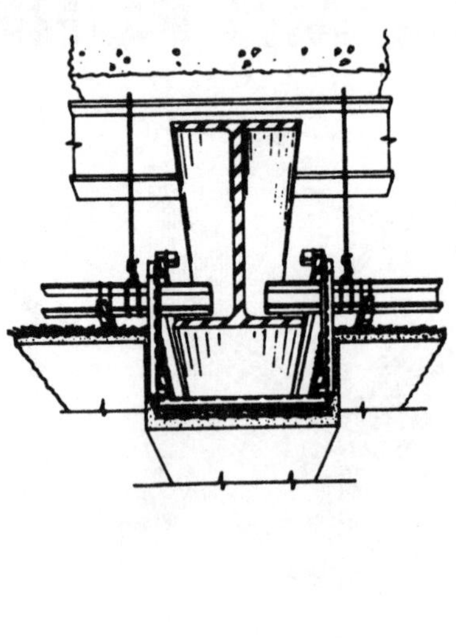

CHANNEL BRACKETS TO MAIN RUNNERS

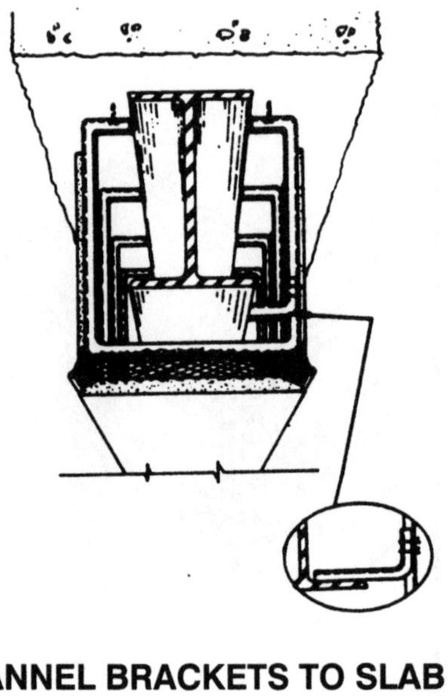

CHANNEL BRACKETS TO SLAB

MAXIMUM SPACING OF SUPPORTS FOR METAL LATH (Inches)

Type Of Lath	Weight of Lath Lb. Per Sq. Yd.	WALLS AND PARTITIONS			CEILINGS	
		Wood Studs	Solid Partitions	Steel Studs Wall Furring, Etc.	Wood or Concrete	Metal
Diamond Mesh (flat expanded)	2.5	16	16	13 1/2	12	12
	3.4	16	16	16	16	
Flat Rib	2.75	16	16	16	16	16
	3.4	19	24 (3)	19	19	19
3/8" Rib (1) (2)	3.4	24	(4)	24	24	24
	4.0	24	(4)	24	24	24
3/4" Rib	5.4	-	(4)	24 (5)	36 (6)	36 (6)
Sheet Lath	4.5	24	(4)	24	24	24

NOTE: Weights are exclusive of paper, fiber or other backing.
(1) 3.4 lb. 3/8" Rib Lath is permissible under Concrete Joists at 27" c.c.
(2) These spacings are based on a narrow bearing surface for the lath. When supports with a relatively wide bearing surface are used, these spacings may be increased accordingly, and still assure satisfactory work.
(3) This spacing permissible for Solid Partitions not exceeding 16' in height. For greater heights, permanent horizontal stiffener channels or rods must be provided on channel side of partitions, every 6' vertically, or else spacing shall be reduced 25%.
(4) For studless solid partitions, lath erected vertically.
(5) For interior wall furring or for application over solid surfaces for stucco.
(6) For contact or ceilings only.

TYPES OF LATH-ATTACHMENT TO WOOD AND METAL SUPPORTS

TYPE OF LATH	NAILS Type & Size	NAILS MAXIMUM SPACING Vertical (In Inches)	NAILS MAXIMUM SPACING Horizontal (In Inches)	SCREWS MAXIMUM SPACING Vertical (In Inches)	SCREWS MAXIMUM SPACING Horizontal (In Inches)	STAPLES Round of Flattened Wire: Wire Gauge No.	STAPLES Crown (In Inches)	STAPLES Leg (In Inches)	STAPLES MAXIMUM SPACING Vertical (In Inches)	STAPLES MAXIMUM SPACING Horizontal (In Inches)
1. Diamond Mesh Expanded Metal Lath and Flat Rib Metal Lath	4d blued smooth box 1 ½ No. 14 gauge 7/32" head (clinched) 1" No.11 gauge 7/16" head, barbed 1 ½" No.11 gauge 7/16" head, barbed	6 6 6	- - 6	6	6	16	¾	7/8	6	6
2. 3/8" Rib Metal Lath and Sheet	1 ½" No. 11 ga. 7/16" head, barbed	6	6	6	6	16	¾	1 ½	At Ribs	At Ribs
3. ¾" Rib Metal Lath	4d common 1 ½" No.12 ½ gauge ¼" head 2" No.11 gauge 7/16" head, barbed	At Ribs	- At Ribs	At Ribs	At Ribs	16	¾	1 5/8	At Ribs	At Ribs
4. Wire Fabric	4d blued smooth box (clinched) 1" No.11 gauge 7/16" head, barbed	6 6	- -	6	6	16	¾	7/8	6	6
	1 ½" No.11 gauge 7/16" head, barbed 1 ¼" No. 12 ga. 3/8" head, furring 1" No.12 gauge	6 6 6	6 6			16	7/16	7/8	6	6
5. 3/8" Gypsum Lath	1 1/8" No.13 gauge 12/61" head, blued	8	8	8	8	16	¾	7/8	8	8
6. ½" Gypsum Lath	1 ¼" No.13 gauge 12/61" head, blued	8	8 6	8	8 6	16	¾	1 1/8	8	8 6

STRESS RELIEF (CONTROL JOINTS)

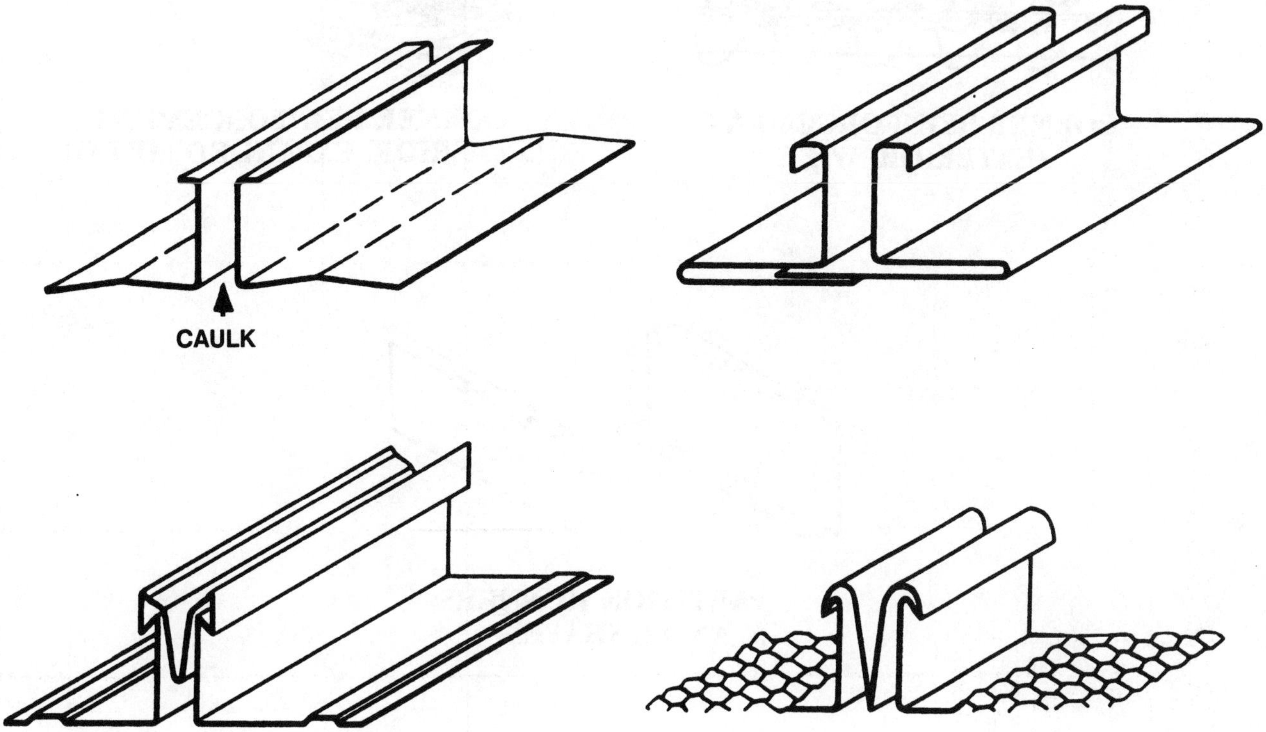

REVEALS

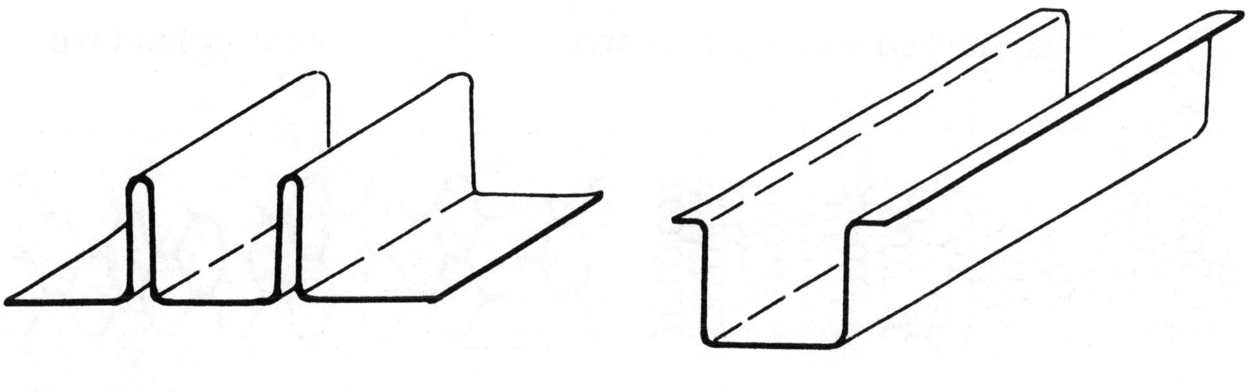

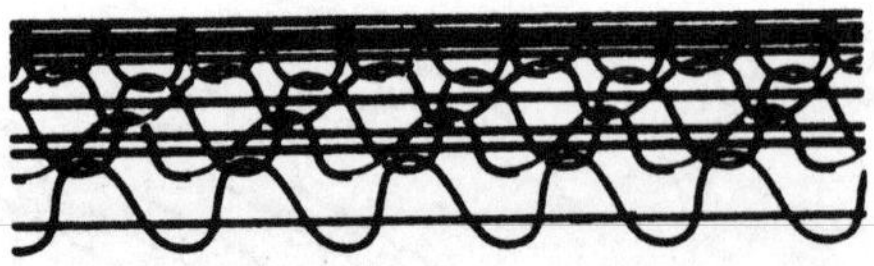

CORNER REINFORCEMENT (EXTERIOR) WIRE

CORNER REINFORCEMENT (EXTERIOR) EXPANDED METAL

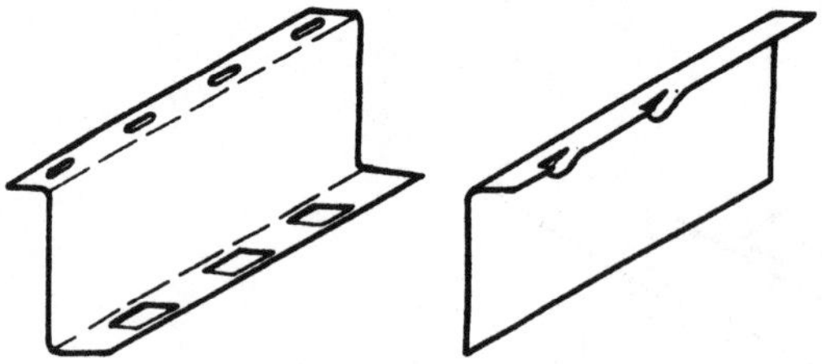

PARTITION RUNNERS (Z AND L SHAPE)

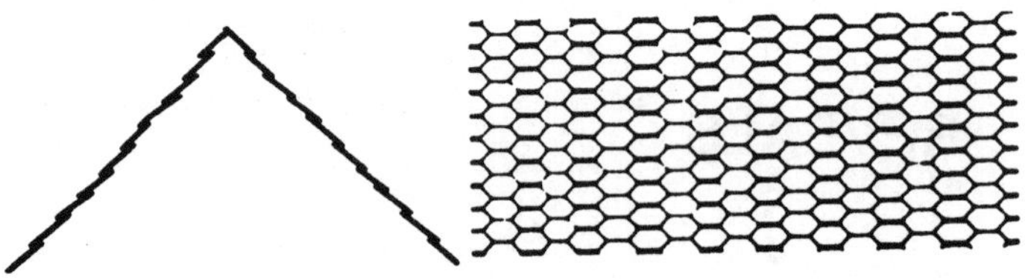

EXPANDED METAL CORNERITE

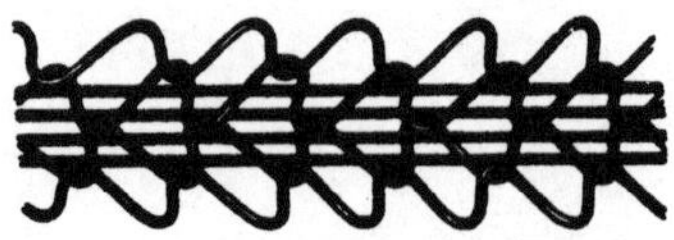

WIRE CORNERITE

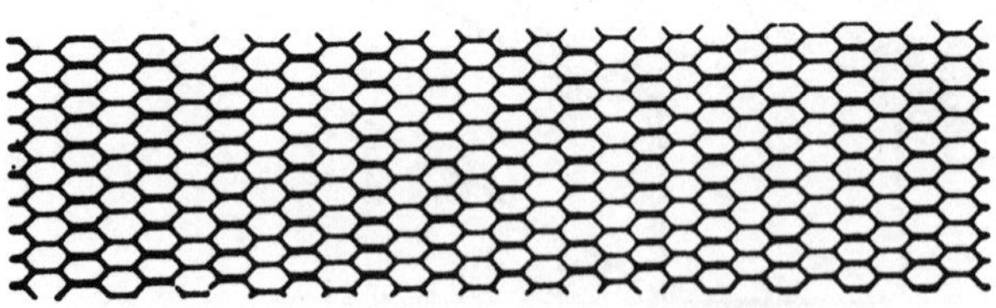

STRIP REINFORCEMENT (EXPANDED METAL)

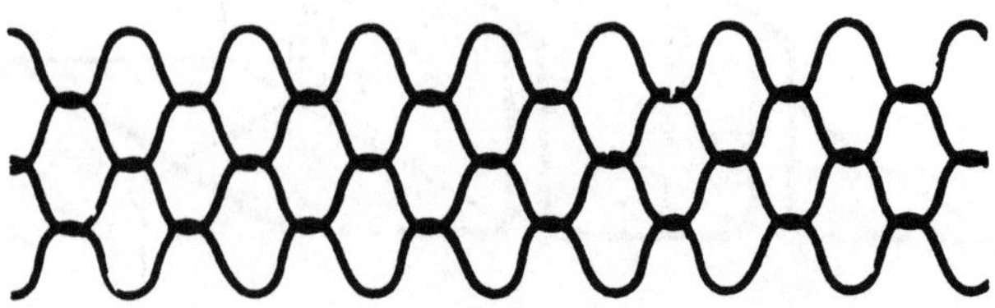

STRIP REINFORCEMENT (WIRE)

FINISHES /PLASTER ACCESSORIES | 09205

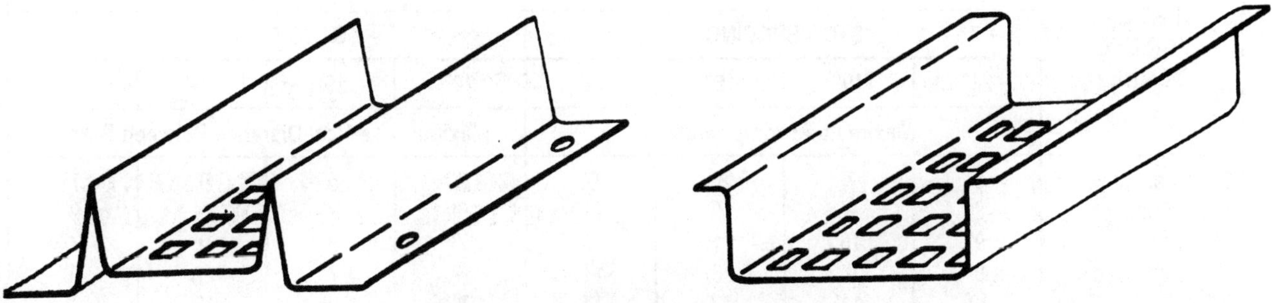

VENTILATING SCREEDS

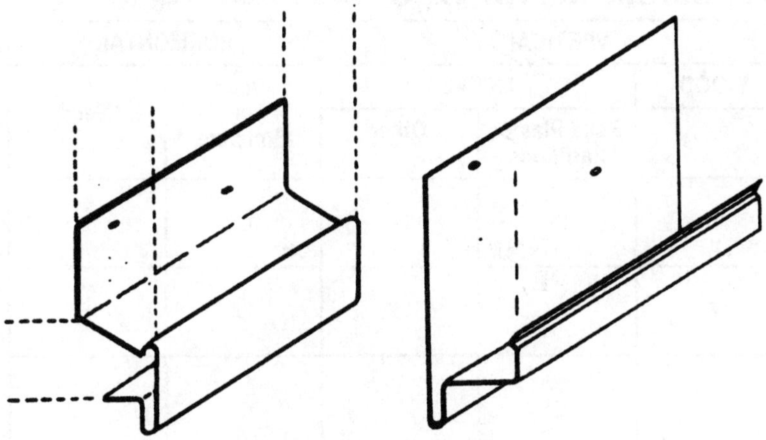

DRIP SCREEDS

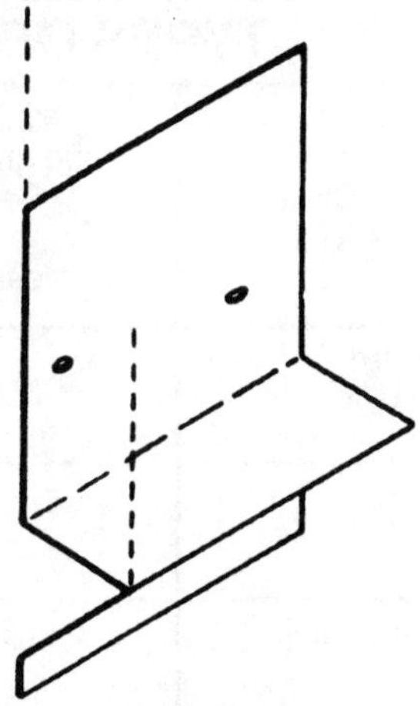

WEEP SCREED
(Also Available with Perforations)

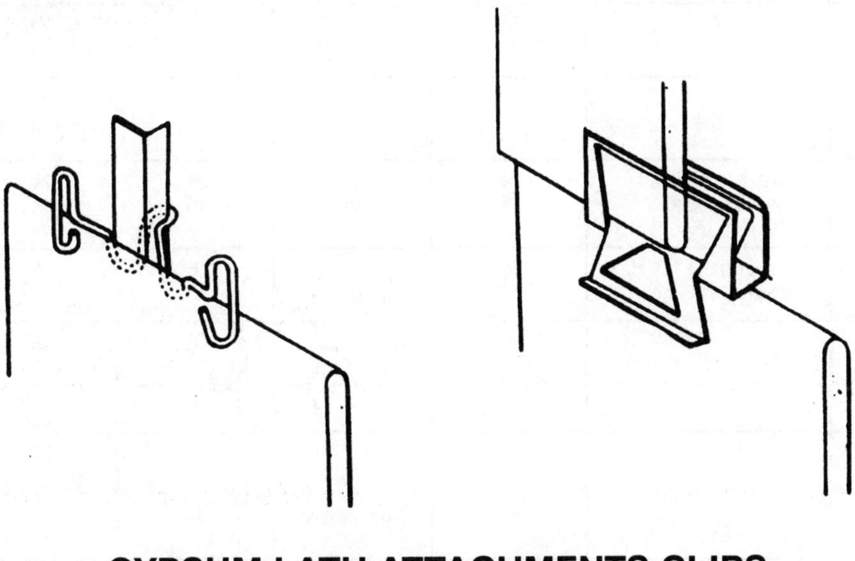

GYPSUM LATH ATTACHMENTS CLIPS

FINISHES / FURRING & LATHING | 09205

VERTICAL FURRING

VERTICAL FURRING MEMBER	UNBRACED				BRACED			
	STUD SPACING				STUD SPACING			
	24"	19"	16"	12"	24"	19"	16"	12"
	Maximum Furring Heights				Maximum Vertical Distance Between Braces			
3/4" Channel	6'	7'	8'	9'	5'	5'	6'	7'
1 1/2" Channel	8'	9'	10'	12'	6'	7'	8'	9'
2" Channel	9'	10'	11'	13'	7'	8'	9'	10'
2" Prefab. Stud	8'	9'	10'	11'	6'	7'	8'	9'
2 1/2" Prefab. Stud	10'	11'	12'	14'	8'	9'	10'	11'
3 1/4" Prefab.Stud	14'	16'	17'	20'	11'	13'	14'	16'

TYPES OF LATH—MAXIMUM SPACING OF SUPPORTS

TYPE OF LATH		Minimum Weight (psy), Gauge & Mesh Size	VERTICAL			HORIZONTAL	
			WOOD	METAL		Wood or Concrete	Metal
				Solid Plaster Partitions	Other		
Expanded Metal Lath		2.5	16"	16"	12"	12"	12"
Flat rib Expanded		2.75	16"	16"	16"	16"	16"
Stucco Mesh Expanded		1.8 and	16"	-	-	-	-
3/8" Rib Expanded		3.4	24"	-	24"	24"	24"
Sheet Lath		4.5	24"	-	24"	24"	24"
3/4" Rib Expanded Metal Lath		5.4	-	-	-	36"	36"
Wire Fabric Lath	Welded	1.95 lbs.,11 ga.,2"x2" 1.4 lbs.,16 ga.,2"x2"	24" 16"	24" 16"	24" 16"	24" 16"	24" 16"
	Woven	1.4 lbs.,17 ga.,1 1/2" Hex. 1.4 lbs.,18 ga.,1" Hex.	24" 24"	16" 16"	16" 16"	24" 24"	16" 16"
3/8" Gypsum Lath		-	16"		16"	16"	16"
(Large Size)		-	16"	-	16"	16"	16"
1/2" Gypsum Lath			24"		24"	24"	24"
(Large Size)			24"	No supports;	24"	24"	16"
5/8" Gypsum Lath (Large Size)		-	24"	No supports; Erected vertically	24"	24"	16"

PLASTERING TABLES

THICKNESS OF PLASTER

PLASTER	FINISHED THICKNESS OF PLASTER FROM FACE OF LATH, MASONRY, CONCRETE	
	Gypsum Plaster	Portland Cement Plaster
Expanded Metal Lath	5/8" minimum	5/8" minimum
Wire Fabric Lath	5/8" minimum	3/4" minimum (interior) 7/8" minimum (exterior)
Gypsum Lath	1/2" minimum	
Gypsum Veneer Base	1/16" minimum	1/2" minimum
Masonry Walls	1/2" minimum	7/8" maximum
Monolithic Concrete Walls	5/8" maximum	1/2" maximum

GYPSUM PLASTER PROPORTIONS

NUMBER OF COATS	COAT	PLASTER BASE OR LATH	MAXIMUM VOLUME AGGREGATE PER 100# NEAT PLASTER (CUBIC FEET)	
			Damp Loose Sand	Perlite or Vermiculite
Two-Coat Work	Basecoat Basecoat	Gypsum Lath Masonry	2 1/2 3	2 1/2 3
Three-Coat Work	First Coat Second Coat First &	Lath Lath	2 3	2 3

PORTLAND CEMENT PLASTER

COAT	VOLUME CEMENT	MAXIMUM WEIGHT (OR VOLUME) LIME PER VOLUME CEMENT	MAXIMUM VOLUME SAND PER VOLUME CEMENT	APPROXIMATE MINIMUM THICKNESS	MINIMUM PERIOD MOIST CURING	MINIMUM INTERVAL BETWEEN COATS
First	1	20 lbs.	4	3/8"	48 Hours	48 Hours
Second	1	20 lbs.	5	1st & 2nd Coats total	48 Hours	7 Days
Finish	1	1	3	1st, 2nd & Finish	-	

PORTLAND CEMENT - LIME PLASTER

COAT	VOLUME CEMENT	MAXIMUM WEIGHT (OR VOLUME) LIME PER VOLUME CEMENT	MAXIMUM VOLUME SAND PER VOLUME CEMENT	APPROXIMATE MINIMUM THICKNESS	MINIMUM PERIOD MOIST CURING	MINIMUM INTERVAL BETWEEN COATS
First	1	1	4	3/8"	48 Hours	48 Hours
Second	1	1	4 1/2	1st & 2nd Coats total	48 Hours	7 Days
Finish	1	1	3	1st, 2nd & Finish	-	

METAL STUD CONSTRUCTION

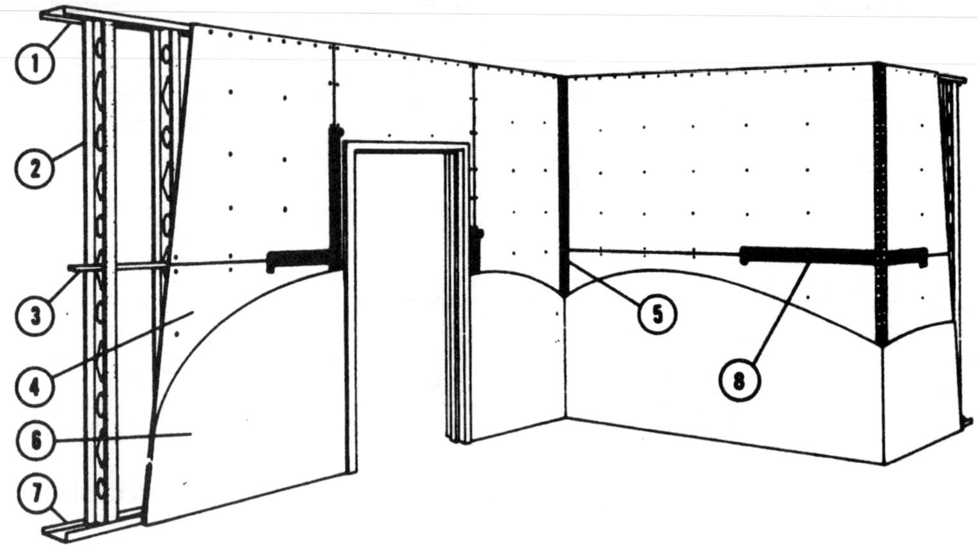

(1) Ceiling Runner Track
(2) Metal Stud (nailable or screw)
(3) Horizontal Stiffener
(4) Large Size Lath
(5) Angle Reinforcement
(6) Veneer Plaster 1/16 to 1/8 inch thick)
(7) Floor Runner Track
(8) Joint Reinforcement

WOOD STUD CONSTRUCTION

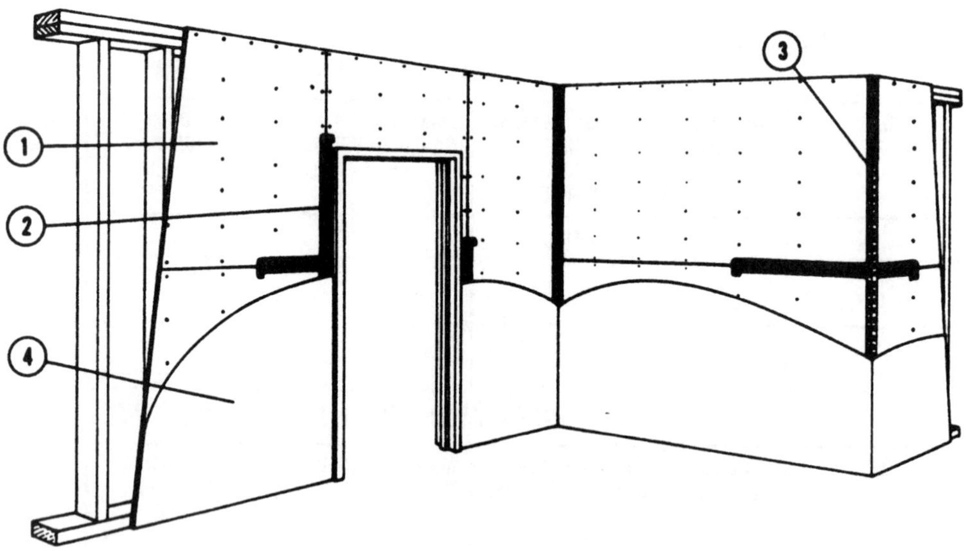

(1) Large Size Lath
(2) Joint Reinforcement
(3) Corner Bead
(4) Veneer Plaster (1/16 to 1/8 inch thick)

EXTERIOR LATH AND PLASTER

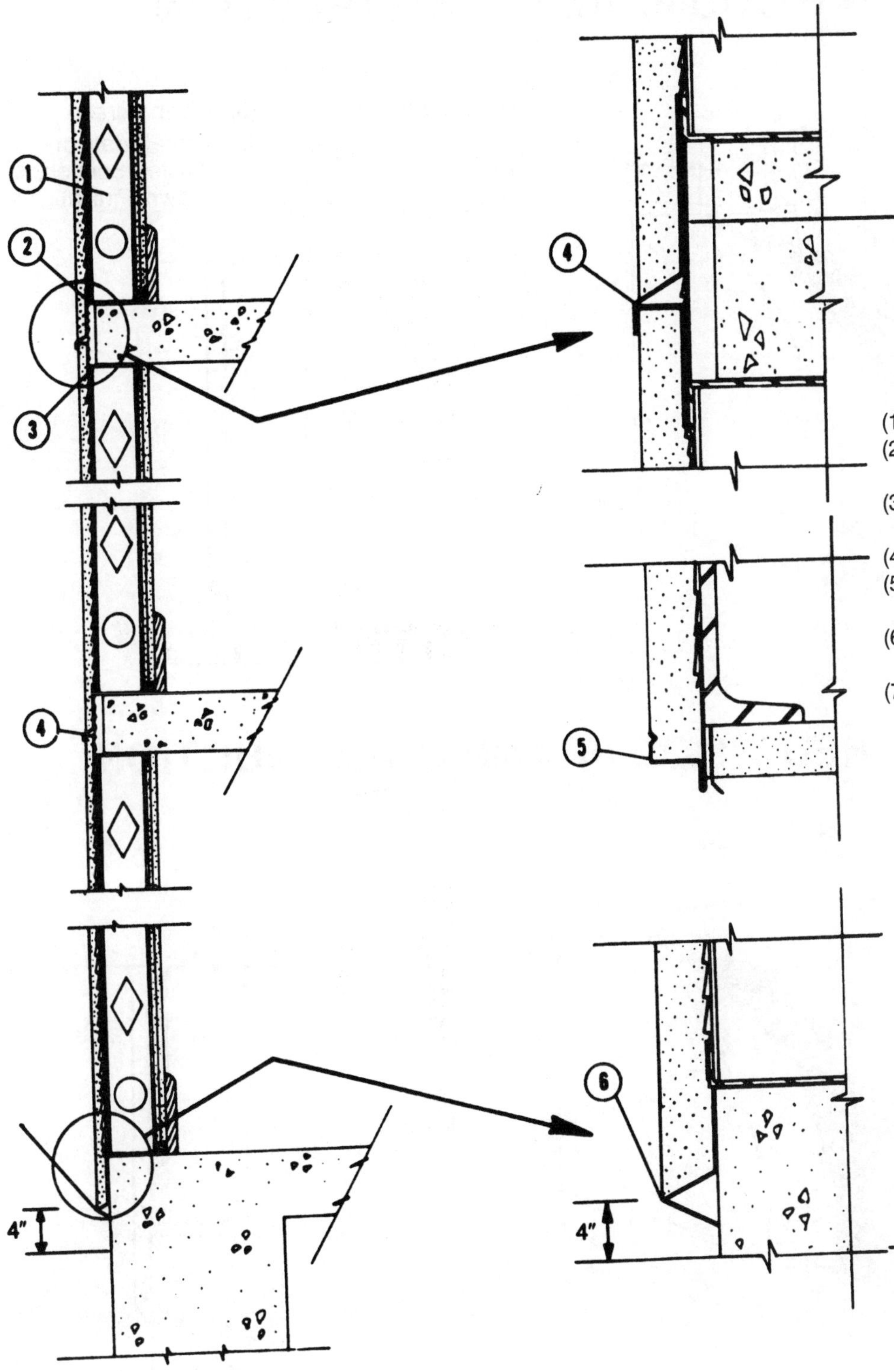

(1) Metal Stud
(2) Floor Runner Track
(3) Ceiling Runner Track
(4) Control Joint
(5) Drip Screed (at soffit)
(6) Weep Screed (at slab or foundation)
(7) Building Paper, Continuous

EXTERIOR LATH AND PLASTER

OPEN WOOD FRAME CONSTRUCTION

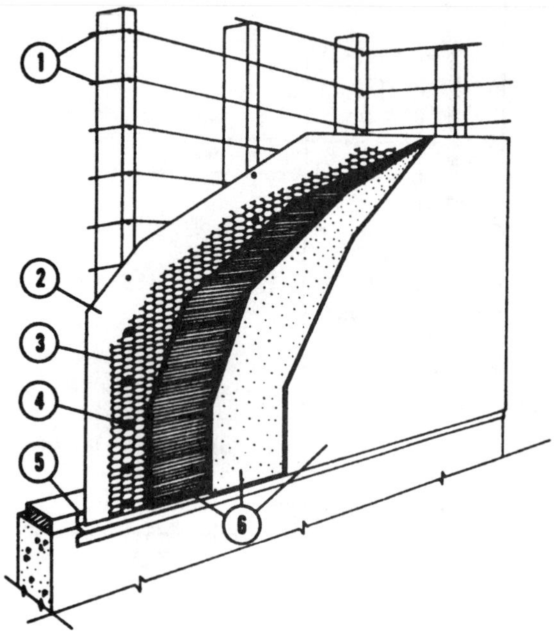

(1) Wire Backing
(2) Building Paper
(3) Wire Fabric Lath
(4) Approved Fasteners
(5) Weep Screed
(6) Three Coats of Plaster (Scratch, Brown, Finish)

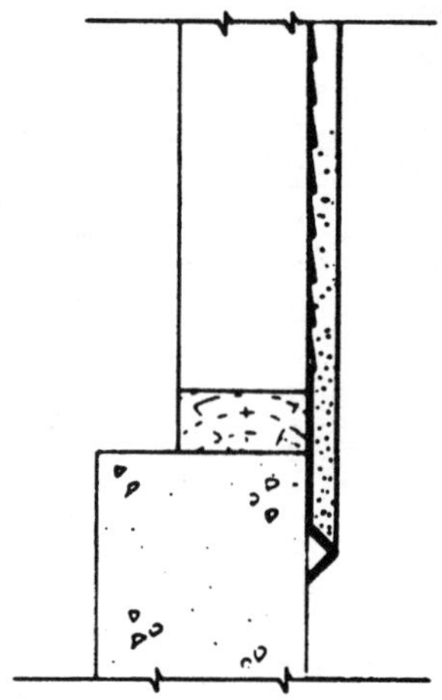

SHEATHED WOOD FRAME CONSTRUCTION

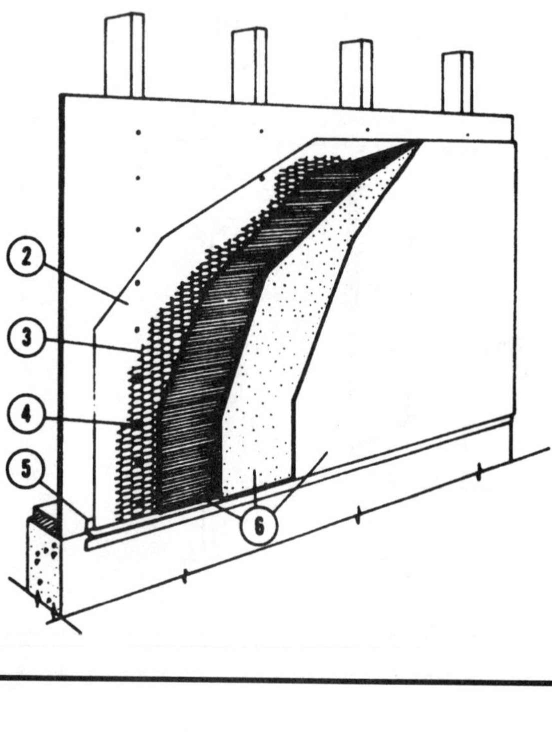

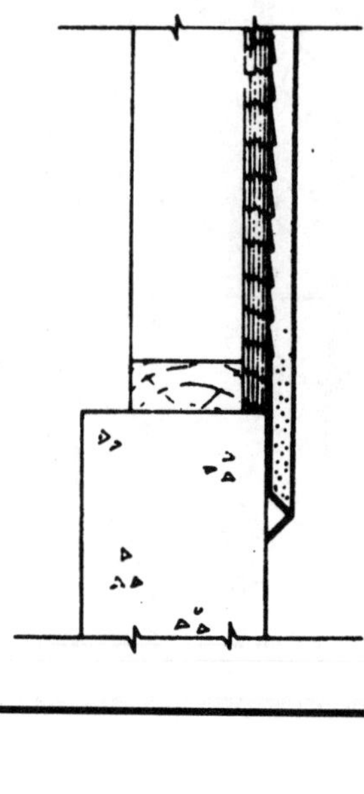

PIPE WEIGHTS

CAST IRON PIPE

SERVICE WEIGHT			
Size, Inches	Weight of Pipe	Weight of Water	Total Weight—Lbs.
2"	3.8	1.45	5.3
3"	5.6	3.2	8.8
4"	7.5	5.5	13.0
5"	9.8	8.7	18.5
6"	12.4	12.5	24.9
8"	18.5	21.7	40.2

EXTRA HEAVY			
Size, Inches	Weight of Pipe	Weight of Water	Total Weight — Lbs.
2"	4.3	1.45	5.8
3"	8.3	3.2	11.5
4"	10.8	5.5	16.3
5"	13.3	8.7	22.0
6"	16.0	12.5	28.5
8"	26.5	21.7	48.2

STEEL PIPE

Pipe Size	W/40	H_2O/Lbs.	Total Lbs./L.F.
2"	3.65	1.45	5.1
2 1/2"	5.79	2.07	7.86
3"	7.57	3.2	10.77
3 1/2"	9.11	4.28	13.39
4"	10.8	5.51	16.31
5"	14.6	8.66	23.26
6"	18.0	12.5	30.5
8"	28.6	21.66	50.26
10"	40.5	34.15	74.65
12"			

SPRINKLER AREA CALCULATIONS

Typical maximum floor area allowed per system riser:

Light Hazard	Ordinary Hazard	Extra Hazard
52,000 S.F.	40,000-52,000 S.F.	25,000 S.F.

Typical maximum floor area coverage allowed per sprinkler head:

Light Hazard	Ordinary Hazard	Extra Hazard
130-200 S.F.	100-130 S.F.	90 S.F.

Typical maximum spacing between lines and sprinkler heads:

Light Hazard	Ordinary Hazard	Extra Hazard
12-15 feet	12-15 feet	12 feet

Note: This data is for estimating purposes only. Check all applicable codes and regulations for specific requirements.

SPRINKLER HEAD CALCULATIONS

Typical maximum quantity of sprinkler heads allowed by pipe size.

Light Hazard:

For sprinklers below ceiling:

Steel		Copper	
1 in. pipe	2 sprinklers	1 in. tube	2 sprinklers
1 ¼ in. pipe	3 sprinklers	1 ¼ in. tube	3 sprinklers
1 ½ in. pipe	5 sprinklers	1 ½ in. tube	5 sprinklers
2 in. pipe	10 sprinklers	2 in. tube	12 sprinklers
2 ½ in. pipe	30 sprinklers	2 ½ in. tube	40 sprinklers
3 in. pipe	60 sprinklers	3 in. tube	65 sprinklers
3 ½ in. pipe	100 sprinklers	3 ½ in. tube	115 sprinklers

For sprinklers above and below ceiling:

Steel		Copper	
1 in.	2 sprinklers	1 in.	2 sprinklers
1 ¼ in.	4 sprinklers	1 ¼ in.	4 sprinklers
1 ½ in.	7 sprinklers	1 ½ in.	7 sprinklers
2 in.	15 sprinklers	2 in.	18 sprinklers
2 ½ in.	50 sprinklers	2 ½ in.	65 sprinklers

Ordinary Hazard:

For sprinklers above ceiling:

Steel		Copper	
1 in. pipe	2 sprinklers	1 in. tube	2 sprinklers
1 ¼ in. pipe	3 sprinklers	1 ¼ in. tube	3 sprinklers
1 ½ in. pipe	5 sprinklers	1 ½ in. tube	5 sprinklers
2 in. pipe	10 sprinklers	2 in. tube	12 sprinklers
2 ½ in. pipe	20 sprinklers	2 ½ in. tube	25 sprinklers
3 in. pipe	40 sprinklers	3 in. tube	45 sprinklers
3 ½ in. pipe	65 sprinklers	3 ½ in. tube	75 sprinklers
4 in. pipe	100 sprinklers	4 in. tube	115 sprinklers
5 in. pipe	160 sprinklers	5 in. tube	180 sprinklers
6 in. pipe	275 sprinklers	6 in. tube	300 sprinklers

For sprinklers above and below ceiling:

Steel		Copper	
1 in.	2 sprinklers	1 in.	2 sprinklers
1 ¼ in.	4 sprinklers	1 ¼ in.	4 sprinklers
1 ½ in.	7 sprinklers	1 ½ in.	7 sprinklers
2 in.	15 sprinklers	2 in.	18 sprinklers
2 ½ in.	30 sprinklers	2 ½ in.	40 sprinklers
3 in.	60 sprinklers	3 in.	65 sprinklers

Extra Hazard:

For sprinklers below ceiling:

Steel		Copper	
1 in. pipe	1 sprinkler	1 in. tube	1 sprinkler
1 ¼ in. pipe	2 sprinklers	1 ¼ in. tube	2 sprinklers
1 ½ in. pipe	5 sprinklers	1 ½ in. tube	5 sprinklers
2 in. pipe	8 sprinklers	2 in. tube	8 sprinklers
2 ½ in. pipe	15 sprinklers	2 ½ in. tube	20 sprinklers
3 in. pipe	27 sprinklers	3 in. tube	30 sprinklers
3 ½ in. pipe	40 sprinklers	3 ½ in. tube	45 sprinklers
4 in. pipe	55 sprinklers	4 in. tube	65 sprinklers
5 in. pipe	90 sprinklers	5 in. tube	100 sprinklers
6 in. pipe	150 sprinklers	6 in. tube	170 sprinklers

Note: This data is for estimating purposes only. Check all applicable codes and regulations for specific requirements.

SPRINKLER HAZARD OCCUPANCIES

Typical Light Hazard Occupancies:

Churches
Clubs
Eaves and overhangs, if combustible construction with no combustible beneath
Educational
Hospitals
Institutional
Libraries, except large stack rooms
Museums
Nursing or convalescent homes
Office, including data processing
Residential
Restaurant seating areas
Theaters seating areas
Theaters and auditoriums excluding stages and prosceniums
Unused attics

Typical Ordinary Hazard Occupancies (Group 1):

Automobile parking garages
Bakeries
Beverage manufacturing
Canneries
Dairy products manufacturing and processing
Electronic plants
Glass and glass products manufacturing
Laundries
Restaurant service areas

Typical Ordinary Hazard Occupancies (Group 2):

Cereal mills
Chemical plants - ordinary
Cold Storage warehouses
Confectionery products
Distilleries
Leather goods mfg.
Libraries-large stack room areas
Mercantiles
Machine shops
Metal working
Printing and publishing
Textile mfg.
Tobacco Products mfg.
Wood product assembly

Typical Ordinary Hazard Occupancies (Group 3):

Feed mills
Paper and pulp mills
Paper process plants
Piers and wharves
Repair garages
Tire manufacturing
Warehouses (having moderate to higher combustibility of content, such as paper, household furniture, paint, general storage, whiskey, etc.)[1]
Wood machining

Typical Extra Hazard Occupancies (Group 1):

Combustible hydraulic fluid use areas
Die casting
Metal extruding
Plywood and particle board manufacturing
Printing (using inks with below 100°F [37.8°C] flash points)
Rubber reclaiming, compounding, drying, milling, vulcanizing
Saw mills
Textile picking, opening, blending, garnetting, carding, combining of cotton, synthetics, wool shoddy or burlap
Upholstering with plastic foams

Typical Extra Hazard Occupancies (Group 2):

Asphalt saturating
Flammable liquids spraying
Flow coating
Mobile Home or Modular Building assemblies (where finished enclosure is present and has combustible interiors)
Open Oil quenching
Solvent cleaning
Varnish and paint dipping

TYPICAL MINIMUM SIZE OF HORIZONTAL BUILDING STORM DRAINS AND BUILDING STORM SEWERS

Diameter of drain	Maximum projected area in square feet for various slopes		
	1/8 inch per feet slope	1/4 inch per feet slope	1/2 inch per feet slope
3	822	1160	1644
4	1880	2650	3760
5	3340	4720	6680
6	5350	7550	10700
8	11500	16300	23000
10	20700	29200	41400
12	33300	47000	66600
15	59500	84000	119000

MECHANICAL / PLUMBING — 15400

PLUMBING BUDGETS

Budget estimates for plumbing can be determined as a percentage of total building costs depending on building type. For example:

- Apartments........................9 to 12 percent
- Assembly........................4 to 7 percent
- Banks........................3 to 6 percent
- Dormitories........................7 to 10 percent
- Factories........................4 to 8 percent
- Hospitals........................8 to 12 percent
- Motels........................9 to 12 percent
- Office Buildings........................4 to 7 percent
- Retail (small)........................4 to 7 percent
- Retail (large)........................3 to 6 percent
- Schools........................3 to 6 percent
- Warehouses........................3 to 7 percent

MECHANICAL / PLUMBING VENTS — 15410

TYPICAL SIZE & LENGTH OF PLUMBING VENTS

Diameter of soil or waste stack (in.)	Total fixture units connected to stack (dfu)	DIAMETER OF VENT PIPE										
		1-1/4	1	2	2-1/2	3	4	5	6	8	10	12
1-1/4	2	30										
1-1/2	8	50	150									
1-1/2	10	30	100									
2	12	30	75	200								
2	20	26	50	150								
2-1/2	42		30	100	300							
3	10		42	150	360	1040						
3	21		32	110	270	810						
3	53		27	94	230	680						
3	102		25	86	210	620						
4	43			35	85	250	980					
4	140			27	65	200	750					
4	320			23	55	170	640					
4	540			21	50	150	580					
5	190				28	82	320	990				
5	490				21	63	250	760				
5	940				18	49	210	670				
5	1400				16	33	190	590				
6	500					26	130	400	1000			
6	1100					22	100	310	780			
6	2000					20	84	260	660			
6	2900						77	240	600			
8	1800						31	95	240	940		
8	3400						24	73	190	720		
8	5600						20	62	160	610		
8	7600						18	56	140	560		

TYPICAL SIZES OF FIXTURE WATER SUPPLY PIPES

Fixture	Nominal pipe size (inches)
Bath tubs	1/2
Combination sink and tray	1/2
Drinking fountain	3/8
Dishwasher (domestic)	1/2
Kitchen sink, residential	1/2
Kitchen sink, commercial	3/4
Lavatory	3/8
Laundry tray, 1, 2 or 3 compartments	1/2
Shower (single head)	1/2
Sinks (service, slop)	1/2
Sinks flushing rim	3/4
Urinal (flash tank)	3/8
Urinal (direct flush valve)	1
Water closet (tank type)	3/8
Water closet (flush valve type)	1
Hose bibs	1/2
Wall hydrant	1/2

TYPICAL VENTILATION AIR REQUIREMENTS FOR SPECIAL USES

Occupancy Classification	Required ventilation air in cfm per human occupant
Special areas	
Lockers	2 * (or 30 per locker)
Wardrobes	2 *
Public bathrooms	40 **
Private bathrooms	25 **
Swimming pools	15 (per occupant)
Water closet (flush valve type)	1
Exitways and corridors	1 1/2 *

*Per square foot floor area.
**Per water closet or urinal.

TYPICAL VENTILATION AIR REQUIREMENTS FOR RETAIL USES

Occupancy Classification	Required ventilation air in cfm per human occupant
Mercantile	
Sales floors and showrooms (basement & grade floors)	7
Sales and showrooms (upper floors)	7
Storage areas	5
Dressing rooms	7
Malls	7
Shipping areas	15
Elevators	7
Supermarkets	
Meat processing rooms	5
Drugs stores	
Pharmacists' work rooms	20
Specialty shops	
Pet shops	1.0*
Florists	5
Greenhouses	5

*cfm per sq. ft. floor area.

TYPICAL VENTILATION AIR REQUIREMENTS FOR RESIDENTIAL USES

Occupancy Classification	Required ventilation air in cfm per human occupant
Residential	
General living areas	5
Bedrooms	5
Kitchens	20
Basements, utility rooms	5
Mobile homes	5
Hotels, motels	
Bedrooms (single, double)	7
Living rooms (suites)	10
Corridors	5
Lobbies	7
Conference rooms (small)	20
Assembly rooms (large)	15

TYPICAL VENTILATION AIR REQUIREMENTS FOR STORAGE USES

Occupancy Classification	Required ventilation air in cfm per human occupant
Storage	
Garages, service stations, parking garages (enclosed)	1.5*
Auto repair shops	1.5**
Warehouses	
General	7

*cfm/s.f. floor area
**Must have positive engine exhaust system.

TYPICAL VENTILATION AIR REQUIREMENTS FOR FACTORY AND INDUSTRIAL USES

Occupany Classification	Required ventilation air in cfm per human occupant
Factory and industrial	
Metalworking & finishing	35
Automotive engine test	*Require Special Exhaust Systems*
Paint spray booths	*Require Special Exhaust Systems*
Picking, etching & plating lines	*Require Special Exhaust Systems*
Degreasing booths	*Require Special Exhaust Systems*
Sandblasting booths	*Require Special Exhaust Systems*
Chemicals and pharmaceuticals	
Dusty operations	30
Rooms containing potential gas emitters	20
Drying oven rooms	15
Fermentation rooms	15
Pillmaking booths	10
Packaging areas	10
Utility rooms	7
Computer rooms	7
Textiles-clothes manufacturer	15
Electronics & aerospace circuit board & soldering rooms	20
Wood products, papermaking	20
Brewing, distilling, wineries, bottling	20*
Food processing	20
Tobacco processing	20
Power plants	
Control rooms	10
Boiler rooms	35
Generator rooms	20
Sewage treatment plants	
Control rooms	10
Compressor/blower motor rooms	20
Glass & ceramic manufacturer	20
Agricultural	20

TYPICAL VENTILATION AIR REQUIREMENTS FOR BUSINESS USES

Occupany Classification	Required ventilation air in cfm per human occupant
Business	
Banks	
(see offices)	
Vaults	5
Barber, beauty and health services	
Beauty shops (hair dressers)	25
Reducing salons	25
Sauna baths, steam rooms	5
Barber shops	7
Photo studios	
Camera rooms, stages	5
Dark rooms	
Shoe repair shops	
Workrooms/trade areas	10
Offices	
General office space and showrooms	15
Conference rooms	25
Drafting/art rooms	7
Doctor's consultation rooms	10
Waiting rooms	10
Lithographing rooms	7
Diazo printing rooms	7
Computer rooms	5
Keypunch rooms	7
Communication	
TV/radio broadcasting booths, studios	30
Motion picture and TV stages	30
Pressrooms	15
Composing rooms	7
Engraving rooms	7
Telephone switchboard rooms (manual)	7
Telephone switchgear rooms (automatic)	7
Teletypewriter/facsimile rooms	5
Research institutes	
Laboratories:	
Light duty; non-chemical	15
Chemical	15
Heavy-duty	15
Radioisotope, chemical & biologically toxic	15
Machine shops	15
Dark rooms, spectroscopy rooms	10
Animal rooms	40
Veterinary hospitals	
Kennels, stalls	25
Operating rooms	25
Reception rooms	10

TYPICAL VENTILATION AIR REQUIREMENTS FOR INSTITUTIONAL USES

Occupancy Classification	Required ventilation air in cfm per human occupant
Institutional	
Prisons	
Cell blocks	7
Eating halls	15
Guard stations	7

AIR CONDITIONING

RECOMMENDED SHEET METAL GAUGES AND CONSTRUCTION FOR RECTANGULAR DUCT

LOW PRESSURE — LOW VELOCITY = 2" W.G. MAX									
				AT JOINTS					
PLATE NO.	DIMENSION OF LONGEST SIDE OF DUCT	Steel Metal Gauges: Steel	Steel Metal Gauges: Aluminum	Plain "S" Slip (B) Pocket Lock (K) Drive Slip (A) 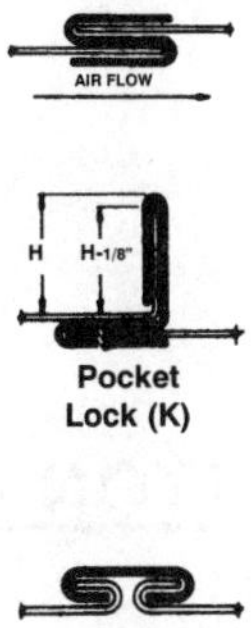	Hemmed "S" Slip (C) Bar Slip (E) Standing Seam (1) 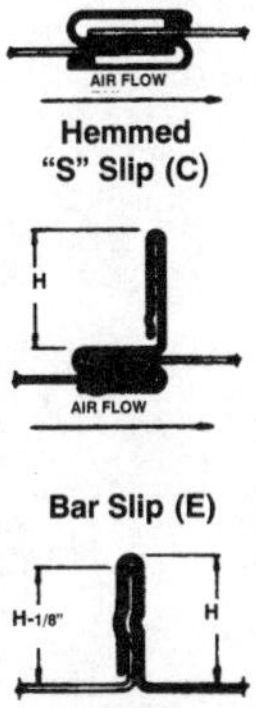	Reinforced Bar Slip (G) 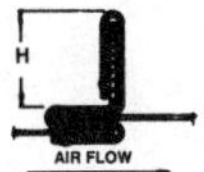	Angle Slip (H) Alternate Bar Slip (F) Angle RFD Pocket (L) 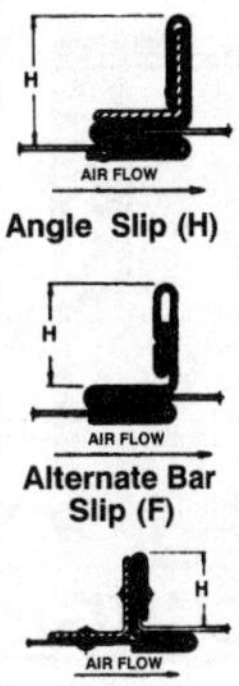	Companion Angles (M) Angle Reinforced Standing Seam (J) 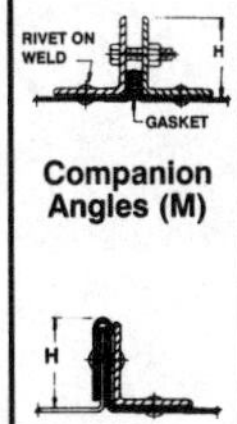	Reinforcing Between Joints
6	Thru 12"	26	24 (.020)	A-B-K	——	——	——	——	
6	13" thru 18"	24	22 (.025)	A-B-K	——	——	——	——	
7 7A	19" thru 30"	24	22 (.025)	K @ 5' cc A	C-E- @ 5' cc C-E- @ 10' cc	——	——	——	1" x 1" x 1/8" @ 5' cc
8	31" thru 42"	22	20 (.032)	K @ 5' cc	E-G-K @ 5' cc E-G-K @ 10' cc	——	——	——	1" x 1" x 1/8" @ 5' cc
9	43" thru 54"	22	20 (.032)	K @ 4' cc K @ 8' cc	E-@ 4' cc E-@ 8' cc	G- @ 4' cc G- @ 8' cc	——	——	1½" x 1½" x 1/8" @ 4' cc
9	55" thru 60"	20	18 (.040)	K @ 4' cc K @ 8' cc	E-@ 4' cc E-@ 8' cc	G- @ 4' cc G- @ 8' cc	——	——	1½" x 1½" x 1/8" @ 4' cc
10	61" thru 84"	20	18 (.040)	——	——	G- @ 4' cc G- @ 5' cc	H- @ 4' cc F- @ 4' cc L- @ 4' cc H- @ 5' cc F- @ 5' cc L- @ 5' cc	J- @ 2' cc	1½" x 1½" x 1/8" @ 2' cc 1½" x 1½" x 1/8" @ 2'- 6" cc
11	85" thru 96"	18	16 (.051)	——	——	——	H- @ 4' cc L- @ 4' cc H- @ 5' cc L- @ 5' cc	M- @ 4' cc M- @ 5' cc J- @ 2' cc	1½" x 1½" x 3/16" @ 2' cc 1½" x 1½" x 3/16" @ 2'- 6" cc 1½" x 1½" x 3/16" @ 2' cc
12	Over 96"	18	16 (.051)	——	——	——	H- @ 4' cc L- @ 4' cc H- @ 5' cc L- @ 5' cc	M- @ 4' cc M- @ 5' cc J- @ 2' cc	2" x 2" x ¼ @ 2' cc 2" x 2" x ¼ @ 2'- 6" cc 2" x 2" x ¼ @ 2' cc

H (height dimension) — up to 42" = 1"
H (height dimension) — 43" to 96" = 1½"
H (height dimension) — over 96" = 2"

AIR CONDITIONING (Cont.)

LONGITUDINAL SEAMS
FOR SHEET METAL DUCTWORK

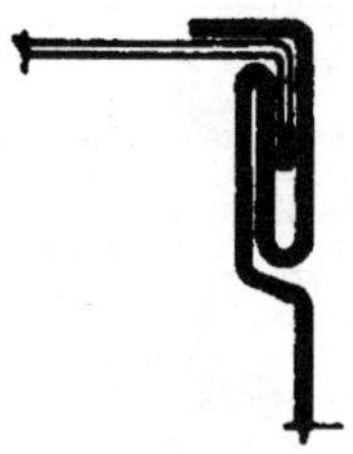

Fig. "N"
PITTSBURGH LOCK

Fig. "Z"
BUTTON PUNCH SNAP LOCK

Fig. "O"
ACME LOCK-GROOVED SEAM

Fig. "T"
DOUBLE SEAM

Approximately 2" Spacing
Between "Buttons"

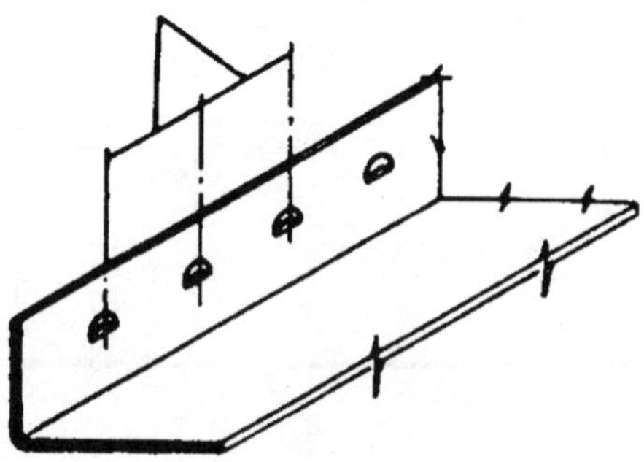

DETAIL NO.1
MALE PIECE-SNAP LOCK

AIR CONDITIONING

TYPICAL DUCT CONNECTIONS

CROSS JOINTS FOR SHEET METAL DUCTWORK

(NOT TO SCALE)

H - HEIGHT REFERRED TO IN DIMESIONS

(A)
DRIVE SLIP

(B)
PLAIN "S" SLIP

(C)
HEMMED "S" SLIP

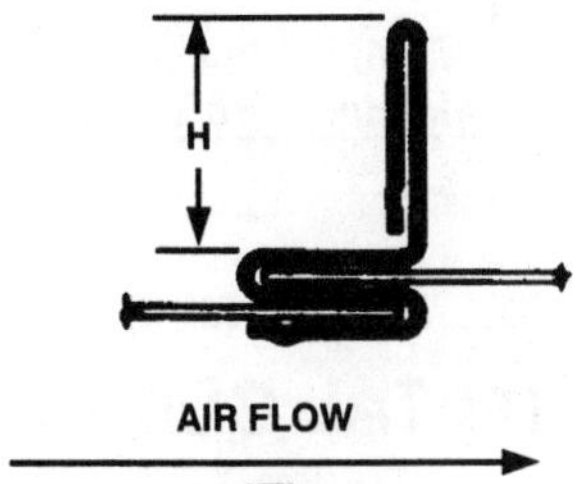

(E)
BAR SLIP

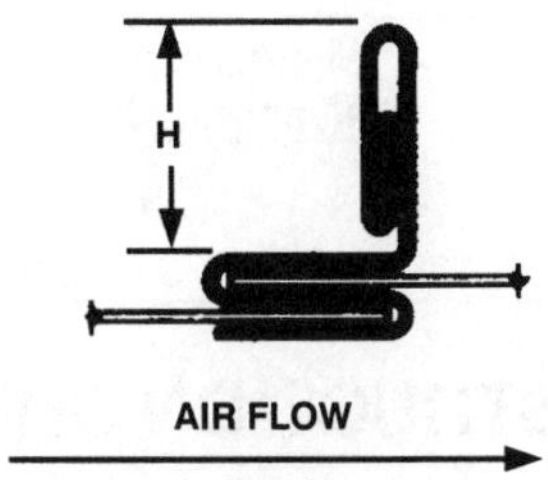

(F)
ALTERNATE BAR SLIP
(STANDARD "S" SLIP)

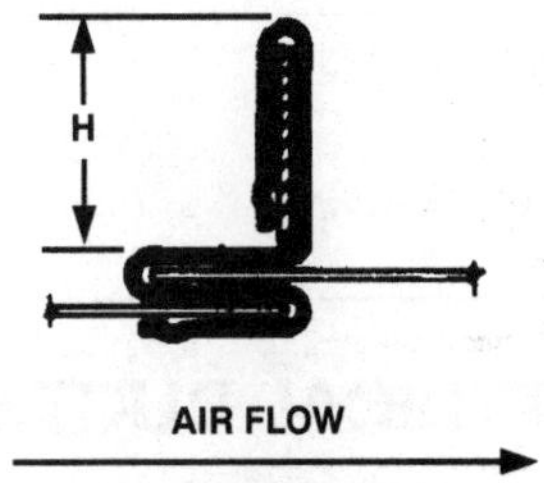

(G)
REINFORCED
BAR SLIP (CLEAT)

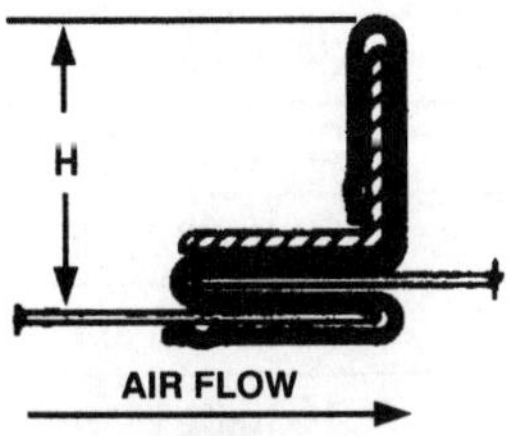

(H)
ANGLE SLIP

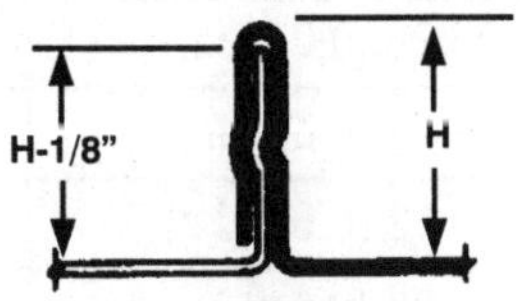

(I)
STANDING SEAM

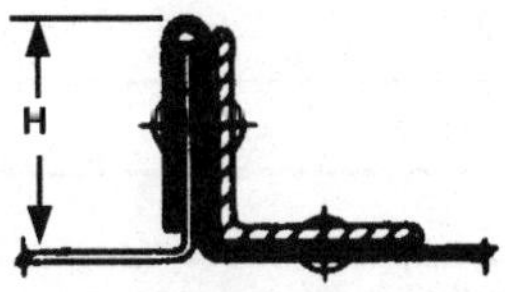

(J)
ANGLE REINFORCED
STANDING SEAM

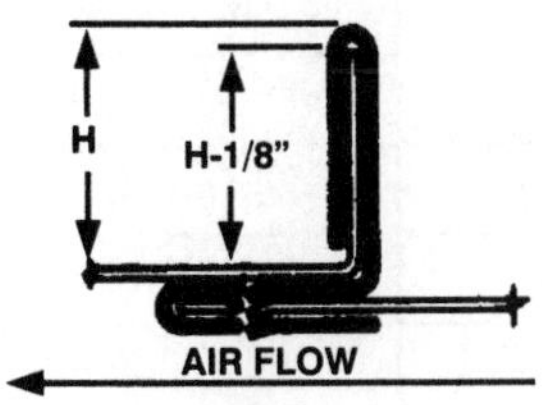

(K)
POCKET LOCK

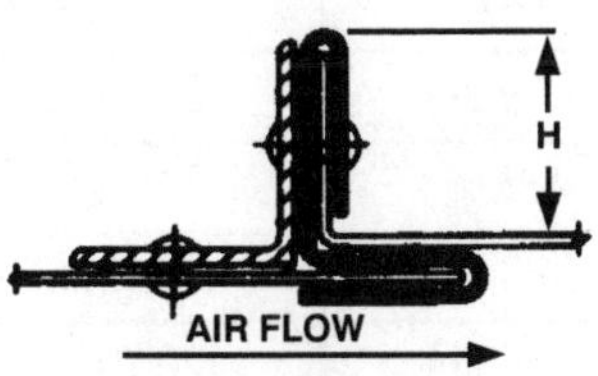

(L)
ANGLE REINFORCED
POCKET LOCK

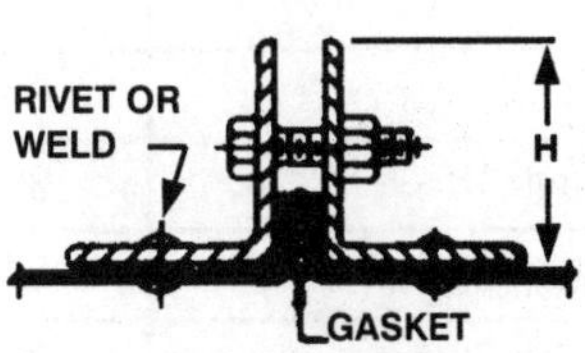

(M)
COMPANION ANGLES
(CAULK OR GASKET)

TYPICAL DUCT CONSTRUCTION SHEET METAL GAGES IN ONE- AND TWO- FAMILY DWELLINGS

Metal Gauges (duct not enclosed in partitions)		
ROUND DUCTS		
Diameter, Inches	**Minimum thickness galvanized sheet gage**	**Minimum thickness aluminum B&S gage**
Less than 12	30	26
12-14	28	26
15-18	26	24
RECTANGULAR DUCTS		
Width, Inches	**Minimum thickness galvanized sheet gage**	**Minimum thickness aluminum B&S gage**
Less than 14	28	24
14-24	26	22
25-30	24	22
Metal Gauges (duct enclosed in partitions)		
Width, Inches	**Minimum thickness galvanized sheet gage**	**Minimum thickness aluminum B&S gage**
14 or less	30	26

TYPICAL DUCT CONSTRUCTION SHEET METAL GLASS

RECTANGULAR DUCTS		
Maximum side inches	**Steel min. Galv. Sheet Gage**	**Aluminum Min. B&S Gage**
Through 12	26 (0.022 in.)	24 (0.020 in.)
13 through 30	24 (0.028 in.)	22 (0.025 in.)
31 through 54	22 (0.034 in.)	20 (0.032 in.)
55 through 84	20 (0.040 in.)	18 (0.040 in.)
Over 84	18 (0.052 in.)	16 (0.051 in.)

ROUND DUCTS			
Diameter Inches	**Spiral seam duct**	**Longitudinal seam duct**	**Fittings**
	Steel min. Galv. Sht. Gage	**Steel min. Galv. Sht. Gage**	**Steel min. Galv. Sht. Gage**
Through 12	28 (0.019 in.)	26 (0.022 in.)	26 (0.022 in.)
13 through 18	26 (0.022 in.)	24 (0.028 in.)	24 (0.028 in.)
19 through 28	24 (0.028 in.)	22 (0.034 in.)	22 (0.034 in.)
29 through 36	22 (0.034 in.)	20 (0.040 in.)	20 (0.040 in.)
37 through 52	20 (0.040 in.)	18 (0.052 in.)	18 (0.052 in.)

APPLICATIONS FOR CONDUCTORS USED FOR GENERAL WIRING

	AMBIENT TEMPERATURE								
	60°C 140°F	75°C 167°F	85°C 185°F	90°C 194°F	110°C 230°F	200°C 392°F	Dry	Dry or Wet	FEATURES
R	X						X		Code Rubber
RH		X					X		Heat Resistant
RHH				X			X		More Heat Resistant
RW	X							X	Moisture Resistant
RH-RW	X							X	Moisture and Heat Resistant
		X					X		Moisture and Heat Resistant
RHW		X						X	Moisture and Heat Resistant
RU	X						X		Latex Rubber
RUH		X					X		Heat Resistant
RUW	X							X	Moisture Resistant
T	X						X		Thermoplastic
TW	X							X	Moisture Resistant
THHN				X			X		Heat Resistant
THW		X						X	Moisture and Heat Resistant
THWN		X						X	Moisture and Heat Resistant
MI			X					X	Mineral Insulated
									Metal Sheathed
V			X				X		Varnished Cambric
AVA					X		X		With Asbestos
AVB				X			X		With Asbestos
AVL					X			X	With Asbestos

This table does not include special condition conductors, thickness of conductor insulation, or reference to all other protective coverings.

GENERAL CLASSIFICATION OF INSULATIONS:

A........... Asbestos
H........... Heat Resistant
MI......... Mineral Insulation
R........... Rubber
RU......... Latex Rubber
V........... Varnished Cambric
T........... Thermoplastic
W.......... (Water) Moisture Resistant

WIRE AND SHEET METAL GAGES
(In Decimals of an Inch)

Name of Gage	American Wire Gage (A.W.G.) (Corresponds to Brown & Sharpe Gage)	Birmingham Iron Wire Gage (B.W.G.)	United States Standard Gage (U.S.S.G.)	
Principal Use	Electrical Wire & Non-Ferrous Sheet Metal	Iron or Steel Wire	Ferrous Sheet Metal	
Gage No.				Gage No.
00 00000				00 00000
0 00000	.5800			0 00000
00000	.5165	.500		00000
0000	.4600	.454		0000
000	.4096	.425		000
00	.3648	.380		00
0	.3249	.340		0
1	.2893	.300		1
2	.2576	.284		2
3	.2294	.259	23.91	3
4	.2048	.238	.2242	4
5	.1819	.220	.2092	5
6	.1620	.203	.1943	6
7	.1443	.180	.1793	7
8	.1285	.165	.1644	8
9	.1144	.148	.1495	9
10	.1019	.134	.1345	10
11	.0907	.120	.1196	11
12	.0808	.109	.1046	12
13	.0720	.095	.0897	13
14	.0641	.083	.0747	14
15	.0571	.072	.0673	15
16	.0508	.065	.0598	16
17	.0453	.058	.0538	17
18	.0403	.049	.0478	18
19	.0359	.042	.0418	19
20	.0320	.035	.0359	20
21	.0285	.032	.0329	21
22	.0253	.028	.0299	22
23	.0226	.025	.0269	23
24	.0201	.022	.0239	24
25	.0179	.020	.0209	25
26	.0159	.018	.0179	26
27	.0142	.016	.0164	27
28	.0126	.014	.0149	28
29	.0113	.013	.0135	29
30	.0100	.012	.0120	30
31	.0089	.010	.0105	31
32	.0080	.009	.0097	32
33	.0071	.008	.0090	33
34	.0063	.007	.0082	34
35	.0056	.005	.0075	35
36	.0050	.004	.0067	36
37	.0045		.0064	37
38	.0040		.0060	38
39	.0035			39
40	.0031			40

Geographic Cost Modifiers

The costs as presented in this book attempt to represent national averages. Costs, however, vary among regions, states and even between adjacent localities.

In order to more closely approximate the probable costs for specific locations throughout the U.S., this table of Geographic Cost Modifiers is provided. These adjustment factors are used to modify costs obtained from this book to help account for regional variations of construction costs and to provide a more accurate estimate for specific areas. The factors are formulated by comparing costs in a specific area to the costs as presented in the Costbook pages. An example of how to use these factors is shown below. Whenever local current costs are known, whether material prices or labor rates, they should be used when more accuracy is required.

Cost Obtained from Costbook Pages **X** Location Cost Adjustment Factor Divided by 100 **=** **Adjusted Cost**

For example, a project estimated to cost \$125,000 using the Costbook pages can be adjusted to more closely approximate the cost in Los Angeles:

$$\$125{,}000 \text{ X } \frac{105}{100} = \$131{,}250$$

GEOGRAPHIC COST MODIFERS

01025

State	Metropolitan Areas	Multiplier
AK	ANCHORAGE	122
AL	ANNISTON	80
	AUBURN-OPELIKA	80
	BIRMINGHAM	79
	DOTHAN	77
	GADSDEN	77
	HUNTSVILLE	79
	MOBILE	82
	MONTGOMERY	77
	TUSCALOOSA	79
AR	FAYETTEVILLE-SPRINGDALE-ROGERS	72
	FORT SMITH	77
	JONESBORO	76
	LITTLE ROCK-NORTH LITTLE ROCK	78
	PINE BLUFF	77
	TEXARKANA	77
AZ	FLAGSTAFF	88
	PHOENIX-MESA	88
	TUCSON	87
	YUMA	89
CA	BAKERSFIELD	100
	FRESNO	102
	LOS ANGELES-LONG BEACH	105
	MODESTO	99
	OAKLAND	108
	ORANGE COUNTY	102
	REDDING	99
	RIVERSIDE-SAN BERNARDINO	100
	SACRAMENTO	102
	SALINAS	104
	SAN DIEGO	101
	SAN FRANCISCO	111
	SAN JOSE	109
	SAN LUIS OBISPO	98
	SANTA CRUZ-WATSONVILLE	104
	SANTA ROSA	105
	STOCKTON-LODI	101
	VALLEJO-FAIRFIELD-NAPA	104
	VENTURA	100
	SANTA BARBARA	103

GEOGRAPHIC COST MODIFERS		01025
State	**Metropolitan Areas**	**Multiplier**
CO	BOULDER-LONGMONT	86
	COLORADO SPRINGS	91
	DENVER	91
	FORT COLLINS-LOVELAND	84
	GRAND JUNCTION	86
	GREELEY	84
	PUEBLO	88
CT	BRIDGEPORT	100
	DANBURY	100
	HARTFORD	99
	NEW HAVEN-MERIDEN	100
	NEW LONDON-NORWICH	98
	STAMFORD-NORWALK	103
	WATERBURY	99
DC	WASHINGTON	94
DE	DOVER	93
	WILMINGTON-NEWARK	94
FL	DAYTONA BEACH	83
	FORT LAUDERDALE	86
	FORT MYERS-CAPE CORAL	79
	FORT PIERCE-PORT ST. LUCIE	85
	FORT WALTON BEACH	87
	GAINESVILLE	81
	JACKSONVILLE	84
	LAKELAND-WINTER HAVEN	81
	MELBOURNE-TITUSVILLE-PALM BAY	88
	MIAMI	86
	NAPLES	88
	OCALA	85
	ORLANDO	85
	PANAMA CITY	75
	PENSACOLA	79
	SARASOTA-BRADENTON	80
	TALLAHASSEE	77
	TAMPA-ST. PETERSBURG-CLEARWATER	83
	WEST PALM BEACH-BOCA RATON	86
GA	ALBANY	76
	ATHENS	78
	ATLANTA	85
	AUGUSTA	74
	COLUMBUS	75
	MACON	78
	SAVANNAH	78

GEOGRAPHIC COST MODIFERS

01025

State	Metropolitan Areas	Multiplier
HI	HONOLULU	121
IA	CEDAR RAPIDS	87
	DAVENPORT	89
	DES MOINES	90
	DUBUQUE	85
	IOWA CITY	89
	SIOUX CITY	85
	WATERLOO-CEDAR FALLS	84
ID	BOISE CITY	88
	POCATELLO	87
IL	BLOOMINGTON-NORMAL	95
	CHAMPAIGN-URBANA	94
	CHICAGO	103
	DECATUR	93
	KANKAKEE	96
	PEORIA-PEKIN	95
	ROCKFORD	95
	SPRINGFIELD	93
IN	BLOOMINGTON	91
	EVANSVILLE	89
	FORT WAYNE	90
	GARY	97
	INDIANAPOLIS	93
	KOKOMO	89
	LAFAYETTE	90
	MUNCIE	90
	SOUTH BEND	90
	TERRE HAUTE	90
KS	KANSAS CITY	88
	LAWRENCE	84
	TOPEKA	83
	WICHITA	83
KY	LEXINGTON	84
	LOUISVILLE	86
	OWENSBORO	84
LA	ALEXANDRIA	79
	BATON ROUGE	82
	HOUMA	82
	LAFAYETTE	81
	LAKE CHARLES	82
	MONROE	79
	NEW ORLEANS	85
	SHREVEPORT-BOSSIER CITY	80

GEOGRAPHIC COST MODIFERS

01025

State	Metropolitan Areas	Multiplier
MA	BARNSTABLE-YARMOUTH	103
	BOSTON	105
	BROCKTON	102
	FITCHBURG-LEOMINSTER	99
	LAWRENCE	99
	LOWELL	98
	NEW BEDFORD	102
	PITTSFIELD	98
	SPRINGFIELD	98
	WORCESTER	98
MD	BALTIMORE	90
	CUMBERLAND	86
	HAGERSTOWN	87
ME	BANGOR	85
	LEWISTON-AUBURN	86
	PORTLAND	88
MI	ANN ARBOR	95
	DETROIT	96
	FLINT	92
	GRAND RAPIDS-MUSKEGON-HOLLAND	89
	JACKSON	91
	KALAMAZOO-BATTLE CREEK	86
	LANSING-EAST LANSING	91
	SAGINAW-BAY CITY-MIDLAND	90
MN	DULUTH	95
	MINNEAPOLIS-ST. PAUL	100
	ROCHESTER	95
	ST. CLOUD	97
MO	COLUMBIA	89
	JOPLIN	85
	KANSAS CITY	91
	SPRINGFIELD	86
	ST. JOSEPH	89
	ST. LOUIS	89
MS	BILOXI-GULFPORT-PASCAGOULA	78
	JACKSON	77
MT	BILLINGS	88
	GREAT FALLS	88
	MISSOULA	86
NC	ASHEVILLE	73
	CHARLOTTE	75

GEOGRAPHIC COST MODIFERS

01025

State	Metropolitan Areas	Multiplier
NC	FAYETTEVILLE	75
	GREENSBORO-WINSTON-SALEM-HIGH POINT	74
	GREENVILLE	74
	HICKORY-MORGANTON-LENOIR	71
	RALEIGH-DURHAM-CHAPEL HILL	74
	ROCKY MOUNT	71
	WILMINGTON	75
ND	BISMARCK	86
	FARGO	89
	GRAND FORKS	87
NE	LINCOLN	82
	OMAHA	86
NH	MANCHESTER	90
	NASHUA	88
	PORTSMOUTH	86
NJ	ATLANTIC-CAPE MAY	99
	BERGEN-PASSAIC	101
	JERSEY CITY	100
	MIDDLESEX-SOMERSET-HUNTERDON	97
	MONMOUTH-OCEAN	99
	NEWARK	102
	TRENTON	99
	VINELAND-MILLVILLE-BRIDGETON	98
NM	ALBUQUERQUE	86
	LAS CRUCES	83
	SANTA FE	90
NV	LAS VEGAS	97
	RENO	96
NY	ALBANY-SCHENECTADY-TROY	91
	BINGHAMTON	90
	BUFFALO-NIAGARA FALLS	97
	ELMIRA	85
	GLENS FALLS	84
	JAMESTOWN	92
	NASSAU-SUFFOLK	103
	NEW YORK	118
	ROCHESTER	95
	SYRACUSE	92
	UTICA-ROME	92

GEOGRAPHIC COST MODIFERS

01025

State	Metropolitan Areas	Multiplier
OH	AKRON	92
	CANTON-MASSILLON	90
	CINCINNATI	90
	CLEVELAND-LORAIN-ELYRIA	93
	COLUMBUS	90
	DAYTON-SPRINGFIELD	88
	LIMA	90
	MANSFIELD	87
	STEUBENVILLE	93
	TOLEDO	92
	YOUNGSTOWN-WARREN	87
OK	ENID	79
	LAWTON	80
	OKLAHOMA CITY	81
	TULSA	80
OR	EUGENE-SPRINGFIELD	93
	MEDFORD-ASHLAND	91
	PORTLAND	95
	SALEM	94
PA	ALLENTOWN-BETHLEHEM-EASTON	94
	ALTOONA	92
	ERIE	93
	HARRISBURG-LEBANON-CARLISLE	90
	JOHNSTOWN	93
	LANCASTER	91
	PHILADELPHIA	102
	PITTSBURGH	93
	READING	93
	SCRANTON-WILKES-BARRE-HAZLETON	91
	STATE COLLEGE	88
	WILLIAMSPORT	89
	YORK	92
RI	PROVIDENCE	98
SC	AIKEN	81
	CHARLESTON-NORTH CHARLESTON	76
	COLUMBIA	76
	FLORENCE	73
	GREENVILLE-SPARTANBURG-ANDERSON	76
	MYRTLE BEACH	81
SD	RAPID CITY	80
	SIOUX FALLS	80

GEOGRAPHIC COST MODIFERS

01025

State	Metropolitan Areas	Multiplier
TN	CHATTANOOGA	79
	JACKSON	78
	JOHNSON CITY	75
	KNOXVILLE	79
	MEMPHIS	83
	NASHVILLE	82
TX	ABILENE	77
	AMARILLO	78
	AUSTIN-SAN MARCOS	78
	BEAUMONT-PORT ARTHUR	79
	BROWNSVILLE-HARLINGEN-SAN BENITO	80
	BRYAN-COLLEGE STATION	78
	CORPUS CHRISTI	76
	DALLAS	83
	EL PASO	76
	FORT WORTH-ARLINGTON	82
	GALVESTON-TEXAS CITY	81
	HOUSTON	84
	LAREDO	70
	LONGVIEW-MARSHALL	75
	LUBBOCK	78
	MCALLEN-EDINBURG-MISSION	75
	ODESSA-MIDLAND	76
	SAN ANGELO	75
	SAN ANTONIO	80
	TEXARKANA	77
	TYLER	77
	VICTORIA	77
	WACO	77
	WICHITA FALLS	77
UT	PROVO-OREM	84
	SALT LAKE CITY-OGDEN	83
VA	CHARLOTTESVILLE	79
	LYNCHBURG	79
	NORFOLK-VIRGINIA BEACH-NEWPORT NEWS	81
	RICHMOND-PETERSBURG	83
	ROANOKE	76
VT	BURLINGTON	91
WA	BELLINGHAM	100
	BREMERTON	98
	OLYMPIA	97

GEOGRAPHIC COST MODIFERS

01025

State	Metropolitan Areas	Multiplier
WA	RICHLAND-KENNEWICK-PASCO	95
	SEATTLE-BELLEVUE-EVERETT	102
	SPOKANE	97
	TACOMA	99
	YAKIMA	95
WI	APPLETON-OSHKOSH-NEENAH	90
	EAU CLAIRE	90
	GREEN BAY	91
	JANESVILLE-BELOIT	90
	KENOSHA	92
	LA CROSSE	89
	MADISON	89
	MILWAUKEE-WAUKESHA	96
	RACINE	92
	WAUSAU	90
WV	CHARLESTON	89
	HUNTINGTON	90
	PARKERSBURG	87
	WHEELING	89
WY	CASPER	87
	CHEYENNE	89

Square Foot Tables

The following Square Foot Tables list hundreds of actual projects for dozens of building types, each with associated building size, total square foot building cost and percentage of project costs for total mechanical and electrical components. This data provides an overview of construction costs by building type. These costs are for actual projects. The variations within similar building types may be due, among other factors, to size, location, quality and specified components, materials and processes. Depending upon all such factors, specific building costs can vary significantly and may not necessarily fall within the range of costs as presented. The data has been updated to reflect current construction costs.

BUILDING CATEGORY | **PAGE**

SQUARE FOOT TABLES

COMMERCIAL

AUTO DEALERSHIP

Project Size Gross S.F.	Project Cost $/S.F.	% Cost Mechanical	% Cost Electrical
7,700	121.30	16.7	9.4
16,100	68.80	10.2	15.9
20,000	72.30	12.9	23.4
26,300	73.70	12.5	22.0
43,600	58.40	19.4	13.2
53,600	97.30	12.5	11.5

BUSINESS CENTER

Project Size Gross S.F.	Project Cost $/S.F.	% Cost Mechanical	% Cost Electrical
3,900	72.60	12.0	9.2
9,900	70.30	9.1	7.6
54,400	47.00	3.4	12.2
135,000	51.00	8.2	1.5

CINEMA

Project Size Gross S.F.	Project Cost $/S.F.	% Cost Mechanical	% Cost Electrical
18,000	166.10	10.9	6.7
22,500 A	102.20	6.6	4.2

MALL/PLAZA

Project Size Gross S.F.	Project Cost $/S.F.	% Cost Mechanical	% Cost Electrical
9,700	51.40	15.0	13.3
10,500	82.20	8.0	13.5
16,300	74.80	9.4	10.6
26,900	74.50	15.0	7.0
36,000	63.10	10.0	11.0
36,300	71.60	12.4	8.2
44,720	74.80	18.3	12.4
59,100	73.40	9.8	9.5
60,000 R	75.20	10.0	9.5
64,100	75.60	22.4	18.4
66,000	86.50	13.5	11.5
67,400	62.90	21.0	15.0
73,500	91.00	14.9	6.9

MALL/PLAZA (Cont.)

Project Size Gross S.F.	Project Cost $/S.F.	% Cost Mechanical	% Cost Electrical
142,000	57.20	7.1	8.0
220,000	151.10	11.0	6.4
223,700	44.40	9.0	9.3
321,200	52.80	7.3	7.4
379,900	68.90	11.2	6.2
405,100	71.10	13.9	6.0
482,000	118.30	10.5	9.4
630,000	79.10	12.2	12.4

RESTAURANT

Project Size Gross S.F.	Project Cost $/S.F.	% Cost Mechanical	% Cost Electrical
4,300 R	130.60	6.9	8.7
4,400 R	192.20	14.6	8.5
5,800	156.20	28.0	10.6
6,800 A	182.60	7.0	11.1
7,360 R	196.60	16.0	6.5
9,600	195.30	24.7	13.1
10,000 R	203.20	21.0	10.0
10,100	175.30	28.6	18.4
10,600	322.80	20.4	6.4
22,900 R	209.80	15.8	16.9

RETAIL STORE

Project Size Gross S.F.	Project Cost $/S.F.	% Cost Mechanical	% Cost Electrical
1,000	196.60	12.8	6.7
3,000 R	176.10	14.3	10.5
12,300	241.30	14.0	10.0
30,000	121.90	15.6	26.2
61,300	63.10	13.3	13.0
115,000	87.30	14.6	11.3
154,700	123.70	11.2	12.4
314,700 R	104.30	13.8	9.4

A = Addition R = Remodel

SQUARE FOOT TABLES

RESIDENTIAL

APARTMENTS

Project Size Gross S.F.	Project Cost $/S.F.	% Cost Mechanical	% Cost Electrical
3,700	73.10	14.9	4.4
13,900	100.10	7.9	7.2
19,200 R	158.90	45.3	7.5
19,700	93.10	7.4	10.4
23,700	102.30	10.6	4.3
26,500	96.80	25.2	12.8
35,100	79.20	16.4	5.6
54,000	137.10	23.3	13.1
62,700	99.60	17.0	9.0
67,300	91.00	13.8	8.4
70,600	51.20	18.1	7.8
75,300	113.00	13.2	8.1
75,600	110.00	14.5	8.9
72,200	95.30	18.4	10.9
77,600	123.40	26.9	14.3
88,100	114.50	15.3	9.3
89,500	107.40	10.7	11.1
94,100	64.70	8.5	6.8
96,000	80.70	17.0	13.3
102,000	140.10	17.8	12.5
103,200	63.70	12.1	8.9
103,600	102.60	19.1	9.0
105,200	130.60	14.9	8.9
106,200	94.10	12.8	9.3
110,900	90.40	15.6	8.4
111,800	128.40	17.3	7.4
115,900	96.30	12.5	8.6
117,200	60.20	12.9	8.0
119,000	82.60	15.6	8.4

APARTMENTS (Cont.)

Project Size Gross S.F.	Project Cost $/S.F.	% Cost Mechanical	% Cost Electrical
119,400	54.10	18.6	8.7
144,300	120.30	15.0	8.6
176,300	102.30	19.1	9.0
192,300	60.20	10.1	6.0
210,900	118.40	19.1	9.5
220,200	121.80	14.8	7.5
253,900	158.30	20.6	7.7
369,500	115.60	15.8	9.0

CONDOS/TOWNHOUSE

Project Size Gross S.F.	Project Cost $/S.F.	% Cost Mechanical	% Cost Electrical
8,600	103.70	8.5	4.4
16,700	83.60	13.2	6.5
18,000	176.10	14.6	9.7
18,400	90.30	12.9	5.8
74,800	81.90	13.8	5.2
111,700	103.40	9.2	7.1
150,300	115.40	15.9	7.9
278,800	180.40	14.1	7.9
1,109,900	83.00	9.8	4.7

SINGLE-FAMILY HOMES

Project Size Gross S.F.	Project Cost $/S.F.	% Cost Mechanical	% Cost Electrical
600 R	74.70	16.3	2.0
900	97.10	33.0	3.0
2,100	112.40	8.8	4.0
2,200	240.80	23.8	3.4
2,500	118.00	6.1	7.1
2,900	113.90	7.7	2.8
3,000	80.00	9.5	6.8
3,100	178.80	15.5	4.6
3,600	112.00	8.5	3.0
3,700	156.50	9.8	3.4

A = Addition R = Remodel

SQUARE FOOT TABLES

RESIDENTIAL (Cont.)

SINGLE-FAMILY HOMES (Cont.)

Project Size Gross S.F.	Project Cost $/S.F.	% Cost Mechanical	% Cost Electrical
4,200	136.10	15.5	5.7
4,600	216.20	9.0	4.7
5,200	280.90	8.3	6.0
5,700	164.60	7.4	3.7
5,700	111.90	8.7	12.6
21,300*	79.80	26.0	4.0
22,700*	74.10	27.0	4.6
45,000*	112.70	7.8	2.5
51,458*	81.70	11.0	5.0

*TOWNHOUSES

EDUCATIONAL

ADMINISTRATION (OFFICES)

Project Size Gross S.F.	Project Cost $/S.F.	% Cost Mechanical	% Cost Electrical
53,700	192.20	15.4	9.0

ATHLETIC FACILITY

Project Size Gross S.F.	Project Cost $/S.F.	% Cost Mechanical	% Cost Electrical
38,100	190.70	16.8	9.8
44,100	176.10	16.8	8.4
100,000	120.30	19.2	5.3
160,000	217.90	13.7	7.9
247,500	182.80	11.2	9.0
271,000	171.70	13.2	7.6
283,100	206.90	14.6	6.0

AUDITORIUM/PERFORMING ARTS

Project Size Gross S.F.	Project Cost $/S.F.	% Cost Mechanical	% Cost Electrical
9,900	308.10	17.5	29.2
17,800	281.70	11.4	16.1
29,200	278.70	16.2	10.5
62,700	202.70	13.0	12.9

CLASSROOM

Project Size Gross S.F.	Project Cost $/S.F.	% Cost Mechanical	% Cost Electrical
35,400	279.50	10.2	6.8
70,000	126.20	24.3	13.1
78,900	234.80	13.3	14.3
80,100	190.70	16.9	10.5
100,000	190.00	21.1	9.8
166,000	131.30	16.1	11.2
298,400	123.20	11.3	11.9

COMPLETE COLLEGE FACILITIES

Project Size Gross S.F.	Project Cost $/S.F.	% Cost Mechanical	% Cost Electrical
95,300	198.10	19.4	11.7
450,000	236.40	18.6	10.8

ELEMENTARY SCHOOL

Project Size Gross S.F.	Project Cost $/S.F.	% Cost Mechanical	% Cost Electrical
18,000	144.90	19.1	13.8
30,800	133.30	16.0	10.0
31,600	111.50	12.1	11.3
35,700	162.90	17.3	8.7
40,000	141.60	18.5	13.4
40,500	124.10	22.1	11.3
57,000	107.30	22.1	7.5
69,700	167.30	20.6	9.3
91,400	134.90	18.2	7.6

HIGH SCHOOL

Project Size Gross S.F.	Project Cost $/S.F.	% Cost Mechanical	% Cost Electrical
116,400	148.10	18.1	12.9
133,000	127.50	17.8	10.7
184,000	224.60	23.0	10.0
217,200 R	121.00	26.4	10.6
254,000 R	93.40	17.2	14.6
431,700	142.30	13.1	9.6

A = Addition R = Remodel

SQUARE FOOT TABLES

EDUCATIONAL (Cont.)

JUNIOR HIGH SCHOOL

Project Size Gross S.F.	Project Cost $/S.F.	% Cost Mechanical	% Cost Electrical
26,000	186.50	9.5	9.3
28,100	115.60	11.9	9.5
52,800	149.60	18.3	8.6
91,600	184.00	21.2	11.0
123,700	150.50	29.1	9.1

LABORATORY/RESEARCH

Project Size Gross S.F.	Project Cost $/S.F.	% Cost Mechanical	% Cost Electrical
9,200	322.80	25.4	4.5
80,300	253.80	20.2	15.5

LIBRARY

Project Size Gross S.F.	Project Cost $/S.F.	% Cost Mechanical	% Cost Electrical
6,900	169.30	17.9	10.3
8,200	155.40	18.8	10.6
12,000	201.40	19.4	15.9
15,000	176.10	15.7	18.7
16,300	149.10	11.3	7.4
28,600	133.50	15.7	9.0
30,100	181.20	13.0	11.0
37,700	142.30	14.6	8.0
43,500	120.30	16.6	6.7
47,900	210.80	29.7	9.0
51,400	215.40	13.1	12.4
63,400 A	162.00	13.5	8.3
64,000	151.60	10.9	11.4
74,000	171.70	17.2	8.0
75,600	228.00	12.5	16.7
176,000	126.30	11.2	9.8

SPECIAL NEEDS FUNCTION

Project Size Gross S.F.	Project Cost $/S.F.	% Cost Mechanical	% Cost Electrical
15,200	143.20	16.6	8.4
27,900	172.50	19.4	11.2

STUDENT CENTER/MULTIPURPOSE

Project Size Gross S.F.	Project Cost $/S.F.	% Cost Mechanical	% Cost Electrical
90,000	217.70	16.2	10.0
49,600	160.80	18.2	9.2
187,700	242.10	15.3	8.8
194,800	127.90	17.2	8.5

HOTEL/MOTEL

CONVENTION/CONFERENCE CENTER

Project Size Gross S.F.	Project Cost $/S.F.	% Cost Mechanical	% Cost Electrical
8,600 A	191.20	20.7	16.3
71,900	212.00	12.5	13.8
433,800	110.80	22.1	8.3

HOTEL

Project Size Gross S.F.	Project Cost $/S.F.	% Cost Mechanical	% Cost Electrical
19,900 A	109.30	16.8	5.8
25,875 R	98.00	11.8	10.5
48,400 A	196.00	23.6	8.2
64,300 R	264.10	20.3	10.1
104,200 A	122.70	15.0	8.8
108,040	104.80	13.0	7.0
110,100	135.60	18.4	10.5
132,000	253.50	16.4	5.4
135,900 A	156.70	15.2	7.7
144,100 A	176.10	19.3	11.5
231,000	190.40	15.2	8.4
449,800 A	110.00	13.1	7.0

HOTEL/INN

Project Size Gross S.F.	Project Cost $/S.F.	% Cost Mechanical	% Cost Electrical
57,400	127.20	13.0	10.0
73,000	83.30	24.4	18.1
75,900	110.80	16.5	7.6
162,000	133.30	17.5	8.0
197,000	127.60	15.7	7.7
277,900	93.70	18.8	9.4

A = Addition R = Remodel

SQUARE FOOT TABLES

INDUSTRIAL

MANUFACTURING

Project Size Gross S.F.	Project Cost $/S.F.	% Cost Mechanical	% Cost Electrical
14,300	108.30	13.0	7.0
18,500	163.30	20.5	13.5
26,600	58.40	6.1	12.1
31,400	129.00	18.9	17.7
33,400	74.50	23.1	15.8
37,300	55.60	3.0	24.0
43,400	74.10	14.7	11.7
45,400	120.30	15.9	15.0
79,800	125.00	16.0	14.5
81,100	133.50	8.2	8.4
137,400	83.80	41.6	13.6
179,600	96.70	26.3	13.6
186,000	90.40	19.5	11.7

RESEARCH AND DEVELOPMENT

Project Size Gross S.F.	Project Cost $/S.F.	% Cost Mechanical	% Cost Electrical
89,140	172.70	20.8	9.0
100,400	228.70	36.1	25.1
114,200	224.00	21.4	9.8
125,000	152.40	20.8	8.4
140,000	212.90	20.1	11.2

WAREHOUSE W/OFFICE

Project Size Gross S.F.	Project Cost $/S.F.	% Cost Mechanical	% Cost Electrical
14,000	49.70	6.5	9.2
19,000	44.50	2.3	1.8
19,700	58.50	9.5	7.0
31,200	50.30	7.0	7.0
40,500	64.20	6.6	10.5
62,000	81.60	10.8	10.0
96,200	47.30	2.2	6.3
105,000	45.90	5.3	11.1
149,800	48.00	14.5	8.6

WAREHOUSE W/OFFICE (Cont.)

Project Size Gross S.F.	Project Cost $/S.F.	% Cost Mechanical	% Cost Electrical
168,600	52.80	11.4	6.3
209,600	48.70	4.9	10.3
402,400	66.20	14.9	8.3

MEDICAL

EDUCATION CENTER

Project Size Gross S.F.	Project Cost $/S.F.	% Cost Mechanical	% Cost Electrical
35,400	296.30	10.2	6.8

HOSPITALS

Project Size Gross S.F.	Project Cost $/S.F.	% Cost Mechanical	% Cost Electrical
9,300 R	292.70	29.5	12.0
15,900 A	216.00	27.4	15.3
16,600 R	77.80	14.7	9.6
22,000	491.50	31.6	17.5
39,100	244.90	23.3	7.9
63,800	197.30	23.4	7.8
98,000 A	283.90	20.5	17.0
100,200 A	452.90	22.2	9.6
103,900 A	251.00	31.1	11.7
109,300	248.70	20.7	15.7
148,700	195.40	25.2	11.8
154,700	317.30	19.3	10.5
165,484	275.80	21.5	14.8
165,700 A	269.20	26.2	17.7
179,400 A	164.00	28.9	20.3
182,800	244.10	19.1	14.0
265,000	291.20	31.2	14.1
281,100	278.90	24.1	14.6
435,000	289.00	21.5	13.7
694,300	160.10	28.9	9.7
772,300	294.90	31.5	13.4

A = Addition   R = Remodel

SQUARE FOOT TABLES

MEDICAL (Cont.)

MEDICAL OFFICES/CENTERS

Project Size Gross S.F.	Project Cost $/S.F.	% Cost Mechanical	% Cost Electrical
3,000	123.80	13.7	11.5
5,500	128.80	8.5	18.7
10,000	155.70	13.9	9.4
10,600	197.20	13.2	8.9
16,300	163.20	21.8	13.2
18,300	89.00	7.1	13.4
20,600	127.90	14.3	6.4
24,900	225.20	21.0	15.1
27,000	227.40	19.7	10.1
28,400	158.50	13.6	9.3
30,500	171.70	17.2	12.7
32.000	136.10	18.1	10.5
44,300	137.30	24.4	16.2
50,200	79.20	9.8	4.9
51,200	200.60	21.0	12.0
64,600	95.50	13.2	6.4
66,000	89.00	9.8	8.8
80,000	76.40	8.2	8.3
137,175	127.60	12.8	6.6

NURSING HOMES

Project Size Gross S.F.	Project Cost $/S.F.	% Cost Mechanical	% Cost Electrical
11,600 A	374.10	53.2	7.9
16,800	243.50	33.3	7.8
31,900 A	183.10	22.0	11.0
64,100	157.00	20.6	11.1
290,000	220.10	16.1	13.6

RESEARCH

Project Size Gross S.F.	Project Cost $/S.F.	% Cost Mechanical	% Cost Electrical
34,600	208.40	18.1	3.0

PUBLIC FACILITIES

ANIMAL CENTER

Project Size Gross S.F.	Project Cost $/S.F.	% Cost Mechanical	% Cost Electrical
20,000	264.10	20.7	4.6
39,100	197.30	22.9	6.8
44,300	188.00	8.1	5.8

AUTO DEALERSHIP

Project Size Gross S.F.	Project Cost $/S.F.	% Cost Mechanical	% Cost Electrical
7,700	123.10	16.7	9.4

BROADCASTING

Project Size Gross S.F.	Project Cost $/S.F.	% Cost Mechanical	% Cost Electrical
20,000	360.00	20.0	13.0
29,500	277.80	16.6	15.0
45,000 R	201.40	15.0	13.0

CIVIC CENTER

Project Size Gross S.F.	Project Cost $/S.F.	% Cost Mechanical	% Cost Electrical
6,000	198.40	9.2	2.8
23,900	249.10	12.3	10.9
34,400	113.50	3.5	17.8
69,800 A	297.50	17.7	8.8
206,500	160.90	15.0	11.0

CORRECTION FACILITIES

Project Size Gross S.F.	Project Cost $/S.F.	% Cost Mechanical	% Cost Electrical
44,600	240.50	20.9	13.1
66,000	146.80	15.0	23.0
257,800 A	254.00	20.9	10.7
360,000	171.10	32.7	13.2

FIRE STATION

Project Size Gross S.F.	Project Cost $/S.F.	% Cost Mechanical	% Cost Electrical
6,900	214.70	12.4	9.7
7,600	164.90	16.0	9.5
8,430	226.80	12.8	9.6
9,600	197.80	13.5	11.8

A = Addition R = Remodel

SQUARE FOOT TABLES

PUBLIC FACILITIES (Cont.)

GOVERMENT BUILDINGS

Project Size Gross S.F.	Project Cost $/S.F.	% Cost Mechanical	% Cost Electrical
12,300	153.90	13.5	10.7
23,500	252.20	11.1	15.6
27,300	199.10	19.5	9.3
31,600	224.20	27.1	11.3
46,600	211.00	17.4	13.8
72,100	229.70	24.4	10.7
78,200	190.60	19.7	15.2
332,900	185.00	16.2	14.2
364,100	242.20	14.6	12.9
771,000	267.60	17.6	11.3

MUSEUM

Project Size Gross S.F.	Project Cost $/S.F.	% Cost Mechanical	% Cost Electrical
27,600	184.80	17.8	14.1
30,100	212.30	18.6	7.9
43,264	197.80	11.1	8.3
63,000	214.10	8.8	18.1

PARKING GARAGE

Project Size Gross S.F.	Project Cost $/S.F.	% Cost Mechanical	% Cost Electrical
66,000	56.50	2.8	3.1
169,000	41.70	10.3	3.7
562,700	37.50	2.4	6.2

TRANSPORTATION

Project Size Gross S.F.	Project Cost $/S.F.	% Cost Mechanical	% Cost Electrical
7,300	339.60	15.1	3.2
14,300	272.90	13.3	16.6
23.000	188.50	9.6	13.7
35,500	157.00	1.0	19.0
49,100	234.20	35.5	11.6

TRANSPORTATION (Cont.)

Project Size Gross S.F.	Project Cost $/S.F.	% Cost Mechanical	% Cost Electrical
288,100	135.00	8.3	13.3
2,160,000	218.60	23.3	11.5

OFFICES

BANKS

Project Size Gross S.F.	Project Cost $/S.F.	% Cost Mechanical	% Cost Electrical
2,900	314.00	5.3	4.3
3,100	140.10	5.9	6.7
3,300	203.80	10.3	16.4
3,600	153.50	10.3	12.2
4,000	174.90	8.0	9.0
4,100	173.10	21.4	13.0
4,200	186.70	8.6	14.21
4,400	215.70	12.2	12.7
4,500	133.90	12.4	13.5
4,900	211.50	11.7	11.2
5,900	157.30	9.3	13.8
6,000	176.80	11.6	7.3
6,100	237.70	11.0	8.0
7,000	319.80	6.0	9.0
7,300	195.60	11.9	11.3
7,700	219.30	8.0	7.5
7,800	240.30	11.0	11.7
8,000	122.10	10.0	14.0
9,200	193.70	9.9	12.1
9,400	140.20	11.7	11.8
10,200	292.50	12.6	12.6
12,600	104.50	7.0	18.0

A = Addition R = Remodel

SQUARE FOOT TABLES

OFFICES (Cont.)

BANKS (Cont.)

Project Size Gross S.F.	Project Cost $/S.F.	% Cost Mechanical	% Cost Electrical
13,300	184.30	9.5	8.3
13,800	164.90	10.00	9.7
15,000	141.30	15.4	12.4
15,200	112.80	9.2	12.9
15,500	144.00	9.8	10.3
16,000	94.50	13.4	23.1
20,100	88.90	13.0	11.0
21,700	188.00	8.8	11.3
44,800	160.60	13.0	8.2
53,200	290.10	14.9	7.2
62,100	166.60	10.1	7.9
95,100	219.20	13.5	4.3

OFFICE BUILDINGS

Project Size Gross S.F.	Project Cost $/S.F.	% Cost Mechanical	% Cost Electrical
2,600	197.00	17.2	9.4
3,400	161.10	10.5	11.3
3,800	158.10	16.4	11.8
4,400	156.20	12.8	8.5
4,500	134.40	13.0	7.0
5,100	97.60	16.6	10.2
5,200	128.10	8.0	5.7
6,700	190.70	16.8	10.4
7,500	205.40	10.1	8.0
7,900	159.90	17.4	9.0
8,100	247.10	10.6	11.0
10,600	142.90	9.9	9.7
10,900	82.30	13.8	10.8

OFFICE BUILDINGS (Cont.)

Project Size Gross S.F.	Project Cost $/S.F.	% Cost Mechanical	% Cost Electrical
11,300	162.40	17.0	6.0
13,000	110.50	15.0	9.0
14,400	145.00	19.8	12.9
14,500	111.20	17.3	12.5
17,000	158.70	14.7	7.9
17,800 A	85.80	10.4	11.0
18,100	166.30	22.5	11.6
19,300	90.30	11.1	8.1
24,600	83.40	18.5	14.1
27,700	100.40	19.6	5.5
27,800	182.10	12.7	5.1
27,800	98.00	17.8	10.3
32,500 R	167.70	13.9	6.4
35,400	97.90	15.0	12.0
36.500	84.80	10.2	10.2
42,300	105.30	10.3	7.7
44,400	143.40	23.5	14.7
44,400	75.40	11.0	5.0
44,500	103.70	11.5	3.1
45,400	89.60	19.1	13.2
47,300	92.20	18.5	8.0
49,700	154.90	22.1	7.2
50,000	147.60	19.4	15.6
50,400	160.20	23.2	7.8
52,200	104.00	18.3	7.8
52,900	121.30	4.4	3.9
53,700	185.50	15.4	9.0

A = Addition R = Remodel

SQUARE FOOT TABLES

OFFICES (Cont.)

OFFICE BUILDINGS (Cont.)

Project Size Gross S.F.	Project Cost $/S.F.	% Cost Mechanical	% Cost Electrical
54,000	75.10	14.4	2.6
56,000	85.20	10.6	6.2
56,500	117.20	19.1	11.0
72,000 R	41.20	12.9	20.2
74,000	81.60	13.5	6.8
80,800	82.50	12.0	6.0
81,800	113.60	21.1	9.5
81,900	95.50	19.9	9.0
82,000	132.80	14.0	4.2
83,100	156.50	16.0	8.6
85,400	137.30	18.7	8.6
86,200	107.50	14.7	11.0
99,900	124.40	16.6	6.6
100,000	139.00	10.6	12.1
100,000	82.00	16.3	10.4
116,400	131.30	12.6	10.0
134,500	267.80	14.1	12.4
140,000	207.60	20.1	11.2
155,700	258.50	11.1	6.1
171,000	121.20	24.0	7.7
174,300	149.20	12.3	9.1
203,300	201.70	19.0	13.2
265,800	279.20	11.0	10.0
287,300	77.30	10.7	4.1
319,800	121.20	13.4	5.6
350,000	179.10	20.0	13.0
360,900	108.40	14.2	11.1
394,000	86.10	22.5	8.7

OFFICE BUILDINGS (Cont.)

Project Size Gross S.F.	Project Cost $/S.F.	% Cost Mechanical	% Cost Electrical
430,000	129.00	21.8	9.3
490,000	137.20	11.2	7.5
588,400	282.00	15.0	9.0
606,000	112.70	9.4	8.8
620,000	311.00	12.6	12.8
733,500	86.10	10.8	4.2

RECREATIONAL

ARENA

Project Size Gross S.F.	Project Cost $/S.F.	% Cost Mechanical	% Cost Electrical
315,200	322.80	10.4	8.00
385,800	184.80	13.2	8.80
727,000	167.10	14.9	7.40

HEALTH CLUB

Project Size Gross S.F.	Project Cost $/S.F.	% Cost Mechanical	% Cost Electrical
15,900	90.30	8.7	9.7
21,800	127.80	10.3	10.3
30,100	197.50	11.4	19.4
66,400	97.30	10.7	8.5

RECREATIONAL CENTER

Project Size Gross S.F.	Project Cost $/S.F.	% Cost Mechanical	% Cost Electrical
9,900	175.50	6.6	9.8
14,000	114.50	11.2	5.0
14,000	137.30	11.3	16.7
15,700	104.50	17.8	14.3
20,000	237.10	19.1	8.9
21,200	133.40	16.0	9.6

A = Addition R = Remodel

SQUARE FOOT TABLES

RECREATIONAL (Cont.)

RECREATIONAL CENTER (Cont.)

Project Size Gross S.F.	Project Cost $/S.F.	% Cost Mechanical	% Cost Electrical
26,000	155.40	9.3	7.0
53,400 A	185.80	11.7	6.4
69,800	301.30	17.7	8.8

RELIGIOUS

CHURCH

Project Size Gross S.F.	Project Cost $/S.F.	% Cost Mechanical	% Cost Electrical
4,100	211.50	8.8	13.5
10,400 R	287.90	12.8	8.1
11,100	126.80	10.9	12.7
13,400	190.00	5.5	6.0
14,500	129.70	12.8	7.4
15,200	168.10	18.3	8.0
15,700	146.40	14.4	8.4

CHURCH (Cont.)

Project Size Gross S.F.	Project Cost $/S.F.	% Cost Mechanical	% Cost Electrical
16,000	238.10	14.1	9.6
20,900	148.50	16.0	14.0
21,500	163.00	7.3	8.0
22,900	147.20	12.1	9.5
30,600	111.30	15.5	8.0
42,700	125.70	18.6	7.7

MULTI-PURPOSE

Project Size Gross S.F.	Project Cost $/S.F.	% Cost Mechanical	% Cost Electrical
4,400	130.00	11.5	17.5
5,800 A	165.70	8.9	7.5
6,400	226.20	15.6	15.8
9,000	110.80	7.7	5.9
9,000	157.30	16.0	6.6
10,100	105.50	11.1	12.0
12,000	208.50	13.1	10.7
18,400	114.30	10.8	10.1
19,500	201.40	16.0	17.3

A = Addition R = Remodel

For more information subscribe to **Design Cost & Data**

BNi Building News

INDEX

- C -

- D -

- E -

- F -

- G -

- H -

- I -

- N -

- O -

- X -

- Y -

- Z -

Notes

Notes

Notes

Notes

Notes

Notes